Introduction to RF Power Amplifier Design and Simulation

Introduction to RF Power Amplifier Design and Simulation

Abdullah Eroglu

INDIANA UNIVERSITY–PURDUE UNIVERSITY
FORT WAYNE, IN, USA

CRC Press is an imprint of the
Taylor & Francis Group, an **informa** business

MATLAB® and Simulink® are trademarks of The MathWorks, Inc. and are used with permission. The MathWorks does not warrant the accuracy of the text or exercises in this book. This book's use or discussion of MATLAB® and Simulink® software or related products does not constitute endorsement or sponsorship by The MathWorks of a particular pedagogical approach or particular use of the MATLAB® and Simulink® software.

CRC Press
Taylor & Francis Group
6000 Broken Sound Parkway NW, Suite 300
Boca Raton, FL 33487-2742

First issued in paperback 2020

ISBN-13: 978-1-4822-3164-9 (hbk)
ISBN-13: 978-0-367-73800-6 (pbk)

Visit the Taylor & Francis Web site at
http://www.taylorandfrancis.com

and the CRC Press Web site at
http://www.crcpress.com

Dedicated to my sons, Duhan and Enes, and daughter Dilem

Contents

Preface

Radio frequency (RF) power amplifiers are used in everyday life for many applications including cellular phones, magnetic resonance imaging, semiconductor wafer processing for chip manufacturing, etc. Therefore, the design and performance of RF amplifiers carry great importance for the proper functionality of these devices. Furthermore, several industrial and military applications require low-profile yet high-powered and efficient power amplifiers. This is a challenging task when several components are needed to be considered in the design of RF power amplifiers to meet the required criteria. As a result, designers are in need of a resource to provide all the essential design components for better-performing, low-profile, high-power, and efficient RF power amplifiers. This book is intended to be the main resource for engineers and students and fill the existing gap in the area of RF power amplifier design by giving a complete guidance with demonstration of the details for the design stages including analytical formulation and simulation. Therefore, in addition to the fact that it can be used as a unique resource for engineers and researchers, this book can also be used as a textbook for RF/microwave engineering students in their senior year at college. Chapter end problems are given to make this option feasible for instructors and students.

Successful realization of RF power amplifiers depends on the transition between each design stage. This book provides practical hints to accomplish the transition between the design stages with illustrations and examples. An analytical formulation to design the amplifier and computer-aided design (CAD) tools to verify the design, have been detailed with a step-by-step design process that makes this book easy to follow. The extensive coverage of the book includes not only an introduction to the design of several amplifier topologies; it also includes the design and simulation of amplifier's surrounding sections and assemblies. This book also focuses on the higher-level design sections and assemblies for RF amplifiers, which make the book unique and essential for the designer to accomplish the amplifier design as per the given specifications.

The scope of each chapter in this book can be summarized as follows. Chapter 1 provides an introduction to RF power amplifier basics and topologies. It also gives a brief overview of intermodulation and elaborates discussion on the difference between linear and nonlinear amplifiers. Chapter 2 gives details on the high-frequency model and transient characteristics of metal–oxide–semiconductor field-effect transistors. In Chapter 3, active device modeling techniques for transistors are detailed. Parasitic extraction methods for active devices are given with application examples. The discussion about network and scattering parameters is also given in this chapter. Resonator and matching networks are critical in amplifier design. The discussion on resonators, matching networks, and tools such as the Smith chart are given in Chapters 4 and 5. Every RF amplifier system has some type of voltage, current, or power-sensing device for control and stability of the amplifier. In Chapter 6, there is an elaborate discussion on power-sensing devices, including four-port directional couplers and new types of reflectometers. RF filter designs for power amplifiers are given in Chapter 7. Several special filter types for amplifiers are discussed, and application examples are presented. In Chapter 8, CAD tools for RF amplifiers are discussed. Unique real-life engineering examples are given. Systematic design techniques using simulation tools are presented and implemented.

Throughout the book, several methods and techniques are presented to show how to blend the theory and practice. In summary, I believe engineers, researchers, and students will greatly benefit from it.

Abdullah Eroglu
Fort Wayne, IN, USA

Acknowledgments

I thank my wife and children for allowing me to write this book instead of spending time with them. I am deeply indebted to their endless support and love. In addition, my students at Indiana University–Purdue University Fort Wayne will always be an inspiration for me to enhance my research in the area of radio frequency/microwave. As usual, special thanks go to my editor, Nora Konopka, for her understanding when I needed more time.

Author

Abdullah Eroglu earned his MSEE in 1999 and PhD in 2004 in electrical engineering from the Electrical Engineering and Computer Science Department of Syracuse University, Syracuse, NY. From 2000 to 2008, he worked as a radio frequency (RF) senior design engineer at MKS Instruments, where he was involved with the design of RF power amplifiers and systems. He is a recipient of the 2013 IPFW Outstanding Researcher Award, 2012 Indiana University-Purdue University Featured Faculty Award, 2011 Sigma Xi Researcher of the Year Award, 2010 College of Engineering, Technology and Computer Science (ETCS) Excellence in Research Award, and the 2004 Outstanding Graduate Student award from the Electrical Engineering and Computer Science Department of Syracuse University. Since 2014, he is a professor of electrical engineering at the Engineering Department of Indiana University–Purdue University, Fort Wayne, IN. He was a faculty Fellow at the Fusion Energy Division of Oak Ridge National Laboratory during the summer of 2009. His teaching and research interests include RF circuit design, microwave engineering, development of nonreciprocal devices, electromagnetic fields, wave propagation, radiation, and scattering in anisotropic and gyrotropic media. Dr. Eroglu has published over 100 peer-reviewed journal and conference papers. He is also the author of four books. He is a reviewer and on the editorial board of several journals.

1 Radio Frequency Amplifier Basics

1.1 INTRODUCTION

Radio frequency (RF) amplifiers are critical components and are widely used in applications including communication systems, radar applications, semiconductor manufacturing, magnetic resonance imaging (MRI), and induction heating. Use of RF amplifier as a core element in conjunction with an antenna in transmitter applications for wireless communication systems is illustrated in Figure 1.1.

The frequency of operation for the amplifiers is based on the application and varies from the very low frequency range to microwave frequencies. In any type of application, when a signal needs to be amplified to a certain level at the frequency of interest, the amplification process of the signal is accomplished using RF power amplifiers (PAs). The power level of the amplifiers also varies, and it can be anywhere from milliwatt to megawatt ranges. The commonly used RF amplifier topologies are A, B, AB, C, D, E, F, and S class. These topologies represent linear or nonlinear amplification of the signal. Linear amplification is realized by using class A, B, or AB amplifier topologies, whereas nonlinear amplification is performed with class C, D, E, F, and S amplifiers. Class D, E, F, and S amplifiers are known as switch-mode amplifiers where the active device or transistor is used as a switch during the operation of the amplifier. The relation between the RF signal waveforms at the input and output of the linear amplifiers can be expressed with the following relation:

$$v_{\mathrm{o}}(t) = \beta v_{\mathrm{i}}(t) \tag{1.1}$$

When the amplifier is operating in nonlinear mode, then the signals at the input and output are expressed using the power series, as given in Equation 1.2:

$$v_{\mathrm{o}}(t) = \alpha_0 + \alpha_1 v_{\mathrm{i}}(t) + \alpha_2 v_{\mathrm{i}}^2(t) + \alpha_3 v_{\mathrm{i}}^3(t) + \ldots \tag{1.2}$$

Equation 1.2 represents weak nonlinearities in the amplifier response. When weak distortion takes place, harmonics disappear as the signal amplitude gets smaller. Coefficients in Equation 1.2 can be found from

$$\alpha_{\mathrm{n}} = \frac{1}{n!} \frac{d^{\mathrm{n}} v_{\mathrm{o}}(t)}{d v_{\mathrm{i}}^{\mathrm{n}}(t)} \bigg|_{v_{\mathrm{i}}=0} \tag{1.3}$$

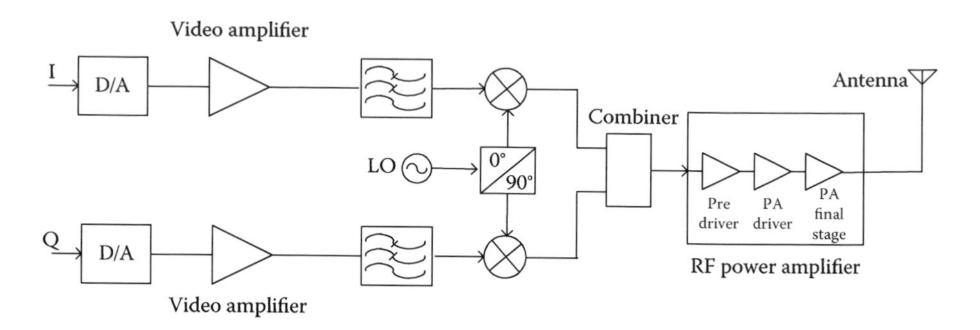

FIGURE 1.1 RF amplifier in transmitter applications for wireless systems.

The frequency components of the signal at the input and output of the amplifier are found from application of the Fourier transform as

$$F(\omega) = \Im\{f(t)\} = \int_{-\infty}^{\infty} f(t)e^{-j\omega t}\,dt \tag{1.4}$$

The energy of the signal can be obtained from Parseval's theorem by assuming that $f(t)$ is either voltage or current across a 1 Ω resistor. Then, the energy associated with $f(t)$ can be found from

$$W = \int_{-\infty}^{\infty} f(t)^2\,dt = \frac{1}{2\pi}\int_{-\infty}^{\infty} |F(\omega)|^2\,d\omega \tag{1.5}$$

Example

Assume that a cosinusoid signal, $v_i(t) = \cos(\omega t)$, with 5-Hz frequency is applied to a linear amplifier, which has output signal, $v_o(t) = 5\cos(\omega t)$, as shown in Figure 1.2. Obtain the time domain representation of the input signal and frequency domain representation of power spectra of the output signal of the amplifier.

Solution

The Fourier transform of the input signal is found from Equation 1.4 as

$$V_i(\omega) = \int_{-\infty}^{\infty} \left(\frac{1}{\pi}\right)\cos(\omega_o t)e^{-j\omega t}\,dt = [\delta(\omega-\omega_o)+\delta(\omega+\omega_o)] \tag{1.6}$$

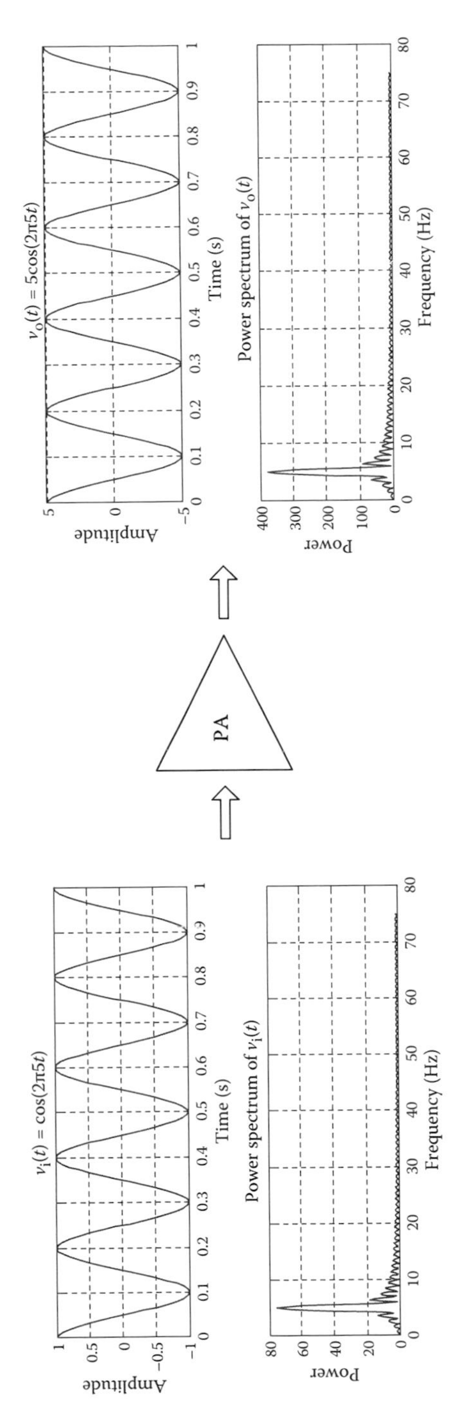

FIGURE 1.2 Linear amplifier input signal and power spectrum of its output signal.

and the output

$$V_o(\omega) = \int_{-\infty}^{\infty} \left(\frac{5}{\pi}\right) \cos(\omega_o t) e^{-j\omega t}\, dt = 5[\delta(\omega - \omega_o) + \delta(\omega + \omega_o)] \tag{1.7}$$

The power spectra of the output signal are obtained with application of Equation 1.5. The input signal and power spectrum of the output signal are illustrated in frequency and time domain in Figure 1.2.

In the amplifier design, there are several parameters that indicate the performance of the amplifier: amplifier gain, output power, stability, linearity, DC supply voltage, efficiency, and ruggedness. The thermal profile of individual transistors and the overall amplifier are also very important to prevent the catastrophic failure of an amplifier.

The basic RF amplifier is illustrated in Figure 1.3. The amplifier mode of operation depends on how the metal–oxide–semiconductor field-effect transistor (MOSFET) is biased based on the voltages applied from gate to source, V_{gs}, and from drain to source, V_{ds}. The MOSFET can be operated either as a dependent current source or as a switch. When it operates as a dependent current source, the ohmic (linear) region becomes the operational region for the MOSFET.

The saturation (active) region is the operational region for the MOSFET when it is operating as a switch. As a result, the operational region of the active device such as the MOSFET shown in Figure 1.3 determines the amplifier mode with variable parameters such as V_{gs} and V_{ds}.

In practice, RF PAs are implemented as part of systems that include several subassemblies such as DC power supply unit for amplifier, line filter, housekeeping power supply, system controller, splitter, combiner, coupler, compensator, and PA modules, as illustrated in Figure 1.4. Several PA modules are combined to obtain higher power levels via combiners. Couplers are used to monitor the reflected and forward power and provide control signal to the system controller. The system controller then adjusts the level of RF input signal to deliver the desired amount of output power.

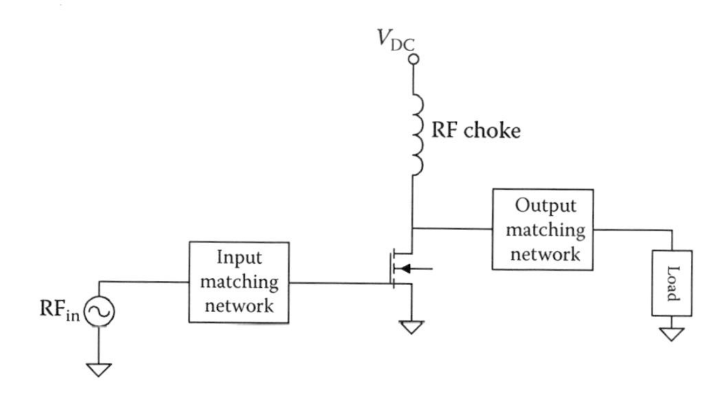

FIGURE 1.3 Basic RF amplifier.

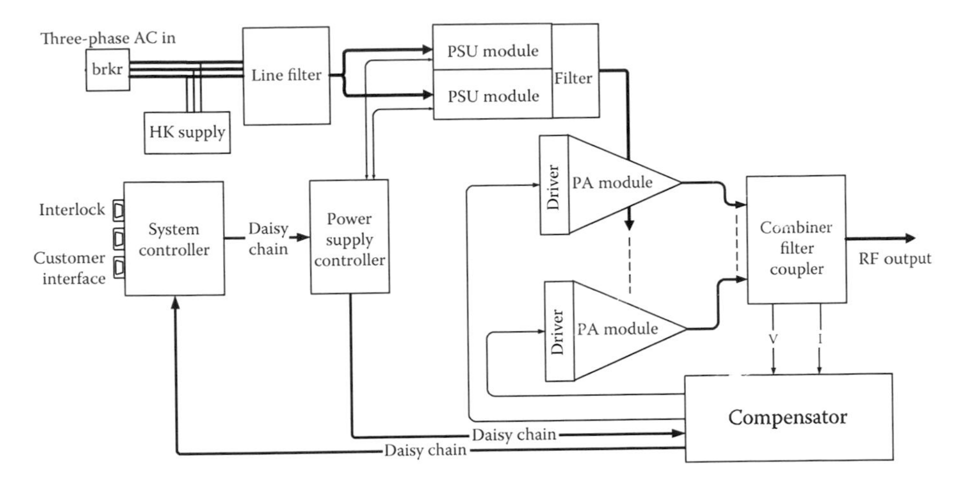

FIGURE 1.4 Typical RF amplifier system architecture.

1.2 RF AMPLIFIER TERMINOLOGY

In this section, some of the common terminologies used in RF PA design will be discussed. For this discussion, consider the simplified RF PA block diagram given in Figure 1.5. In the figure, RF PA is simply considered as a three-port network, where the RF input signal port, the DC input port, and the RF signal output port constitute the ports of the network as illustrated. Power that is not converted to RF output power, P_{out}, is dissipated as heat and designated by P_{diss}, as shown in Figure 1.5. The dissipated power, P_{diss}, is found from

$$P_{diss} = (P_{in} + P_{DC}) - P_{out} \tag{1.8}$$

1.2.1 Gain

RF PA gain is defined as the ratio of the output power to the input power, as given by

$$G = \frac{P_{out}}{P_{in}} \tag{1.9}$$

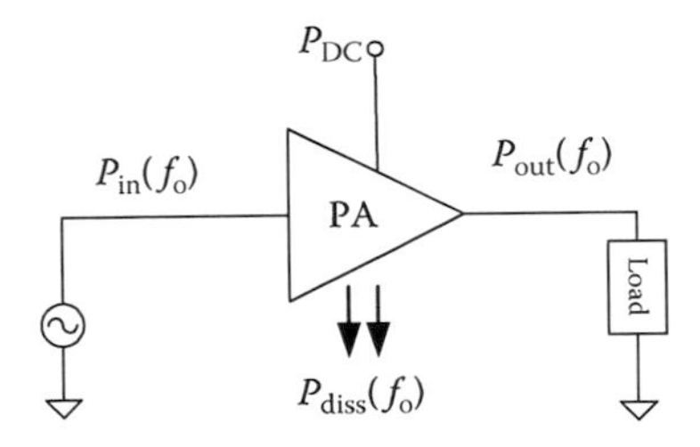

FIGURE 1.5 RF PA as a three-port network.

It can be defined in terms of decibels as

$$G(\mathrm{dB}) = 10\log\left(\frac{P_{\mathrm{out}}}{P_{\mathrm{in}}}\right)[\mathrm{dB}] \tag{1.10}$$

RF PA amplifier gain is higher at lower frequencies. This can be illustrated based on the measured data for a switched-mode RF amplifier operating at high-frequency (HF) range in Figure 1.6 for several applied DC supply voltages.

It is possible to obtain higher gain level when multiple amplifiers are cascaded to obtain multistage amplifier configuration, as shown in Figure 1.7.

The overall gain of the multistage amplifier system for the one shown in Figure 1.7 can then be found from

$$G_{\mathrm{tot}}(\mathrm{dB}) = G_{\mathrm{PA}_1}(\mathrm{dB}) + G_{\mathrm{PA}_2}(\mathrm{dB}) + G_{\mathrm{PA}_3}(\mathrm{dB}) \tag{1.11}$$

The unit of the gain is given in terms of decibels because it is the ratio of the output power to the input power. It is important to note that decibel is not a unit that defines the power. In amplifier terminology, dBm is used to define the power. dBm is found from

$$\mathrm{dBm} = 10\log\left(\frac{P}{1\ \mathrm{mW}}\right) \tag{1.12}$$

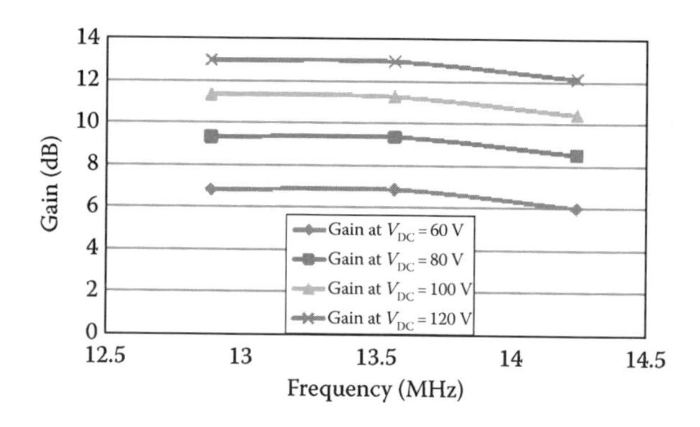

FIGURE 1.6 Measured gain variation vs. frequency for a switched-mode RF amplifier.

P_{in} PA_1 $G_{PA_1}(dB)$ PA_2 $G_{PA_2}(dB)$ PA_3 $G_{PA_3}(dB)$ P_{out}

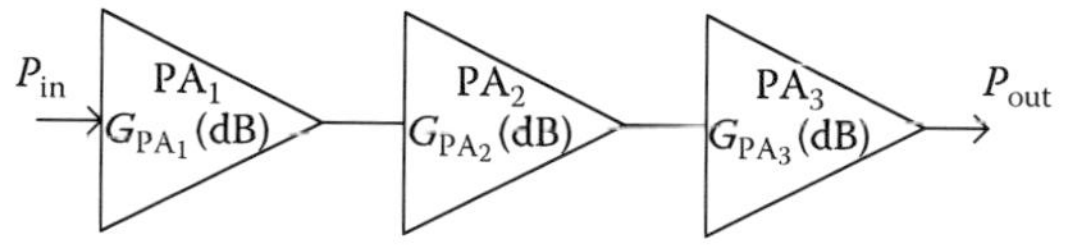

FIGURE 1.7 Multistage RF amplifiers.

Example

In the RF system shown in Figure 1.8, the RF signal source can provide power output from 0 to 30 dBm. The RF signal is fed through a 1-dB T-pad attenuator and a 20-dB directional coupler where the sample of the RF signal is further attenuated by a 3-dB π-pad attenuator before power meter reading in dB. The "through" port of the directional coupler has 0.1 dB of loss before it is sent to PA. RF PA output is then connected to a 6-dB π-pad attenuator. If the power meter is reading 10 dBm, what is the power delivered to the load shown in Figure 1.8 in mW?

Solution

We need to find the power source first. The loss from the power meter to the RF signal source is

$$\text{loss from power meter to source} = 3 \text{ dB} + 20 \text{ dB} + 1 \text{ dB} = 23 \text{ dB}$$

Then, the power at the source is

$$\text{RF source signal} = 23 \text{ dBm} + 10 \text{ dBm} = 33 \text{ dBm}$$

The total loss toward PA is due to the T-pad attenuator (1 dB) and the directional coupler (0.1 dB) = 1.1 dB. So the transmitted RF signal at PA is

$$\text{RF signal at PA} = 33 \text{ dBm} - 1.1 \text{ dBm} = 31.9 \text{ dBm}$$

Hence, the power delivered to the load is found from

$$\text{power delivered to the load} = 31.9 \text{ dBm} - 6 \text{ dBm} = 25.9 \text{ dBm}$$

Power delivered in mW is found from

$$P(\text{mW}) = 10^{\frac{\text{dBm}}{10}} = 10^{\frac{25.9}{10}} = 389.04\,[\text{mW}]$$

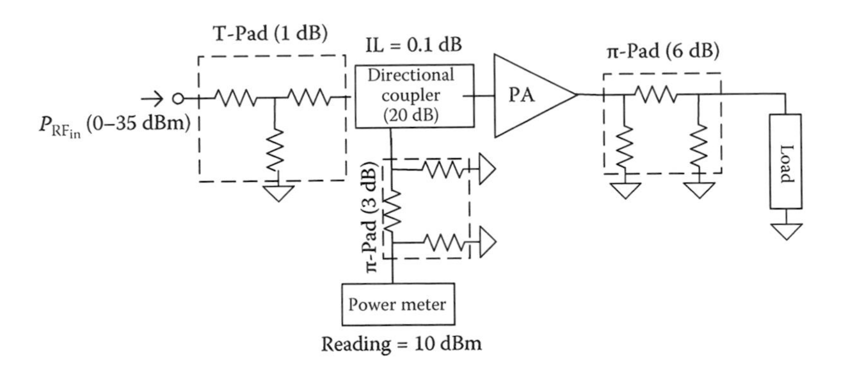

FIGURE 1.8 RF system with coupler and attenuation pads.

1.2.2 Efficiency

In practical applications, RF PA is implemented as a subsystem and consumes most of the DC power from the supply. As a result, minimal DC power consumption for the amplifier becomes important and can be accomplished by having high RF PA efficiency. RF PA efficiency is one of the critical and most important amplifier performance parameters. Amplifier efficiency can be used to define the drain efficiency for MOSFET or collector efficiency for a bipolar junction transistor (BJT). Amplifier efficiency is defined as the ratio of the RF output power to power supplied by the DC source and can be expressed as

$$\eta(\%) = \frac{P_{\text{out}}}{P_{\text{DC}}} \times 100 \tag{1.13}$$

Efficiency in terms of gain can be put in the following form:

$$\eta(\%) = \frac{1}{1 + \left(\frac{P_{\text{diss}}}{P_{\text{out}}}\right) - \left(\frac{1}{G}\right)} \times 100 \tag{1.14}$$

The maximum efficiency is possible when there is no dissipation, i.e., $P_{\text{diss}} = 0$. The maximum efficiency from Equation 1.14 is then equal to

$$\eta(\%) = \frac{1}{1 - \left(\frac{1}{G}\right)} \times 100 \tag{1.15}$$

When RF input power is included in the efficiency calculation, the efficiency is then called as power-added efficiency, η_{PAE}, and found from

$$\eta_{\text{PAE}}(\%) = \frac{P_{\text{out}} - P_{\text{in}}}{P_{\text{DC}}} \times 100 \tag{1.16}$$

or

$$\eta_{\text{PAE}}(\%) = \eta\left(1 - \frac{1}{G}\right) \times 100 \tag{1.17}$$

Example

RF PA delivers 200[W] to a given load. If the input supply power for this amplifier is given to be 240[W], and the power gain of the amplifier is 15 dB, find the (a) drain efficiency and (b) power-added efficiency.

Solution

a. The drain efficiency is found from Equation 1.13 as

$$\eta(\%) = \frac{P_{out}}{P_{DC}} \times 100 = \frac{200}{240} \times 100 = 83.33\%$$

b. Power-added efficiency is found from Equation 1.17 as

$$\eta_{PAE}(\%) = \eta\left(1 - \frac{1}{G}\right) \times 100 = 83\left(1 - \frac{1}{15}\right) = 77.47\%$$

1.2.3 Power Output Capability

The power output capability of an amplifier is defined as the ratio of the output power for the amplifier to the maximum values of the voltage and current that the device experiences during the operation of the amplifier. When there are more than one transistor, or the number of the transistors increases due to amplifier configuration used in such push–pull configuration or any other combining techniques, this is reflected in the denominator of the following equation:

$$c_p = \frac{P_o}{\mathrm{N}I_{max}V_{max}} \tag{1.18}$$

1.2.4 Linearity

Linearity is a measure of the RF amplifier output to follow the amplitude and phase of its input signal. In practice, the linearity of an amplifier is measured in a very different way; it is measured by comparing the set power of an amplifier with the output power. The gain of the amplifier is then adjusted to compensate one of the closed-loop parameters such as gain. The typical closed-loop control system that is used to adjust the linearity of the amplifier through closed-loop parameters is shown in Figure 1.9. When the linearity of the amplifier is accomplished, the linear curve shown in Figure 1.10 is obtained. The experimental setup that is used to calibrate RF PAs to have linear characteristics is given in Figure 1.11.

In Figure 1.11, the RF amplifier output is measured by a thermocouple-based power meter via a directional coupler. The directional coupler output is terminated with 50 Ω load. The set power is adjusted by the user and output forward power, P_{fwr}, and the reverse power is measured with the power meter. If the set power and output power are different, the control closed-loop parameters are then modified.

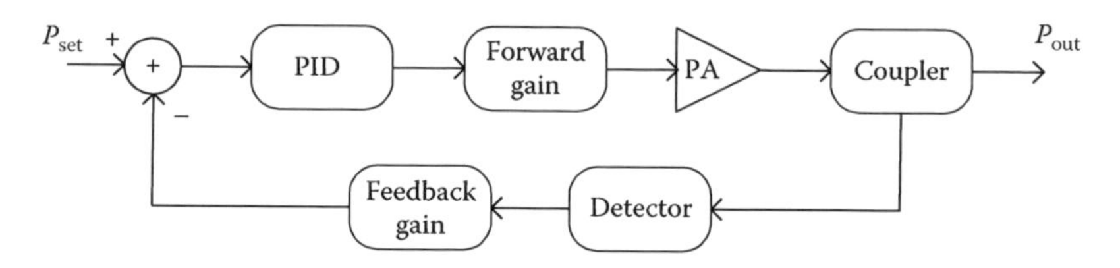

FIGURE 1.9 Typical closed-loop control for RF PA for linearity control.

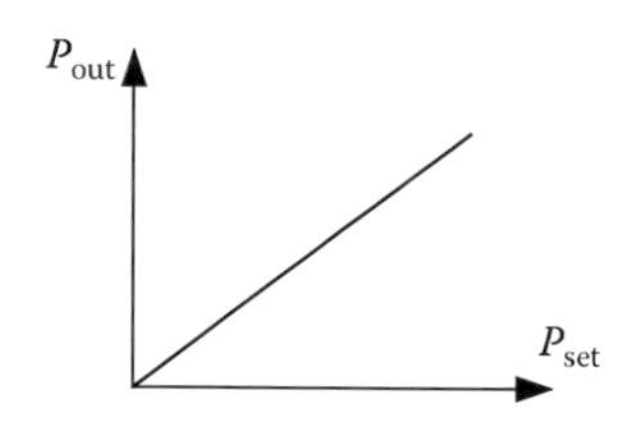

FIGURE 1.10 Linear curve for RF amplifier.

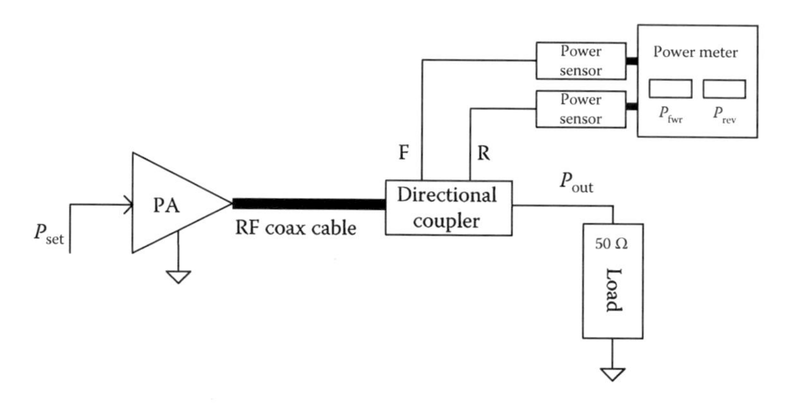

FIGURE 1.11 Experimental setup for linearity adjustment of RF PAs.

1.2.5 1-dB Compression Point

The compression point for an amplifier is the point where the amplifier gain becomes 1 dB below its ideal linear gain, as shown in Figure 1.12. Once the 1-dB compression point is identified for the corresponding input power range, the amplifier can be operated in linear or nonlinear mode. Hence, the 1-dB compression point can also be conveniently used to identify the linear characteristics of the amplifier.

The gain at 1-dB compression point can be found from

$$P_{1dB,out} - P_{1dB,in} = G_{1dB} = G_0 - 1 \tag{1.19}$$

where G_0 is the small signal or linear gain of the amplifier at fundamental frequency. The 1-dB compression point can also be expressed using the input and output

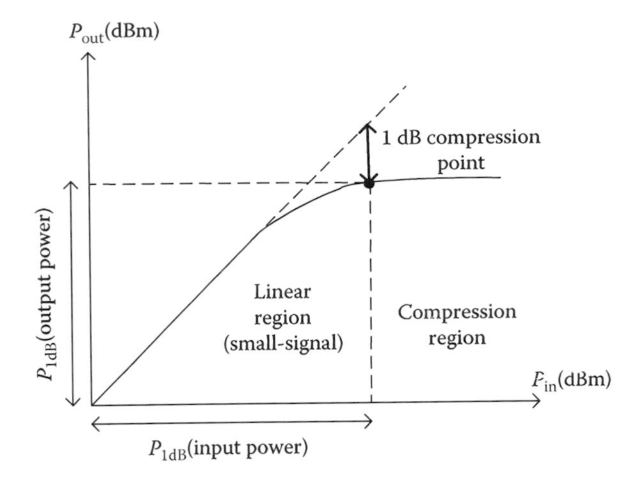

FIGURE 1.12 1-db compression point for amplifiers.

voltages and their coefficients. From Equations 1.1 and 1.2, the gain of the amplifier at the fundamental frequency when $v_i(t) = \beta\cos\omega t$ is found as

$$G_{1dB} = 20\log\left|\alpha_1 + 3\alpha_3\frac{\beta^2}{4}\right| \tag{1.20}$$

$$G_0(\text{linear/small signal gain}) = 20\log|\alpha_1| \tag{1.21}$$

As a result, the 1-dB compression point in Figure 1.12 can be calculated from

$$20\log\left|\alpha_1 + 3\alpha_3\frac{\beta^2_{in,1dB}}{4}\right| = 20\log|\alpha_1| - 1\text{ dB} \tag{1.22}$$

where β_{1dB} is the amplitude of the input voltage at the 1-dB compression point. The solution of Equation 1.20 for β_{1dB} leads to

$$\beta_{1dB} = \sqrt{0.145\left|\frac{\alpha_1}{\alpha_3}\right|} \tag{1.23}$$

1.3 SMALL-SIGNAL VS. LARGE-SIGNAL CHARACTERISTICS

Small-signal analysis is based on the condition that the variation between the active device output and input voltages and currents exhibits small fluctuations such that the device can be modeled using its equivalent linear circuit and analyzed with two-port parameters. The large-signal analysis of the amplifiers is based on the fact that the variation between voltages and currents is large. The small-signal amplifier can then be approximated to have the linear relation given by Equation 1.1, whereas the large-signal amplifier presents the nonlinear characteristics given by Equation 1.2.

1.3.1 Harmonic Distortion

The harmonic distortion (HD) of an amplifier can be defined as the ratio of the amplitude of the $n\omega$ component to the amplitude of the fundamental component. The second- and third-order HDs can then be expressed as

$$\mathrm{HD}_2 = \frac{1}{2}\frac{\alpha_2}{\alpha_1}\beta \tag{1.24}$$

$$\mathrm{HD}_3 = \frac{1}{4}\frac{\alpha_3}{\alpha_1}\beta^2 \tag{1.25}$$

From Equations 1.24 and 1.25, it is apparent that the second HD is proportional to the signal amplitude, whereas the third-order amplitude is proportional to the square of the amplitude. Hence, when the input signal is increased by 1 dB, HD_2 increases by 1 dB, and HD_3 increases by 2 dB. The total HD (THD) in the amplifier can be found from

$$\mathrm{THD} = \sqrt{(\mathrm{HD}_2)^2 + (\mathrm{HD}_3)^2 + \ldots} \tag{1.26}$$

Example

An RF signal, $v_i(t) = \beta\cos\omega t$, is applied to a linear amplifier and then to a nonlinear amplifier given in Figure 1.13 with output response $v_o(t) = \alpha_0 + \alpha_1\beta\cos\omega t + \alpha_2\beta^2\cos^2\omega t + \alpha_3\beta^3\cos^3\omega t$. Assuming that input and output impedances are equal to R, (a) calculate and plot the gain for the linear amplifier; and (b) obtain the second and third HD for the nonlinear amplifier when $\alpha_0 = 0$, $\alpha_1 = 1$, $\alpha_2 = 3$, $\alpha_3 = 1$ and $\beta = 1$, $\beta = 2$. Calculate also the THD for both cases.

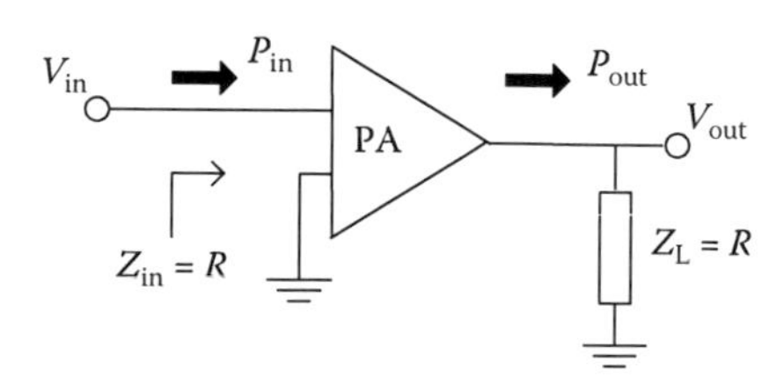

FIGURE 1.13 PA amplifier output response.

Solution

a. For linear amplifier characteristics, the output voltage is expressed using Equation 1.1 as

$$v_o(t) = \beta v_i(t) \tag{1.27}$$

which can also be written as

$$\frac{1}{2R}v_o^2(t) = \beta^2 \frac{1}{2R}v_i^2(t) \tag{1.28}$$

or

$$P_o = \beta^2 P_i \tag{1.29}$$

When power (Equation 1.21) is given in dBm, then Equation 1.21 can be expressed as

$$10\log\left(\frac{P_o}{1\ \text{mW}}\right) = 10\log\left(\beta^2 \frac{P_i}{1\ \text{mW}}\right) \tag{1.30}$$

or

$$P_o(\text{dBm}) = 10\log(\beta^2) + P_{in}(\text{dBm}) \tag{1.31}$$

Then, the power gain is obtained from Equation 1.24 as

$$\text{Gain(dBm)} = G(\text{dBm}) = 10\log(\beta^2) = P_o(\text{dBm}) - P_{in}(\text{dBm}) \tag{1.32}$$

The relation between input and output power is plotted and illustrated in Figure 1.14.

b. The nonlinearity response of the amplifier using third-order polynomial can be expressed using Equation 1.2 as

$$v_o(t) = \alpha_0 + \alpha_1 v_i(t) + \alpha_2 v_i^2(t) + \alpha_3 v_i^3(t) \tag{1.33}$$

or

$$v_o(t) = \alpha_0 + \alpha_1 \beta\cos\omega t + \alpha_2 \beta^2 \cos^2\omega t + \alpha_3 \beta^3 \cos^3\omega t \tag{1.34}$$

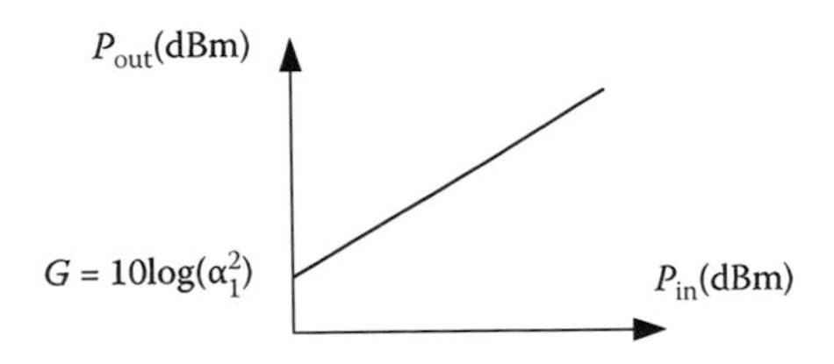

FIGURE 1.14 Power gain for linear operation.

As seen from Equation 1.24, we have fundamental, second-order harmonic, third-order harmonic, and a DC component in the output response of the amplifier. In Equation 1.27, it is also seen that the DC component exists due to the second harmonic content. Equation 1.28 can be rearranged to give the following closed-form relation:

$$v_o(t) = \alpha_0 + \alpha_1\beta\cos\omega t + \frac{\alpha_2\beta^2}{2} + \frac{\alpha_2\beta^2}{2}\cos 2\omega t + 3\frac{\alpha_3\beta^3}{4}\cos\omega t + \frac{\alpha_3\beta^3}{4}\cos 3\omega t \quad (1.35)$$

which can be simplified to

$$v_o(t) = \left(\alpha_0 + \frac{\alpha_2\beta^2}{2}\right) + \left(\alpha_1 + 3\frac{\alpha_3\beta^2}{4}\right)\beta\cos\omega t + \left(\frac{\alpha_2}{2}\right)\beta^2\cos 2\omega t + \frac{\alpha_3\beta^3}{4}\cos 3\omega t \quad (1.36)$$

When $\alpha_1 = 1$, $\alpha_2 = 3$, $\alpha_3 = 1$ and $\beta = 1$, HD_2 and HD_3 are obtained from Equations 1.24 and 1.25 as

$$HD_2 = \frac{1}{2}\frac{\alpha_2}{\alpha_1}\beta = \frac{1}{2}\frac{3}{1}(1) = 1.5 \quad (1.37)$$

$$HD_3 = \frac{1}{4}\frac{\alpha_3}{\alpha_1}\beta^2 = \frac{1}{4}\frac{1}{1}(1)^2 = 0.25 \quad (1.38)$$

When $\alpha_1 = 1$, $\alpha_2 = 3$, $\alpha_3 = 1$ and $\beta = 2$,

$$HD_2 = \frac{1}{2}\frac{\alpha_2}{\alpha_1}\beta = \frac{1}{2}\frac{3}{1}(2) = 3 \quad (1.39)$$

$$HD_3 = \frac{1}{4}\frac{\alpha_3}{\alpha_1}\beta^2 = \frac{1}{4}\frac{1}{1}(2)^2 = 1 \quad (1.40)$$

The THD for this system is found from Equation 1.26 as

$$\text{THD} = \sqrt{(1.5)^2 + (0.25)^2} = 1.5625 \quad (1.41)$$

and

$$\text{THD} = \sqrt{(3)^2 + (1)^2} = 3.16 \quad (1.42)$$

Example

The input of voltage for an RF circuit is given to be $v_{in}(t) = \beta\cos(\omega t)$. The RF circuit generates signal at the third harmonic as $V_3\cos(3\omega t)$. What is the 1-dB compression point?

Solution

Using Equation 1.36, the amplitude of the third harmonic component can be found from

$$\frac{\alpha_3\beta^3}{4}=V_3 \quad \text{or} \quad \alpha_3=\frac{4V_3}{\beta^3} \tag{1.43}$$

Then, the 1-dB compression point, β_{1dB}, is found from Equation 1.23 as

$$\beta_{1dB}=\sqrt{0.145\left|\frac{\alpha_1}{\alpha_3}\right|}=\sqrt{\frac{0.145}{4}\left|\frac{\beta^3\alpha_1}{V_3}\right|}=0.19\sqrt{\left|\frac{\beta^3\alpha_1}{V_3}\right|} \tag{1.44}$$

1.3.2 Intermodulation

When a signal composed of two cosine waveforms with different frequencies

$$v_i(t)=\beta_1\cos\omega_1 t+\beta_2\cos\omega_2 t \tag{1.45}$$

is applied to an input of an amplifier, the output signal consists of components of the self-frequencies and their products created by frequencies by ω_1 and ω_2 given by the following equation:

$$v_o(t)=\alpha_1(\beta_1\cos\omega_1 t+\beta_2\cos\omega_2 t)+\alpha_2(\beta_1\cos\omega_1 t+\beta_2\cos\omega_2 t)^2 \\ +\alpha_3(\beta_1\cos\omega_1 t+\beta_2\cos\omega_2 t)^3 \tag{1.46}$$

or

$$v_o(t)=\alpha_1(\beta_1\cos\omega_1 t+\beta_2\cos\omega_2 t)+\alpha_2\begin{pmatrix}\frac{1}{2}\beta_1^2(1+\cos 2\omega_1 t)+\frac{1}{2}\beta_2^2(1+\cos 2\omega_2 t)\\ +\frac{1}{2}\beta_1\beta_2(\cos(\omega_1+\omega_2)t+\cos(\omega_1-\omega_2)t)\end{pmatrix}$$
$$+\alpha_3\begin{pmatrix}\frac{3}{4}\beta_1^3(\cos\omega_1 t)+\frac{3}{2}\beta_1\beta_2^2(\cos\omega_1 t)+\frac{1}{4}\beta_1^3\cos(3\omega_1 t)+\frac{3}{4}\beta_1\beta_2^2(\cos(\omega_1-2\omega_2)t)\\ +\frac{3}{4}\beta_1^2\beta_2(\cos(2\omega_1-\omega_2)t)+\frac{3}{2}\beta_1^2\beta_2(\cos\omega_2 t)+\frac{3}{4}\beta_2^3(\cos\omega_2 t)+\frac{1}{4}\beta_2^3(\cos 3\omega_2 t)\\ +\frac{3}{4}\beta_1^2\beta_2(\cos(2\omega_1+\omega_2)t)+\frac{3}{4}\beta_1\beta_2^2(\cos(\omega_1+2\omega_2)t)\end{pmatrix} \tag{1.47}$$

In Equations 1.46 and 1.47, the DC component, α_0, is ignored. The components will rise due to combinations of the frequencies, ω_1 and ω_2, as given by Equations 1.46 and 1.47 and shown in Table 1.1. The corresponding frequency components in Table 1.1 are also illustrated in Figure 1.15.

In amplifier applications, intermodulation distortion (IMD) products are undesirable components in the output signal. As a result, the amplifier needs to be tested using an input signal, which is the sum of two cosines to eliminate these side products. This test is also known as a *two-tone test*. This specific test is important for an amplifier specifically when two frequencies, ω_1 and ω_2, are close to each other.

The second-order IMD, IM_2, can be found from Equation 1.26 and Table 1.1 when $\beta_1 = \beta_2 = \beta$. It is the ratio of the components at $\omega_1 \pm \omega_2$ to the fundamental components at ω_1 or ω_2.

$$IM_2 = \frac{\alpha_2}{\alpha_1}\beta \tag{1.48}$$

The third-order distortion, IM_3, can be found from the ratio of the component at $2\omega_2 \pm \omega_1$ (or $2\omega_1 \pm \omega_2$) to the fundamental components at ω_1 or ω_2.

TABLE 1.1
Intermodulation (IM) Frequencies and Corresponding Amplitudes

$\omega = \omega_1$	$\left(\alpha_1\beta_1 + \frac{3}{4}\alpha_3\beta_1^3 + \frac{3}{2}\alpha_3\beta_1\beta_2^2\right)\cos(\omega_1 t)$
$\omega = \omega_2$	$\left(\alpha_1\beta_2 + \frac{3}{4}\alpha_3\beta_2^3 + \frac{3}{2}\alpha_3\beta_2\beta_1^2\right)\cos(\omega_2 t)$
$\omega = \omega_1 + \omega_2$	$\frac{1}{2}(\alpha_2\beta_1\beta_2)\cos(\omega_1 + \omega_2)t$
$\omega = \omega_1 - \omega_2$	$\frac{1}{2}(\alpha_2\beta_1\beta_2)\cos(\omega_1 - \omega_2)t$
$\omega = 2\omega_1 + \omega_2$	$\left(\frac{3}{4}\alpha_3\beta_1^2\beta_2\right)\cos(2\omega_1 + \omega_2)t$
$\omega = 2\omega_1 - \omega_2$	$\left(\frac{3}{4}\alpha_3\beta_1^2\beta_2\right)\cos(2\omega_1 - \omega_2)t$
$\omega = 2\omega_2 + \omega_1$	$\left(\frac{3}{4}\alpha_3\beta_1^2\beta_2\right)\cos(2\omega_2 + \omega_1)t$
$\omega = 2\omega_2 - \omega_1$	$\left(\frac{3}{4}\alpha_3\beta_1^2\beta_2\right)\cos(2\omega_2 - \omega_1)t$

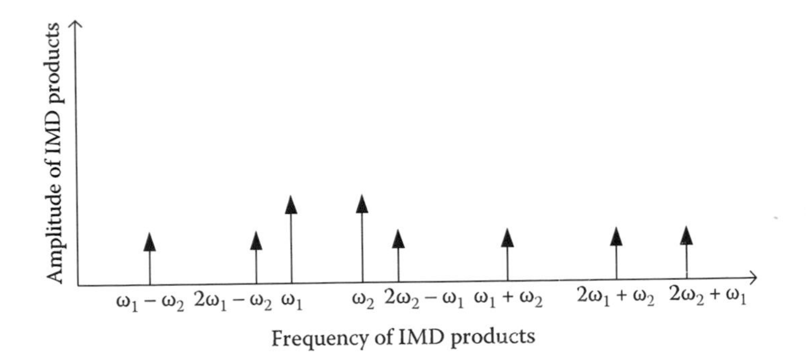

FIGURE 1.15 Illustration of IMD frequencies and products.

$$IM_3 = \frac{3}{4}\frac{\alpha_3}{\alpha_1}\beta^2 \tag{1.49}$$

The IM product frequencies are summarized in Table 1.2.

If Equations 1.24 and 1.25 and Equations 1.48 and 1.49 are compared, IM products can be related to HD products as

$$IM_2 = 2HD_2 \tag{1.50}$$

$$IM_3 = 3HD_3 \tag{1.51}$$

IM_3 distortion components at frequencies $2\omega_1 - \omega_2$ and $2\omega_2 - \omega_1$ are very close to the fundamental components. It is why the IM_3 signal is measured most of the time for IMD characterization of the amplifier. The simplified measurement setup for IMD testing is shown in Figure 1.16.

The point where the output components at the fundamental frequency and IM_3 intersect is called as an intercept point or IP_3. At this point, $IM_3 = 1$, and IP_3 is found from Equation 1.49 as

$$IM_3 = 1 = \frac{3}{4}\frac{\alpha_3}{\alpha_1}(IP_3)^2 \tag{1.52}$$

TABLE 1.2
Summary of IM Product Frequencies

IM_2 frequencies	$\omega_1 \pm \omega_2$	
IM_3 frequencies	$2\omega_1 \pm \omega_2$	$2\omega_2 \pm \omega_1$
IM_5 frequencies	$3\omega_1 \pm 2\omega_1$	$3\omega_2 \pm 2\omega_1$

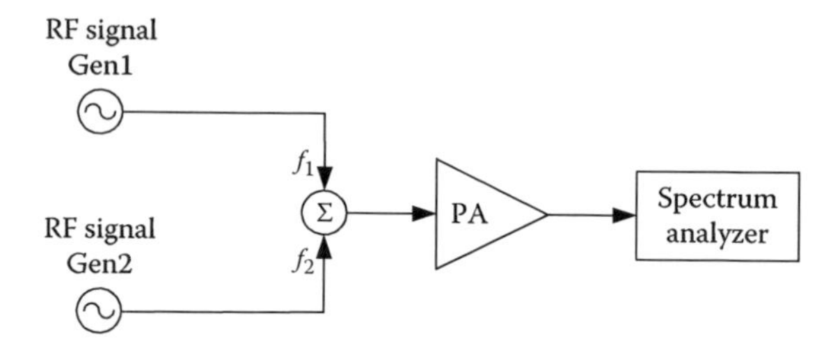

FIGURE 1.16 Simplified IMD measurement setup.

or

$$\mathrm{IP}_3 = \sqrt{\frac{4}{3}\frac{\alpha_1}{\alpha_3}} \tag{1.53}$$

which can also be written as

$$\mathrm{IP}_3 = \frac{V_{\mathrm{in}}}{\sqrt{\mathrm{IM}_3}} \tag{1.54}$$

where V_{in} is the input voltage. Equation 1.54 can be expressed in terms of dB by taking the log of both sides in Equation 1.54 as

$$\mathrm{IP}_3(\mathrm{dB}) = V_{\mathrm{in}}(\mathrm{dB}) - \frac{1}{2}\mathrm{IM}_3(\mathrm{dB}) \tag{1.55}$$

The dynamic range, DR, is measured to understand the level of the output noise and is defined as

$$\mathrm{DR} = \alpha_1 \frac{V_{\mathrm{in}}}{V_{\mathrm{Nout}}} = \frac{V_{\mathrm{in}}}{V_{\mathrm{Nin}}} \tag{1.56}$$

where input noise is related to output noise by

$$V_{\mathrm{Nin}} = \frac{V_{\mathrm{Nout}}}{\alpha_1} \tag{1.57}$$

So,

$$\mathrm{DR}(\mathrm{dB}) = V_{\mathrm{in}}(\mathrm{dB}) - V_{\mathrm{Nin}}(\mathrm{dB}) \tag{1.58}$$

Intermodulation free dynamic range, $IMFDR_3$, is defined as the largest DR possible with no IM_3 product. For the third-order IMD, V_{Nout} is defined by

$$V_{Nout} = \frac{3}{4}\alpha_3 V_{in}^3 \tag{1.59}$$

We can obtain

$$V_{in} = \sqrt[3]{\frac{4}{3\alpha_3} V_{Nout}} \tag{1.60}$$

Substitution of Equation 1.59 into Equation 1.56 gives $IMFDR_3$ as

$$DR = IMFDR_3 = \alpha_1 \frac{V_{in}}{V_{Nout}} = V_{in} = \sqrt[3]{\frac{4}{3}\frac{\alpha_1^3}{\alpha_3}\frac{1}{V_{Nout}^2}} \tag{1.61}$$

Since $V_{Nout} = \alpha_1 V_{Nin}$ from Equation 1.54, Equation 1.56 can be written in terms of input noise as

$$IMFDR_3 = \alpha_1 \frac{V_{in}}{V_{Nout}} = V_{in} = \sqrt[3]{\frac{4}{3}\frac{\alpha_1}{\alpha_3}\frac{1}{V_{Nin}^2}} \tag{1.62}$$

When Equations 1.53 and 1.62 are compared, $IMFDR_3$ can also be expressed using IP_3 as

$$IMFDR_3 = \left(\frac{IP_3}{V_{Nin}}\right)^{\frac{2}{3}} \tag{1.63}$$

or in terms of dB, Equation 1.63 can also be given by

$$IMFDR_3(dB) = \frac{2}{3}(IP_3(dB) - V_{Nin}(dB)) \tag{1.64}$$

The relationship between the fundamental component and the third-order distortion component via input and output voltages is illustrated in Figure 1.17. In the figure, –1-dB compression point is used to characterize IM_3 product. The –1-dB compression point can be defined as the value of the input voltage, V_{in}, which is designated by $V_{in,1dBc}$, where the fundamental component is reduced by 1 dB. $V_{in,1dBc}$ can be defined by

$$V_{in,1dBc} = \sqrt{(0.122)\left(\frac{4}{3}\right)\left|\frac{\alpha_1}{\alpha_3}\right|} \tag{1.65}$$

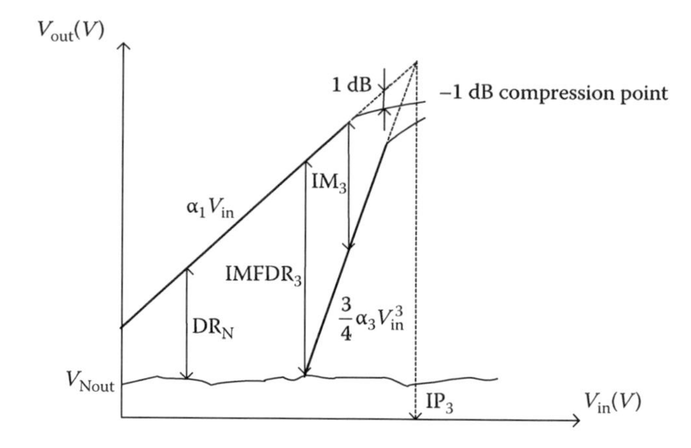

FIGURE 1.17 Illustration of the relation between the fundamental components and IM_3.

which is also equal to

$$V_{in,1dBc} = \sqrt{(0.122)IP_3} \tag{1.66}$$

Equation 1.66 can be expressed in dB as

$$V_{in,1dBc}(dB) = IP_3(dB) - 9.64\ dB \tag{1.67}$$

As a result, once IP_3 is determined, Equation 1.66 can be used to calculate the −1-dB compression point for the amplifier.

Example

Assume that a sinusoid signal, $v_i(t) = \sin(\omega t)$, with 5-Hz frequency is applied to a nonlinear amplifier, which has an output signal, (a) $v_o(t) = 10\sin(\omega t) + 2\sin^2(\omega t)$, (b) $v_o(t) = 10\sin(\omega t) - 3\sin^3(\omega t)$, and (c) $v_o(t) = 10\sin(\omega t) + 2\sin^2(\omega t) - 3\sin^3(\omega t)$. Obtain the time domain representation of the input signal and frequency domain representation of power spectra of the output signal of the amplifier.

Solution

a. The frequency spectrum for the input signal and the power spectrum for the output signal of the amplifier are obtained using the MATLAB® script given in the following. Based on the results shown in Figure 1.18, the amplifier output has components at DC, *f*, and 2*f*.

```
% Example for Figure 1.18
fs = 150; % Assign Sampling frequency
t = 0:1/fs:1; % Create time vector
f = 5; % Frequency in Hz.
vin = sin(2*pi*t*f); % input voltage
```

```
vout = 10*sin(2*pi*t*f)+2*(sin(2*pi*t*f)).^2; % output
  voltage
lfft = 1024; % length of FFT
Vin = fft(vin,lfft);% Take FFT
Vin = Vin(1:lfft/2); % FFT is symmetric, dont need second
                     % half
Vout = fft(vout,lfft);% Repeat it for output

Vout = Vout(1:lfft/2);
magvin = abs(Vin); % Magnitude of FFT of vin
magvout = abs(Vout);% Magnitude of FFT of vout
f = (0:lfft/2-1)*fs/lfft; % Create frequency vector
figure(1) % Plotting begins
subplot(2,1,1)
plot(t,vin);
title('v_i(t) = sin(2{\pi}5t)','fontsize',12)
xlabel('Time (s)');
ylabel('Amplitude');
grid on
subplot(2,1,2)
plot(f,magvin);
title('Power Spectrum of v_i(t)','fontsize',12);
xlabel('Frequency (Hz)');
ylabel('Power');
grid on

% Obtain the Output Waveforms

figure(2)
subplot(2,1,1)
plot(t,vout);
title('v_o(t) = 10sin(2{\pi}5t)+2sin^2(2{\pi}5t)',
  'fontsize',12)
xlabel('Time (s)');
ylabel('Amplitude');
grid on
subplot(2,1,2)
plot(f,magvout);
title('Power Spectrum of v_o(t)','fontsize',12);
xlabel('Frequency (Hz)');
ylabel('Power');
grid on
```

b. The MATLAB script in part (a) is modified for input and output voltage to obtain the third-order response shown in Figure 1.19. As seen from Figure 1.19, the output signal does not have a DC component anymore. The third-order effect shows itself as clipping in the time domain signal and fundamental and third-order components at the output power spectra of the signal.
c. Using the modified MATLAB script in parts (a) and (b), the time domain and frequency domain signals are obtained and illustrated in Figure 1.20.

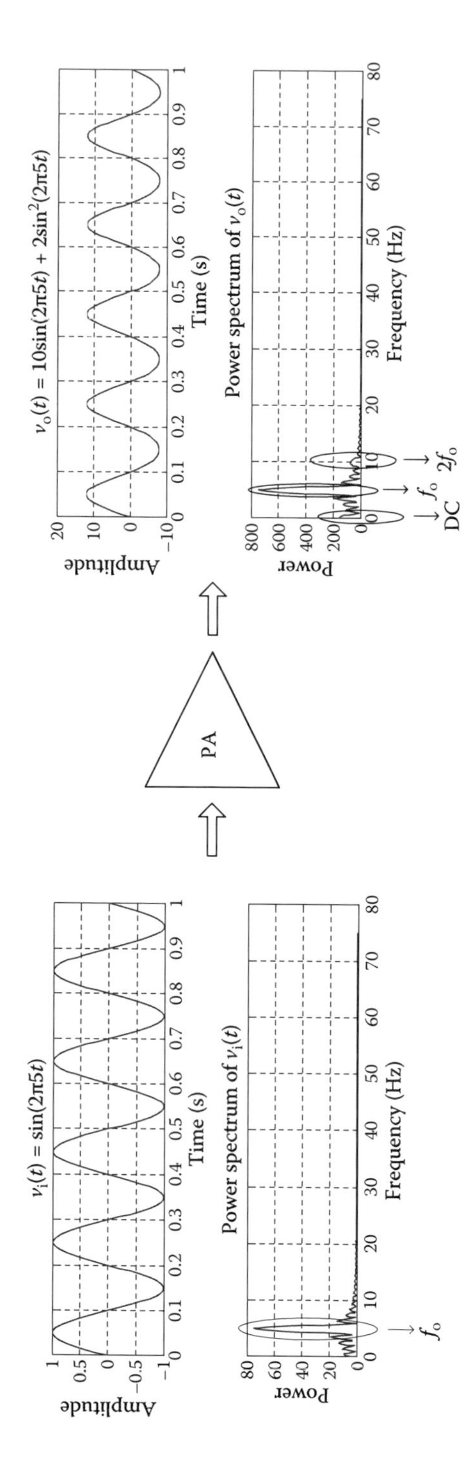

FIGURE 1.18 Second-order nonlinear amplifier output response, which has components at DC, f, and $2f$.

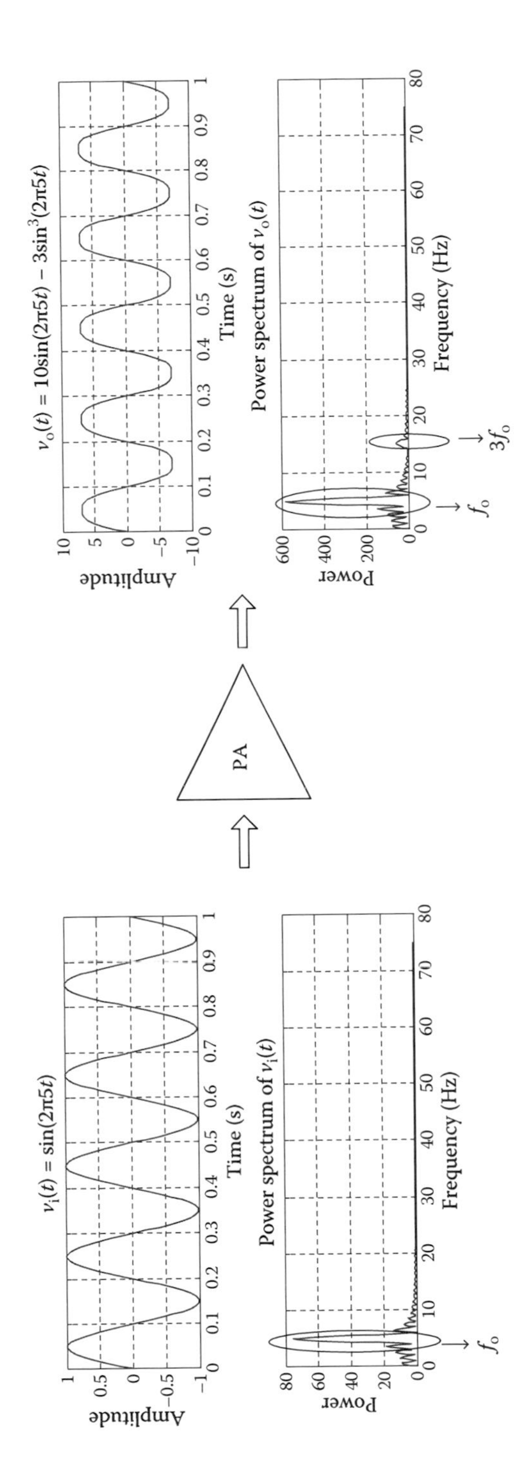

FIGURE 1.19 Third-order nonlinear amplifier output response, which has components at f and $3f$.

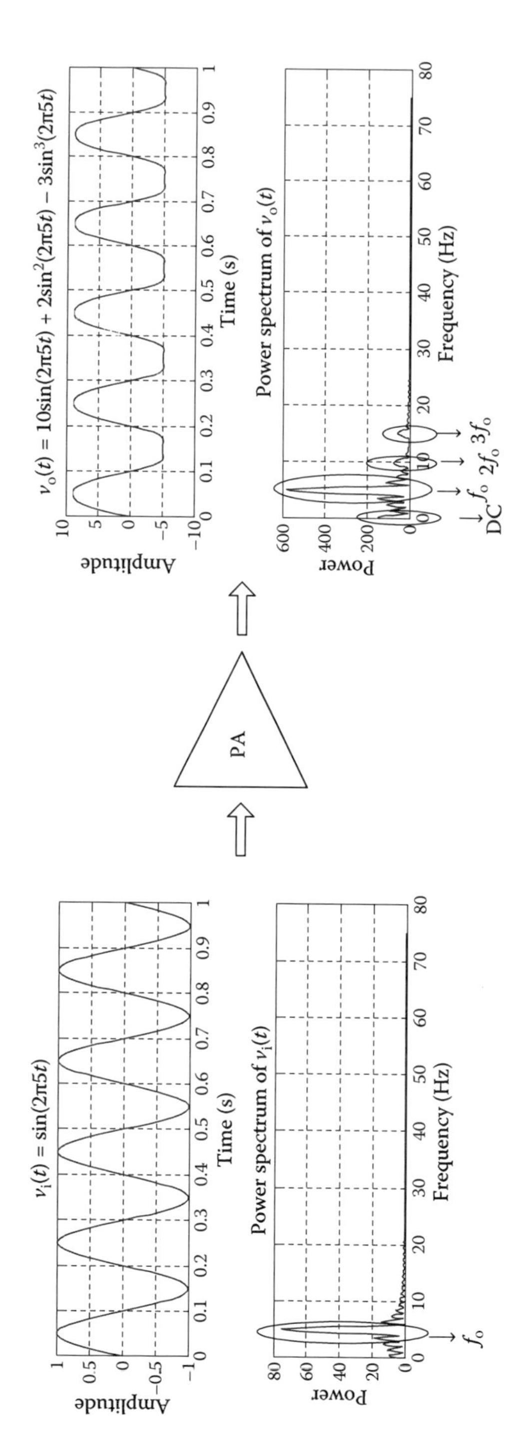

FIGURE 1.20 Nonlinear amplifier response, which has second- and third-order nonlinearities.

As illustrated, the output response has components at DC, f, $2f$, and $3f$. Overall, the level of the nonlinearity response of the amplifier strongly depends on the coefficients of the output signal.

1.4 RF AMPLIFIER CLASSIFICATIONS

In this section, amplifier classes such as classes A, B, and AB for linear mode of operation and classes C, D, E, F, and S for nonlinear mode of operation will be discussed. Classes D, E, F, and S are also known as switch-mode amplifiers.

When the transistor is operated as a dependent current source, the conduction angle, θ, is used to determine the class of the amplifier. The conduction angle varies up to 2π based on the amplifier class. The use of transistor as a dependent current source represents linear mode of operation, which is shown in Figure 1.21. When the transistor is used as a switch, then the amplifier operates in nonlinear mode of operation, and it can be illustrated with the equivalent circuit in Figure 1.22.

The conduction angles, bias, and quiescent points for linear amplifier are shown in Figure 1.23 and illustrated in Table 1.3. Conduction angle, θ, is defined as the duration of the period in which the given transistor is conducting. The full cycle of

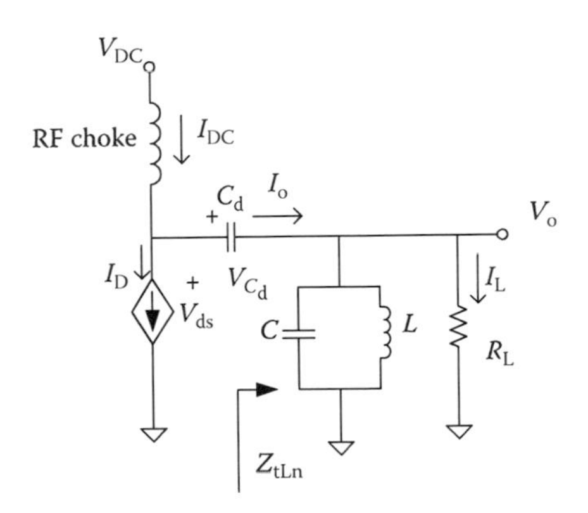

FIGURE 1.21 Equivalent circuit representation of linear amplifier mode of operation.

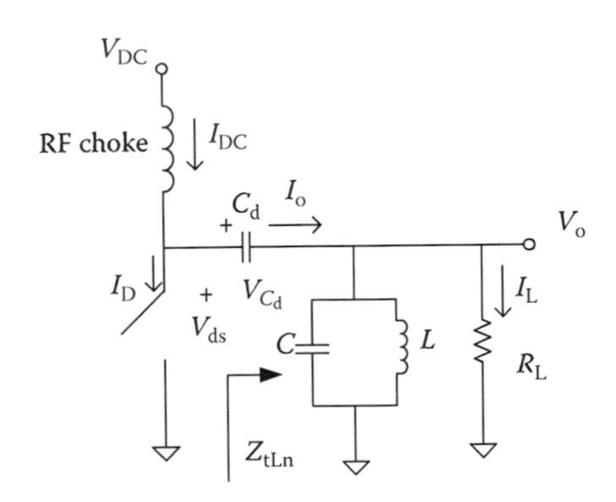

FIGURE 1.22 Equivalent circuit representation of nonlinear amplifier mode of operation.

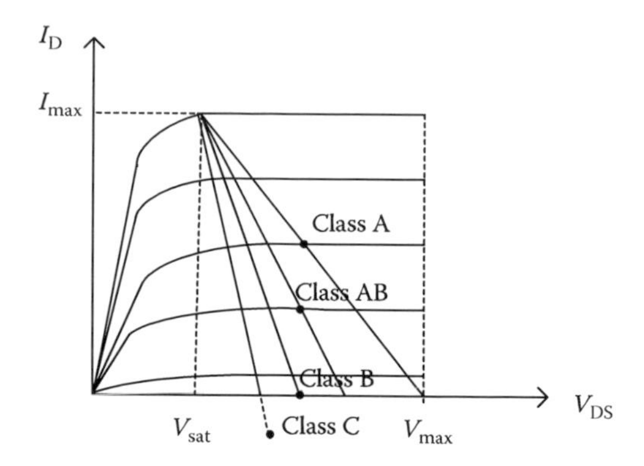

FIGURE 1.23 Load lines and bias points for linear amplifiers.

TABLE 1.3
Conduction Angles, Bias, and Quiescent Points for Linear Amplifiers

Class	Bias Point	Quiescent Point	Conduction Angle
A	0.5	0.5	2π
AB	0–0.5	0–0.5	π–2π
B	0	0	π
C	<0	0	0–π

conduction is considered to be 360°. The points of intersection with the load line are known as the "quiescent" conditions or "Q points" or the DC bias conditions for the transistor and represent the operational device voltages and drain current, as shown in Figure 1.23.

In our analysis, MOSFET is used as an active device. The ideal MOSFET transfer characteristics are illustrated in Figure 1.24. There is a maximum drain current level for a corresponding gate–source voltage, V_{GS}, that a MOSFET conducts. In the cutoff region, the gate–source voltage, V_{GS}, is less than the threshold voltage or saturation voltage, V_{sat}, and the device is an open circuit or off. In the ohmic region, the device acts as a resistor with an almost constant on-resistance R_{DSon} and is equal to the ratio of drain voltage, V_{DS}, and the drain current I_D. In the linear mode of operation, the device operates in the active region where I_D is a function of the gate–source voltage V_{GS} and is defined by

$$I_D = K_n(V_{GS} - V_{th})^2 = g_m(V_{GS} - V_{th}) \tag{1.68}$$

where K_n is a parameter depending on the temperature and device geometry, and g_m is the current gain or transconductance. When V_{DS} is increased, the positive drain potential opposes the gate–voltage bias and reduces the surface potential in the

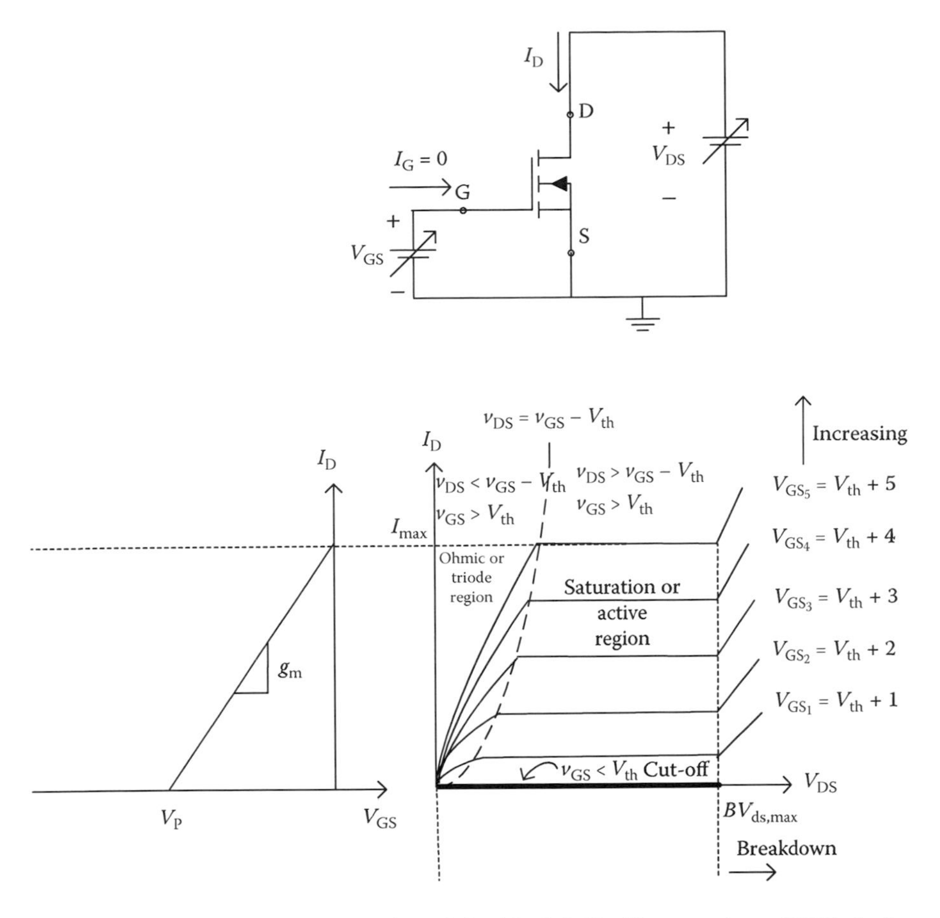

FIGURE 1.24 Transfer characteristics of the ideal field effect transistor (FET) device, N-type metal oxide semiconductor (NMOS).

channel. The channel inversion-layer charge decreases with increasing V_{DS} and, ultimately, becomes zero when the drain voltage is equal to $V_{GS} - V_{th}$. This point is called the "channel pinch-off point," where the drain current becomes saturated.

When the operational drain current, I_D, at a given V_{GS} goes above the ohmic or linear region "knee or saturation point," any further increase in drain current results in a significant rise in the drain–source voltage, V_{DS}, as shown in Figure 1.24. This results in a rise in conduction loss. If power dissipation is high and over the limit that the transistor can handle, then the device may catastrophically fail. When V_{GS} is less than the threshold or pinch-off voltage, V_P, then the device does not conduct, and I_D becomes 0. As V_{GS} increases, the transistor enters the saturation or active region, and I_D increases in a nonlinear fashion. It will remain almost constant until the transistor gets into the breakdown region. The characteristics of I_D vs. V_{GS} in saturation region are illustrated in Figure 1.25. The simplified device model for MOSFET in each region is illustrated in Figure 1.26.

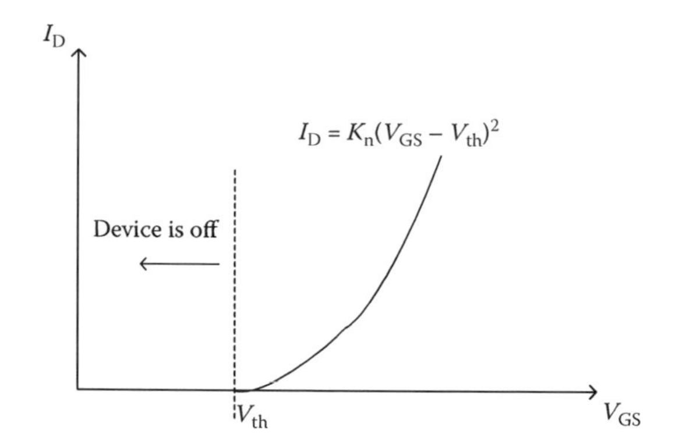

FIGURE 1.25 I_D vs. V_{GS} in the saturation region for the ideal NMOS device.

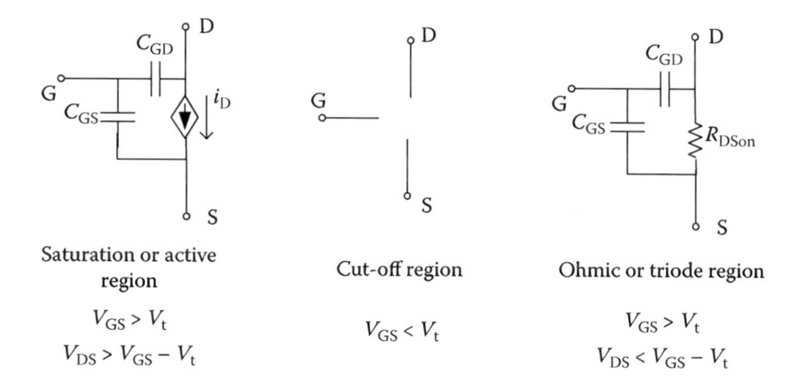

FIGURE 1.26 Simplified large-signal model for the NMOS FET device.

Example

(a) State if the following transistors in Figure 1.27 are operating in the saturation region, and (b) what should be the value of the gate voltage, V_G, for the transistor to operate in the saturation region? Assume that $V_t = 1$ V.

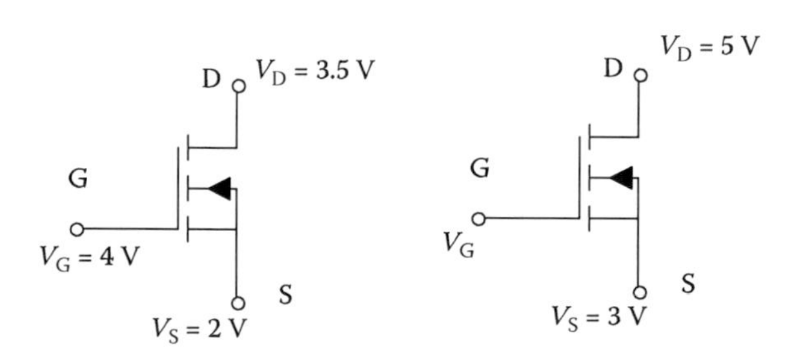

FIGURE 1.27 Operational region of transistors.

Solution

The condition to operate in the saturation region is

$$V_{GS} > V_t$$

$$V_{DS} > V_{GS} - V_t$$

a. $V_{GS} = 4 - 2 = 2$ [V] $> V_t = 1$ [V] and $V_{DS} = 3.5 - 2 = 1.5$ [V] $> V_{GS} - V_t = 2 - 1 = 1$ [V]. So, the first transistor is operating in the saturation region.
b. From the first condition, for the transistor to operate in the saturation region, $V_{GS} > 1$ [V]. That requires gate voltage $V_G > 4$ [V]. In addition, it is required that $V_{GS} < V_{DS} + V_t = 2 + 1 = 3$ [V]. The overall solution is

$$1 < V_{GS} < 3$$

Since the source voltage is $V_S = 2$ V, then,

$$1 < V_G - 2 < 3 \quad \text{or} \quad 3 < V_G < 5$$

1.4.1 Conventional Amplifiers—Classes A, B, and C

In amplifier classes A, B, and C, the transistor can be modeled as the voltage-dependent current source and is shown in Figure 1.28. Most of the time, the class of these amplifiers is also known as conventional amplifiers.

The current i_D flowing through the device, the drain-to-source voltage, and the voltage applied at the gate to source of the transistor in Figure 1.28 are

$$i_D = I_{DC} + I_m\cos(\theta) \tag{1.69}$$

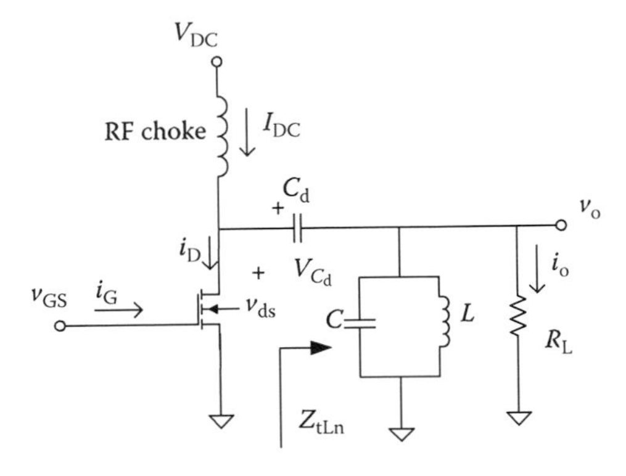

FIGURE 1.28 Conventional RF PA classes: classes A, B, and C.

$$v_{DS} = V_{DC} - V_m \cos(\theta) \tag{1.70}$$

$$v_{GS} = V_t + V_{gsm} \cos(\theta) \tag{1.71}$$

where $\theta = \omega t$. The DC component of the drain current and the drain-to-source voltage are equal to I_{DC} and V_{DC}, and the AC component is $I_m \cos(\theta)$, $-V_m \cos(\theta)$ as given by

$$I_D = I_{DC}, \quad V_{DS} = V_{DC} \tag{1.72}$$

$$i_D = I_m \cos(\theta), \quad v_{DS} = -V_m \cos(\theta) \tag{1.73}$$

Fourier integrals can be used to determine the DC power and output of the amplifier. The drain current, i_D, can be represented using Fourier series expansion using sine or cosine functions, which are harmonically related as

$$i_D(t) = I_o + \sum_{n=1}^{\infty} I_{an} \cos n\omega_o t + I_{bn} \sin n\omega_o t \tag{1.74}$$

$$v_{DS}(t) = V_o + \sum_{n=1}^{\infty} V_{an} \cos n\omega_o t + V_{bn} \sin n\omega_o t \tag{1.75}$$

The odd and even harmonic coefficients can be combined into a single cosine (or sine) to give

$$i_D(t) = I_o + \sum_{n=1}^{\infty} I_n \cos(n\omega_o t + \alpha_n) \tag{1.76}$$

$$v_{DS}(t) = V_o + \sum_{n=1}^{\infty} V_n \cos(n\omega_o t + \beta_n) \tag{1.77}$$

where

$$I_n = \sqrt{I_{an}^2 + I_{bn}^2} \tag{1.78}$$

and

$$V_n = \sqrt{V_{an}^2 + V_{bn}^2} \tag{1.79}$$

and

$$\alpha_n = \tan^{-1}\left(\frac{I_{bn}}{I_{an}}\right) \tag{1.80}$$

and

$$\beta_n = \tan^{-1}\left(\frac{V_{bn}}{V_{an}}\right) \tag{1.81}$$

The impedance at the transistor load line of the transistor is found from

$$Z_{tLn} = \frac{V_n e^{j\alpha}}{I_n e^{j\beta}} = \left|Z_{tLn}\right| e^{j\phi_n} \tag{1.82}$$

where $\phi_n = (\alpha_n - \beta_n)$. The Fourier coefficients I_o, I_{an}, and I_{bn} for the drain current are calculated from

$$I_o = \frac{1}{T}\int_{-T/2}^{T/2} i_D(t)\,dt \tag{1.83}$$

$$I_{an} = \frac{2}{T}\int_{-T/2}^{T/2} i_D(t)\cos(k\omega_o t)\,dt \tag{1.84}$$

$$I_{bn} = \frac{2}{T}\int_{-T/2}^{T/2} i_D(t)\sin(k\omega_o t)\,dt \tag{1.85}$$

where $\omega_o = 2\pi/T$. The fundamental component in Equation 1.82 is obtained when $n = 1$, and the DC components are found from Equation 1.83 as follows:

$$I_o = \frac{1}{2\pi}\int_{-\pi}^{\pi} (I_{DC} + I_m\cos(\theta))\,d\theta = I_{DC} \tag{1.86}$$

$$I_{an} = \frac{1}{\pi}\int_{-\pi}^{\pi}(I_{DC} + I_m \cos(\theta))\cos(n\theta)\,d\theta = I_m \quad \text{when } n = 1 \text{ otherwise } I_{an} = 0 \tag{1.87}$$

$$I_{bn} = \frac{1}{\pi}\int_{-\pi}^{\pi}(I_{DC} + I_m \cos(\theta))\sin(n\theta)\,d\theta = 0 \text{ for all } n \tag{1.88}$$

The same analysis and derivation can also be repeated for the drain-to-source voltage. Hence, the DC and the fundamental components of the drain current and the drain-to-source voltage at the resonant frequency, f_o, are

$$I_o = I_{DC}, \quad V_o = V_{DC} \tag{1.89}$$

$$I_1 = I_m, \quad V_1 = V_m \tag{1.90}$$

DC power, P_{DC}, from supply is then calculated from

$$P_{DC} = V_{DC} I_{DC} \tag{1.91}$$

The power delivered from the device to the output is represented by P_o and calculated at the fundamental frequency, which is the resonant frequency of the LC network and is obtained from Equation 1.90 as

$$P_o = \frac{1}{2} V_m I_m \tag{1.92}$$

The more general expression when operational frequency is not equal to resonant frequency for the output power can be found at the fundamental and harmonic frequencies as

$$P_{on} = \frac{1}{2} V_n I_n \cos(\phi_n), \quad n = 1, 2, \ldots \tag{1.93}$$

The transistor dissipation is calculated using

$$P_{diss} = \frac{1}{T}\int_{0}^{T} v_{DS}(t) i_D(t)\,dt \tag{1.94}$$

which is also equal to

$$P_{\text{diss}} = P_{\text{DC}} - \sum_{n=1}^{\infty} P_{o,n} \tag{1.95}$$

The drain efficiency is the ratio of the output power to DC supply power and calculated using Equations 1.91 and 1.93 as

$$\eta = \frac{P_{o,n}}{P_{\text{DC}}} = \frac{P_{o,n}}{P_{\text{diss}} + P_{o,n}} \tag{1.96}$$

The maximum efficiency is obtained when

$$P_{\text{diss}} + \sum_{n=2}^{\infty} P_{o,n} = 0$$

Then, the maximum efficiency from Equation 1.96 is found to be $\eta_{\text{max}} = 100\%$.

The drive input power is calculated using Equation 1.71 for gate-to-source voltage, v_{GS}, and gate current, i_G, from

$$P_G = \frac{1}{T}\int_0^T v_{GS}(t) i_G(t)\,\mathrm{d}t \tag{1.97}$$

1.4.1.1 Class A

The bias point in class A mode of operation is selected at the center of the *I–V* curve between the saturation voltage, V_{sat}, and the maximum operational transistor voltage, V_{max}, as shown in Figure 1.29, whereas the DC current for the class A amplifier is biased between 0 and the maximum allowable current, I_{max}. The conduction angle for the transistor for class A operation is 2π, which means that the transistor conducts the entire RF cycle. The typical load line and drain-to-source voltage and drain current waveforms are illustrated in Figures 1.29 and 1.30, respectively.

The maximum efficiency for class A amplifier happens when the drain voltage swings from 0 to 2 V_{DC}. The DC power for this case is

$$P_{\text{DC}} = V_{\text{DC}} I_{\text{DC}} = \frac{V_{\text{DC}}^2}{R_L} \tag{1.98}$$

and the RF output power is

$$P_o = \frac{1}{2}\frac{V_{\text{DC}}^2}{R_L} \tag{1.99}$$

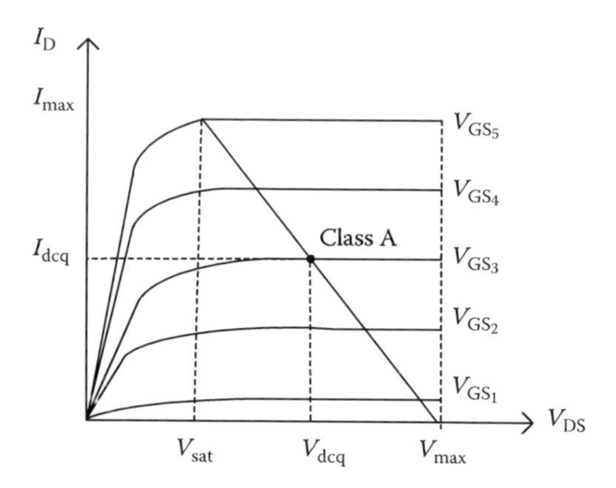

FIGURE 1.29 Typical load line for class A amplifiers.

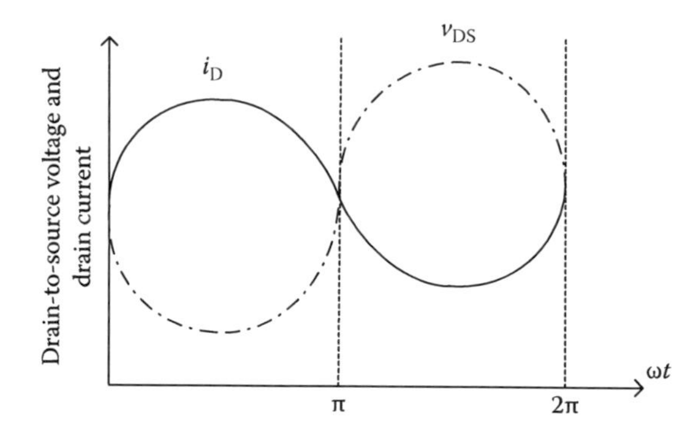

FIGURE 1.30 Typical drain-to-source voltage and drain current for class A amplifiers.

Then, the efficiency from Equations 1.13, 1.98, and 1.99 is

$$\eta_{max} = \frac{P_{out}}{P_{DC}} = 0.5 \tag{1.100}$$

1.4.1.2 Class B

Transistor power dissipation due to its 360° conduction angle for class A amplifiers significantly limits the amplifier RF output power capacity. The power dissipation in the active device can be reduced if the device is biased to conduct less than the full RF period. The transistor is turned on only one-half of the cycle, and as a result, the conduction angle for class B amplifiers is $\theta = 180°$. Class B amplifiers can be implemented as a single-ended amplifier when narrow band is required or transformed coupled push–pull configuration when high linear output power is desired. The typical load line and drain-to-source voltage and drain current waveforms are illustrated in Figures 1.31 and 1.32, respectively.

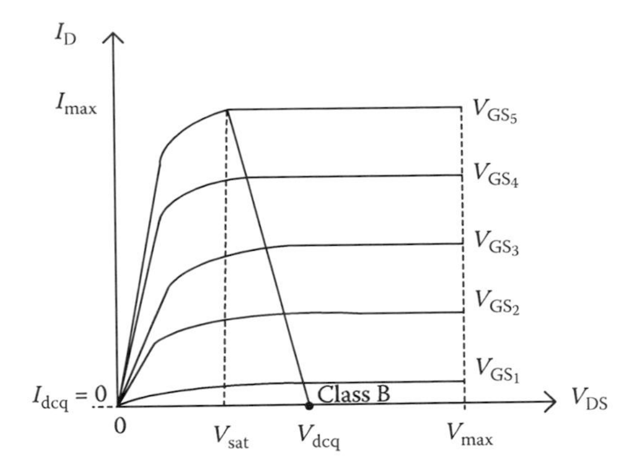

FIGURE 1.31 Typical load line for class B amplifiers.

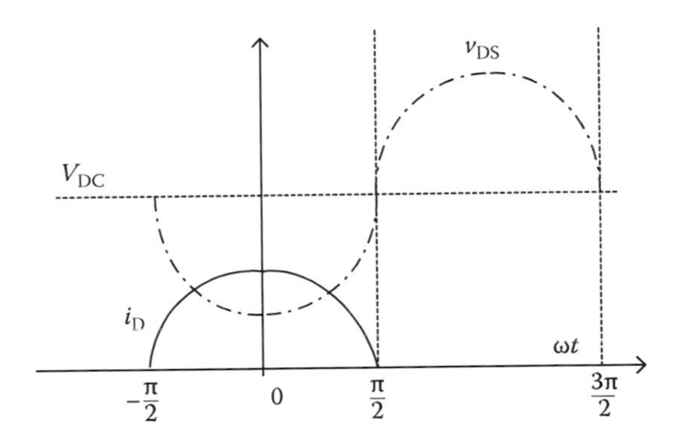

FIGURE 1.32 Typical drain-to-source voltage and drain current for class B amplifiers.

The maximum efficiency for class B amplifier occurs when $V_m = V_{DC}$. Under this condition, the DC or the input power is

$$P_{DC} = \frac{2}{\pi} I_m V_{DC} \tag{1.101}$$

and the RF output power is

$$P_o = \frac{1}{2} I_m V_m = \frac{1}{2} I_m V_{DC} \tag{1.102}$$

So, the maximum efficiency is equal to

$$\eta_{max} = \frac{P_{out}}{P_{DC}} = \frac{\pi}{4} \approx 0.7853 \tag{1.103}$$

1.4.1.3 Class AB

In class B mode of operation, amplifier efficiency is sacrificed for linearity. When it is desirable to have an amplifier with better efficiency than the class A amplifier, and yet better linearity than the class B amplifier, then class AB is chosen as a compromise. The conduction angle for the class AB amplifier is between 180° and 360°. As a result, the bias point for class AB amplifiers is chosen between the bias points for class A and class B amplifiers. Class AB amplifiers are widely used in RF applications when linearity and efficiency together become requirements. The ideal efficiency of class AB amplifiers is between 50% and 78.53%. The typical drain-to-source voltage and drain current waveforms are illustrated in Figure 1.33.

1.4.1.4 Class C

Class A, B, and AB amplifiers are considered to be linear amplifiers where the phase and amplitude of the output signal are linearly related to the amplitude and phase of the input signal. If efficiency is a more important parameter than linearity, then nonlinear amplifier classes such as class C, D, E, or F can be used. The conduction angle for class C amplifier is less than 180°, which makes this amplifier class have higher efficiencies than class B amplifiers. The typical drain-to-source voltage and drain current waveforms for class C amplifiers are illustrated in Figure 1.34.

The drain efficiency for PA classes A, B, and C can also be calculated using the conduction angle, θ, as

$$\eta = \frac{\theta - \sin\theta}{4[\sin(\theta/2) - (\theta/2)\cos(\theta/2)]} \tag{1.104}$$

The maximum drain efficiency for the class C amplifier is obtained when θ = 0°. However, the output power decreases very quickly when the conduction angle approaches as shown by

$$P_o \propto \frac{\theta - \sin(\theta)}{1 - \cos(\theta/2)} \tag{1.105}$$

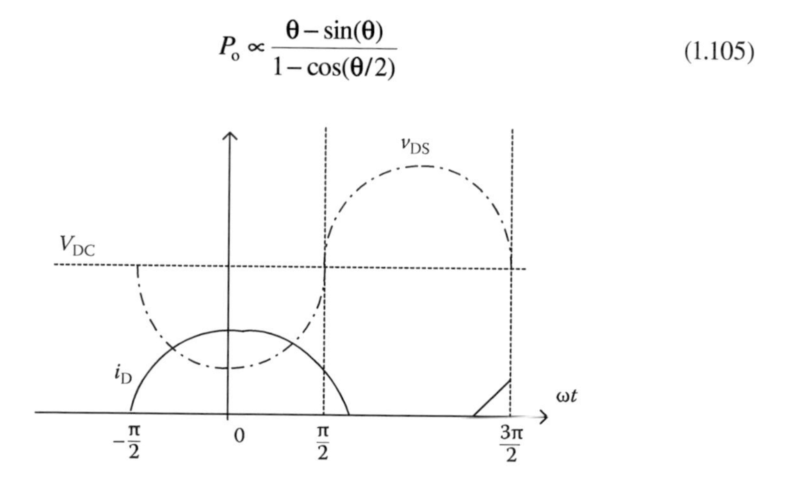

FIGURE 1.33 Typical drain-to-source voltage and drain current for class AB amplifiers.

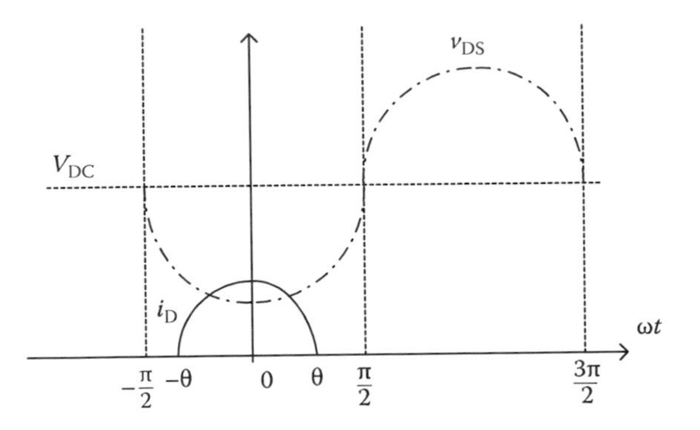

FIGURE 1.34 Typical drain-to-source voltage and drain current for class C amplifiers.

As a result, it is not feasible to obtain 100% efficiency with class C amplifiers. The typical class C amplifier efficiency in practice is between 75% and 80%.

The efficiency distribution and power capacity of conventional amplifiers, classes A, B, AB, and C, are illustrated in Figures 1.35 and 1.36, respectively.

1.4.2 Switch-Mode Amplifiers—Classes D, E, and F

Class A, AB, and B amplifiers have been used for linear applications where amplitude modulation (AM), single-sideband modulation, and quadrate AM might be required. Classes C, D, E, and F are usually implemented for narrow band-tuned amplifiers when high efficiency is desired with high power. Classes A, B, AB, and C are operated as transconductance amplifiers, and the mode of operation depends on the conduction angle. In switch-mode amplifiers such as classes D, E, and F, the active device is intentionally driven into the saturation region, and it is operated as a switch rather than a current source unlike class A, AB, B, or C amplifiers, as shown in Figure 1.22. In theory, power dissipation in the transistor can be totally eliminated, and hence, 100% efficiency can be achieved for switching-mode amplifiers.

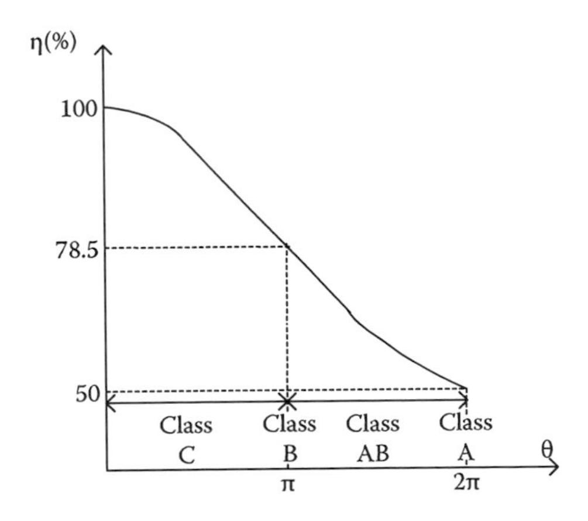

FIGURE 1.35 Efficiency distribution of conventional amplifiers.

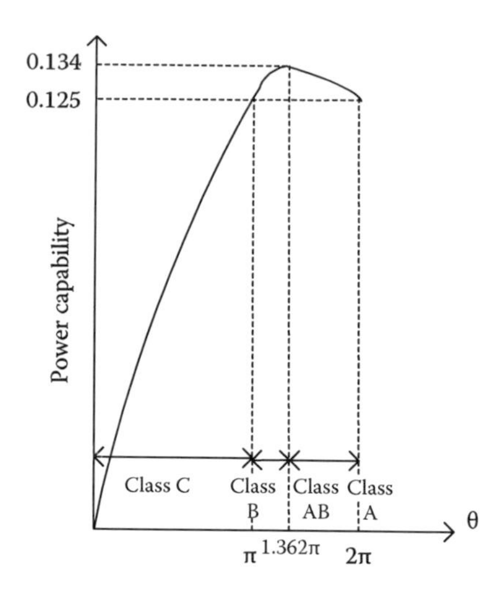

FIGURE 1.36 Power capability of conventional amplifiers.

1.4.2.1 Class D

Class D amplifiers have two pole-switching operations of transistors, either in voltage-mode (VM) configuration that uses a series resonator or current-mode (CM) configuration that uses a parallel resonator circuit. There are several implementation methods for class D amplifiers. The complementary version of the voltage switching mode class D amplifier is shown in Figure 1.37. In the operation of the class D amplifier, transistors act as switches, and they turn on and off alternately. The series resonator circuit composed of L_o and C_o resonates at the operational frequency and tunes the amplifier output circuit to provide sinusoidal output current waveform. The voltage applied to the resonator circuit, $v_d(t)$, and the output sinusoidal current, $i_o(t)$, flowing through R_L can be represented as

$$v_d(t) = \begin{cases} V_{DC}, & 0 \le \theta \le \pi \\ 0 & \pi \le \theta \le 2\pi \end{cases} \tag{1.106}$$

$$i_o(t) = I_m \sin(\theta) = \frac{2}{\pi}\frac{V_{DC}}{R_L}\sin(\theta) \tag{1.107}$$

$$v_o(t) = V_m \sin(\theta) = \frac{2}{\pi} V_{DC} \sin(\theta) \tag{1.108}$$

where $\theta = \omega t$ and

$$I_m = I_{DC} = \frac{2}{\pi}\frac{V_{DC}}{R_L} \tag{1.109}$$

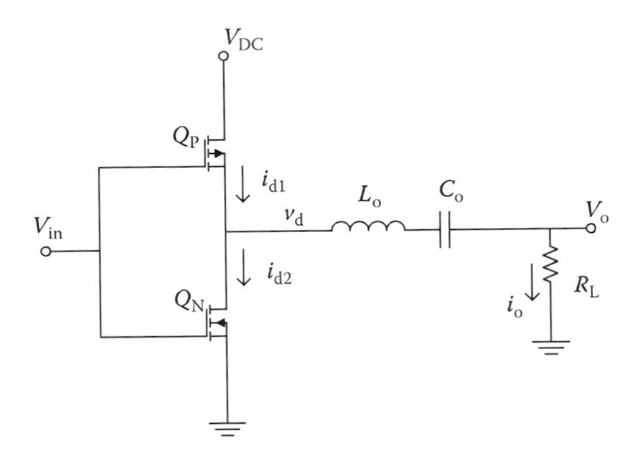

FIGURE 1.37 Complementary voltage switching class D amplifier.

The voltage and current waveforms are illustrated in Figure 1.38. The output and input powers are calculated from Equations 1.107 through 1.109 as

$$P_o = v_o(t)i_o(t) = \frac{2}{\pi^2}\frac{V_{DC}^2}{R_L} \tag{1.110}$$

$$P_{DC} = V_{DC}I_{DC} = \frac{2}{\pi^2}\frac{V_{DC}^2}{R_L} \tag{1.111}$$

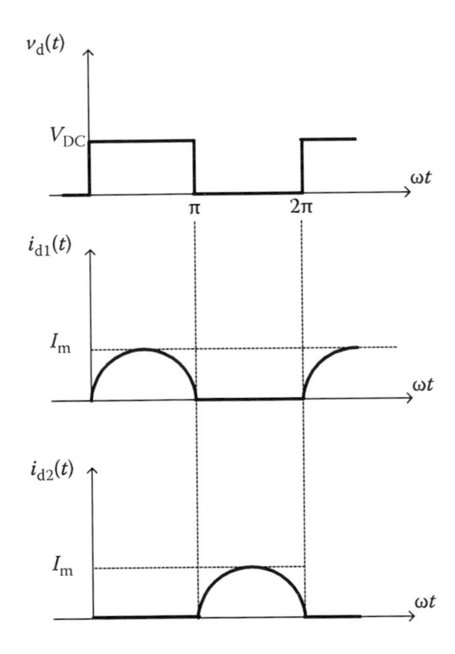

FIGURE 1.38 Voltage and current waveforms for complementary voltage switching class D amplifier.

As a result, in theory, for class D amplifiers, $P_o = P_{DC}$, so the efficiency is equal to 100%, as shown in the following equation:

$$\eta(\%) = \frac{P_o}{P_{DC}} \times 100 = 100\% \tag{1.112}$$

1.4.2.2 Class E

The basic analysis of the class D amplifiers shows that it is possible to obtain 100% efficiency in theory modeling the active devices as ideal switches. However, this is not accurate specifically at higher frequencies as device parasitics such as device capacitances play an important role in determining the amplifier performance, which makes the class D amplifier mode of operation challenging. This challenge can be overcome by making parasitic capacitance of the transistor as part of the tuning network as in class E amplifiers, as shown in Figure 1.39. The class E amplifier shown in Figure 1.39a consists of a single transistor that acts as a switch S, RF choke, a parallel-connected capacitance C_p, a resonator circuit L–C, and a load R_L, as shown in Figure 1.39b. With the application of the input signal, switch S is turned

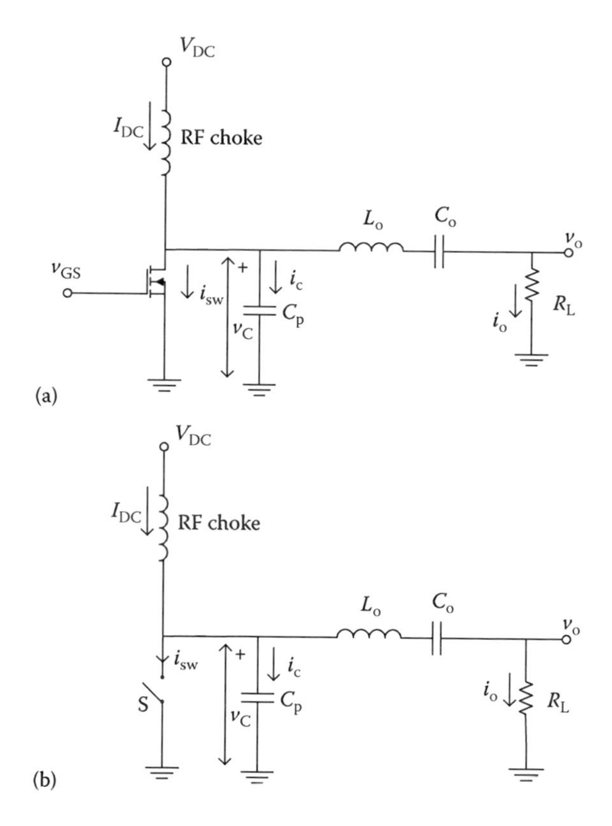

FIGURE 1.39 (a) Simplified circuit of class E amplifier. (b) Simplified class E amplifier with switch S.

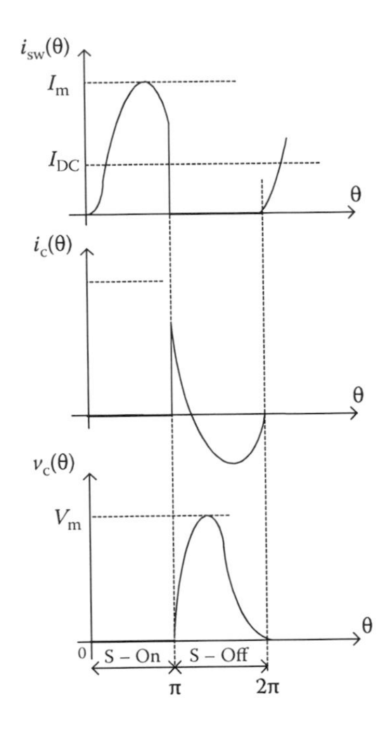

FIGURE 1.40 Voltage and current waveforms for basic class E amplifier.

on in half of the period, and off in the other half. When S is on, the voltage across S is zero, and when it is off, the current through S is zero. A high Q resonator circuit, LC network, produces a sinusoidal output signal at the output of the amplifier. The current and voltage waveforms for the basic class E amplifier in Figure 1.39 are illustrated in Figure 1.40.

1.4.2.3 Class DE

In practical applications, class D amplifiers suffer from drastic switching loss due to the fact that device capacitances are charged and discharged every switching cycle and dissipate energy. This results in significant reduction of efficiency for class D amplifiers. Class E amplifiers overcome this problem by utilizing the device capacitance and making it part of the tuning network so that zero-voltage switching is accomplished by turning the switch on only when the shunt capacitance connected has been discharged. However, one of the disadvantages of class E amplifiers is the higher voltage stress on the switches. In ideal class E operation, the peak voltage on the device is 3.6 times more than class D when the same DC supply voltage is applied.

Class DE PAs bring the advantages of both class D and E amplifiers. It is free of switching losses and has increased efficiency vs. class D amplifiers, and it has less voltage stress on its transistors in comparison to class E amplifiers. However, it is a challenge to drive class DE amplifiers with rectangular pulse signal due to the necessity of control of accurate dead time at high frequencies; as a result, the sinusoidal input signal is preferred. The voltage and current waveform for the class DE amplifier shown in Figure 1.41 with duty ratio, $D = 0.25$, are given in Figure 1.42.

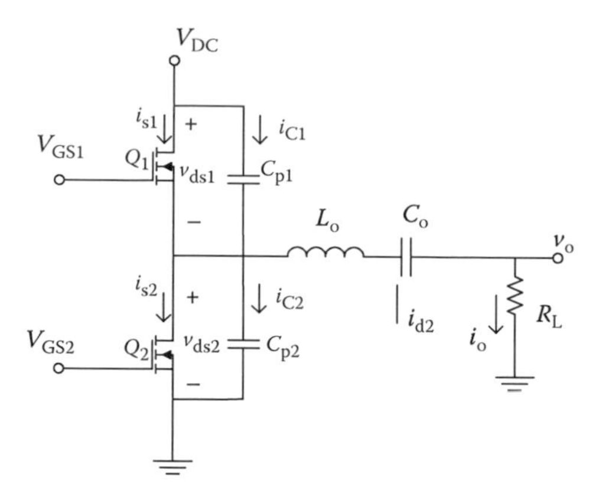

FIGURE 1.41 Class DE amplifier circuit.

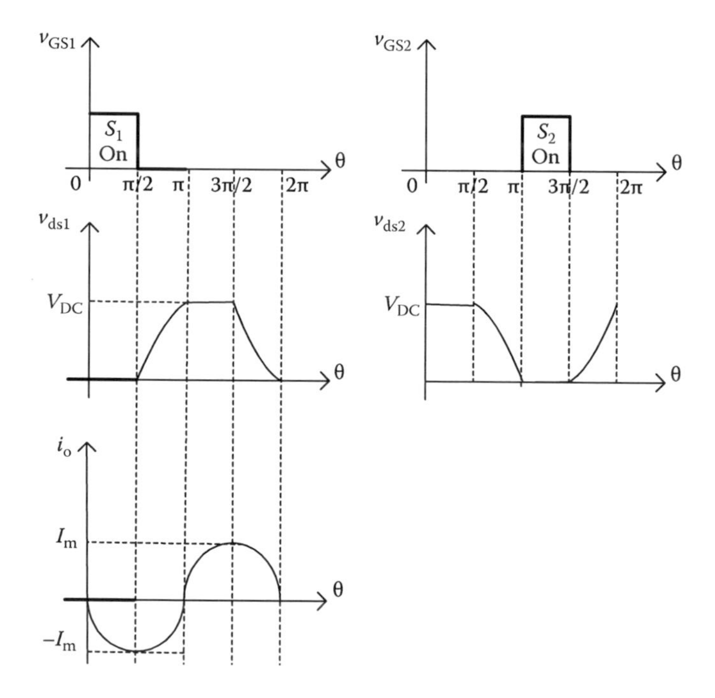

FIGURE 1.42 Class DE amplifier voltage and current waveforms.

1.4.2.4 Class F

Class F amplifiers carry similar characteristics to class B or C amplifiers, which use a single-resonant load network to produce simple sinusoid at the resonant frequency. The power capacity and efficiency obtained for class B and C amplifiers can be improved by introducing the harmonic terminations of the load network as in class F amplifiers. So the load network in the class F amplifier resonates at the operational frequency as well as one or more harmonic frequencies. The multiresonant load network in the class F amplifier helps in controlling the harmonic contents of the drain voltage and current and wave shapes to minimize the overlap region between them to reduce the transistor power dissipation. This results in improvement of both efficiency and power capacity. In theory, an ideal class F amplifier can control an infinite number of harmonics; it has a square voltage waveform and can give 100% efficiency. However, in practical applications, it is very difficult to control more than the fifth harmonics with class F amplifiers. The schematics of the basic class F amplifier with the third harmonic called class F_3 peaking and its voltage and current waveforms are given in Figures 1.43 and 1.44, respectively. The class F amplifier shown

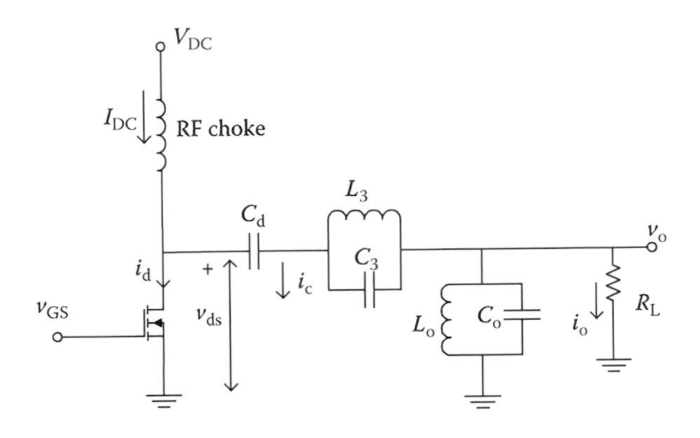

FIGURE 1.43 Class F amplifier with third harmonic peaking.

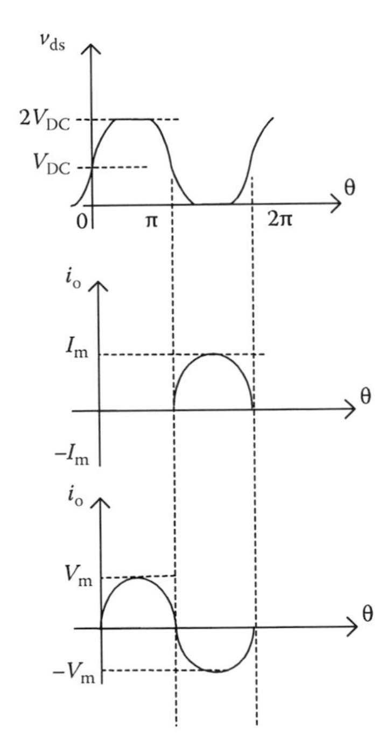

FIGURE 1.44 Voltage and current waveforms for class F amplifier with third harmonic peaking.

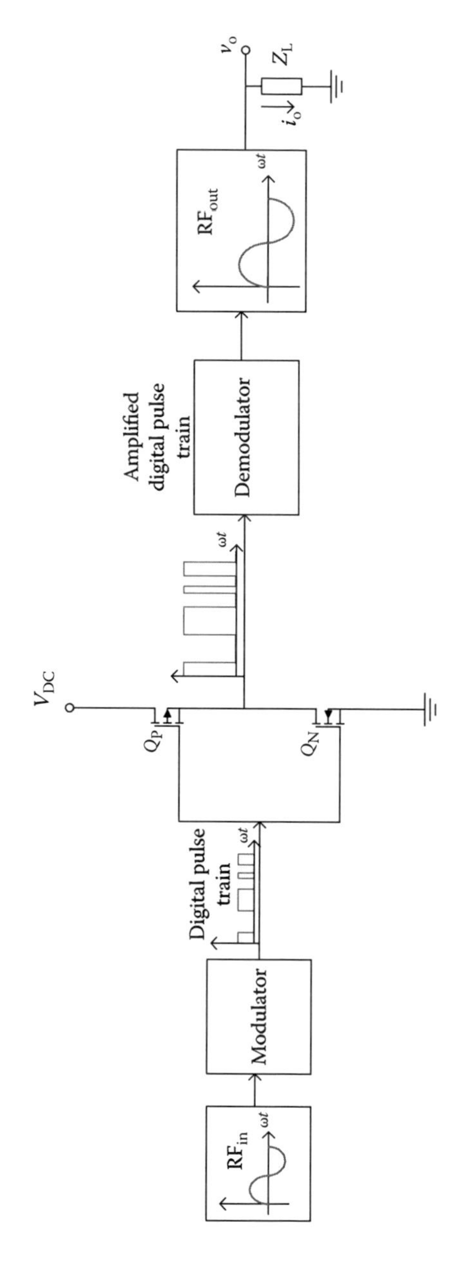

FIGURE 1.45 VM class S amplifier configuration.

in Figure 1.43 has two parallel LC resonators, which are tuned to center frequency, f_o, and third harmonic frequency $3f_o$.

1.4.2.5 Class S

The class S amplifier is based on switching of two transistors similar to the class D amplifier concept. The main difference between class D and S amplifiers is the way in which the amplifier is driven. According to the class S amplifier operation, the analog input signal is converted into a digital pulse train via a modulator instead of the signal being alternately switched at the carrier frequency with a constant duty cycle. The fully digital pulse train then feeds the power-switching final-stage amplifiers, which in turn amplify it to the proper power level. In the ideal case, no overlapping occurs between current and voltage waveforms, and hence no power loss exists, which leads to a 100% efficiency independently of the power back-off. A demodulator is required at the output network to pick the required signal frequency and to restore the analog input signal. The amplifier can be implemented in VM or CM configurations. The VM configuration of class S amplifiers with waveforms is given in Figure 1.45.

The summary of some of the basic linear and nonlinear amplifier performance parameters including efficiency, normalized RF power, normalized maximum drain voltage swing, and power capability is given in Table 1.4. Table 1.5 compares each amplifier class based on transistor operation and application and gives the advantages and disadvantages of each class.

1.5 HIGH-POWER RF AMPLIFIER DESIGN TECHNIQUES

RF amplifier output power capacity can be increased in several ways. One of the ways is to use higher voltage and current-rated active devices for the frequency of operation. However, this has many implications such as availability of devices, increased size, and other implementation issues. When use of higher voltage and current-rated devices is not feasible, the output power capacity can also be increased

TABLE 1.4
Summary of Basic Amplifier Performance Parameters

Amplifier Class	Max Efficiency $\eta_{max}(\%)$	Normalized RF Output Power $\frac{P_{o,max}}{V_{dc}^2/2R_L}$	Normalized $V_{ds,max}$ $\frac{V_m}{V_{dc}}$	Normalized $I_{d,max}$ $\frac{I_m}{I_{dc}}$	Power Capability $\frac{P_{o,max}}{V_m I_m}$
A	50	1	2	2	0.125
B	78.5	1	2	$\pi = 3.14$	0.125
C	86 ($\theta = 71°$)	1	2	3.9	0.11
D	100	$16/\pi^2 = 1.624$	2	$\pi/2 = 1.57$	$1/\pi = 0.318$
E	100	$4/(1 + \pi^2/4) = 1.154$	3.6	2.86	0.098
F	100	$16/\pi^2 = 1.624$	2	$\pi = 3.14$	$1/2\pi = 0.318$

TABLE 1.5
Summary of Basic Amplifier Performance Parameters

Amplifier Class	Mode	Transistor (Q) Operation	Pros	Cons
A	Linear	Always conducting	Most linear, lowest distortion	Poor efficiency
B	Linear	Each device is on half cycle	$\eta_B > \eta_A$	Worse linearity than class A
AB	Linear	Mid-conduction	Improved linearity with respect to class B	Power dissipation for low signal levels higher than class B
C	Nonlinear	Each device is on half cycle	High P_o	Inherent harmonics
D	Switch mode	Q_1 and Q_2 switched on/off alternately	Max efficiency and best power	Device parasitics are issued at high frequencies
E	Switch mode	Transistor is switched on/off	Max efficiency, no loss due to parasitics	High voltage stress on transistor
F	Switch mode	Transistor is switched on/off	Max efficiency and no harmonic power delivered	Power loss due to discharge of output capacitance
S	Switch mode	Q_1 and Q_2 are switched on/off with a modulated signal	Wider DR and high efficiency	Upper frequency range is limited

by using push–pull configuration and/or parallel transistor configuration. The following discussion describes how these two techniques can be implemented.

1.5.1 Push–Pull Amplifier Configuration

Practical RF PA design uses push–pull configuration widely to meet with the demand of high output power. The input drive signal for push–pull amplifiers is outphased by 180° using transformers such as input balun. Output balun is used at the output to combine the outphased amplifier output signal and double the RF power. The RF input voltage signal at operational frequency, ignoring the phases and assuming ideal conditions for simplicity, can be expressed as

$$v_s(t) = 4V_m \sin(\omega t) \tag{1.113}$$

Assuming the matched impedance case, the signal at the input of the balun is then

$$v_{in}(t) = \frac{v_s(t)}{2} = 2V_m \sin(\omega t) \tag{1.114}$$

The signal at the output of the balun will be equally split and phased by 180°. These signals at the input of the amplifiers can be represented as

$$v_1(t) = V_m \sin(\omega t) \tag{1.115}$$

$$v_2(t) = V_m \sin(\omega t + \theta) = -V_m \sin(\omega t) \tag{1.116}$$

where $\theta = 180°$.

The signals $v_1(t)$ and $v_2(t)$ will be amplified by the amplifiers by their corresponding gains, A_1 and A_2, respectively. The signals at the output of the amplifiers are then equal to

$$v_1(t) = A_1 V_m \sin(\omega t) \tag{1.117}$$

$$v_2(t) = -A_2 V_m \sin(\omega t) \tag{1.118}$$

The amplifier output signals are then combined via the output balun. The final load signal is

$$v_L(t) = v_1(t) - v_2(t) = (A_1 + A_2)V_m \sin(\omega t) \tag{1.119}$$

The illustration of the push–pull amplifier with waveforms is given in Figure 1.46.

1.5.2 Parallel Transistor Configuration

RF power output in the amplifier system can be increased by also paralleling the transistors. When transistors are paralleled, the current is increased proportional to the number of transistors used under ideal conditions. This can be illustrated in Figure 1.47.

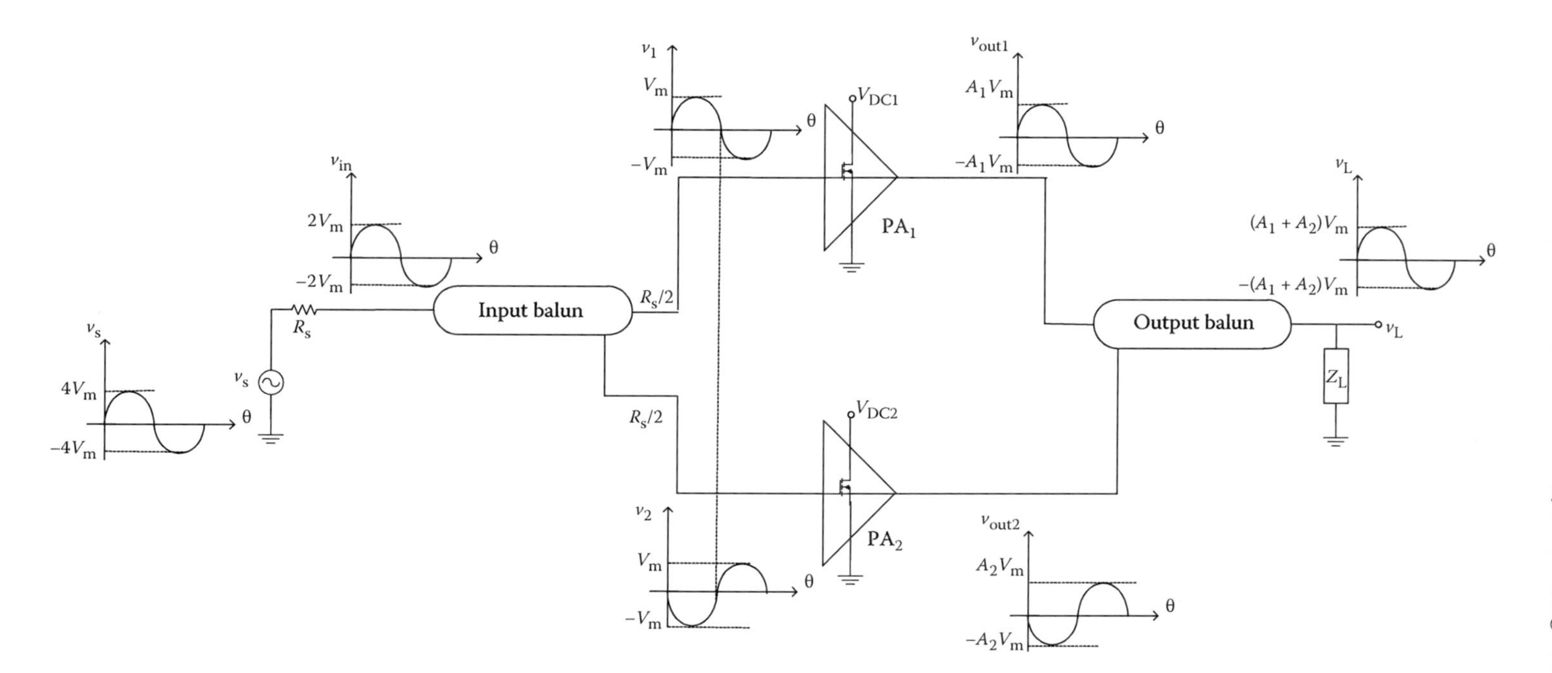

FIGURE 1.46 Implementation of push–pull amplifiers.

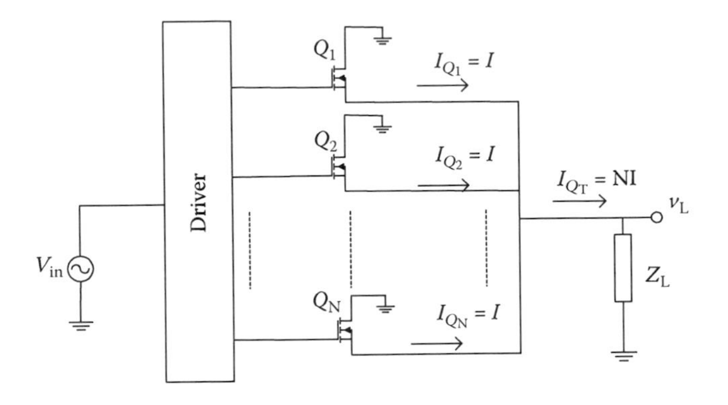

FIGURE 1.47 Parallel connection of transistors for RF amplifiers.

1.5.3 PA Module Combiners

RF output power can be increased further up to several kilowatts by combining individual identical PA modules. These PA modules are usually combined via a Wilkinson-type power combiner. The typical PA module combining technique is shown via a two-way power combiner in Figure 1.48.

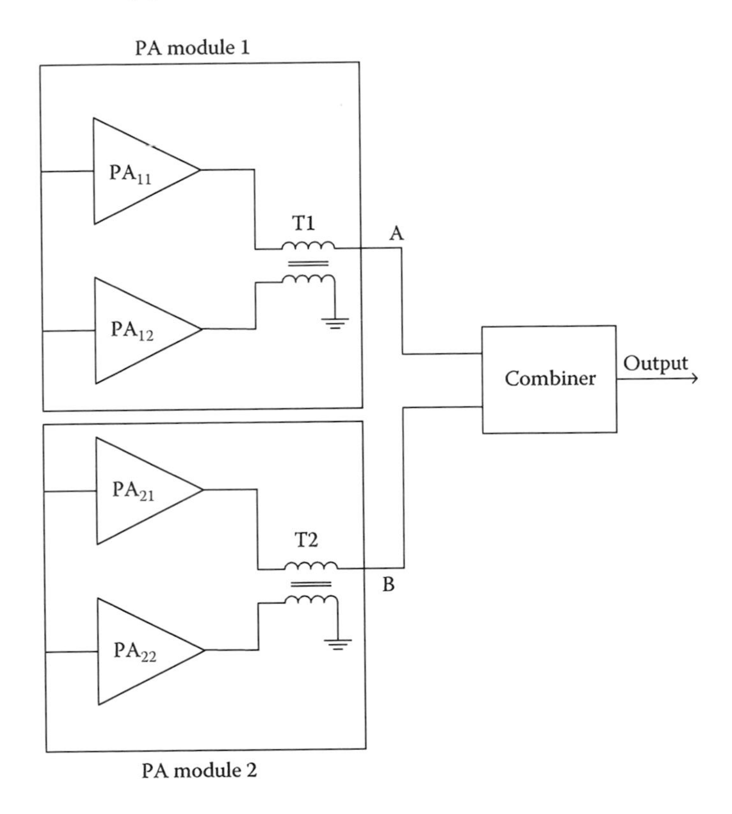

FIGURE 1.48 Power combiner for PA modules.

1.6 RF POWER TRANSISTORS

The selection of a transistor for the amplifier being designed is critical as it will impact the performance of the amplifier parameters including efficiency, dissipation, power delivery, stability, and linearity. Once the transistor is selected for the corresponding amplifier topology, the size of the transistor, die placement, bond pads, bonding of the wires, and lead connections will determine the layout of the amplifier and the thermal management of the system. RF power transistors are fabricated using silicon (Si), gallium arsenide (GaAs), and related compound semiconductors. There is intense research into the development of high-power density devices using wide-bandgap materials such as silicon carbide (SiC) and gallium nitride (GaN). RF power transistors and their major applications and frequency of operation are given in Table 1.6. The material properties of the semiconductor-based compounds used for device manufacturing are given in Table 1.7 [1].

TABLE 1.6
RF Power Transistors and Their Applications and Frequency of Operations

RF Transistor	Drain BV [V]	Frequency (GHz)	Major Applications
RF power FET	65	0.001–0.4	VHF PA
GaAs MesFET	16–22.60	1–30	Radar, satellite, defense
SiC MesFET	100	0.5–2.3	Base station
GaN MesFET	160	1–30	Replacement for GaAs
Si LDMOS (FET)	65	0.5–2	Base station
Si VDMOS (FET)	65–1200	0.001–0.5	HF power amplifier and FM Broadcasting and MRI

Note: LDMOS, laterally diffused metal oxide semiconductor; VDMOS, vertical diffusion metal-oxide semiconductor.

TABLE 1.7
RF High-Power Transistor Material Properties

RF High-Power Material	μ (cm^2/V_s)	ε_r	E_g (eV)	Thermal Conductivity (W/cmK)	E_{br} (MV/cm)	JM = $E_{br}v_{sat}/2\pi$	T_{max} (°C)
Si	1350	11.8	1.1	1.3	0.3	1.0	300
GaAs	8500	13.1	1.42	0.46	0.4	2.7	300
SiC	700	9.7	3.26	4.9	3.0	20	600
GaN	1200 (bulk) 2000 (2DEG)	9.0	3.39	1.7	3.3	27.5	700

1.7 CAD TOOLS IN RF AMPLIFIER DESIGN

CAD tools have been widely used in engineering applications [2–5] and in RF PA systems to expedite the design process, increase the system performance, and reduce the associated cost by eliminating the need for several prototypes before the implementation stage [6–8]. RF PAs are simulated with nonlinear circuit simulators, which use large-signal equivalent models of the active devices. The passive components used in RF PA simulation are usually modeled as ideal components and hence do not include the frequency characteristics. Furthermore, it is rare to include the electromagnetic effects such as coupling between traces and leads, parasitic effects, current distribution, and radiation effects that exist among the components in RF PA simulation. This is partly due to the requirement in expertise in both nonlinear circuit simulators and electromagnetic simulators. However, use of electromagnetic simulators in simulation of RF PAs will drastically improve the accuracy of the results by taking into account the coupling and radiation effects.

Nonlinear circuit simulation of RF PAs can be done using a harmonic balance technique with the application of Krylov subspace methods in the frequency domain or nonlinear differential algebraic equations using the integration methods, Newton's method, or sparse matrix solution techniques in the time domain [9–11]. Time domain methods are preferred over frequency domain methods because of their advantage in providing accurate solutions using the transient response of the circuits. This is specifically valid for the frequencies where industrial, scientific, and medical (ISM) applications take place.

Several different simulator types will be used throughout this textbook for simulation of the PA circuits with its surrounding components that will also be able to take electromagnetic effects. In addition, use of MATLAB/Simulink in conjunction with Orcad/PSpice will be illustrated. This is a unique technique that is needed specifically when advanced signal processing and control of the PA output signal are required.

Example

Model a nonlinear, voltage-controlled capacitor that can be used to imitate the behavior of the drain-to-source capacitor, C_{ds}, of MOSFET by simulating the capacitive input half rectifier circuit shown in Figure 1.49 with Orcad/PSpice.

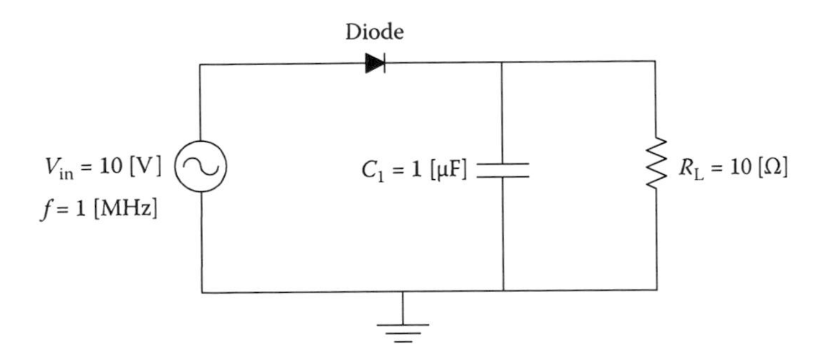

FIGURE 1.49 Rectifier circuit for voltage-controlled capacitor example.

Solution

We begin with simulating the circuit in Figure 1.49, as shown in Figure 1.50, following capacitive input half rectifier circuit with PSpice.

The simulation results for the output voltage are shown in Figure 1.51.

From the graph, $V_m = 8.5095$ V, which can be also found from

$$V_r = V_m \frac{T}{\tau} = V_m \frac{T}{RC} \tag{1.120}$$

where $T = 1$ μs

$$\tau = RC = 10 \times 1e-6 = 10\ \mu s$$

$$V_r = V_m \frac{T}{RC} = 0.85095\ \text{V}$$

$$V_m - V_r = 8.5095 - 0.85095 = 7.65855\ \text{V}$$

The PSpice-measured value is 7.8434 V, which is very close to the calculated value above.

Now, let us simulate the same circuit given in Figure 1.49 by using a nonlinear voltage-controlled capacitor, as shown in Figure 1.52.

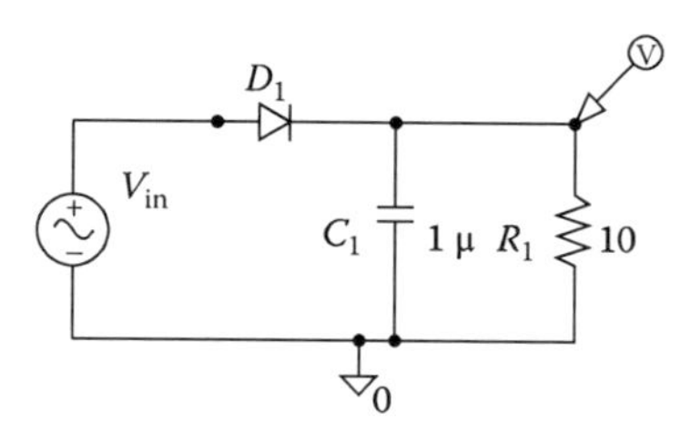

FIGURE 1.50 Rectifier circuit simulation of voltage-controlled capacitor example.

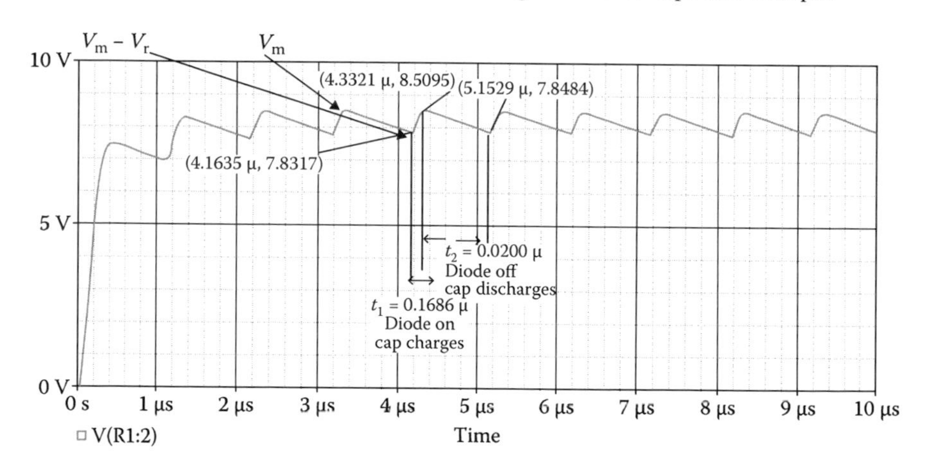

FIGURE 1.51 Rectifier circuit simulation results of voltage-controlled capacitor.

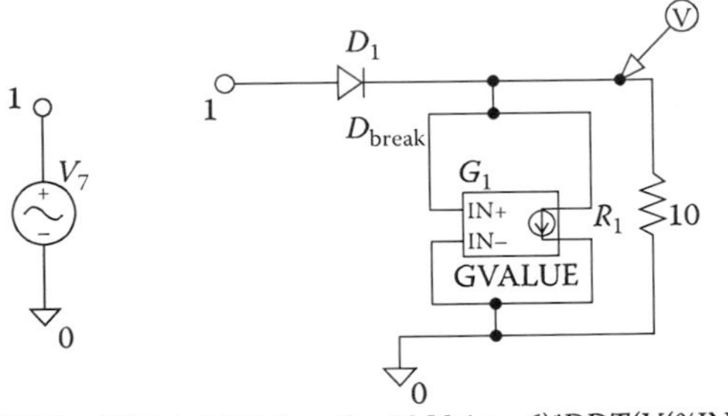

FIGURE 1.52 Modeling nonlinear voltage-controlled capacitor with PSpice.

In this circuit, GVALUE, which is an analog behavior modeling (ABM) component, is used to model the nonlinear behavior in the capacitor. The TABLE in Figure 1.52 is incorporated into the model to look up the measured values of cap values vs. voltage values. This model has a fixed value of 1 μF. However, it can also be used for variable values. If the cap values are different, then they linearly interpolate the cap value over the range. The idea is to replace the capacitor by a controlled current source, G_{out}, whose current is defined by

$$I = C(V)\frac{dV}{dt} \tag{1.121}$$

The time derivative (DDT) can be defined by the DDT() function. The simulation result showing the output response of the equivalent circuit in Figure 1.52 with a nonlinear voltage-controlled capacitor is shown in Figure 1.53.

From the graph, $V_m = 8.4701$ V, which can be calculated similarly when $T = 1$ μs as

$$\tau = RC = 10 \times 1e - 6 = 10\ \mu s$$

$$V_r = V_m \frac{T}{RC} = 0.84701\,\text{V}$$

$$V_m - V_r = 8.4701 - 0.84701 = 7.62309\ \text{V}$$

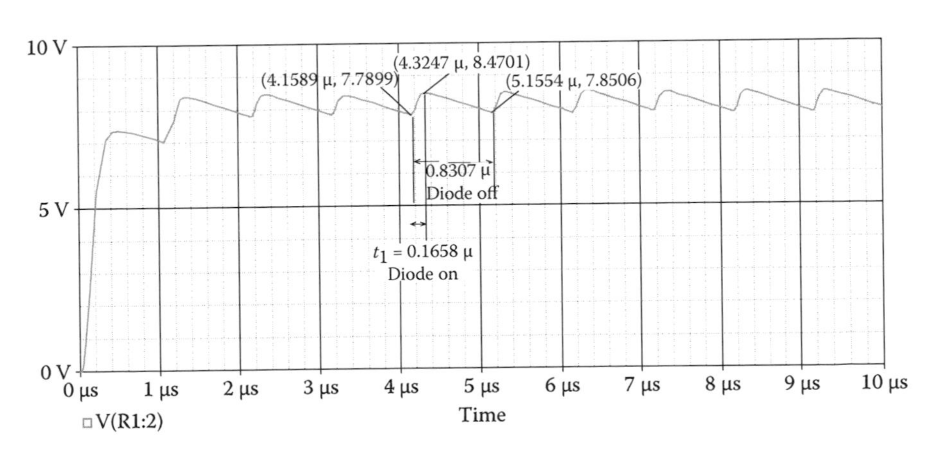

FIGURE 1.53 Equivalent circuit simulation results using a voltage-controlled capacitor.

The PSpice-measured value is 7.8506 V, which matches the previous result. Hence, the nonlinear capacitor is accurate based on the PSpice measurements using PSpice ABM models. They also correlate well with the analytical results.

Example

Model a transistor as an ideal switch such that it will be on half of the period, $T/2$, and it will be off the other half. Assume that the operational frequency is f = 13.56 MHz.

Solution

The period $T = \frac{1}{f} = \frac{1}{13.56 \times 10^6} = 73.746\,\text{ns}$

This can be accomplished using switch from the parts list. The PSpice circuit is shown in Figure 1.54.

In the simulation, V_{control} (V_{pulse}) parameters:

$V_1 = 0$, $V_2 = 6$, TD = 0.0001u, TR = 0.0001u, TF = 0.0001u, PW = 36.873n, PER = 73.746n

Switch, S_1, parameters:

$$R_{\text{off}} = 1\text{e+},\ R_{\text{on}} = 10n,\ V_{\text{off}} = 0,\ V_{\text{on}} = 5$$

The output response of the simulation is an ideal switch response of the transistor, as shown in the simulation results in Figure 1.55.

Example

The large-signal model of a diode is shown in Figure 1.56 using PSpice.

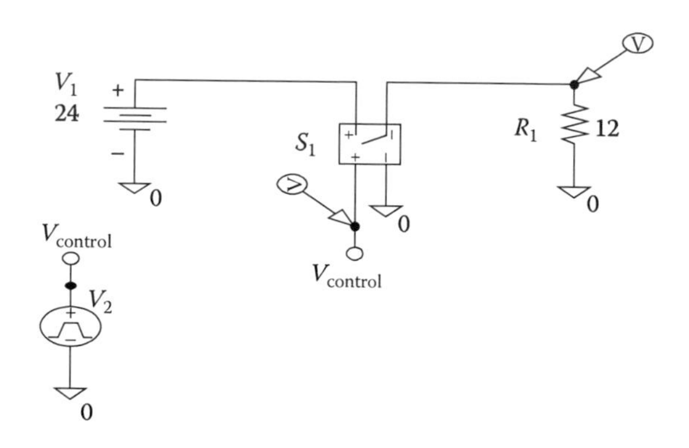

FIGURE 1.54 Modeling transistor as a switch using PSpice.

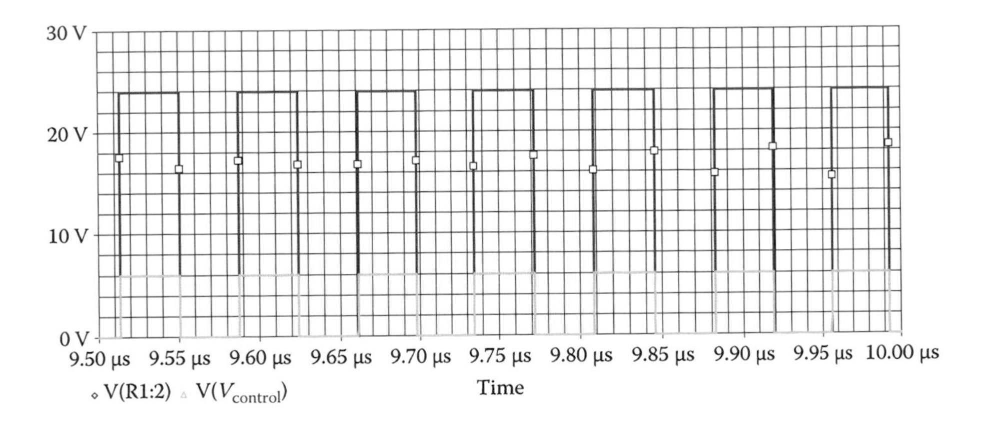

FIGURE 1.55 Output simulation response of an ideal switch.

A Diode K

FIGURE 1.56 Diode for large-signal modeling.

Solution

The simplified large-signal model of the diode is shown in Figure 1.57.

A diode contains two capacitances. One of them is voltage dependent and is called junction capacitance, C_j, and the other is called diffusion capacitance, C_d. The junction capacitance is a function DC bias of the diode. C_d is the capacitance associated with the PN junction, which is only observable under transient conditions. This transient condition is for the diode to be under the forward bias condition and quickly switched from forward bias to reverse bias.

So when the diode is off, i.e., $V < V_d$, then the diode capacitance can be approximated by

$$C = C_j(V_R) + C_d \tag{1.122}$$

where

$$C_j(V_R) = \frac{C_{jo}}{\left(1 - \frac{V_d}{V_j}\right)^M} \tag{1.123}$$

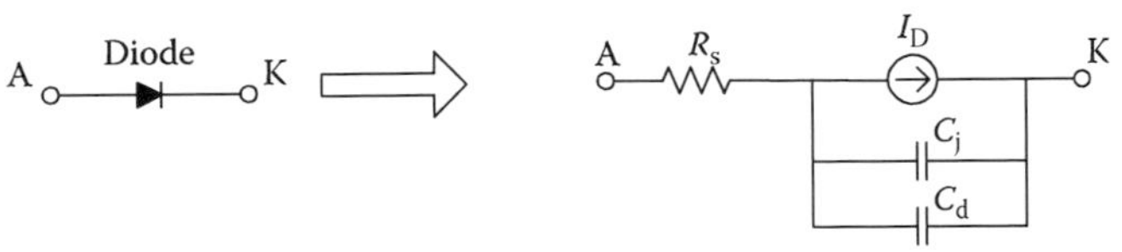

A Diode K ⇒ A R_s I_D C_j C_d K

FIGURE 1.57 The simplified large-signal model for diode.

where C_{jo} is the initial PN junction capacitance, M is the grading coefficient, V_j is the built-in junction voltage, and V_d is the forward bias voltage. The parameters in Equations 1.122 and 1.123 can be calculated from

$$C_d = TT.g_d = TT.\frac{dI_d}{dV_d} = TT.\frac{Is}{NV_{th}}e^{\frac{V_d}{NV_{th}}} \tag{1.124}$$

$$V_{th} = \frac{kT}{q} = \frac{(1.38\times10^{-23})\times(300)}{(1.6\times10^{-19})} = 0.0258 \tag{1.125}$$

where TT is the transit time of minority carriers, g_d is the conductance, N is the emission, and Is is the saturation current. Hence, in the reverse bias mode, C_d is very small and can be neglected. Also, the overall capacitance of the diode is equal to C_j. In the forward bias mode, C_j is very small (because V_d is small), and C_d is large. It effects diode turn OFF properties and delays diode's turn OFF.

PROBLEMS

1. Assume that a sinusoid signal, $v_i(t) = 12\sin(\omega t)$, with 60-Hz frequency is applied to a linear amplifier that has an output signal, $v_o(t) = 60\sin(\omega t)$, as shown in Figure 1.58. Obtain the time domain representation of the input signal and the frequency domain representation of power spectra of the output signal of the amplifier.
2. Calculate the output power for the RF system shown in Figure 1.59.

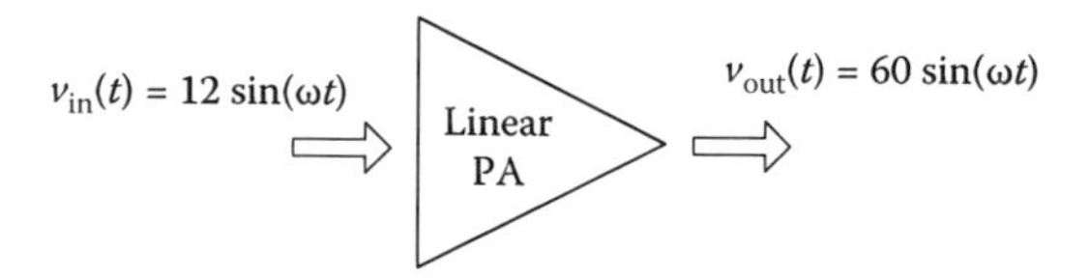

FIGURE 1.58 Linear amplifier.

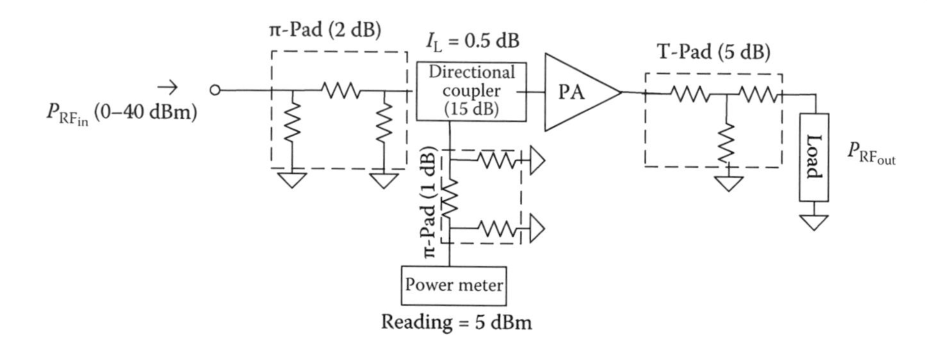

FIGURE 1.59 RF system power calculation.

3. RF PA delivers 600 [W] to a given load. If the input supply power for this amplifier is given to be 320 [W], and the power gain of the amplifier is 13 dB, find the (a) drain efficiency and (b) power-added efficiency.
4. RF signal $v_i(t) = \beta \cos \omega t$ is applied to a linear amplifier and then to a nonlinear amplifier shown in Figure 1.60 with output response $v_o(t) = \alpha_0 + \alpha_1\beta\cos\omega t + \alpha_2\beta^2\cos^2\omega t$. Assume that input and output impedances are equal to R. (a) Calculate and plot the gain for the linear amplifier. (b) Obtain the second and third HD for the nonlinear amplifier when $\alpha_0 = 0$, $\alpha_1 = 2$, $\alpha_2 = 1$ and $\beta = 2$, $\beta = 4$. Calculate also the THD for both cases.
5. In the RF amplifier circuit given in Figure 1.61, calculate the 1-dB compression point.
6. Find the region of operation for each of the transistors in Figure 1.62 when $V_t = 1$ V.
7. For the given amplifier configuration in Figure 1.63, find the impedance, voltage, and power at each point shown on the figure.

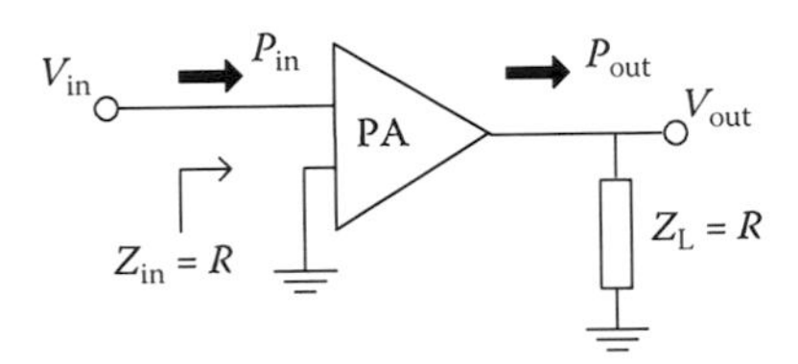

FIGURE 1.60 THD calculation for RF amplifier.

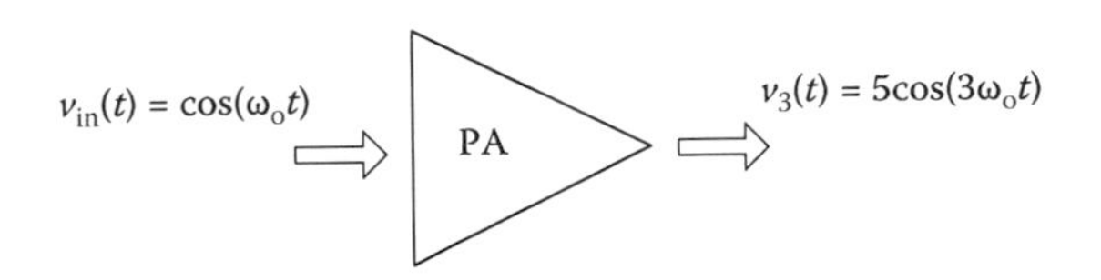

FIGURE 1.61 1-dB compression point for RF amplifier.

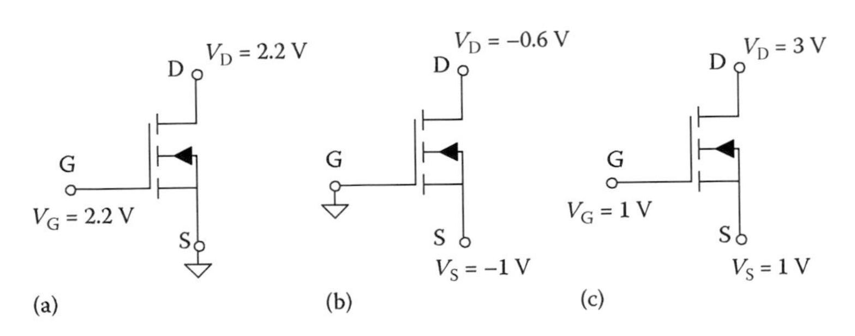

FIGURE 1.62 Operational region for transistors: (a) transistor 1, (b) transistor 2, and (c) transistor 3.

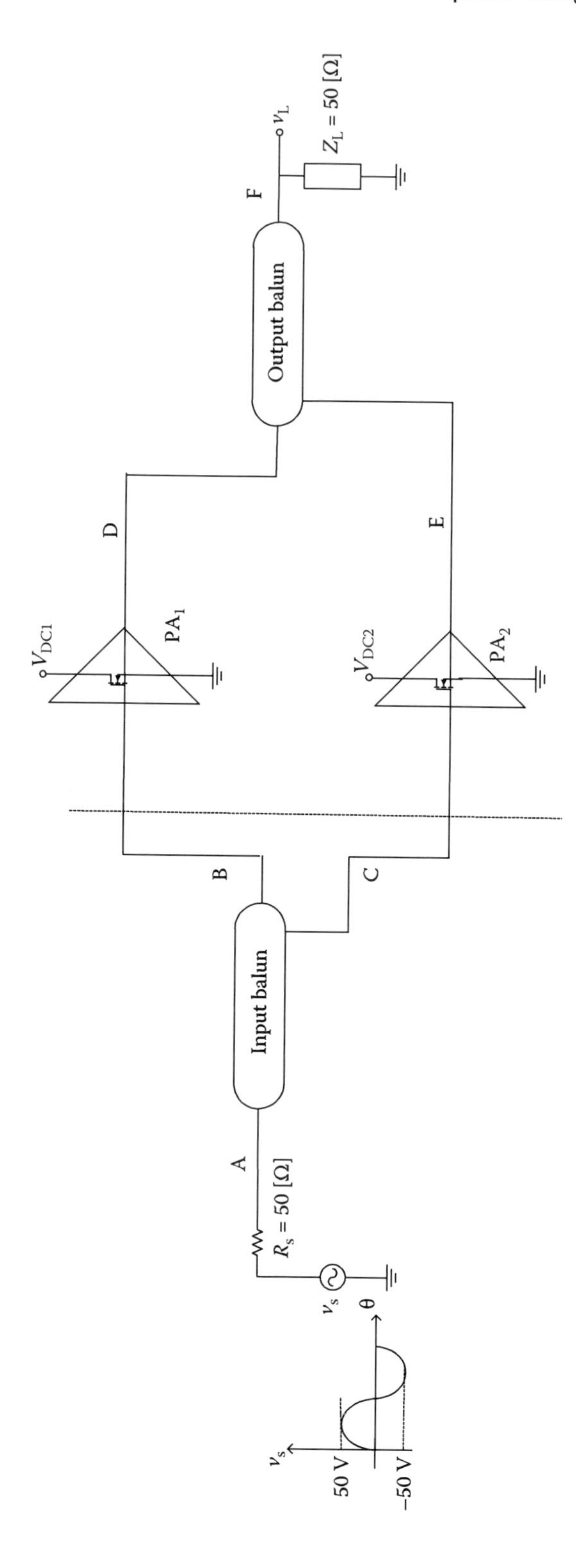

FIGURE 1.63 Push–pull amplifier configuration.

REFERENCES

1. U.K. Mishra, L. Shen, T.E. Kazior, and Y.-F. Wu. 2008. GaN-based RF power devices and amplifiers. *Proceedings of the IEEE*, Vol. 96, No. 2, pp. 287–305, February.
2. S. El Alimi, C. Münch, A. Azzi, P. Bégou, and S.B. Nasrallah. 2009. Large eddy simulation of a compressible flow in a locally heated square duct. *International Review of Modeling and Simulations (IREMOS)*, Vol. 2, No. 1, pp. 93–97, February.
3. O. Chocron, and H. Mangel. 2011. Models and simulations for reconfigurable magnetic-coupling thrusters technology. *International Review of Modeling and Simulations (IREMOS)*, Vol. 4, No. 1, pp. 325–334, February.
4. A. Milovanovic, B. Koprivica, and M. Bjekic. 2010. Application of the charge simulation method to the calculation of the characteristic parameters of printed transmission lines. *International Review of Electrical Engineering (IREE)*, Vol. 5, No. 6, Pt. A, pp. 2722–2726, December.
5. M. Heidari, R. Kianinezhad, S.Gh. Seifossadat, M. Monadi, and D. Mirabbasi. 2011. Effects of distribution network unbalance voltage types with identical unbalance factor on the induction motors simulation and experimental. *International Review of Electrical Engineering (IREE)*, Vol. 6, No. 1, Pt. A, pp. 223–228, February.
6. P.H. Aaen, J.A. Pla, and C.A. Balanis. 2006. Modeling techniques suitable for CAD-based design of internal matching networks of high-power RF/microwave transistors. *IEEE Transactions on Microwave Theory and Techniques*, Vol. 54, No. 7, pp. 3052–3059, July.
7. R. Mittra, and V. Veremey. 2000. Computer-aided design of RF circuits. *2000 IEEE AP-S*, Vol. 2, pp. 596–599, Salt Lake City.
8. K.C. Gupta. 1998. Emerging trends in millimeter-wave CAD. *IEEE Transactions on Microwave Theory and Techniques*, Vol. 46, No. 6, pp. 747–755, June.
9. R.J. Gilmore, and M.B. Steer. 1991. Nonlinear circuit analysis using the method of harmonic balance—A review of the art. Part I. Introductory concepts. *International Journal of Microwave and Millimeter-Wave Computer-Aided Engineering*, Vol. 1, No. 1, pp. 22–37.
10. R.J. Gilmore, and M.B. Steer. 1991. Nonlinear circuit analysis using the method of harmonic balance—A review of the art. Part II. Advanced concepts. *International Journal of Microwave and Millimeter-Wave Computer-Aided Engineering*, Vol. 1, No. 2, pp. 159–180.
11. K.S. Kundert. 1998. *The Designer's Guide to Spice and Spectre*. Kluwer Academic Publishers, Boston.

2 Radio Frequency Power Transistors

2.1 INTRODUCTION

Radio frequency (RF) power transistors are the most critical components in the amplifier system since their characteristics have direct implications in amplifier response including power, stability, linearity, and the profile of the amplifier itself. The parameters of the transistor such as gain, intrinsic parameters such as capacitances, and extrinsic parameters such as lead inductances are determining factors of the amplifier design and performance. Transistors have small-signal and large-signal models. Small-signal models assume small variation between voltages and currents and hence do not provide an accurate design for amplifiers when large variations take place between voltages and currents. It is also very challenging to obtain accurate large-signal models for active devices. Small- and large-signal models are provided by manufacturers and implemented in simulators such as advanced design system (ADS) and Orcad/PSpice environment. When the transistor model is available as an equivalent circuit, the transistor then can be treated as a two-port device, as shown in Figure 2.1, and network analysis is performed to produce analytical and numerical results.

2.2 HIGH-FREQUENCY MODEL FOR MOSFETs

Metal–oxide–semiconductor field-effect transistors (MOSFETs) are the preferred active device that will be used throughout this book for illustration of the majority of the concepts and examples in RF power amplifier design. There are several different versions of MOSFETs, which are manufactured using different processes, as illustrated in Figure 2.2.

MOSFET parameters are identified by manufacturers at different static and dynamic conditions. Hence, each MOSFET device has been manufactured with different characteristics. A designer makes the selection of the appropriate device for the specific circuit under consideration. One of the standard ways commonly used by designers for the selection of the right MOSFET device is called figure of merit (FOM) [1]. Although there are different types of FOMs that are used, FOM in its simplest form compares the gate charge, Q_g, against R_{dson}. The multiplication of gate charge and drain to source on resistance relates to a certain device technology as it can be related to the required Q_g and R_{dson} to achieve the right scale for MOSFETs. The challenge is the relation between Q_g and R_{dson} because MOSFETs have inherent trade-offs between the ON resistance and gate charge, i.e., the lower the R_{dson} is,

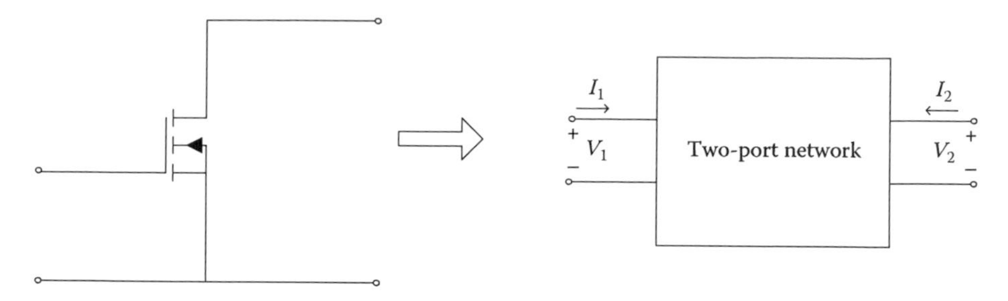

FIGURE 2.1 RF power transistor as a two-port network.

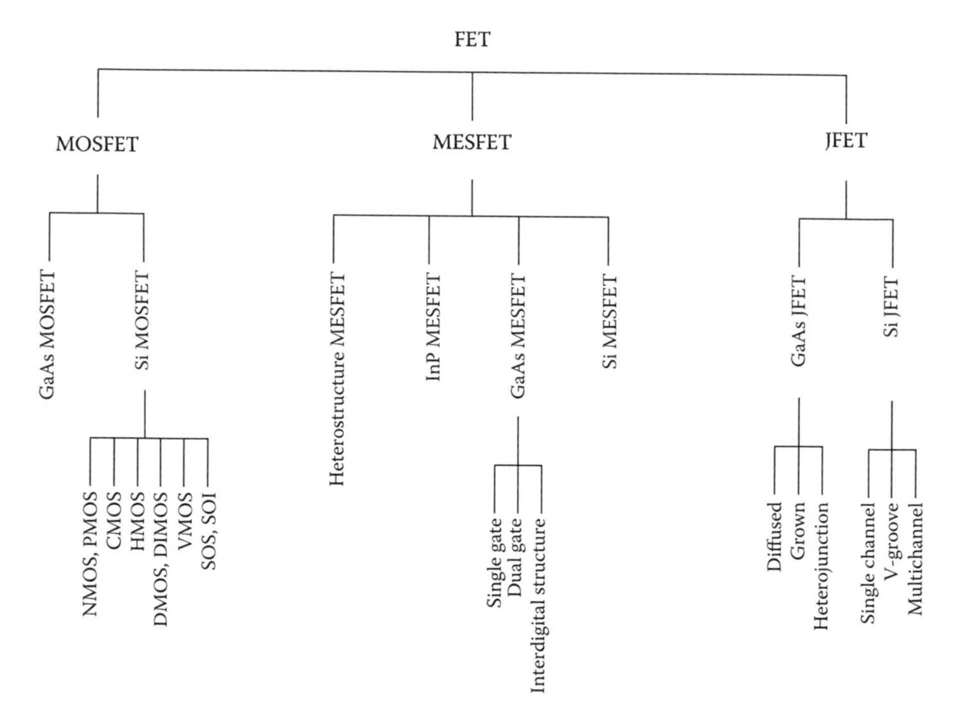

FIGURE 2.2 Types of MOSFETs.

the higher the gate charge will be. In device design, this translates into conduction loss vs. switching loss trade-off. The new-generation MOSFETs are manufactured to have an improved FOM [2–4]. The comparison of FOM on MOSFETs manufactured with different processes can be illustrated on planar MOSFET structure and trench MOSFET structure. MOSFETs with trench structure have seven times better FOM vs. planar structure, as shown in Figure 2.3.

Two variations of trench power MOSFETs are shown in Figure 2.4. The trench technology has the advantage of higher cell density but is more difficult to manufacture than the planar device.

There is a best die size for MOSFET devices for a given output power. The optimum die size for minimal power loss, P_{loss}, depends on load impedance, rated power,

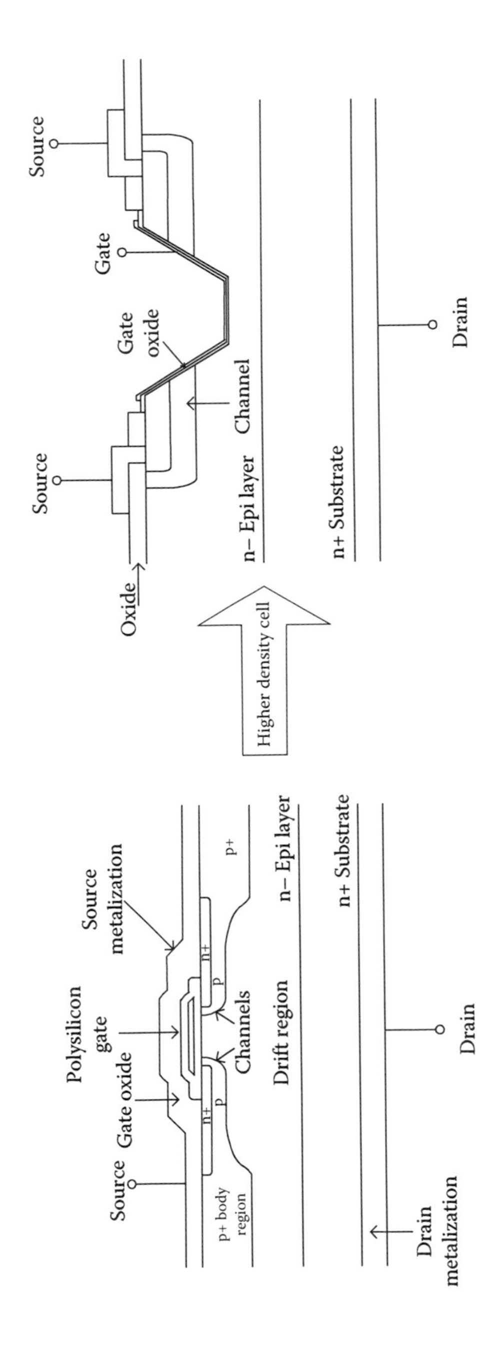

FIGURE 2.3 FOM comparison of planar and trench MOSFET structures.

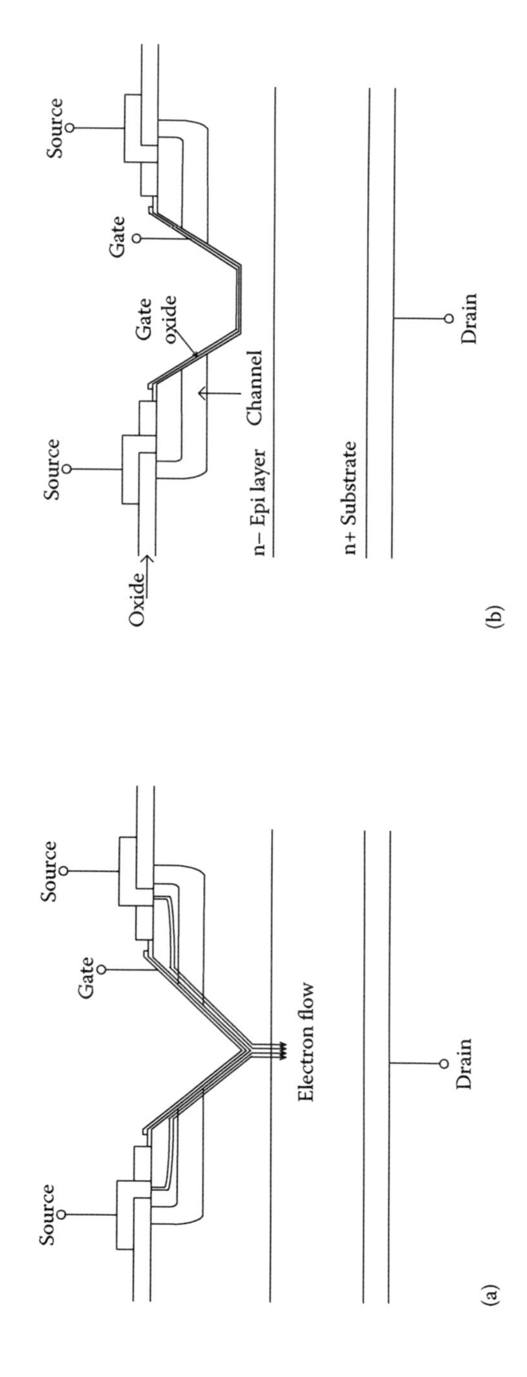

FIGURE 2.4 Trench MOSFETs: (a) current crowding in V-groove trench MOSFETs and (b) truncated V-groove.

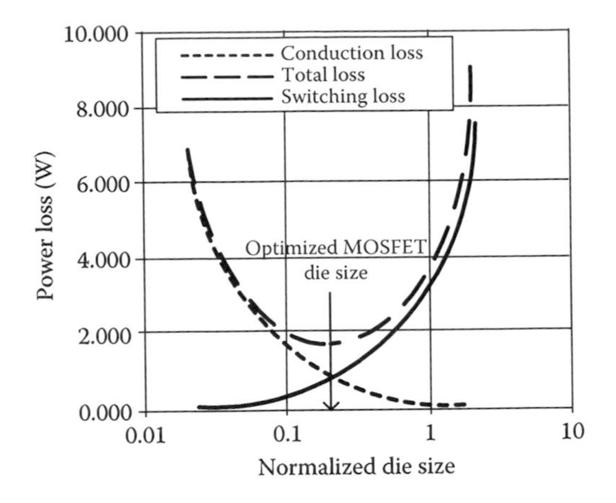

FIGURE 2.5 Die size vs. power loss.

and switching frequency. The relation between power loss and optimized die size is illustrated in Figure 2.5.

The typical MOSFET structure with internal capacitances is illustrated in Figure 2.6.

In MOSFET structures, considerable capacitance is observed at the input due to the oxide layer. The simplest view of an *n*-channel MOSFET is shown in Figure 2.7, where the three capacitors, C_{gd}, C_{ds}, and C_{gs}, represent the parasitic capacitances. These values can be manipulated to form the input capacitance, output capacitance, and transfer capacitance. C_{gs} is the capacitance due to the overlap of the source and channel regions by the polysilicon gate. It is independent of the applied voltage. C_{gd} consists of the part associated with the overlap of the polysilicon gate and the silicon underneath the junction gate field effect transistor (JFET) region, which is also independent of the applied voltage and the capacitance associated with the depletion region immediately under the gate, which is a nonlinear function of the applied voltage. This capacitance provides a feedback loop between the output and input circuit. C_{gd} is also called the Miller capacitance because it causes the total dynamic input capacitance to become greater than the sum of the static capacitors. C_{ds} is the capacitance associated with the body drift diode. It varies inversely with the square root

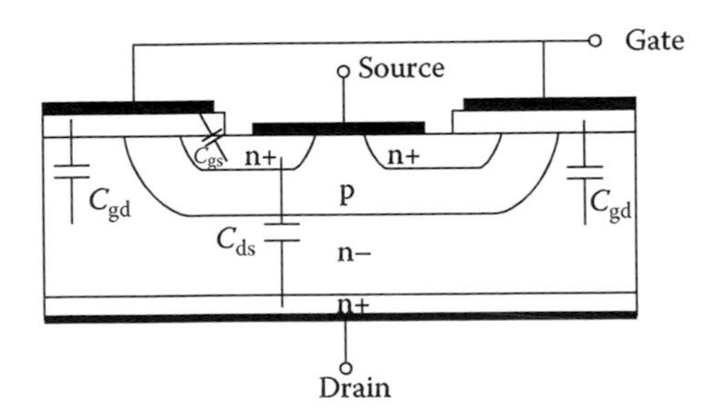

FIGURE 2.6 MOSFET structure capacitance illustration.

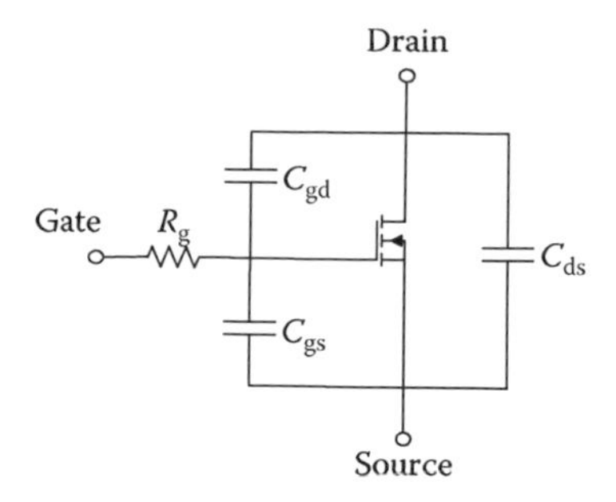

FIGURE 2.7 Simplest view of MOSFETs with intrinsic capacitances.

of the drain source bias voltage. In the manufacturer data sheet, input capacitance, C_{iss}, and output capacitance, C_{oss}, information is usually given. C_{iss} is made up of the gate-to-drain capacitance, C_{gd}, in parallel with the gate-to-source capacitance, C_{gs}, or

$$C_{iss} = C_{gs} + C_{gd} \tag{2.1}$$

The input capacitance must be charged to the threshold voltage before the device begins to turn on and discharged to the plateau voltage before the device turns off. Therefore, the impedance of the drive circuitry and C_{iss} have a direct effect on the turn-on and turn-off delays. C_{oss} is made up of the drain-to-source capacitance, C_{ds}, in parallel with the gate-to-drain capacitance, C_{gd}, or

$$C_{oss} = C_{ds} + C_{gd} \tag{2.2}$$

Common MOSFET packages are identified as transistor outline (TO), small outline transistor (SOT), and small outline package (SOP). The early TO package specifications such as TO-92, TO-92L, TO-220, TO-247, TO-252, etc., are plug-in package design. In recent years, market demand has increased for surface mount, and TO packages also progressed to the surface-mount package. SOT packages are a lower-power SMD transistor package than the small TO package, generally used for small-power MOSFETs. The common SOT packages are SOT-23, SOT-89, and SOT-236. SOP is a surface-mount package. The pin from the package was gull wing leads on both sides (L-shaped). MOSFET manufacturers are trying to improve chip production technology and processes to have better packaging technology. MOSFET package has important characteristics that will have limitations in device performance. The characteristics of the package include package resistance, package inductance, and thermal impedance. Package resistance depends on bonding wire resistance based on the bonding wire type and length and lead frame resistance. The bonding wire can be Cu, Al, Al ribbon, or Cu clip based. A lead frame-based package with internal wire bonds introduces parasitic inductance on the gate, source, and drain terminals. During current switching, this inductance produces a large Ldi/dt effect to slow down the turn-on and turn-off of the device. This effect will significantly hinder the performance at high switching frequencies. Parasitic inductances directly affect body diode reverse recovery characteristics and peak voltage spikes. Thermal impedance of the package consists of thermal resistances due to the junction to the printed circuit board (PCB) and the junction to the case.

Equivalent circuits are used to model and represent the characteristics of transistors. The equivalent circuits for transistors are physically measured for electrical parameters using specific measurement setups and fixtures. The measured results are then transformed into a circuit that mimics the electrical behavior of the transistor. This is the common procedure that is applied for any active device used in power amplifiers. The commonly used high-frequency, small-signal model for MOSFETs is given in Figure 2.8. This figure also illustrates the intrinsic parameters based on the measured parameters for MOSFETs. The transistors consist of die, exterior package, bonding wires, bonding pads, etc., designed for specific ratings and applications as discussed before. The die has to be placed inside a package, and internal and external connections are established with bonding wires and vias. The effects of the transistor packages such as lead inductances and package capacitances are called extrinsic parasitics. The complete MOSFET small-signal, high-frequency model including extrinsic and intrinsic parameters is shown in Figure 2.9.

When there is no feedback capacitance nor any resistance, the high-frequency model in Figure 2.8 simplifies to the one with a load resistor, as shown in Figure 2.10.

The circuit parameters such as current, i_o, and voltage, v_o, can be expressed as

$$i_o = \frac{g_m v_{GS}}{\sqrt{1+\omega^2 C_{ds}^2 R_L^2}} \tag{2.3}$$

$$v_o = \frac{g_m v_{GS} R_L}{\sqrt{1+\omega^2 C_{ds}^2 R_L^2}} \tag{2.4}$$

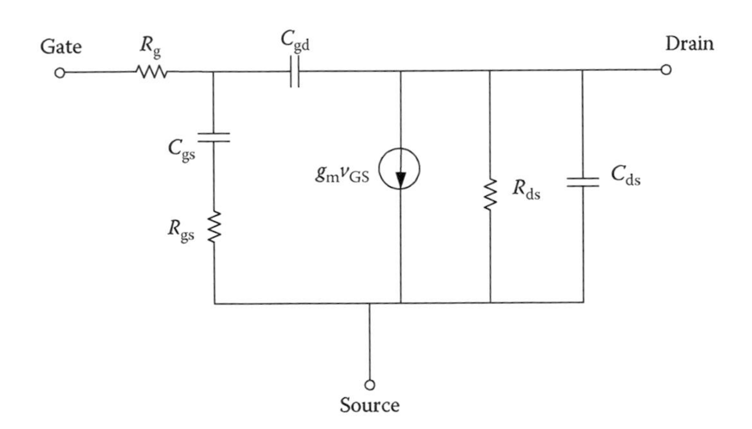

FIGURE 2.8 High-frequency, small-signal model for MOSFET transistor.

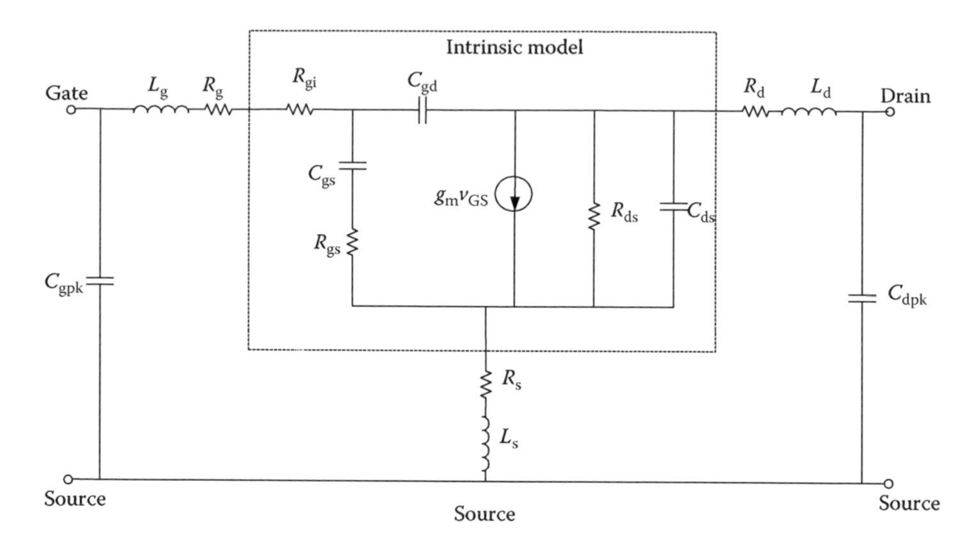

FIGURE 2.9 High-frequency, small-signal model for MOSFET transistor with extrinsic parameters.

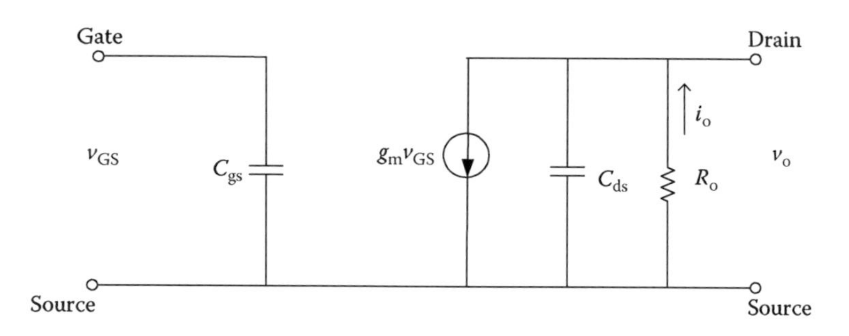

FIGURE 2.10 Simplified high-frequency model for MOSFETs.

The output power is then equal to

$$P_o = \frac{g_m^2 v_{GS}^2 R_L}{1 + \omega^2 C_{ds}^2 R_L^2} \tag{2.5}$$

The corresponding current and voltage gains are

$$A_I = \frac{g_m}{\omega C_{gs}\sqrt{1 + \omega^2 C_{ds}^2 R_L^2}} \tag{2.6}$$

$$G = \frac{g_m^2}{\omega C_{gs}\left(1 + \omega^2 C_{ds}^2 R_L^2\right)} \tag{2.7}$$

Example

Obtain the I–V characteristics of an *n*-channel MOSFET for a given spice.lib file using Orcad/PSpice. The spice.lib file of MOSFET is

```
.SUBCKT MOSFETex1 1 2 3
* External Node Designations
* Node 1 -> Drain
* Node 2 -> Gate
* Node 3 -> Source
M1 9 7 8 8 MM L=100u W=100u
.MODEL MM NMOS LEVEL=1 IS=1e-32
+VTO=4.54 LAMBDA=0.00633779 KP=7.09
+CGSO=2.17164e-05 CGDO=3.39758e-07
RS 8 3 0.0001
D1 3 1 MD
.MODEL MD D IS=5.2e-09 RS=0.00580776 N=1.275 BV=1000
+IBV=10 EG=1.061 XTI=2.999 TT=3.28994e-05
+CJO=2.95707e-09 VJ=1.57759 M=0.9 FC=0.1
RDS 3 1 1e+06
RD 9 1 0.95
RG 2 7 0.4
D2 4 5 MD1
* Default values used in MD1:
*   RS=0 EG=1.11 XTI=3.0 TT=0
*   BV=infinite IBV=1mA
.MODEL MD1 D IS=1e-32 N=50
+CJO=2.05889e-09 VJ=0.5 M=0.9 FC=1e-08
D3 0 5 MD2
* Default values used in MD2:
*   EG=1.11 XTI=3.0 TT=0 CJO=0
*   BV=infinite IBV=1mA
.MODEL MD2 D IS=1e-10 N=0.4 RS=3.00001e-06
RL 5 10 1
FI2 7 9 VFI2 -1
VFI2 4 0 0
EV16 10 0 9 7 1
CAP 11 10 2.05889e-09
FI1 7 9 VFI1 -1
VFI1 11 6 0
RCAP 6 10 1
D4 0 6 MD3
* Default values used in MD3:
*   EG=1.11 XTI=3.0 TT=0 CJO=0
*   RS=0 BV=infinite IBV=1mA
.MODEL MD3 D IS=1e-10 N=0.4
.ENDS MOSFETex1
```

Solution

PSpice schematics are shown in Figure 2.11. A DC sweep has to be set up in conjunction with nested sweep so that for each value of the nested sweep variable, V_{gs}, from 0 to 10 V, the drain current, I_d, is calculated by varying the DC sweep variable, V_{ds}, from 0 to 100 V. The setup required for obtaining I–V curves is outlined next.

The analysis setup is

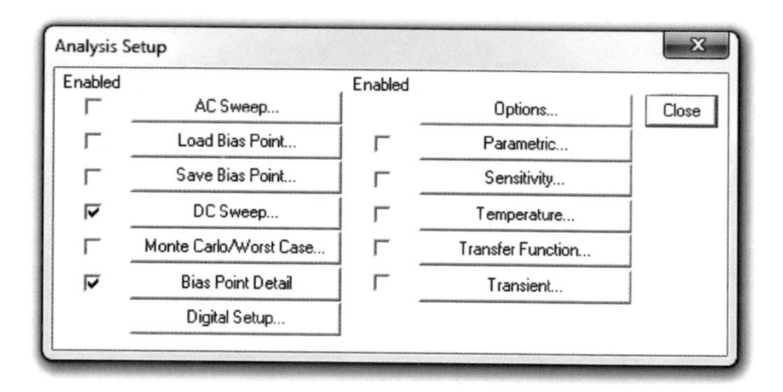

The DC sweep setup is

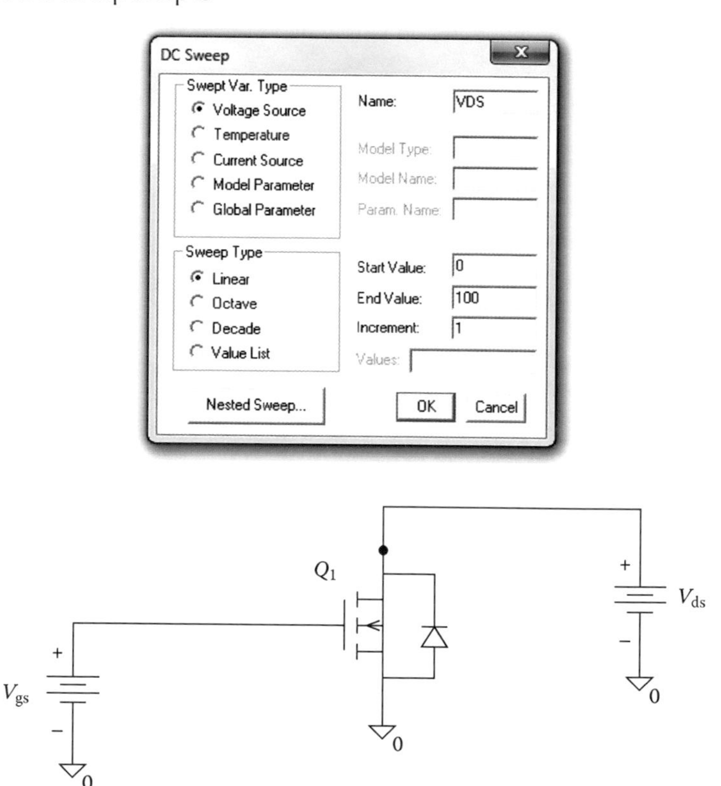

FIGURE 2.11 I–V characterization of MOSFETs via PSpice simulation.

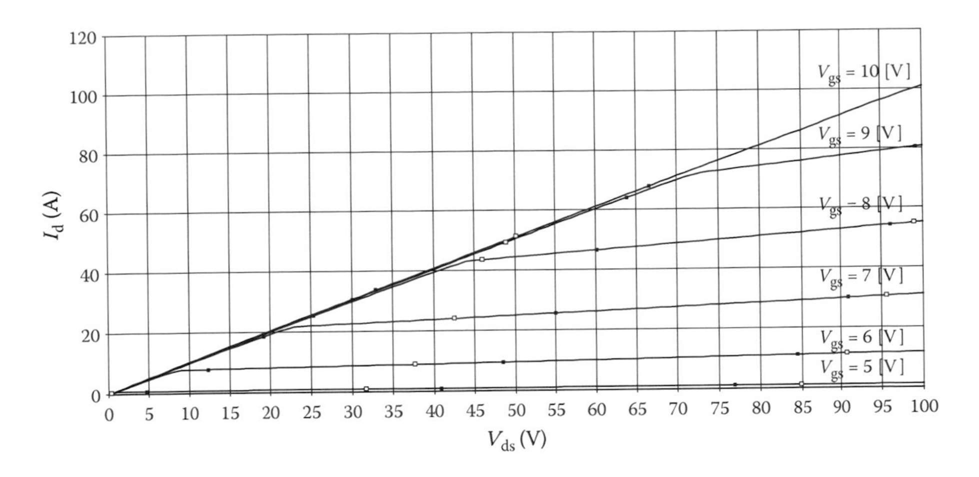

FIGURE 2.12 PSpice simulation results for I–V curves of an enhancement MOSFET.

The nested sweep setup is

DC Nested Sweep

Swept Var. Type
- (•) Voltage Sourc
- ( ) Temperature
- ( ) Current Source
- ( ) Model Paramete
- ( ) Global Parameter

Name: VGS
Model Type:
Model Name:
Param. Name:

Sweep Type
- (•) Linear
- ( ) Octave
- ( ) Decad
- ( ) Value List

Start Value: 0
End Value: 10
Increment: 1
Values:

Main Sweep... [✓] Enable Nested Sweep

OK Cancel

When simulation is run, the I–V characteristics are obtained from the Probe interface. As seen from simulation results shown in Figure 2.12, $V_{gs} > V_{th} = 4.54$ [V] as specified in the lib file that MOSFET conducts.

2.3 USE OF SIMULATION TO OBTAIN INTERNAL CAPACITANCES OF MOSFETs

2.3.1 Finding C_{iss} with PSpice

The C_{iss} given by Equation 2.1 for any MOSFET is obtained using the PSpice circuit given in Figure 2.13 by following the steps outlined below.

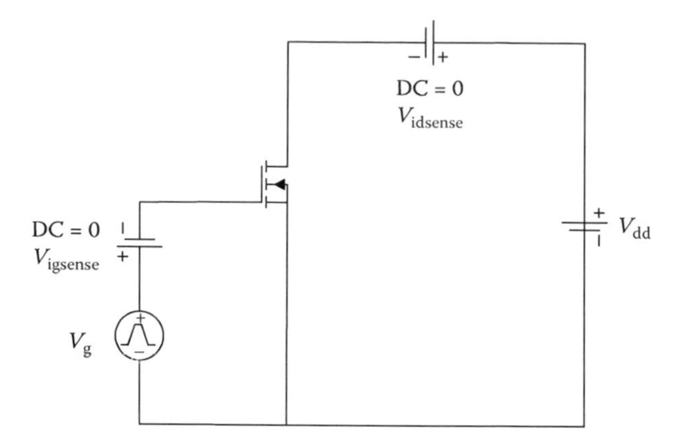

FIGURE 2.13 PSpice circuit to obtain C_{iss} for a given MOSFET.

In the circuit given in Figure 2.13, it is assumed that

- V_{gs} increases with constant d*v*/d*t* of 1 V/μs.
- $|I_g(t)|$ in milliamperes vs. V_{gs} is equal to C_{iss} in nanofarads vs. V_{gs} when $V_{ds} = 0$. $V_{gs} < V_{th}$ is valid only when the curve is obtained vs. V_{ds}.
- $|I_g(t)| - |I_d(t)|$ in milliamperes vs. V_{gs} is equal to C_{gs} in nanofarads when $V_{ds} = 0$ and is only valid for $0 < V_{gs} < V_{th}$. For $V_{gs} > V_{th}$, $i_d(t)$ has contributions not only from C_{gd} but also from the dependent current source in the MOSFET model activated by V_{gs}.
- $|I_d(t)|$ in milliamperes vs. V_{gs} is equal to C_{gd} in nanofarads vs. V_{gs} when $V_{ds} = 0$ and is only valid for $0 < V_{gs} < V_{th}$. For $V_{gs} > V_{th}$, $i_d(t)$ has contributions not only from C_{gd} but also from the dependent current source in the MOSFET model activated by V_{gs}.

The parameters for V_{pulse} in the simulation are suggested to be

$$\text{PULSE [1} = 0\text{ V, V2} = 4\text{ V, TD} = 0\text{, TR} = 0.4e - 5\text{, TF} = 0\text{,}\\ \text{PW} = 1e - 4\text{, PER} = 1e - 4]$$

This should be kept constant during the test. The purpose is to make a 1-V change in 1 μs. For instance, d*v*/d*t* = 4 V/*0.4*e* – 5 = 1 V/1 μs. Also, PW and PER have to be large enough.

2.3.2 Finding C_{OSS} and C_{RSS} with PSpice

The C_{oss} given by Equation 2.2 and C_{rss} (or C_{gd}) for any MOSFET is obtained using the PSpice circuit given in Figure 2.14 by following the steps below.

In the circuit given in Figure 2.14, it is assumed that

- V_{ds} falls with constant d*v*/d*t* of 1 V/1 μs.
- $|I_d(t)|$ in milliamperes vs. V_{ds} is equal to C_{oss} in nanofarads vs. V_{ds}.

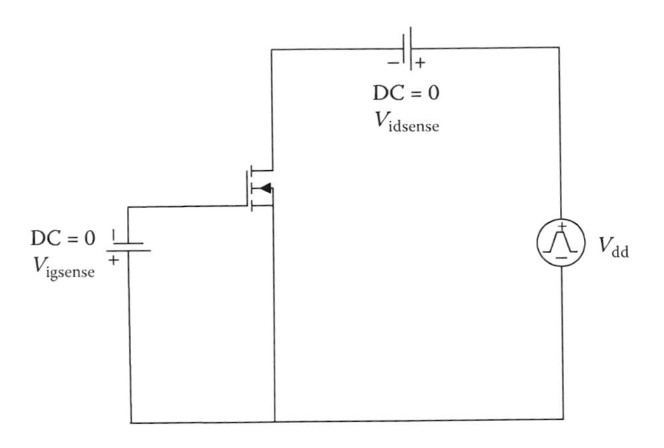

FIGURE 2.14 PSpice circuit to obtain C_{oss} and C_{rss} (or C_{gd}) for the given MOSFET.

- $|I_g(t)|$ in milliamperes vs. V_{ds} is equal to C_{gd} (or C_{rss}) in nanofarads vs. V_{ds}.
- $|I_d(t)| - |I_g(t)|$ in milliamperes vs. V_{ds} is equal to C_{ds} in nanofarads vs. V_{ds}.

The parameters for V_{pulse} in the simulation are suggested to be

$$\text{PULSE } [V_1 = 200 \text{ V}, V_2 = 0 \text{ V}, \text{TD} = 0, \text{TR} = 20e - 5, \text{TF} = 0,$$
$$\text{PW} = 3e - 4, \text{PER} = 3e - 4]$$

This should be kept constant during the test. The purpose is to make a 1-V change in 1 μs. For instance $dv/dt = 200 \text{ V}/{*}20e - 5 = 1 \text{ V}/1 \text{ μs}$. Also, PW and PER have to be large enough.

Example

A high-power very high frequency (VHF) MOSFET large-signal spice model lib file is given by the manufacturer. (a) Obtain the nonlinear circuit representation of the MOSFET. (b) Find the C_{iss}, C_{oss}, and C_{rss} of the given MOSFET with PSpice. The large signal spice.lib file of the MOSFET is

```
.SUBCKT MOSFETex2 1 2 3
* External Node Designations
* Node 1 -> Drain
* Node 2 -> Gate
* Node 3 -> Source
M1 9 7 8 8 MM L=100u W=100u
.MODEL MM NMOS LEVEL=1 IS=1e-32
+VTO=3.74315 LAMBDA=0.0111194 KP=1000
+CGSO=3.52512e-05 CGDO=3.35356e-07
RS 8 3 0.0803342
D1 3 1 MD
.MODEL MD D IS=3.68875e-13 RS=0.016113 N=0.888321 BV=1100
+IBV=10 EG=1.07489 XTI=4 TT=1.999e-07
+CJO=1.9261e-09 VJ=0.5 M=0.62478 FC=0.1
RDS 3 1 6e+07
```

```
RD 9 1 0.755455
RG 2 7 2.3488
D2 4 5 MD1
* Default values used in MD1:
*   RS=0 EG=1.11 XTI=3.0 TT=0
*   BV=infinite IBV=1mA
.MODEL MD1 D IS=1e-32 N=50
+CJO=3.1316e-09 VJ=0.5 M=0.9 FC=1e-08
D3 0 5 MD2
* Default values used in MD2:
*   EG=1.11 XTI=3.0 TT=0 CJO=0
*   BV=infinite IBV=1mA
.MODEL MD2 D IS=1e-10 N=0.4 RS=3e-06
RL 5 10 1
FI2 7 9 VFI2 -1
VFI2 4 0 0
EV16 10 0 9 7 1
CAP 11 10 9.84747e-09
FI1 7 9 VFI1 -1
VFI1 11 6 0
RCAP 6 10 1
D4 0 6 MD3
* Default values used in MD3:
*   EG=1.11 XTI=3.0 TT=0 CJO=0
*   RS=0 BV=infinite IBV=1mA
.MODEL MD3 D IS=1e-10 N=0.4
.ENDS MOSFETex2
```

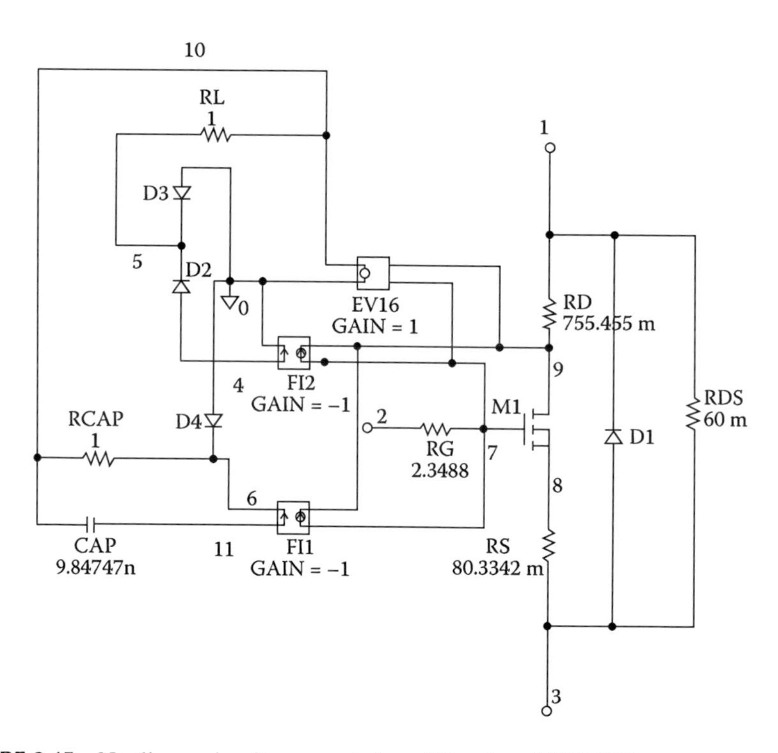

FIGURE 2.15 Nonlinear circuit representation of the given MOSFET.

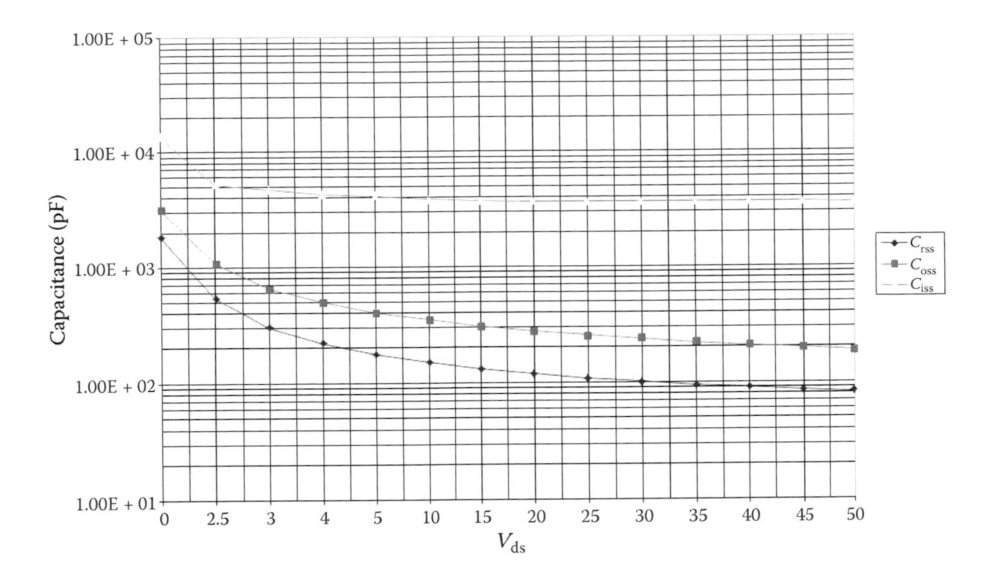

FIGURE 2.16 C–V curves obtained using PSpice for the given MOSFET.

Solution

a. The nonlinear circuit model is constructed using the spice.lib file given in the question and shown in Figure 2.15.
b. The C_{iss}, C_{oss}, and C_{rss} of the given MOSFET are obtained using the PSpice simulation circuit that is established using the steps and circuits given in Sections 2.3.1 and 2.3.2. The C–V curves are illustrated in Figure 2.16.

2.4 TRANSIENT CHARACTERISTICS OF MOSFET

The transient characteristics of a MOSFET are a function of many factors including DC voltage, switching frequency, and internal and extrinsic parameters. Typical MOSFET turn-on and turn-off characteristics showing the behavior of drain current, i_d, drain-to-source voltage, V_{ds}, gate current, i_g, and gate-to-source voltage, V_{gs}, during turn-on and turn-off transients are illustrated in Figures 2.17 and 2.18.

Assume that step voltage is applied to the gate of the simplified MOSFET equivalent circuit shown in Figure 2.19.

2.4.1 During Turn-On

Initially, MOSFET is turned off with the following initial conditions: $V_{gg} = 0$ V, $V_{ds} = V_{in}$, and $i_o = i_{diode}$. At $t = t_o$, V_{gg} is applied, and internal (and externally added) capacitors across the gate–source and gate–drain start to be charged through the gate resistor R_g. The gate–source voltage, V_{gs}, increases exponentially with the time constant $\tau = R_g(C_{gs} + C_{gd})$. The drain current remains zero until V_{gs} reaches the threshold voltage. This behavior can be mathematically expressed as [5]

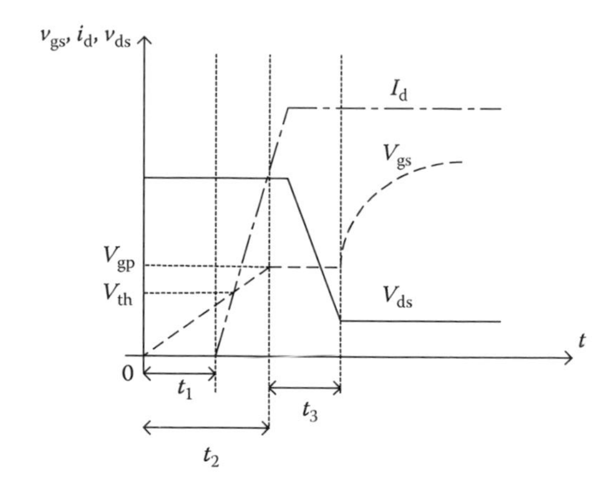

FIGURE 2.17 MOSFET turn-on characteristics.

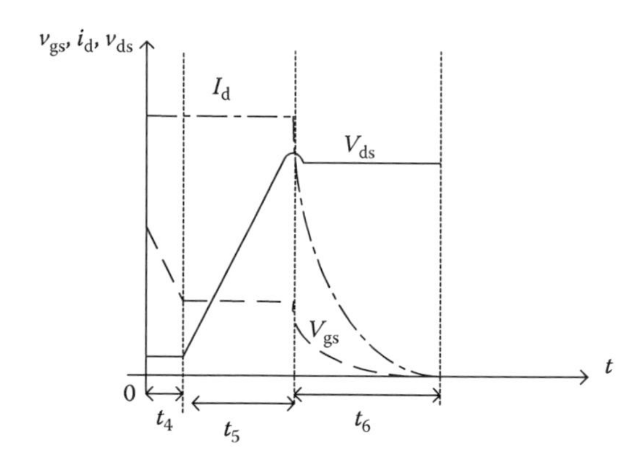

FIGURE 2.18 MOSFET turn-off characteristics.

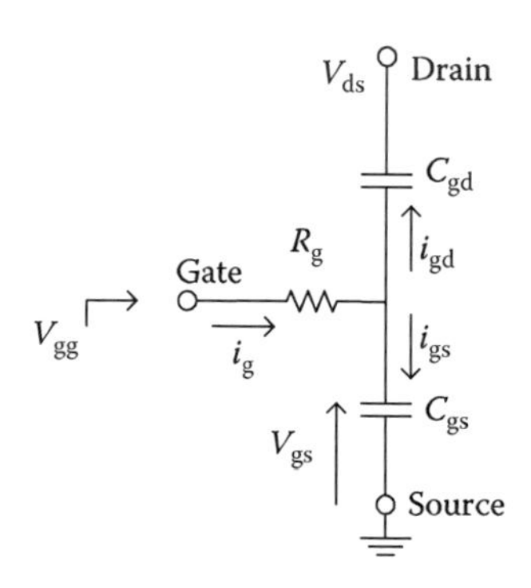

FIGURE 2.19 Simplified MOSFET equivalent circuit for switching analysis.

$$i_g = \frac{V_{gg} - V_{gs}}{R_g} \tag{2.8}$$

or

$$i_g = i_{gs} + i_{gd} \tag{2.9}$$

where

$$i_{gs} = C_{gs} \frac{dV_{gs}}{dt} \tag{2.10}$$

$$i_{gd} = C_{gd} \frac{dV_{gs}}{dt} \tag{2.11}$$

So,

$$\frac{V_{gg} - V_{gs}}{R_g} = C_{gs} \frac{dV_{gs}}{dt} + C_{gd} \frac{dV_{gs}}{dt} \tag{2.12}$$

Rearranging Equation 2.12 gives

$$\frac{dV_{gs}}{(V_{gg} - V_{gs})} = \frac{dt}{(C_{gs} + C_{gd})R_g} \tag{2.13}$$

Integrating both sides of Equation 2.13 gives

$$-\ln(V_{gg} - V_{gs}) = \frac{t}{(C_{gs} + C_{gd})R_g} + K \tag{2.14}$$

Then,

$$V_{gs} = V_{gg} - Ke^{-\frac{t}{(C_{gs}+C_{gd})R_g}} \tag{2.15}$$

In Equation 2.15, time constant τ is defined as

$$\tau = (C_{gs} + C_{gd})R_g \tag{2.16}$$

K in Equation 2.15 is found by application of the initial condition at $t = 0$. At $t = 0$, $V_{gs} = 0$. So $K = V_{gg}$. Hence, Equation 2.15 can be expressed as

$$V_{gs} = V_{gg}\left(1 - e^{-\frac{t}{(C_{gs}+C_{gd})R_g}}\right) = V_{gg}\left(1 - e^{-\frac{t}{\tau}}\right) \tag{2.17}$$

Equation 2.17 is used to determine the amount of time it takes to reach the gate-to-source threshold voltage, V_{th}, to turn the device on. In our analysis, the following time durations, t_1 to t_6, in Figures 2.17 and 2.18 are derived by designating V_{th} as the threshold voltage, V_{gp}, as the gate plateau voltage, V_f as the voltage across MOSFET with full current, and V_{ds} as the drain-to-source voltage when the device is off. One important note is on the equation for t_s. t_s depends on the gate-to-source voltage, which changes with the applied voltage. t_1 and t_2 can be accurately calculated with the parameters given by the manufacturer data sheet.

$$t_1 = R_g(C_{gs} + C_{gd})\ln\left(\frac{1}{1 - \frac{V_{th}}{V_{gg}}}\right) \tag{2.18}$$

t_1 is also equal to delay time, which is the time it takes for the gate-to-source voltage to reach the threshold voltage and is expressed by

$$t_d = t_1 = -R_g(C_{gs} + C_{gd})\ln\left(1 - \frac{V_{th}}{V_{gg}}\right) = -\tau\ln\left(1 - \frac{V_{th}}{V_{gg}}\right) \tag{2.19}$$

$$t_2 = R_g(C_{gs} + C_{gd})\ln\left(\frac{1}{1 - \frac{V_{gp}}{V_{gg}}}\right) \tag{2.20}$$

and

$$t_3 = \frac{(V_{gs} - V_f)R_g C_{gd}}{(V_{gg} - V_{gp})} \tag{2.21}$$

The drain current can be calculated from the gate-to-source voltage as

$$i_d(t) = g_m(V_{gg} - V_{th}) - g_m V_{gg} e^{-(t-t_1)/\tau} \tag{2.22}$$

When the drain current reaches the value of the load current, the gate-to-source voltage, V_{gs}, and the gate current, i_g, remain constant. So the gate current is found from

$$i_g = i_{gd} = -C_{gd}\frac{d(V_{gg} - V_{in})}{dt} = -C_{gd}\frac{dV_{ds}}{dt} \tag{2.23}$$

Hence, the drain-to-source voltage is

$$V_{ds}(t) = -\frac{(V_{gg} - V_{th})}{R_g C_{gd}}(t - t_2) + V_{in} \tag{2.24}$$

The dV/dt that happens during turn-on is an important parameter that impacts the performance of the amplifier. dV/dt can be calculated from Equation 2.23 as

$$\frac{dV_{ds}}{dt} = -\frac{i_g}{C_{gd}} \tag{2.25}$$

Substituting Equation 2.8 into Equation 2.25 gives

$$\frac{dV_{ds}}{dt} = -\frac{i_g}{C_{gd}} = -\frac{V_{gg} - V_{gs}}{R_g C_{gd}} \tag{2.26}$$

The value of the gate-to-source voltage, V_{gs}, in Equation 2.26 is calculated from the values given in the manufacturer data sheet for MOSFETs as

$$V_{gs} = \frac{I_o}{g_m} + V_{th} \tag{2.27}$$

So, the maximum value of dV/dt during turn-on is found from

$$\left|\frac{dV_{ds}}{dt}\right|_{max} = \frac{1}{R_g C_{gd}} \left| V_{gg} - \left(\frac{I_o}{g_m} + V_{th} \right) \right| \tag{2.28}$$

2.4.2 During Turn-Off

The initial conditions for turn-off are $V_{ds,onstate} = R_{ds} I_o$, $i_g = 0$, $V_{gs} = V_{gg}$, and $i_d = I_o$. The analysis performed during turn-on can be applied for derivation of the time duration turn-off. The transition time during turn-off is

$$t_4 = R_g (C_{gs} + C_{gd}) \ln\left(\frac{V_{gg}}{V_{gp}} \right) \tag{2.29}$$

At the time $t = t_4$, the gate-to-source and gate-to-drain capacitors discharge through R_g, and the gate current can be represented by

$$i_g = -\frac{V_{gg}}{R_g} \tag{2.30}$$

Also,

$$i_g = i_{gs} + i_{gd} = (C_{gs} + C_{gd}) \frac{dV_{gs}}{dt} \tag{2.31}$$

which leads to

$$V_{gs} = V_{gg} e^{-\frac{t}{\tau}} \tag{2.32}$$

V_{gs} decreases until it reaches the constant value at the plateau, whereas the drain current, i_d, remains constant at $i_d = I_o$. The gate-to-source voltage, V_{gs}, at $i_d = I_o$ is found from

$$V_{gs} = \frac{I_o}{g_m} + V_{th} \tag{2.33}$$

From Equations 2.32 and 2.33, we can obtain the time it takes for V_{gs} to reach the plateau voltage, or the constant value can be found as

$$V_{gs} = \tau \frac{V_{gg}}{\frac{I_o}{g_m} + V_{th}} \tag{2.34}$$

Time duration t_5 can be calculated from

$$t_5 = R_g C_{gd} \ln\left(\frac{V_{ds} - V_f}{V_{gp}}\right) \tag{2.35}$$

Since V_{gs} is constant during t_5, all the current at the gate is due to the gate-to-drain capacitance and can be expressed as

$$i_g = C_{gd} \frac{dv_{gd}}{dt} \tag{2.36}$$

or

$$i_g = C_{gd} \frac{d(v_{gs} - v_{ds})}{dt} = -C_{gd} \frac{dv_{ds}}{dt} \tag{2.37}$$

Hence,

$$i_g = \frac{1}{R_g}\left(\frac{I_o}{g_m} + V_{th}\right) \tag{2.38}$$

From Equations 2.37 and 2.38, we can find the drain-to-source voltage as

$$V_{ds} = V_{ds,on} + \frac{1}{R_g C_{gd}}\left(\frac{I_o}{g_m} + V_{th}\right)(t - t_1) \tag{2.39}$$

The final time duration t_6 is calculated from

$$t_6 = R_g(C_{gs} + C_{gd})\ln\left(\frac{V_{gp}}{V_{th}}\right) \tag{2.40}$$

When t_6 begins, the free-wheeling diode in the MOSFET structure turns on, and the drain current starts falling. The drain current can then be represented as

$$i_d = g_m(V_{gs} - V_{th}) \tag{2.41}$$

where

$$V_{gs} = \left(\frac{I_o}{g_m} + V_{th}\right)e^{-\frac{(t-t_2)}{\tau}} \tag{2.42}$$

The gate current during this cycle is found from

$$i_g = -\frac{1}{R_g}\left(\frac{I_o}{g_m} + V_{th}\right)e^{-\frac{(t-t_2)}{\tau}} \tag{2.43}$$

The turn-off time period is completed when the gate-to-source voltage becomes zero, $V_{gs} = 0$.

The MOSFET turn-on and turn-off times are illustrated in Figure 2.20. They can be related to the transition times defined by Equations 2.18 through 2.32 as follows. Turn-on delay time, t_d(on), is the time for the gate voltage, V_{gs}, to reach the threshold voltage, V_{th}. The input capacitance during this period is $C_{iss} = C_{gs} + C_{gd}$. This means that this is the charging period to bring up the capacitance to the threshold voltage.

It is expressed as

$$t_d(\text{on}) \approx t_1 + t_{ir} \tag{2.44}$$

where t_{ir} is the current rise time and is defined by

$$t_{ir} = t_2 - t_1 \tag{2.45}$$

or

$$t_{ir} = R_g C_{iss} \ln\left(\frac{g_m(V_{gg} - V_{th})}{g_m(V_{gg} - V_{th}) - I_o}\right) \tag{2.46}$$

Rise time is the period after V_{gs} reaches V_{th} to complete the transient. During the rise time, as both the high voltage and the high current exist in the device, high power dissipation occurs. The rise time should be reduced by reducing the gate series resistance and the drain–gate capacitance, C_{gd}. After this, the gate voltage continues

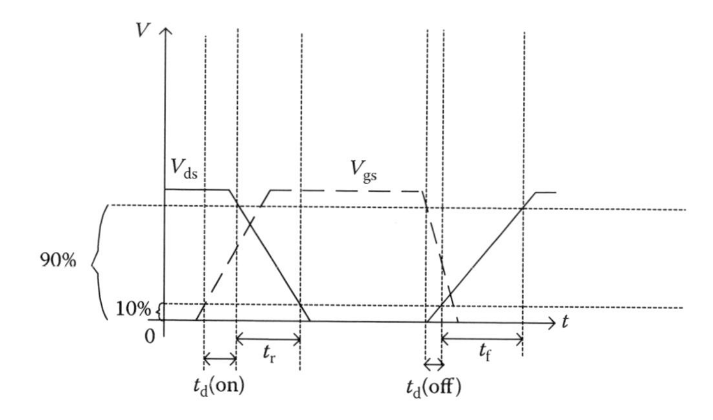

FIGURE 2.20 Illustration of the MOSFET turn-on and turn-off times.

to increase up to the supplied voltage level, but as the drain voltage and the current are already in steady state, they are not affected during this region. Rise time can be expressed as

$$t_r \approx t_{vf} \tag{2.47}$$

where t_{vf} is the voltage fall time and is defined by

$$t_{vf} = t_3 \tag{2.48}$$

or

$$t_{vf} = \frac{Q_{gd\_d}(V_{ds} - V_f)R_g}{\left(V_{ds\_d} - V_{f\_d}\right)\left(V_{gg} - \left(V_{th} + \frac{I_o}{g_m}\right)\right)} \tag{2.49}$$

In Equation 2.49, $(V_{ds\_d} - V_{f\_d})$ is the voltage swing at the drain of the MOSFET, and Q_{gd_d} is the corresponding gate charge. Then, the MOSFET turn-on time, t_{on}, is defined by

$$t_{on}(\text{MOSFET}) = t_d(\text{on}) + t_r \tag{2.50}$$

Turn-off delay time, t_d(off), is the time for the gate voltage to reach the point where it is required to make the drain current become saturated at the value of the load current. During this time, there are no changes to the drain voltage and the current. It is defined by

$$t_d(\text{off}) \approx t_4 \tag{2.51}$$

where t_4 is given in Equation 2.29. Then, MOSFET turn-off time is defined as

$$t_{off}(\text{MOSFET}) = t_d(\text{off}) + t_f \tag{2.52}$$

where t_f is the fall time where the gate voltage reaches the threshold voltage after t_d(off).

There is a lot of power dissipation in the t_f region during the turn-off state similar to the turn-on state. Hence, t_f must be reduced as much as possible. After this, the gate voltage continues to decrease until it reaches zero. As the drain voltage and the current are already in steady state, they are not affected during this region. t_f is defined by

$$t_f \approx t_{vr} \tag{2.53}$$

where t_{vr} is the voltage rise time and defined by

$$t_{vr} = t_5 \tag{2.54}$$

or

$$t_{vr} = \frac{Q_{gd\_d}(V_{ds} - V_f)R_g}{\left(V_{ds\_d} - V_{f\_d}\right)\left(V_{th} + \dfrac{I_o}{g_m}\right)} \tag{2.55}$$

The current fall time can also be expressed as

$$t_{if} = R_g C_{iss} \ln\left(\frac{V_{th} + \dfrac{I_o}{g_m}}{V_{th}}\right) \tag{2.56}$$

The rate change of current is calculated from

$$\frac{di_d}{dt} = \frac{i_d}{t_{if}} \tag{2.57}$$

2.5 LOSSES FOR MOSFET

The power losses in MOSFETs are mainly due to conduction losses, P_c, and switching losses, P_{sw} [6–8]. Hence, the overall power loss for MOSFETs can be approximated as

$$P_{loss} \approx P_c + P_{sw} \tag{2.58}$$

The instantaneous value of the conduction losses for MOSFETs can be calculated using

$$P_c(t) = V_d(t) i_d(t) = R_{ds} i_d^2(t) \tag{2.59}$$

The average value of the conduction loss in Equation 2.59 is calculated from

$$P_c = \frac{1}{T}\int_0^T P_c(t)\,dt = \frac{1}{T}\int_0^T R_{ds} i_d^2(t)\,dt = R_{ds} I_d^2 \tag{2.60}$$

where T represents the time period for switching. The switching loss for MOSFETs is

$$P_{sw} = (E_{on} + E_{off}) f_{sw} \tag{2.61}$$

where E_{on} and E_{off} are the turn-on and turn-off energies for MOSFETs, respectively, and calculated from

$$E_{on} = \int_0^{t_{ri}+t_{fv}} V_d(t) i_d(t)\,dt = V_{dd} I_{don} \frac{t_{ri} + t_{fv}}{2} + Q_{rr} V_{dd} \tag{2.62}$$

and

$$E_{on} = \int_0^{t_{rv}+t_{fi}} V_d(t) i_d(t)\,dt = V_{dd} I_{doff} \frac{t_{rv} + t_{fi}}{2} \tag{2.63}$$

where Q_{rr} is the reverse recovery charge. So the overall loss in MOSFETs from Equation 2.58 can be expressed as

$$P_{loss} = P_c + P_{sw} = R_{dson} I_{drms}^2 + (E_{on} + E_{off}) f_{sw} \tag{2.64}$$

2.6 THERMAL CHARACTERISTICS OF MOSFETs

The thermal profile of MOSFETs is important in device performance. When power loss occurs, it is turned into heat and increases the junction temperature. This degrades the device characteristics and can cause device failure. As a result, it is crucial to lower the junction temperature by transferring heat from the chip junction to ambient via the cold plate. The thermal path of MOSFETs illustrating die, case, junctions, cold plate, and ambient is shown in Figure 2.21. In the figure, T_j is used for junction temperature, T_c represents the case temperature at a point of the package that has the semiconductor die inside, T_s is used for heat sink or cold plate

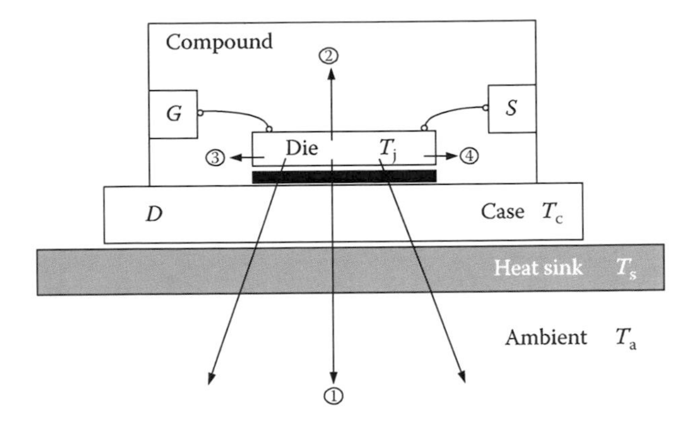

FIGURE 2.21 Illustration of thermal path for MOSFETs.

temperature, and T_a represents the ambient temperature of the surrounding environment where the device is located. $R_{\theta jc}$ shows the junction-to-case thermal resistance, $R_{\theta cs}$ is used for the case-to-heat sink thermal resistance, and $R_{\theta sa}$ shows the heat sink-to-ambient thermal resistance.

The heat produced at the die junction commonly radiates over 80% in the direction of 1 and about 20% in the direction of 2, 3, and 4. The equivalent electrical circuit for the thermal path shown in Figure 2.21 is established and given in Figure 2.22. Thermal capacitance effect of the structure is ignored for simplicity of the analysis in Figure 2.22. Die junction-to-ambient thermal, $R_{\theta ja}$, can be expressed as

$$R_{\theta ja} = R_{\theta jc} + R_{\theta cs} + R_{\theta sa} \tag{2.65}$$

In Equation 2.65, junction-to-case thermal resistance, $R_{\theta jc}$, is the internal thermal resistance from the die junction to the case package. If the size of the die is known, this thermal resistance of pure package is determined by the package design and lead frame material and is expressed by

$$R_{\theta jc} = \frac{T_j - T_c}{P_d} \,[°\text{C/W}] \tag{2.66}$$

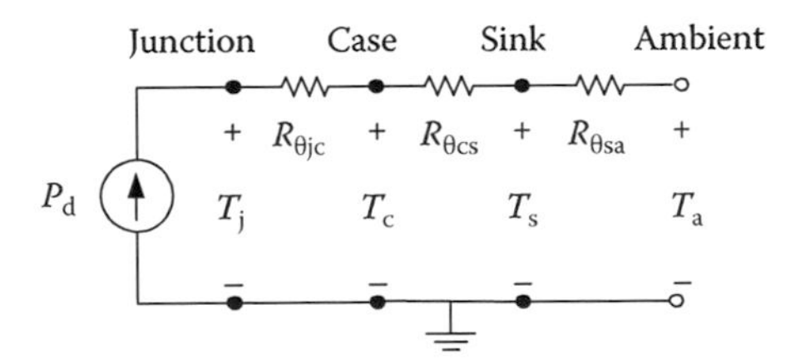

FIGURE 2.22 Electrical equivalent circuit for thermal path.

Consider the power dissipation profile of MOSFETs shown in Figure 2.23 for a pulsing RF amplifier. The junction temperature varies accordingly and can be expressed as

$$T_{\text{jmax}} - T_{\text{c}} = R_{\theta\text{jc}} P_{\text{d}} \tag{2.67}$$

The time-dependent thermal impedance for repetitive power pulses having a constant duty factor (D) is calculated using

$$Z_{\theta\text{jc}}(t) = R_{\theta\text{jc}} D + (1 - D) S_{\theta\text{jc}}(t) \tag{2.68}$$

where $S_{\theta\text{jc}}(t)$ is the thermal impedance for a single pulse. The drain current, I_{d}, rating of the device is found from

$$I_{\text{d}}(T_{\text{C}}) = \sqrt{\frac{T_{\text{jmax}} - T_{\text{C}}}{R_{\text{dson}}(T_{\text{jmax}}) R_{\theta\text{jc}}}} \tag{2.69}$$

When MOSFETs are pulsed, the pulsed drain current rating is obtained from

$$I_{\text{dpulse}}(T_{\text{c}}) = 4 I_{\text{d}}(T_{\text{c}}) \tag{2.70}$$

The crossover point is an important point that identifies the safe operational region for MOSFETs. It is obtained by finding the intersection point of I–V curves of MOSFETs for extreme and nominal junction temperatures, T_{j}. The I–V characteristics of MOSFETs vs. junction temperature, T_{j}, giving the crossover point for MOSFETs are shown in Figure 2.24. MOSFETs are inherently stable when they are operated above the crossover point because hot spotting van occurs below the crossover point.

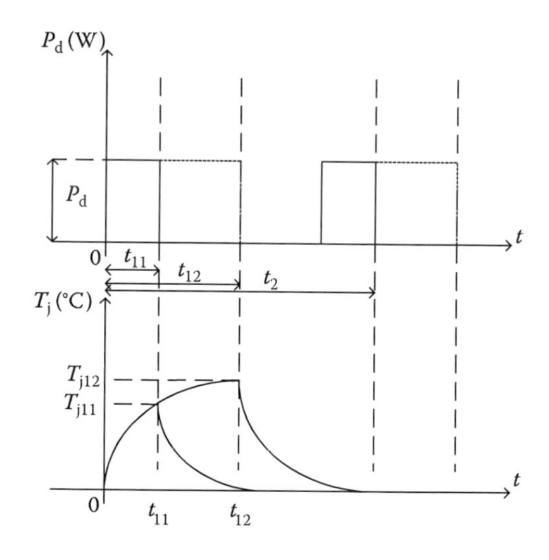

FIGURE 2.23 Power dissipation profile and the corresponding junction temperature.

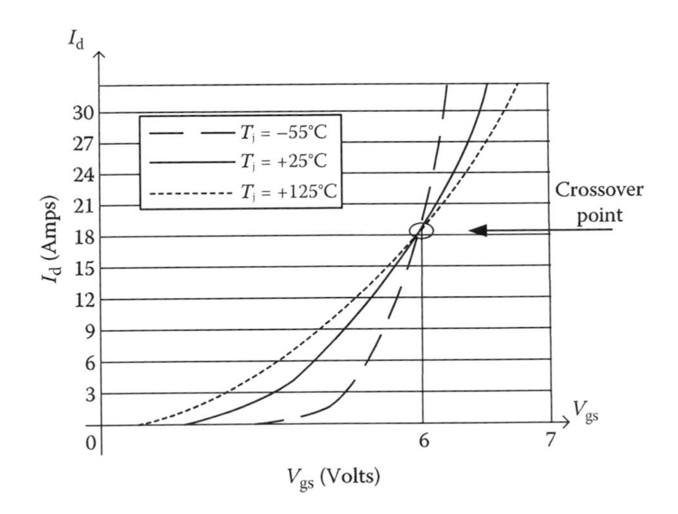

FIGURE 2.24 Transfer characteristics of MOSFETs vs. junction temperature, T_j.

2.7 SAFE OPERATING AREA FOR MOSFETs

Safe operating area (SOA) is the region identified by the maximum value of the drain-to-source voltage and drain current that guarantees safe operation when the device is at the forward bias. The SOA for MOSFETs can be expressed by a constant-power line, which is limited by thermal resistance with pulse width as a parameter. MOSFETs can be safely operated over a very wide range within the breakdown voltage between the drain and the source without narrowing the high-voltage area because secondary breakdown does not occur in the high-voltage area. The typical SOA for a MOSFET is illustrated in Figure 2.25. The relationship between SOA and device parameters such as R_{dson} and P_{dmax} is given in Figure 2.26. In Figure 2.26, the actual response of device dissipation characteristics, which has steeper slope when it is over a certain value, is assumed.

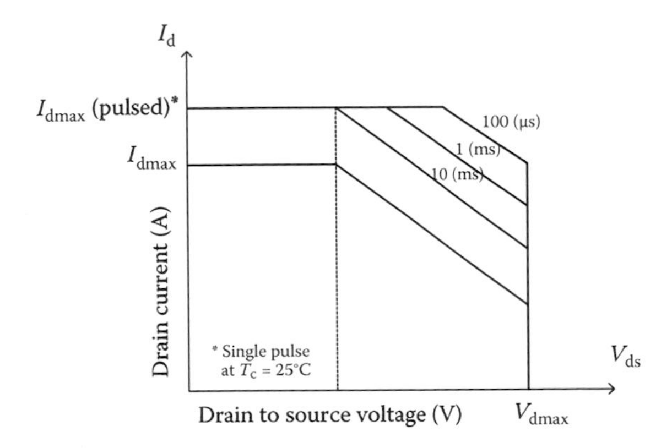

FIGURE 2.25 Typical SOA curves for MOSFETs.

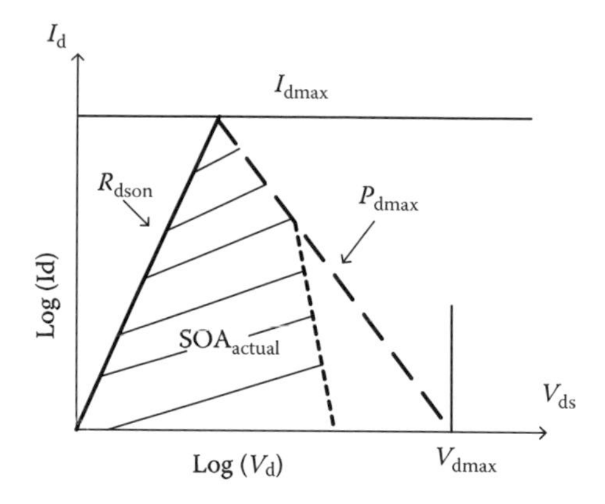

FIGURE 2.26 SOA vs. device parameters R_{dson} and P_{dmax}.

2.8 MOSFET GATE THRESHOLD AND PLATEAU VOLTAGE

MOSFET gate threshold, V_{th}, and Miller plateau voltage, V_{gp}, values are required to calculate the switching times for MOSFETs given in Section 2.3. However, these values are not well defined in the manufacturer data sheet. As a result, they need to be obtained using the measured transfer characteristics of MOSFETs. The plateau voltage, V_{gp}, is determined by finding two different values of drain current at the same temperature to identify the gate-to-source voltages, V_{gs} for the given load current, I_o. The drain current when the MOSFET is in the active region was given in Chapter 1 and found from

$$I_{d1} = K(V_{gs1} - V_{th}) \tag{2.71}$$

$$I_{d2} = K(V_{gs2} - V_{th}) \tag{2.72}$$

Solving Equations 2.71 and 2.72 for V_{th} gives

$$V_{th} = \frac{V_{gs1}\sqrt{I_{d2}} - V_{gs2}\sqrt{I_{d1}}}{\sqrt{I_{d2}} - \sqrt{I_{d1}}} \tag{2.73}$$

The constant K is found from Equation 2.71 or 2.72 as

$$K = \frac{I_{d1}}{(V_{gs1} - V_{th})^2} \tag{2.74}$$

The plateau voltage, V_{gp}, is then equal to

$$V_{gp} = V_{th} + \sqrt{\frac{I_o}{K}} \tag{2.75}$$

Example

Assume that the measured transfer characteristics for a MOSFET are given by the manufacturer in Figure 2.27 when T_j = 25°C and T_j = 125°C. It is communicated that the full load current is equal to I_o = 5 [A]. Find the threshold voltage and the gate plateau voltage using the transfer curves.

Solution

The transfer curve when T_j = 125°C is chosen for analysis. From the curve, we obtain the values of drain currents and gate-to-source voltages as

$$I_{d1} = 3\,[A], \qquad V_{gs1} = 4.13\,[V]$$
$$I_{d1} = 20\,[A], \qquad V_{gs2} = 5.67\,[V]$$

Using Equation 2.72, the threshold voltage, V_{th}, is calculated as

$$V_{th} = \frac{V_{gs1}\sqrt{I_{d2}} - V_{gs2}\sqrt{I_{d1}}}{\sqrt{I_{d2}} - \sqrt{I_{d1}}} = \frac{4.13\sqrt{20} - 5.67\sqrt{3}}{\sqrt{20} - \sqrt{3}} = 3.16\,[V]$$

The constant K is found from Equation 2.74 as

$$K = \frac{I_{d1}}{(V_{gs1} - V_{th})^2} = \frac{3}{(4.13 - 3.16)^2} = 3.19$$

Then, the gate plateau voltage, V_{gp}, from Equation 2.75 is

$$V_{gp} = V_{th} + \sqrt{\frac{I_o}{K}} = 3.16 + \sqrt{\frac{5}{3.19}} = 4.41\,[V]$$

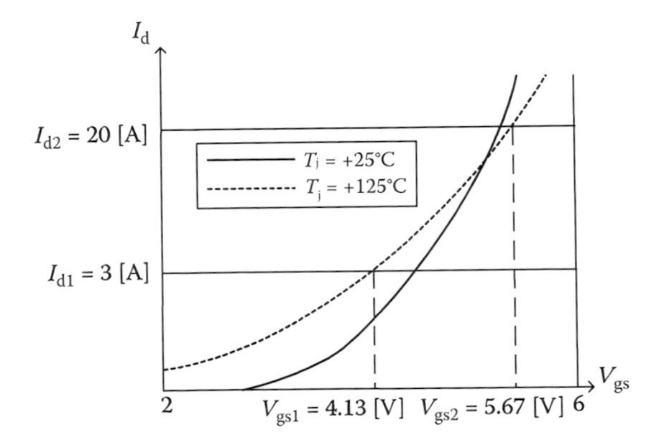

FIGURE 2.27 Transfer curves for the MOSFET at T_j = 25°C and T_j = 125°C.

PROBLEMS

1. Obtain the I–V characteristics of the MOSFET given in the example in Section 2.2 using ADS and MATLAB®/Simulink, and compare your results with the results obtained with PSpice in the example.
2. An RF amplifier used IRF440 MOSFET as an active device. It is rated for 500 V at 8 A with TO-3 package. The following information is given from the data sheet:

$$R_{\theta cs} = 0.2°C/W,\ R_{\theta ss} = 1°C/W,\ R_{\theta jc} = 1°C/W$$

$$R_{dson} \text{ at } 25°C = 0.8\ \Omega,\ V_{gs} = 10\ V,\ T_a = 50°C,\ I_{Drms} = 3\ A,\ P_l = 5\ W$$

Calculate the total die dissipation, P_t, and junction temperature, T_j, using the transient thermal model with

$$T_j = T_a + R_{\theta ja}\left[P_l + I^2_{D(rms)}R_{ds(25°C)}R_{dson}\right]$$

$$P_t = P_l + I^2_{D(rms)}R_{ds(25°C)}R_{dson}$$

and $R_{dson} = 0.4\ \Omega$ at 67.3°C, $R_{dson} = 2\ \Omega$ at 92.7°C.

3. The measurement of important parameters of a specific MOSFET has been done, and the results obtained are tabulated in Table 2.1. Calculate the turn-on and turn-off times shown in Figure 2.20.

TABLE 2.1
Measured Value of MOSFET Parameters

	Min	Type	Max	Unit
R_g	0.6	0.8	1	Ω
R_{g_app}	5.4	6	6.6	Ω
C_{iss} at V_{ds}	620	775	930	pF
C_{iss} at 0 V	880	1100	1320	pF
g_{fs}	21.6	27	32.4	S
V_{gs_app}	9	10	11	V
V_{th}	0.8	1.4	1.8	V
I_{ds}	0.9	1	1.1	A
Q_{gd_d}	2.8	3.5	4.2	nC
V_{ds_d}	13.5	15	16.5	V
I_{ds_d}	11.2	12.4	13.6	A
R_{dson}	0.008	0.01	0.012	Ω
V_f	0.0072	0.01	0.0132	V
V_{f_d}	0.09	0.12	0.16	V
V_{ds}	13.5	15	16.5	V

REFERENCES

1. B.J. Baliga. 2010. *Advanced Power MOSFET Concepts*. Springer, San Francisco.
2. M. Deboy, N. Marz, J.-P. Stengl, H. Strack, J. Tihanyi, and H. Weber. 1998. A new generation of high voltage MOSFETs breaks the limit line of silicon. *IEDM '98 Electron Devices Meeting, IEDM '98, Technical Digest*, pp. 683–685.
3. G. Sabui, and Z.J. Shen. 2014. On the feasibility of further improving Figure of Merits (FOM) of low voltage power MOSFETs. *Proceedings of the 26th International Symposium on Power Semiconductor Devices & IC's*, June 15–19, Waikoloa, Hawaii.
4. S. Xu et al. 2009. NexFET: A new power device. *Proceedings of the International Electron Devices Meeting*, pp. 1–4.
5. B.J. Baliga. 1995. *Power Semiconductor Devices*. PWS Pub. Co, Boston.
6. M.H. Rashid. 2011. *Power Electronics Handbook: Devices, Circuits, and Applications*. Academic Press, Burlington, MA.
7. N. Mohan. 1995. *Power Electronics: Converters, Applications, and Design*. John Wiley and Sons, Hoboken, NJ.
8. D. Graovac, M. Pürschel, and A. Kiep. 2006. *MOSFET Power Losses Calculation Using the Data-Sheet Parameters*. Infineon, Neubiberg, Germany.

3 Transistor Modeling and Simulation

3.1 INTRODUCTION

In this chapter, network parameters will be introduced and used to obtain response of the available electrical equivalent circuit models for transistors. In transistor modeling, network parameters will be used as a mathematical tool for designers to model and characterize critical parameters of devices by establishing relations between voltages and currents. Important transistor parameters such as voltage and current gains can be obtained with the application of these parameters. They can also be applied in small-signal power amplifier design, and parameters such as overall system gain and loss and several other responses can be obtained.

3.2 NETWORK PARAMETERS

Network parameters are analyzed and studied using two-port networks in Ref. [1]. The two-port network shown in Figure 3.1 is described by a set of four independent parameters, which can be related to voltage and current at any port of the network. As a result, the two-port network can be treated as a black box modeled by the relationships between the four variables. There exist six different ways to describe the relationships between these variables, depending on which two of the four variables are given, whereas the other two can always be derived. All voltages and currents are complex variables and represented by phasors containing both magnitude and phase. Two-port networks are characterized by using two-port network parameters such as Z-impedance, Y-admittance, h-hybrid, and $ABCD$. High-frequency networks are characterized by S-parameters. They are usually expressed in matrix notation, and they establish relations between the following parameters: input voltage V_1, output voltage V_2, input current I_1, and output current I_2.

3.2.1 *Z*-Impedance Parameters

The voltages are represented in terms of currents through Z-parameters as follows:

$$V_1 = Z_{11}I_1 + Z_{12}I_2 \tag{3.1}$$

$$V_2 = Z_{21}I_1 + Z_{22}I_2 \tag{3.2}$$

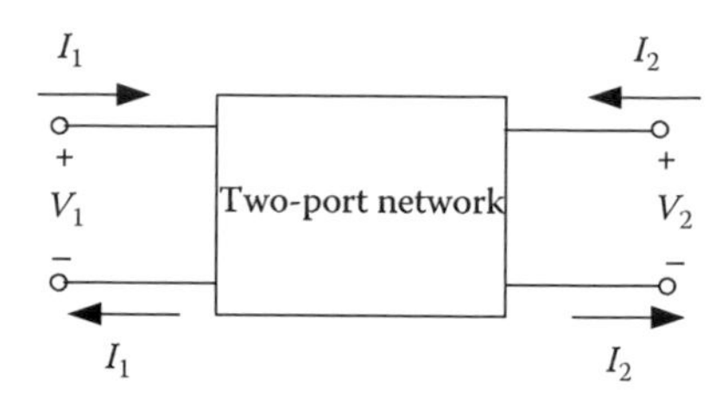

FIGURE 3.1 Two-port network representation.

In matrix form, Equations 3.1 and 3.2 can be combined and written as

$$\begin{bmatrix} V_1 \\ V_2 \end{bmatrix} = \begin{bmatrix} Z_{11} & Z_{12} \\ Z_{21} & Z_{22} \end{bmatrix} \begin{bmatrix} I_1 \\ I_2 \end{bmatrix} \tag{3.3}$$

The Z-parameters for a two-port network are defined as

$$\begin{aligned} Z_{11} &= \left.\frac{V_1}{I_1}\right|_{I_2=0} & Z_{12} &= \left.\frac{V_1}{I_2}\right|_{I_1=0} \\ Z_{21} &= \left.\frac{V_2}{I_1}\right|_{I_2=0} & Z_{22} &= \left.\frac{V_2}{I_2}\right|_{I_1=0} \end{aligned} \tag{3.4}$$

The formulation in Equation 3.4 can be generalized for an N-port network as

$$Z_{\mathrm{nm}} = \left.\frac{V_{\mathrm{n}}}{I_{\mathrm{m}}}\right|_{I_{\mathrm{k}}=0(k\neq m)} \tag{3.5}$$

Z_{nm} is the input impedance seen looking into port n when all other ports are open circuited. In other words, Z_{nm} is the transfer impedance between ports n and m when all other ports are open. It can be shown that for reciprocal networks,

$$Z_{\mathrm{nm}} = Z_{\mathrm{mn}} \tag{3.6}$$

3.2.2 *Y*-Admittance Parameters

The currents are related to voltages through Y-parameters as follows:

$$I_1 = Y_{11}V_1 + Y_{12}V_2 \tag{3.7}$$

$$I_2 = Y_{21}V_1 + Y_{22}V_2 \tag{3.8}$$

In matrix form, Equations 3.6 and 3.7 can be written as

$$\begin{bmatrix} I_1 \\ I_2 \end{bmatrix} = \begin{bmatrix} Y_{11} & Y_{12} \\ Y_{21} & Y_{22} \end{bmatrix} \begin{bmatrix} V_1 \\ V_2 \end{bmatrix} \tag{3.9}$$

The Y-parameters in Equation 3.9 can be defined as

$$\begin{aligned} Y_{11} &= \left.\frac{I_1}{V_1}\right|_{V_2=0} & Y_{12} &= \left.\frac{I_1}{V_2}\right|_{V_1=0} \\ Y_{21} &= \left.\frac{I_2}{V_1}\right|_{V_2=0} & Y_{22} &= \left.\frac{I_2}{V_2}\right|_{V_1=0} \end{aligned} \tag{3.10}$$

The formulation in Equation 3.10 can be generalized for an N-port network as

$$Y_{\mathrm{nm}} = \left.\frac{I_{\mathrm{n}}}{V_{\mathrm{m}}}\right|_{V_{\mathrm{k}}=0(k\neq m)} \tag{3.11}$$

Y_{nm} is the input impedance seen looking into port n, when all other ports are short circuited. In other words, Y_{nm} is the transfer admittance between ports n and m when all other ports are short. It can be shown that

$$Y_{\mathrm{nm}} = Y_{\mathrm{mn}} \tag{3.12}$$

In addition, it can be further proved that the impedance and admittance matrices are related through

$$[Z] = [Y]^{-1} \tag{3.13}$$

or

$$[Y] = [Z]^{-1} \tag{3.14}$$

3.2.3 *ABCD*-Parameters

$ABCD$-parameters relate the voltages to current in the following form for a two-port network:

$$V_1 = AV_1 - BI_2 \tag{3.15}$$

$$I_1 = CV_1 - DI_2 \tag{3.16}$$

which can be put in matrix form as

$$\begin{bmatrix} V_1 \\ I_1 \end{bmatrix} = \begin{bmatrix} A & B \\ C & D \end{bmatrix} \begin{bmatrix} V_1 \\ -I_2 \end{bmatrix} \tag{3.17}$$

The $ABCD$ parameters in Equation 3.17 are defined as

$$\begin{aligned} A &= \left.\frac{V_1}{V_2}\right|_{I_2=0} & B &= \left.\frac{V_1}{-I_2}\right|_{V_2=0} \\ C &= \left.\frac{I_1}{V_2}\right|_{I_2=0} & D &= \left.\frac{I_1}{-I_2}\right|_{V_2=0} \end{aligned} \tag{3.18}$$

It can be shown that

$$AD - BC = 1 \tag{3.19}$$

for the reciprocal network and $A = D$ for the symmetrical network. The $ABCD$ network is useful in finding the voltage or current gain of a component or the overall gain of a network. One of the great advantages of $ABCD$ parameters is their use in cascaded network or components. When this condition exists, the overall $ABCD$ parameter of the network simply becomes the matrix product of an individual network and a component. This can be generalized for an N-port network shown in Figure 3.2 as

$$\begin{Bmatrix} v_1 \\ i_1 \end{Bmatrix} = \left(\begin{bmatrix} A_1 & B_1 \\ C_1 & D_1 \end{bmatrix} \cdots \begin{bmatrix} A_n & B_n \\ C_n & D_n \end{bmatrix} \right) \begin{Bmatrix} v_2 \\ -i_2 \end{Bmatrix} \tag{3.20}$$

3.2.4 *h*-Hybrid Parameters

Hybrid parameters relate voltage and current in a two-port network as

$$V_1 = h_{11}I_1 + h_{12}V_2 \tag{3.21}$$

$$I_2 = h_{21}I_1 + h_{22}V_2 \tag{3.22}$$

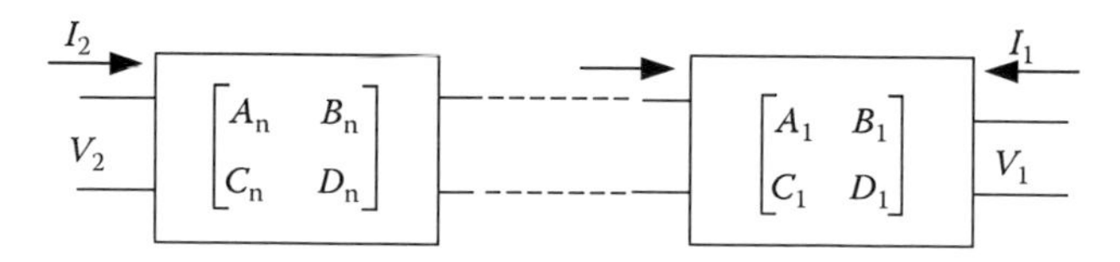

FIGURE 3.2 *ABCD*-parameter of cascaded networks.

Equations 3.21 and 3.22 can be put in matrix form as

$$\begin{bmatrix} V_1 \\ I_2 \end{bmatrix} = \begin{bmatrix} h_{11} & h_{12} \\ h_{21} & h_{22} \end{bmatrix} \begin{bmatrix} I_1 \\ V_2 \end{bmatrix} \tag{3.23}$$

The hybrid parameters in Equation 3.23 can be found from

$$\begin{aligned} h_{11} &= \left.\frac{V_1}{I_1}\right|_{V_2=0}, & h_{12} &= \left.\frac{V_1}{V_2}\right|_{I_1=0} \\ h_{21} &= \left.\frac{I_2}{I_1}\right|_{V_2=0} & h_{22} &= \left.\frac{I_2}{V_2}\right|_{I_1=0} \end{aligned} \tag{3.24}$$

Hybrid parameters are preferred for components such as transistors and transformers since they can be measured with ease in practice.

Example

Obtain the h-parameter of the circuit shown in Figure 3.3 if $L_1 = L_2 = M = 1$ H.

Solution

There are two methods to solve this example.

- *First method.* From KVL on the primary side,

$$V_1 = sL_1I_1^x + sMI_2 \tag{3.25}$$

Application of KCL gives

$$I_1^x = I_1 - V_1 \tag{3.26}$$

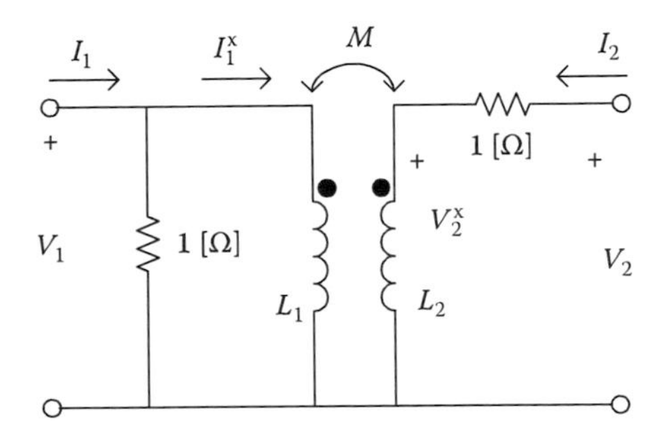

FIGURE 3.3 Coupling transformer example.

Substitution of I_1^x into the above equation gives

$$(1 + sL_1)V_1 - sMI_2 = sL_1I_1 \quad (3.27)$$

From KVL on the secondary side,

$$V_2 = V_2^x + I_2 \quad (3.28)$$

Also,

$$V_2^x = sL_2I_2 + sMI_1^x \quad (3.29)$$

Substitution of I_1^x into the above leads to

$$V_2^x = sL_2I_2 + sM(I_1 - V_1) \quad (3.30)$$

When V_2^x is inserted in V_2, we obtain

$$V_2 = (1 + sL_2)I_2 + sM(I_1 - V_1) \quad (3.31)$$

or

$$sMV_1 - (1 + sL_2)I_2 = sMI_1 - V_2 \quad (3.32)$$

Equation 3.32 can be written in matrix form as

$$\begin{bmatrix} 1+sL_1 & -sM \\ SM & -(1+sL_2) \end{bmatrix} \begin{bmatrix} V_1 \\ I_2 \end{bmatrix} = \begin{bmatrix} sL_1 & 0 \\ sM & -1 \end{bmatrix} \begin{bmatrix} I_1 \\ V_2 \end{bmatrix} \quad (3.33)$$

or

$$\begin{bmatrix} V_1 \\ I_2 \end{bmatrix} = \begin{bmatrix} 1+sL_1 & -sM \\ sM & -(1+sL_2) \end{bmatrix}^{-1} \begin{bmatrix} sL_1 & 0 \\ sM & -1 \end{bmatrix} \begin{bmatrix} I_1 \\ V_2 \end{bmatrix} \quad (3.34)$$

When $L_1 = L_2 = M = 1$ H, Equation 3.34 becomes

$$\begin{bmatrix} V_1 \\ I_2 \end{bmatrix} = \begin{bmatrix} 1+s & -s \\ s & -(1+s) \end{bmatrix}^{-1} \begin{bmatrix} s & 0 \\ s & -1 \end{bmatrix} \begin{bmatrix} I_1 \\ V_2 \end{bmatrix} \quad (3.35)$$

or

$$\begin{bmatrix} V_1 \\ I_2 \end{bmatrix} = \frac{1}{(2s+1)} \begin{bmatrix} s & s \\ -s & s+1 \end{bmatrix} \begin{bmatrix} I_1 \\ V_2 \end{bmatrix} \quad (3.36)$$

Hence,

$$[h] = \begin{bmatrix} h_{11} & h_{12} \\ h_{21} & h_{22} \end{bmatrix} = \begin{bmatrix} \frac{s}{2s+1} & \frac{s}{2s+1} \\ -\frac{s}{2s+1} & \frac{s+1}{2s+1} \end{bmatrix} \quad (3.37)$$

- *Second method.* The mutual inductance equivalent circuit can be converted into an equivalent circuit with the transformation parameters shown in Figure 3.4.

So, the original circuit can then be translated to that shown in Figure 3.5.
The *ABCD* parameter of network 1, N_1, is

$$ABCD_{N_1} = \begin{bmatrix} 1 & 0 \\ Y & 1 \end{bmatrix} = \begin{bmatrix} 1 & 0 \\ 1 & 1 \end{bmatrix} \quad (3.38)$$

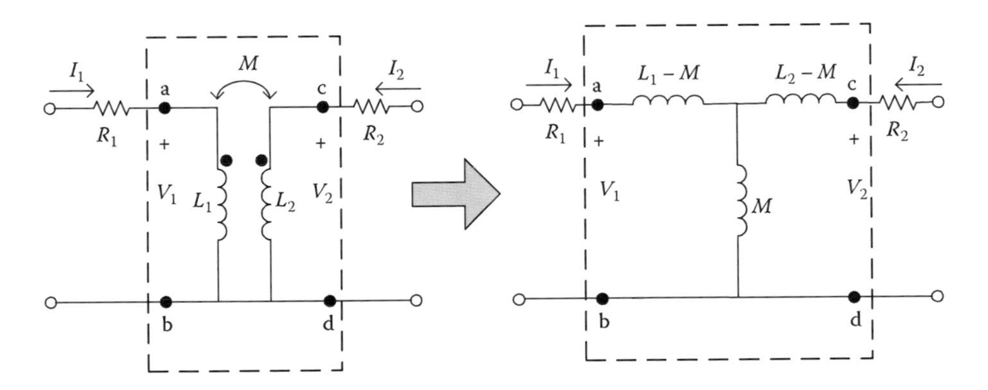

FIGURE 3.4 Conversion of transformer coupling circuit to equivalent circuit.

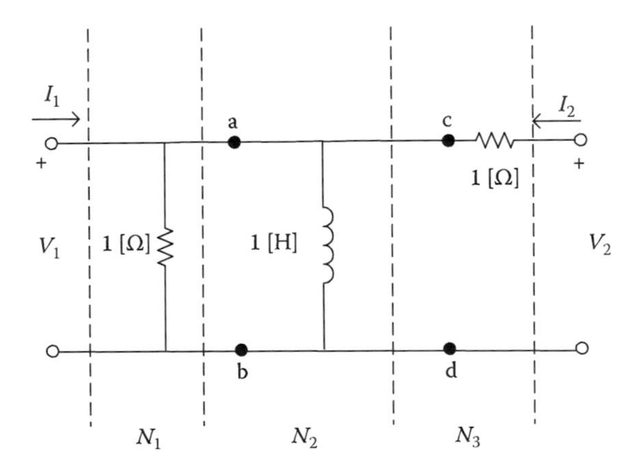

FIGURE 3.5 Transformer coupling circuit to equivalent circuit.

The *ABCD* parameter of network 2, N_2, is

$$ABCD_{N_2} = \begin{bmatrix} 1 & 0 \\ Y & 1 \end{bmatrix} = \begin{bmatrix} 1 & 0 \\ \dfrac{1}{j\omega} & 1 \end{bmatrix} \quad (3.39)$$

The *ABCD* parameter of network 3, N_3, is

$$ABCD_{N_3} = \begin{bmatrix} 1 & Z \\ 0 & 1 \end{bmatrix} = \begin{bmatrix} 1 & 1 \\ 0 & 1 \end{bmatrix} \quad (3.40)$$

The *ABCD* parameter of the overall network is

$$ABCD = (ABCD_{N_1})(ABCD_{N_2})(ABCD_{N_3}) = \begin{bmatrix} 1 & 0 \\ 1 & 1 \end{bmatrix} \begin{bmatrix} 1 & 0 \\ \dfrac{1}{j\omega} & 1 \end{bmatrix} \begin{bmatrix} 1 & 1 \\ 0 & 1 \end{bmatrix}$$

$$= \begin{bmatrix} 1 & 1 \\ 1+\dfrac{1}{j\omega} & 2+\dfrac{1}{j\omega} \end{bmatrix} \quad (3.41)$$

So, the hybrid parameters are

$$h = \begin{bmatrix} \dfrac{B}{D} & \dfrac{\Delta}{D} \\ -\dfrac{1}{D} & \dfrac{C}{D} \end{bmatrix} = \begin{bmatrix} \dfrac{j\omega}{2j\omega+1} & \dfrac{j\omega}{2j\omega+1} \\ -\dfrac{j\omega}{2j\omega+1} & \dfrac{j\omega+1}{2j\omega+1} \end{bmatrix} \quad (3.42)$$

where $\Delta = 1$. The same result is obtained.

Example

Find the (a) impedance, (b) admittance, (c) *ABCD*, and (d) hybrid parameters of the *T*-network given in Figure 3.6.

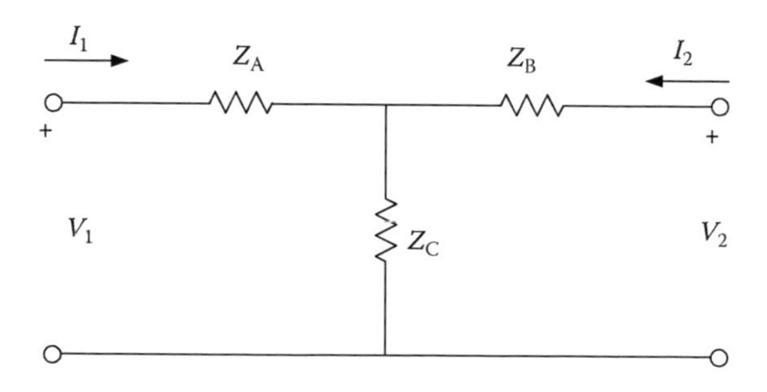

FIGURE 3.6 *T*-network configuration.

Solution

a. Z-parameters are found with application of Equation 3.4 by opening all the other ports except the measurement port. This leads to

$$Z_{11} = \left.\frac{V_1}{I_1}\right|_{I_2=0} = Z_A + Z_C \qquad Z_{21} = \left.\frac{V_2}{I_1}\right|_{I_2=0} = Z_C$$

$$Z_{12} = \left.\frac{V_1}{I_2}\right|_{I_1=0} = \frac{V_2}{I_2}\frac{Z_C}{Z_B+Z_C} = (Z_B+Z_C)\frac{Z_C}{Z_B+Z_C} = Z_C \qquad Z_{22} = \left.\frac{V_2}{I_2}\right|_{I_1=0} = Z_B + Z_C$$

The Z-matrix is then constructed as

$$Z = \begin{bmatrix} Z_A + Z_C & (Z_B+Z_C)\dfrac{Z_C}{Z_B+Z_C} = Z_C \\ (Z_B+Z_C)\dfrac{Z_C}{Z_B+Z_C} = Z_C & Z_B + Z_C \end{bmatrix}$$

b. Y-parameters are found from Equation 3.10 by shorting all the other ports except the measurement port. Y_{11} and Y_{21} are found when port 2 is shorted as

$$Y_{11} = \left.\frac{I_1}{V_1}\right|_{V_2=0} \rightarrow I_1 = \frac{V_1}{Z_A + (Z_B // Z_C)} = V_1\left(\frac{Z_B+Z_C}{Z_AZ_B + Z_AZ_C + Z_BZ_C}\right) \rightarrow$$

$$Y_{11} = \left(\frac{Z_B+Z_C}{Z_AZ_B + Z_AZ_C + Z_BZ_C}\right)$$

$$Y_{21} = \left.\frac{I_2}{V_1}\right|_{V_2=0} \rightarrow I_2 = \frac{-V_1}{(Z_A + (Z_B // Z_C))}\frac{Z_C}{(Z_C+Z_B)} \rightarrow Y_{21} = \left(\frac{-Z_C}{Z_AZ_B + Z_AZ_C + Z_BZ_C}\right)$$

Similarly, Y_{12} and Y_{22} are found when port 1 is shorted as

$$Y_{12} = \left.\frac{I_1}{V_2}\right|_{V_1=0} \rightarrow I_1 = \frac{-V_2}{(Z_B + (Z_A // Z_C))}\frac{Z_C}{(Z_A+Z_C)} \rightarrow Y_{12} = \left(\frac{-Z_C}{Z_AZ_B + Z_AZ_C + Z_BZ_C}\right)$$

$$Y_{22} = \left.\frac{I_2}{V_2}\right|_{V_1=0} \rightarrow I_2 = \frac{V_2}{Z_B + (Z_A // Z_C)} = V_1\left(\frac{Z_A+Z_C}{Z_AZ_B + Z_AZ_C + Z_BZ_C}\right) \rightarrow Y_{22} = \left(\frac{Z_A+Z_C}{Z_AZ_B + Z_AZ_C + Z_BZ_C}\right)$$

Y-parameters can also be found by just inverting the Z-matrix, as given by Equation 3.14 as

$$[Y] = [Z]^{-1} = \frac{1}{(Z_AZ_B + Z_AZ_C + Z_BZ_C)}\begin{bmatrix} Z_B + Z_C & -Z_C \\ -Z_C & Z_A + Z_C \end{bmatrix}$$

So, the *Y*-matrix for the *T*-network is then

$$Y=\begin{bmatrix} \left(\dfrac{Z_B+Z_C}{Z_AZ_B+Z_AZ_C+Z_BZ_C}\right) & \left(\dfrac{-Z_C}{Z_AZ_B+Z_AZ_C+Z_BZ_C}\right) \\ \left(\dfrac{-Z_C}{Z_AZ_B+Z_AZ_C+Z_BZ_C}\right) & \left(\dfrac{Z_A+Z_C}{Z_AZ_B+Z_AZ_C+Z_BZ_C}\right) \end{bmatrix}$$

As seen from the results of parts a and b, the network is reciprocal since

$$Z_{12}=Z_{21} \quad \text{and} \quad Y_{12}=Y_{21}$$

c. Hybrid parameters are found using Equation 3.24. Parameters h_{11} and h_{21} are obtained when port 2 is shorted as

$$h_{11}=\left.\frac{V_1}{I_1}\right|_{V_2=0} \rightarrow V_1=I_1(Z_A+(Z_B//Z_C))=I_1\left(\frac{Z_AZ_B+Z_AZ_C+Z_BZ_C}{Z_B+Z_C}\right)\rightarrow h_{11}$$

$$=\left(\frac{Z_AZ_B+Z_AZ_C+Z_BZ_C}{Z_B+Z_C}\right)$$

and

$$h_{21}=\left.\frac{I_2}{I_1}\right|_{V_2=0} \rightarrow I_2=-I_1\left(\frac{Z_C}{Z_B+Z_C}\right)\rightarrow h_{21}=-\left(\frac{Z_C}{Z_B+Z_C}\right)$$

Parameters h_{12} and h_{22} are obtained when port 1 is open circuited as

$$h_{12}=\left.\frac{V_1}{V_1}\right|_{I_1=0} \rightarrow V_1=V_2\left(\frac{Z_C}{Z_B+Z_C}\right)\rightarrow h_{12}=\left(\frac{Z_C}{Z_B+Z_C}\right)$$

and

$$h_{22}=\left.\frac{I_2}{V_2}\right|_{I_1=0} \rightarrow I_2=V_2\left(\frac{1}{Z_B+Z_C}\right)\rightarrow h_{22}=\left(\frac{1}{Z_B+Z_C}\right)$$

The hybrid matrix for the *T*-network can now be constructed as

$$h=\begin{bmatrix} \left(\dfrac{Z_AZ_B+Z_AZ_C+Z_BZ_C}{Z_B+Z_C}\right) & \left(\dfrac{Z_C}{Z_B+Z_C}\right) \\ -\left(\dfrac{Z_C}{Z_B+Z_C}\right) & \left(\dfrac{1}{Z_B+Z_C}\right) \end{bmatrix}$$

d. *ABCD* parameters are found using Equations 3.1 through 3.18. Parameters *A* and *C* are determined when port 2 is open circuited as

$$A = \left.\frac{V_1}{V_2}\right|_{I_2=0} \rightarrow V_2 = \frac{Z_C}{Z_C + Z_A} V_1 \rightarrow A = \left(\frac{Z_C + Z_A}{Z_C}\right)$$

and

$$C = \left.\frac{I_1}{V_2}\right|_{I_2=0} \rightarrow I_1 = V_2\left(\frac{1}{Z_C}\right) \rightarrow C = \left(\frac{1}{Z_C}\right)$$

Parameters *B* and *D* are determined when port 2 is short circuited as

$$B = \left.\frac{V_1}{-I_2}\right|_{V_2=0} \rightarrow I_2 = \frac{-V_1}{Z_A + (Z_B // Z_C)} \frac{Z_C}{(Z_B + Z_C)} \rightarrow B = \left(\frac{Z_A Z_B + Z_A Z_C + Z_B Z_C}{Z_C}\right)$$

and

$$D = \left.\frac{-I_1}{I_2}\right|_{V_2=0} \rightarrow I_2 = -I_1\left(\frac{Z_C}{Z_B + Z_C}\right) \rightarrow D = \left(\frac{Z_B + Z_C}{Z_C}\right)$$

So, the *ABCD* matrix is

$$ABCD = \begin{bmatrix} \left(\frac{Z_C + Z_A}{Z_C}\right) & \left(\frac{Z_A Z_B + Z_A Z_C + Z_B Z_C}{Z_C}\right) \\ \left(\frac{1}{Z_C}\right) & \left(\frac{Z_B + Z_C}{Z_C}\right) \end{bmatrix}$$

It can be proven that *Z*, *Y*, *h*, and *ABCD* parameters are related using the relations given in Table 3.1.

3.3 NETWORK CONNECTIONS

Networks and components in engineering applications can be connected in different ways to perform certain tasks. The commonly used network connection methods are series, parallel, and cascade connections. The series connection of two networks is shown in Figure 3.7. Since the networks are connected in series, currents are the same and voltages are added across ports of the network to find the overall voltage at the ports of the combined network. This can be represented by impedance matrices as

$$[Z] = [Z^x] + [Z^y] = \begin{bmatrix} Z_{11}^x & Z_{12}^x \\ Z_{21}^x & Z_{22}^x \end{bmatrix} + \begin{bmatrix} Z_{11}^y & Z_{12}^y \\ Z_{21}^y & Z_{22}^y \end{bmatrix} = \begin{bmatrix} Z_{11}^x + Z_{11}^y & Z_{12}^x + Z_{12}^y \\ Z_{21}^x + Z_{21}^y & Z_{22}^x + Z_{22}^y \end{bmatrix} \tag{3.43}$$

TABLE 3.1
Network Parameter Conversion Table

	[Z]	[Y]	[ABCD]	[h]
[Z]	$\begin{bmatrix} Z_{11} & Z_{12} \\ Z_{21} & Z_{22} \end{bmatrix}$	$\begin{bmatrix} \frac{Y_{22}}{\Delta Y} & -\frac{Y_{12}}{\Delta Y} \\ -\frac{Y_{21}}{\Delta Y} & \frac{Y_{11}}{\Delta Y} \end{bmatrix}$	$\begin{bmatrix} \frac{A}{C} & -\frac{\Delta_{ABCD}}{C} \\ \frac{1}{C} & \frac{D}{C} \end{bmatrix}$	$\begin{bmatrix} \frac{\Delta_h}{h_{22}} & \frac{h_{12}}{h_{22}} \\ -\frac{h_{21}}{h_{22}} & \frac{1}{h_{22}} \end{bmatrix}$
[Y]	$\begin{bmatrix} \frac{Z_{22}}{\Delta Z} & -\frac{Z_{12}}{\Delta Z} \\ -\frac{Z_{21}}{\Delta Z} & \frac{Z_{11}}{\Delta Z} \end{bmatrix}$	$\begin{bmatrix} Y_{11} & Y_{12} \\ Y_{21} & Y_{22} \end{bmatrix}$	$\begin{bmatrix} \frac{D}{B} & -\frac{\Delta_{ABCD}}{B} \\ -\frac{1}{B} & \frac{A}{B} \end{bmatrix}$	$\begin{bmatrix} \frac{1}{h_{11}} & -\frac{h_{12}}{h_{11}} \\ \frac{h_{21}}{h_{11}} & \frac{\Delta_h}{h_{11}} \end{bmatrix}$
[ABCD]	$\begin{bmatrix} \frac{Z_{11}}{Z_{21}} & \frac{\Delta Z}{Z_{21}} \\ \frac{1}{Z_{21}} & \frac{Z_{22}}{Z_{21}} \end{bmatrix}$	$\begin{bmatrix} -\frac{Y_{22}}{Y_{21}} & -\frac{1}{Y_{21}} \\ -\frac{\Delta Y}{Y_{21}} & -\frac{Y_{11}}{Y_{21}} \end{bmatrix}$	$\begin{bmatrix} A & B \\ C & D \end{bmatrix}$	$\begin{bmatrix} -\frac{\Delta h}{h_{21}} & -\frac{h_{11}}{h_{21}} \\ -\frac{h_{22}}{h_{21}} & -\frac{1}{h_{21}} \end{bmatrix}$
[h]	$\begin{bmatrix} \frac{\Delta_Z}{Z_{22}} & \frac{Z_{12}}{Z_{22}} \\ \frac{Z_{21}}{Z_{22}} & \frac{1}{Z_{22}} \end{bmatrix}$	$\begin{bmatrix} \frac{1}{Y_{11}} & -\frac{Y_{12}}{Y_{11}} \\ \frac{Y_{21}}{Y_{11}} & \frac{\Delta_Y}{Y_{11}} \end{bmatrix}$	$\begin{bmatrix} \frac{B}{D} & \frac{\Delta_{ABCD}}{D} \\ -\frac{1}{D} & \frac{C}{D} \end{bmatrix}$	$\begin{bmatrix} h_{11} & h_{12} \\ h_{21} & h_{22} \end{bmatrix}$

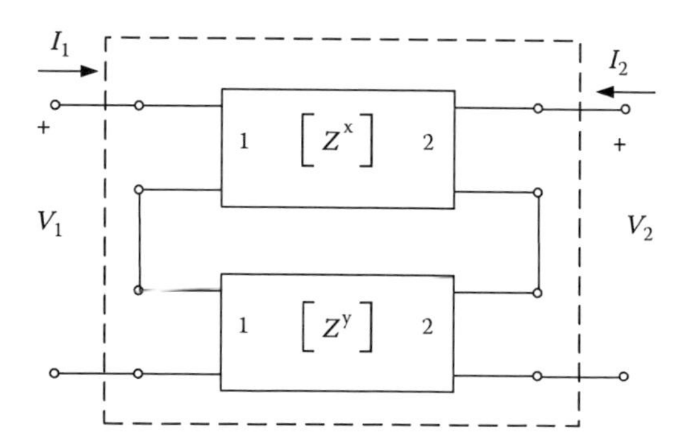

FIGURE 3.7 Series connection of two-port networks.

So,

$$\begin{bmatrix} V_1 \\ V_2 \end{bmatrix} = \begin{bmatrix} Z_{11}^x + Z_{11}^y & Z_{12}^x + Z_{12}^y \\ Z_{21}^x + Z_{21}^y & Z_{22}^x + Z_{22}^y \end{bmatrix} \begin{bmatrix} I_1 \\ I_2 \end{bmatrix} \tag{3.44}$$

The parallel connection of two-port networks is illustrated in Figure 3.8. In parallel-connected networks, voltages are the same across ports and currents are added to find the overall current flowing at the ports of the combined network. This can be represented by Y-matrices as

$$[Y] = [Y^x] + [Y^y] = \begin{bmatrix} Y_{11}^x & Y_{12}^x \\ Y_{21}^x & Y_{22}^x \end{bmatrix} + \begin{bmatrix} Y_{11}^y & Y_{12}^y \\ Z_{21}^y & Y_{22}^y \end{bmatrix} = \begin{bmatrix} Y_{11}^x + Y_{11}^y & Y_{12}^x + Y_{12}^y \\ Y_{21}^x + Y_{21}^y & Y_{22}^x + Y_{22}^y \end{bmatrix} \tag{3.45}$$

As a result,

$$\begin{bmatrix} I_1 \\ I_2 \end{bmatrix} = \begin{bmatrix} Y_{11}^x + Y_{11}^y & Y_{12}^x + Y_{12}^y \\ Y_{21}^x + Y_{21}^y & Y_{22}^x + Y_{22}^y \end{bmatrix} \begin{bmatrix} V_1 \\ V_2 \end{bmatrix} \tag{3.46}$$

The cascade connection of two-port networks is shown in Figure 3.9. In cascade connection, the magnitude of the current flowing at the output of the first network is equal to the current at the input port of the second network. The voltages at the

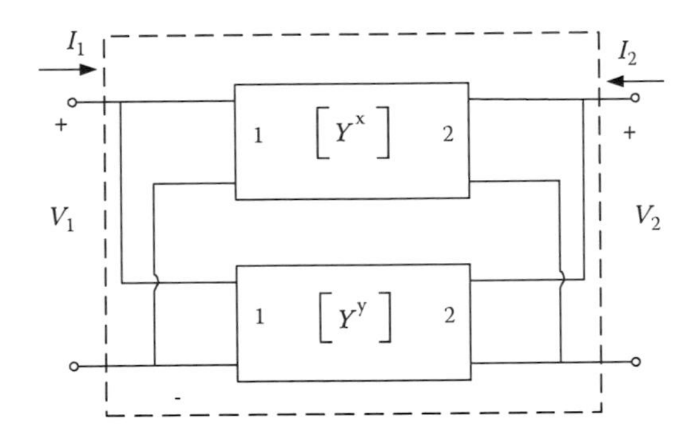

FIGURE 3.8 Parallel connection of two-port networks.

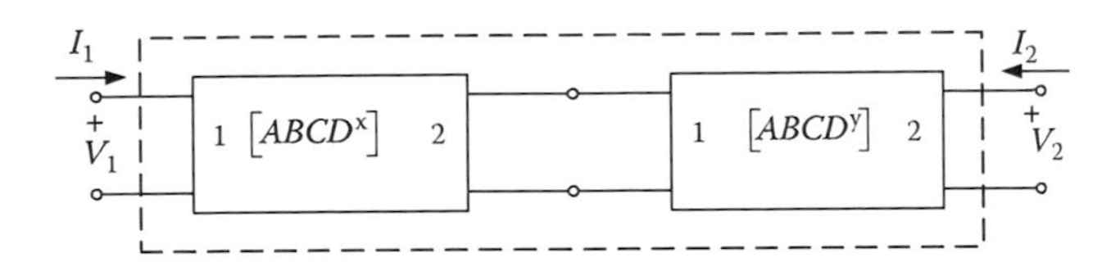

FIGURE 3.9 Cascade connection of two-port networks.

output of the first network is also equal to the voltage across the input of the second network. This can be represented by using *ABCD* matrices as

$$[ABCD] = [ABCD^x][ABCD^y] = \begin{bmatrix} A^x & B^x \\ C^x & D^x \end{bmatrix} \begin{bmatrix} A^y & B^y \\ C^y & D^y \end{bmatrix}$$

$$= \begin{bmatrix} A^x A^y + B^x C^y & A^x B^y + B^x B^y \\ C^x A^y + D^x C^y & C^x B^y + D^x D^y \end{bmatrix} \quad (3.47)$$

Example

Consider the radio frequency (RF) amplifier given in Figure 3.10. It has feedback network for stability and input and output matching networks. The transistor used is NPN BJT, and its characteristic parameters are given by $r_{BE} = 400\ \Omega$, $r_{CE} = 70\ k\Omega$, $C_{BE} = 15$ pF, $C_{BC} = 2$ pF, and $g_m = 0.2$ S. Find the voltage and current gain of this amplifier when $L = 2$ nH, $C = 12$ pF, $l = 5$ cm, and $v_p = 0.65$ c.

Solution

The high-frequency characteristics of the transistor are modeled using the hybrid parameters given by

$$h_{11} = h_{ie} = \frac{r_{BE}}{1 + j\omega(C_{BE} + C_{BC})r_{BE}} \quad (3.48)$$

$$h_{12} = h_{re} = \frac{j\omega C_{BC} r_{BE}}{1 + j\omega(C_{BE} + C_{BC})r_{BE}} \quad (3.49)$$

$$h_{21} = h_{fe} = \frac{r_{BE}(g_m - j\omega C_{BC})}{1 + j\omega(C_{BE} + C_{BC})r_{BE}} \quad (3.50)$$

$$h_{22} = h_{oe} = \frac{[1 + j\omega(C_{BE} + C_{BC})r_{BE}] + [(1 + r_{BE}g_m + j\omega C_{BE}r_{BE})]r_{CE}}{1 + j\omega(C_{BE} + C_{BC})r_{BE}} \quad (3.51)$$

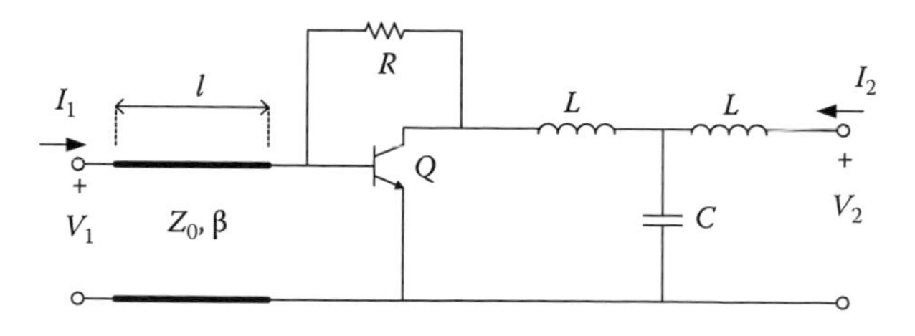

FIGURE 3.10 RF amplifier analysis by network parameters.

The amplifier network shown in Figure 3.9 is a combination of four networks that are connected in parallel and cascade. The overall network first has to be partitioned. This can be illustrated as shown in Figure 3.11.

In the partitioned amplifier circuit, networks N_2 and N_3 are connected in parallel, as shown in Figure 3.12. Then, the parallel-connected network, Y, can be represented by an admittance matrix. The admittance matrix of network 3 is

$$Y^{y} = \begin{bmatrix} \frac{1}{R} & -\frac{1}{R} \\ -\frac{1}{R} & \frac{1}{R} \end{bmatrix} \tag{3.52}$$

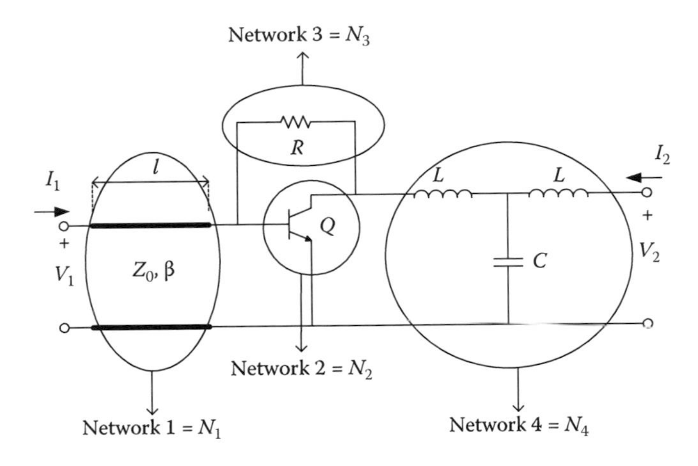

FIGURE 3.11 Partition of amplifier circuit for network analysis.

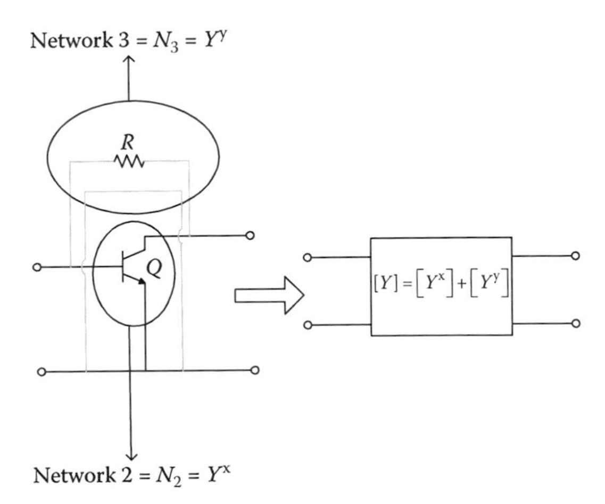

FIGURE 3.12 Illustration of parallel connection between networks 2 and 3.

The admittance matrix for the transistor can be obtained by using the conversion table given in Table 3.1 since the hybrid parameters for it are available. This can be done by using

$$Y^x = \begin{bmatrix} \dfrac{1}{h_{11}} & -\dfrac{h_{12}}{h_{11}} \\ \dfrac{h_{21}}{h_{11}} & \dfrac{\Delta h}{h_{11}} \end{bmatrix} \tag{3.53}$$

Then, the overall admittance matrix is found as

$$[Y]=[Y^x]+[Y^Y]=\begin{bmatrix} \dfrac{1}{R}+\dfrac{1}{h_{11}} & -\dfrac{1}{R}-\dfrac{h_{12}}{h_{11}} \\ -\dfrac{1}{R}+\dfrac{h_{21}}{h_{11}} & \dfrac{1}{R}+\dfrac{\Delta h}{h_{11}} \end{bmatrix} \tag{3.54}$$

where Δ is for the determinant of the corresponding matrix. At this point, it is now clearer that networks 1, *Y*, and 4 are cascaded. We need to determine the *ABCD* matrix of each network in this connection, as shown in Figure 3.13. The first step is then to convert the admittance matrix in Equation 3.54 to an *ABCD* parameter using the conversion table. The conversion table gives the relation as

$$ABCD^Y = \begin{bmatrix} \dfrac{Y_{22}}{Y_{21}} & -\dfrac{1}{Y_{21}} \\ \dfrac{\Delta Y}{Y_{21}} & \dfrac{Y_{11}}{Y_{21}} \end{bmatrix} \tag{3.55}$$

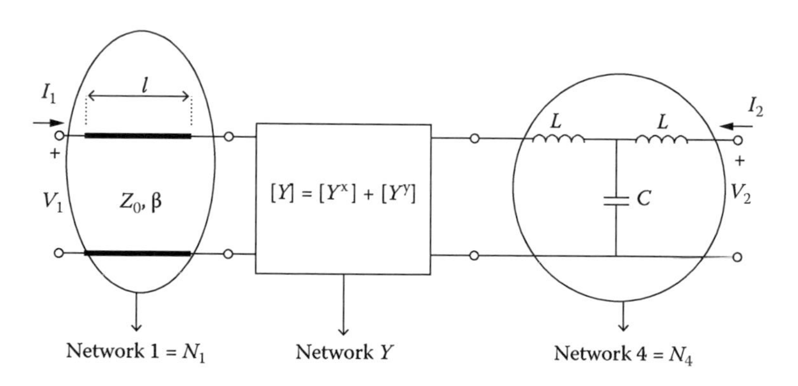

FIGURE 3.13 Cascade connection of the final circuit.

The *ABCD* matrices for networks 1 and 4 are obtained as

$$ABCD^{N_1} = \begin{bmatrix} \cos(\beta l) & jZ_0 \sin(\beta l) \\ \dfrac{j\sin(\beta l)}{Z_0} & \cos(\beta l) \end{bmatrix} \tag{3.56}$$

$$ABCD^{N_4} = \begin{bmatrix} 1-\omega^2 LC & j\omega L(2-\omega^2 LC) \\ j\omega C & 1-\omega^2 LC \end{bmatrix} \tag{3.57}$$

The overall *ABCD* parameter of the combined network shown in Figure 3.13 is

$$ABCD = ABCD^{N_1}(ABCD^{Y})ABCD^{N_4} \tag{3.58}$$

MATLAB® Script for Network Analysis of RF Amplifier

```
Zo=50;
l=0.05;
L=2e-9;
C=12e-12;
rbe=400;
rce=70e3;
Cbe=15e-12;
Cbc=2e-12;
gm=0.2;
VGain=zeros(5,150);
IGain=zeros(5,150);
freq=zeros(1,150);
R=[200 300 500 1000 10000];

for i=1:5
for t=1:150;

f=10^((t+20)/20);
freq(t)=f;
lambda=0.65*3e8/(f);
bet=(2*pi)/lambda;
w=2*pi*f;
N1=[cos(bet*l) 1j*Zo*sin(bet*l);1j*(1/Zo)*sin(bet*l) cos(bet*l)];
Y1=[1/R(i) -1/R(i);-1/R(i) 1/R(i)];
k=(1+1j*w*rbe*(Cbc+Cbe));
h=[(rbe/k) (1j*w*rbe*Cbc)/k;(rbe.*(gm-1j*w*Cbc))/k ((1/rce)+(1j*w*Cbc*
  (1+gm*rbe+1j*w*Cbe*rbe)/k))];
Y2=[1/h(1,1) -h(1,2)/h(1,1);h(2,1)/h(1,1) det(h)/h(1,1)];
Y=Y1+Y2;
N23=[-Y(2,2)/Y(2,1) -1/Y(2,1);(det(Y)/Y(2,1)) -Y(1,1)/Y(2,1)];
N4=[(1-(w^2)*L*C) (2j*w*L-1j*(w^3)*L^2*C);1j*(w*C) (1-(w^2)*L*C)];
NT=N1*N23*N4;
VGain(i,t)=20*log10(abs(1/NT(1,1)));
IGain(i,t)=20*log10(abs(-1/NT(2,2)));

end
end
```

```
figure
semilogx(freq,(IGain))
axis([10^4 10^9 20 50]);
ylabel('I_{Gain} (I_2/I_1) (dB)');
xlabel('Freq (Hz)');
legend('R=200Ohm','R=300Ohm','R=500Ohm','R=1000Ohm','R=10000Ohm')
figure
semilogx(freq,(VGain))
axis([10^4 10^9 20 80]);
ylabel('V_{Gain} (V_2/V_1) (dB)');
xlabel('Frcq (Hz)');
legend('R=200Ohm','R=300Ohm','R=500Ohm','R=1000Ohm','R=10000Ohm')
```

$$ABCD = \begin{bmatrix} \cos(\beta l) & jZ_0 \sin(\beta l) \\ \dfrac{j\sin(\beta l)}{Z_0} & \cos(\beta l) \end{bmatrix} \begin{bmatrix} \dfrac{Y_{22}}{Y_{21}} & -\dfrac{1}{Y_{21}} \\ \dfrac{\Delta Y}{Y_{21}} & \dfrac{Y_{11}}{Y_{21}} \end{bmatrix} \begin{bmatrix} 1-\omega^2 LC & j\omega L(2-\omega^2 LC) \\ j\omega C & 1-\omega^2 LC \end{bmatrix} \tag{3.59}$$

Voltage and current gains from *ABCD* parameters are found using

$$V_{\text{gain}} = 20\log\left(\left|\frac{1}{A}\right|\right)(\text{dB}) \tag{3.60}$$

$$I_{\text{gain}} = 20\log\left(\left|\frac{1}{D}\right|\right)(\text{dB}) \tag{3.61}$$

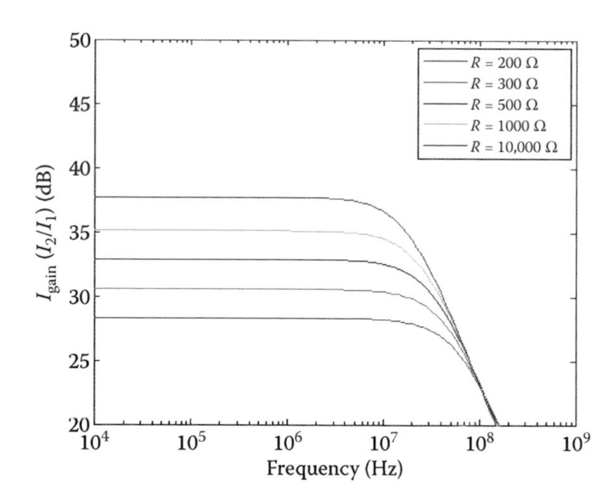

FIGURE 3.14 Current gain of RF amplifier vs. feedback resistor values and frequency.

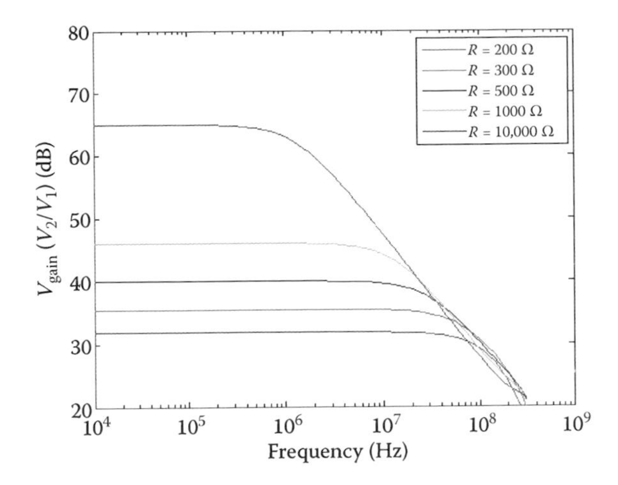

FIGURE 3.15 Voltage gain of RF amplifier vs. feedback resistor values and frequency.

The MATLAB script has been written to obtain the voltage and current gains. The script that can be used for analysis of any other amplifier network is given for reference. The voltage and current gains that are obtained by MATLAB versus various feedback resistor values and frequency are shown in Figures 3.14 and 3.15. This type of analysis gives the designer the effect of several parameters on output response in an amplifier circuit including feedback, matching networks, and parameters of the transistor.

3.3.1 MATLAB Implementation of Network Parameters

Network parameters can be easily implemented by MATLAB as illustrated, and the amount of computational time can be reduced. MATLAB scripts and functions then can be used for calculation of two-port parameters and conversion between them for the networks. The MATLAB programs below are given to systematize these techniques to increase the computational time. This will be illustrated in the upcoming examples.

```
%%%%%%%%%%%%%%%%%%%%%%%%%%%%%%%%%%%%%%%%%%%%%%%%%%%%%%%%%%%%%%%%
This m-file is function program to add two series connected network
%%%%%%%%%%%%%%%%%%%%%%%%%%%%%%%%%%%%%%%%%%%%%%%%%%%%%%%%%%%%%%%%

function [z11,z12,z21,z22]=SERIES(zx11,zx12,zx21,zx22,zw11,zw12,
zw21,zw22)
z11=zx11 + zw11;
z12=zx12 + zw12;
z21=zx21 + zw21;
z22=zx22 + zw22;
Z=[z11 z12;z21 z22]
end
```

```
%%%%%%%%%%%%%%%%%%%%%%%%%%%%%%%%%%%%%%%%%%%%%%%%%%%%%%%%%%%%%%%%%%
This m-file is function program to add two parallel connected network
%%%%%%%%%%%%%%%%%%%%%%%%%%%%%%%%%%%%%%%%%%%%%%%%%%%%%%%%%%%%%%%%%%%

function [y11,y12,y21,y22]=PARALLEL(yx11,yx12,yx21,yx22,yw11,
yw12,yw21,yw22)
y11=yx11 + yw11 ;
y12=yx12 + yw12 ;
y21=yx21 + yw21 ;
y22=yx22 + yw22 ;
Y=[y11 y12;y21 y22]
end

%%%%%%%%%%%%%%%%%%%%%%%%%%%%%%%%%%%%%%%%%%%%%%%%%%%%%%%%%%%%%%%%%%%
This m-file is function program to add two cascaded connected network
%%%%%%%%%%%%%%%%%%%%%%%%%%%%%%%%%%%%%%%%%%%%%%%%%%%%%%%%%%%%%%%%%%%

function [a11,a12,a21,a22]=CASCADE(ax11,ax12,ax21,ax22,aw11,
aw12,aw21,aw22)
a11=ax11.*aw11 + ax12.*aw21;
a12=ax11.*aw12 + ax12.*aw22;
a21=ax21.*aw11 + ax22.*aw21;
a22=ax21.*aw12 + ax22.*aw22;
A=[a11 a12;a21 a22]
end

%%%%%%%%%%%%%%%%%%%%%%%%%%%%%%%%%%%%%%%%%%%%%%%%%%%%%%%%%%%%%%%%%%%
This m-file is function program to convert Z Parameters to Y Parameters
%%%%%%%%%%%%%%%%%%%%%%%%%%%%%%%%%%%%%%%%%%%%%%%%%%%%%%%%%%%%%%%%%%%

function [y11,y12,y21,y22]=Z2Y(z11,z12,z21,z22)
DET=z11.*z22-z21.*z12;
y11=z22./DET;
y12=-z12./DET;
y21=-z21./DET;
y22=z11./DET;
end

%%%%%%%%%%%%%%%%%%%%%%%%%%%%%%%%%%%%%%%%%%%%%%%%%%%%%%%%%%%%%%%%%%%
This m-file is function program to convert Y Parameters to Z Parameters
%%%%%%%%%%%%%%%%%%%%%%%%%%%%%%%%%%%%%%%%%%%%%%%%%%%%%%%%%%%%%%%%%%%

function [z11,z12,z21,z22]=Y2Z(y11,y12,y21,y22)
DET=y11.*y22-y21.*y12;
z11=y22./DET;
z12=-y12./DET;
z21=-y21./DET;
z22=y11./DET;
end
```

```
%%%%%%%%%%%%%%%%%%%%%%%%%%%%%%%%%%%%%%%%%%%%%%%%%%%%%%%%%%%%%%%
This m-file is function program to convert Z Parameters to A Parameters
%%%%%%%%%%%%%%%%%%%%%%%%%%%%%%%%%%%%%%%%%%%%%%%%%%%%%%%%%%%%%%%

function [a11,a12,a21,a22]=Z2A(z11,z12,z21,z22)
DET=z11.*z22-z21.*z12;
a11=z11./z21;
a12=DET./z21;
a21=1./z21;
a22=z22./z21;
end


%%%%%%%%%%%%%%%%%%%%%%%%%%%%%%%%%%%%%%%%%%%%%%%%%%%%%%%%%%%%%%%
This m-file is function program to convert Y Parameters to A Parameters
%%%%%%%%%%%%%%%%%%%%%%%%%%%%%%%%%%%%%%%%%%%%%%%%%%%%%%%%%%%%%%%

function [a11,a12,a21,a22]=Y2A(y11,y12,y21,y22)
DET=y11.*y22-y21.*y12;
a11=-y22./y21;
a12=-1./y21;
a21=-DET./y21;
a22=-y11./y21;
end
```

The following MATLAB program uses the menu option and asks the user to enter the two-port network parameters when networks are connected in series or in parallel or are cascaded, and outputs the desired results.

```
%%%%%%%%%%%%%%%%%%%%%%%%%%%%%%%%%%%%%%%%%%%%%%%%%%%%%%%%%%%%%%%%%%%
This m-file is a script program that calculates the final 2-port%
%parameters when networks are connected in series, parallel or cascaded
%%%%%%%%%%%%%%%%%%%%%%%%%%%%%%%%%%%%%%%%%%%%%%%%%%%%%%%%%%%%%%%%%%%
clear;

M = menu('Network Analysis','2 Network in Series', '2 Network
in Parallel','2-Network in Cascade');

switch M
       case 1
         zx11 = input('enter Z1_11: ');
         zx12 = input('enter Z1_12: ');
         zx21 = input('enter Z1_21: ');
         zx22 = input('enter Z1_22: ');
         zw11 = input('enter Z2_11: ');
         zw12 = input('enter Z2_12: ');
         zw21 = input('enter Z2_21: ');
         zw22 = input('enter Z2_22: ');
         SERIES(zx11,zx12,zx21,zx22,zw11,zw12,zw21,zw22);
```

```
    case 2
       yx11 = input('enter Y1_11: ');
       yx12 = input('enter Y1_12: ');
       yx21 = input('enter Y1_21: ');
       yx22 = input('enter Y1_22: ');
       yw11 = input('enter Y2_11: ');
       yw12 = input('enter Y2_12: ');
       yw21 = input('enter Y2_21: ');
       yw22 = input('enter Y2_22: ');
       PARALLEL(yx11,yx12,yx21,yx22,yw11,yw12,yw21,yw22);
    case 3
       ax11 = input('enter A1_11: ');
       ax12 = input('enter A1_12: ');
       ax21 = input('enter A1_21: ');
       ax22 = input('enter A1_22: ');
       aw11 = input('enter A2_11: ');
       aw12 = input('enter A2_12: ');
       aw21 = input('enter A2_21: ');
       aw22 = input('enter A2_22: ');
       CASCADE(ax11,ax12,ax21,ax22,aw11,aw12,aw21,aw22);

    otherwise
          disp('ERROR: invalid entry')
end
```

When the program is run, the following window appears for the user. The user then specifies how the networks are connected. Then, the network parameters of the two networks can be manually entered from MATLAB Command Window for execution.

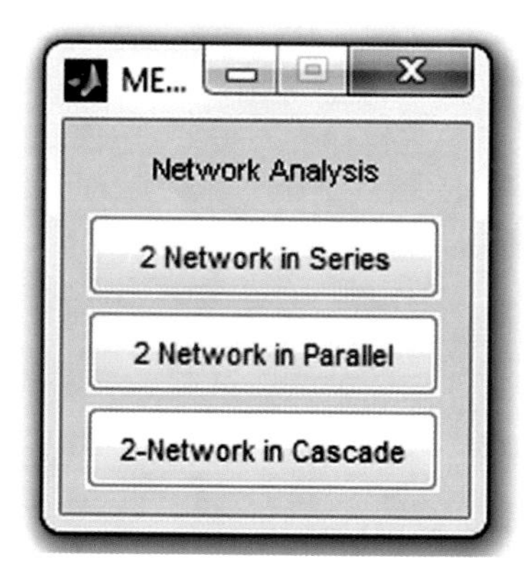

Example

The *Z*-parameters of the two-port network *N* in Figure 3.16a are $Z_{11} = 4s$, $Z_{12} = Z_{21} = 3s$, and $Z_{22} = 9s$ where $s = j\omega$. (a) Replace network *N* by its *T*-equivalent. (b) Use part (a) to find and input current $i_1(t)$ for $v_s = \cos 1000t$ (V), and write a MATLAB script to compute the equivalent network parameters of the circuit for (a) and (b). Your program should make the conversion from the two-port network to the

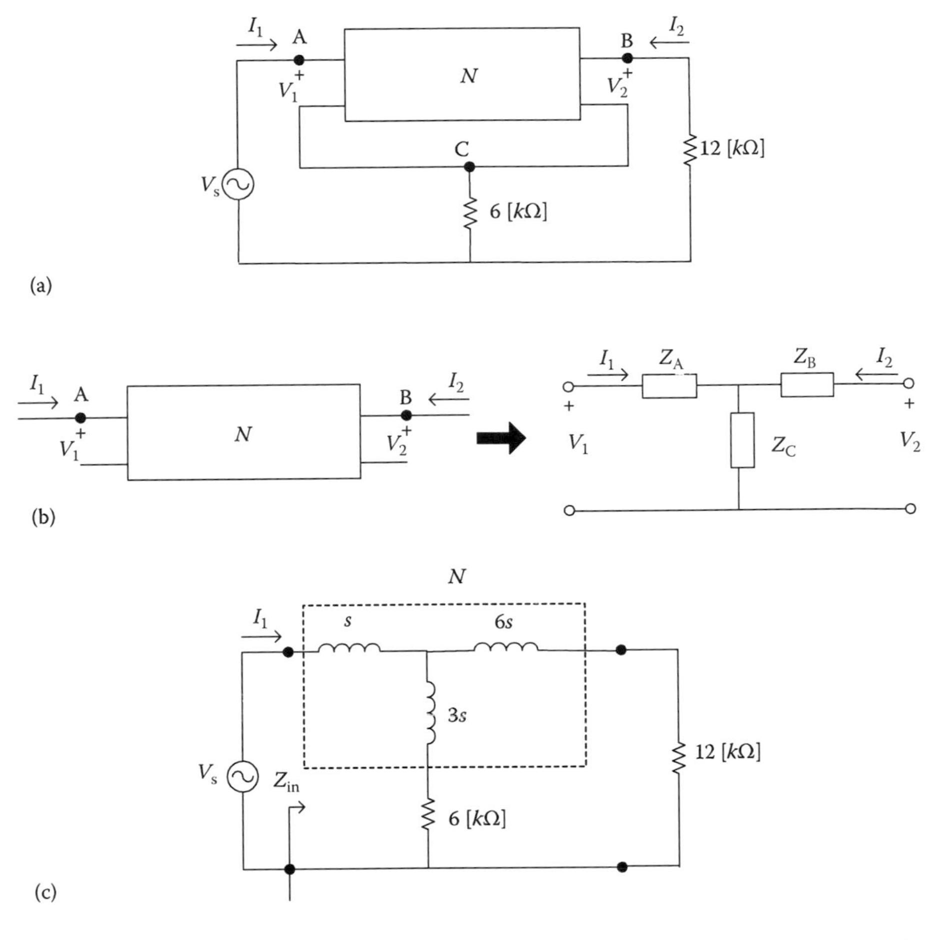

FIGURE 3.16 (a) *N*-network for analysis. (b) Network transformation to *T*-network. (c) Equivalent *T*-network.

T-equivalent network by checking if the two-port network is reciprocal. Execute your program, plot *i(t)*, and confirm your results.

Solution

a. Any two networks can be converted to their equivalent *T*-network shown in Figure 3.16b if it is reciprocal.

   The transformation of the network to the *T*-network shown in Figure 3.16b is valid with the following relations:

$$Z_a = Z_{11} - Z_{12}$$

$$Z_b = Z_{22} - Z_{21}$$

$$Z_c = Z_{12} = Z_{21}$$

Based on the given information, the network is reciprocal because $Z_{12} = Z_{21}$. So, we can convert the network to its *T*-network equivalent and obtain

$$Z_a = Z_{11} - Z_{12} = 4s - 3s = s$$

$$Z_b = Z_{22} - Z_{21} = 9s - 3s = 6s$$

$$Z_c = Z_{12} = Z_{21} = 3s$$

Hence, this simplified network can now be analyzed by establishing the relations between voltage and current.

$$V_1 = Z_{11}I_1 + Z_{12}I_2$$

$$V_2 = Z_{21}I_1 + Z_{22}I_2$$

b. From the final circuit, we obtain Z_{in} as

$$Z_{in}(s) = V_s/I_{in} = s + \frac{(3s+6)(6s+12)}{9s+18} = 3s + 4 = 3j + 5 = 5\angle 36.9°$$

So, the current is

$$i(t) = 0.2\cos(1000t - 36.9°)\ [\text{mA}]$$

This operation can be implemented by MATLAB simply.

```
%%%%%%%%%%%%%%%%%%%%%%%%%%%%%%%%%%%%%%%%%%%%%%%%%%%%%%%%%%%%%%%%%%%%%%%%%%%
This m-file is a script checks if a network is reciprocal network and
then converts it to its equivalent T-network and calculates current
%%%%%%%%%%%%%%%%%%%%%%%%%%%%%%%%%%%%%%%%%%%%%%%%%%%%%%%%%%%%%%%%%%%%%%%%%%%
clear;
w = input('Circuit frequency: ');
s=sqrt(-1)*w;

'Enter Z parameters for network N in figure 2:'
z11 = input('Enter Z_11: ');
z12 = input('Enter Z_12: ');
z21 = input('Enter Z_21: ');
z22 = input('Enter Z_22: ');
 if (z12==z21)
  za=z11-z12;
  zb=z22-z21;
  zc=z12;
  'The T equivalent of network N is:'
  za
  zb
  zc
  zeq=((6000+zc)*(12000+zb))/((6000+zc)+(12000+zb))+za;
  t=0:0.0001:0.1;
  v=cos(1000*t);
```

```
  i=1/(abs(zeq))*cos(w*t-angle(zeq));
  plot(t,i)
  title('i(t) for problem 2b');
  xlabel('t');
  ylabel('Amplitude');
else
  'Network N is not reciprocal'
end
```

Example

a. Obtain a small-signal model of the MOSFET using *Y*-parameters for the equivalent shown in Figure 3.17. (b) Use MATLAB to compute the voltage gain and phase of the voltage gain of the model when $R_g = 5\ \Omega$, $C_{gs} = 10 \times 10^{-12}$, $R_{gs} = 0.5\ \Omega$, $C_{gd} = 100 \times 10^{-12}$ F, $C_{ds} = 2 \times 10^{-12}$ F, $g_m = 20 \times 10^{-3}$ S, $R_{ds} = 70 \times 10^3\ \Omega$, $R_S = 3\ \Omega$, $R_{highL} = 10 \times 10^3\ \Omega$, and the connected load is $R_L = 10 \times 10^3\ \Omega$.

Solution

a. The equivalent circuit in Figure 3.17 is simplified and shown in Figure 3.18. When Figures 3.17 and 3.18 are compared, the following can be written:

$$Z_g = R_g, \quad Z_1 = R_{gs} - j\frac{1}{\omega C_{gs}}, \quad Z_2 = -j\frac{1}{\omega C_{gd}},$$

$$Z_3 = R_{ds}, Z_4 = R_{highL} - j\frac{1}{\omega C_{ds}}, Z_s = R_s, Z_L = R_L, Z' = \frac{Z_3 Z_4}{Z_3 + Z_4}$$

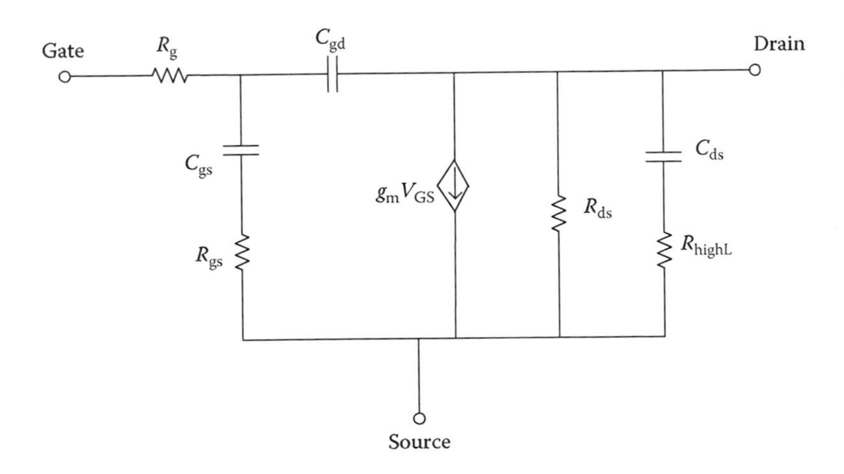

FIGURE 3.17 Small-signal MOSFET model.

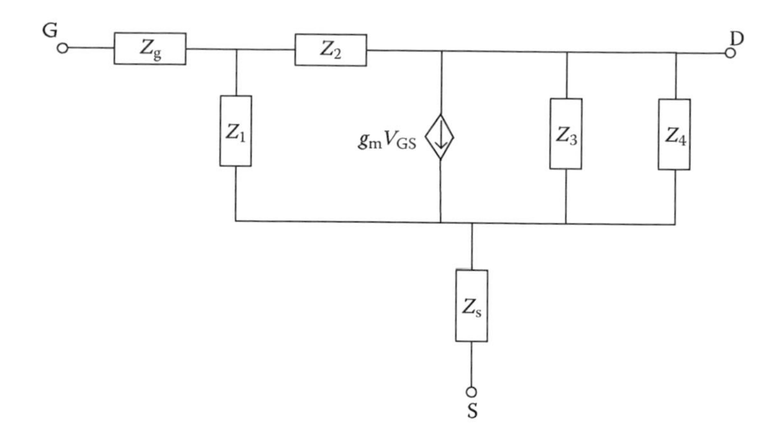

FIGURE 3.18 The simplified equivalent circuit for the MOSFET small-signal model.

The small-signal MOSFET model circuit given in Figure 3.17 can be analyzed when each of the components is represented as a network with the connection that they are introduced in the circuit, as shown in Figure 3.19. From Figure 3.19, it is seen that network 2, N_2, and network 3, N_3, are connected in parallel. The overall parallel-connected network, N_{23}, can be found from

$$Y_{23} = Y_2 + Y_3 \rightarrow N_{23}$$

Now, network 4, N_4, and the resultant parallel network, N_{23}, are connected in series. The combination of these two networks can be obtained from

$$Z_{234} = Z_{23} + Z_4 \rightarrow N_{234}$$

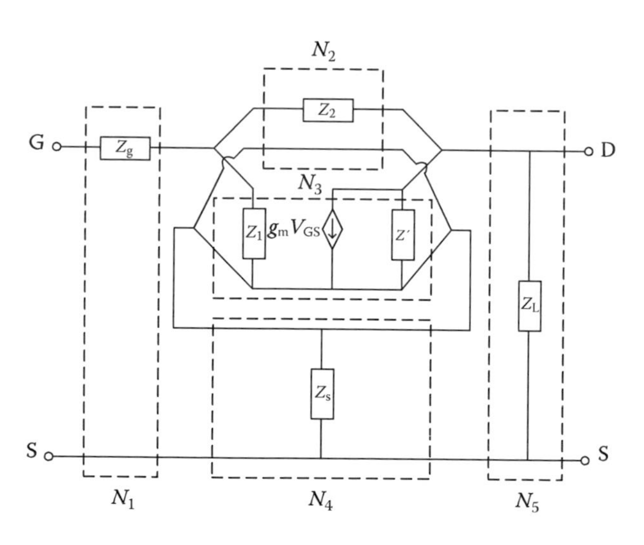

FIGURE 3.19 The network equivalent circuit for the MOSFET small-signal model.

where $Z_{23} = (Y_{23})^{-1}$. From Figure 3.19, it is also observed that networks, N_1, N_{234}, and N_5 are cascaded. Hence, the overall network parameters can now be found from *ABCD* parameters:

$$[ABCD]_{\text{Network}} = [ABCD]_{N_1}[ABCD]_{N_{234}}[ABCD]_{N_5}$$

- The *ABCD* parameters for networks 1 and 5 and the *Y* parameters for network 2 are

$$[ABCD]_{N_1} = \begin{bmatrix} 1 & Z_g \\ 0 & 1 \end{bmatrix}, [Y]_{N_2} = \begin{bmatrix} Y_{C_{gd}} & -Y_{C_{gd}} \\ -Y_{C_{gd}} & Y_{C_{gd}} \end{bmatrix}, \text{ and } [ABCD]_{N_5} = \begin{bmatrix} 1 & 0 \\ 1/Z_L & 1 \end{bmatrix}$$

- The *Y*-parameters for network 3 are found using Figure 3.20. The *Y*-parameters are found using Figure 3.20 as

$$Y_{11} = \left.\frac{I_1}{V_1}\right|_{V_2=0} = Y_{GS} \qquad Y_{12} = \left.\frac{I_1}{V_2}\right|_{V_1=0} = 0$$

$$Y_{21} = \left.\frac{I_2}{V_1}\right|_{V_2=0} = g_m \qquad Y_{22} = \left.\frac{I_2}{V_2}\right|_{V_1=0} = \frac{1}{Z_3}$$

So,

$$[Y]_{N_3} = \begin{bmatrix} Y_{GS} & 0 \\ g_m & 1/Z' \end{bmatrix}$$

- The parallel-connected network, N_{23}, is now found from

$$[Y]_{N_{23}} = [Y]_{N_3} + [Y]_{N_2} = \begin{bmatrix} Y_{C_{gd}} & -Y_{C_{gd}} \\ -Y_{C_{gd}} & Y_{C_{gd}} \end{bmatrix} + \begin{bmatrix} Y_{GS} & 0 \\ g_m & 1/Z' \end{bmatrix}$$

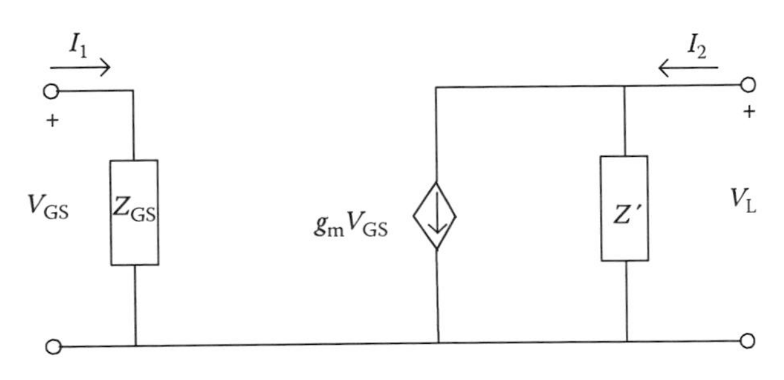

FIGURE 3.20 The equivalent circuit for network 3.

Hence,

$$[Y]_{N_{23}} = \begin{bmatrix} Y_{C_{gd}} + Y_{GS} & -Y_{C_{gd}} \\ -Y_{C_{gd}} + g_m & Y_{C_{gd}} + 1/Z' \end{bmatrix}$$

- The Z-parameters for network 4 are calculated using the circuit shown in Figure 3.21a as

$$Z_{11} = \left.\frac{V_1}{I_1}\right|_{I_2=0} = Z_s \qquad Z_{12} = \left.\frac{V_1}{I_2}\right|_{I_1=0} = Z_s$$

$$Z_{21} = \left.\frac{V_2}{I_1}\right|_{I_2=0} = Z_s \qquad Z_{22} = \left.\frac{V_2}{I_2}\right|_{I_1=0} = Z_s$$

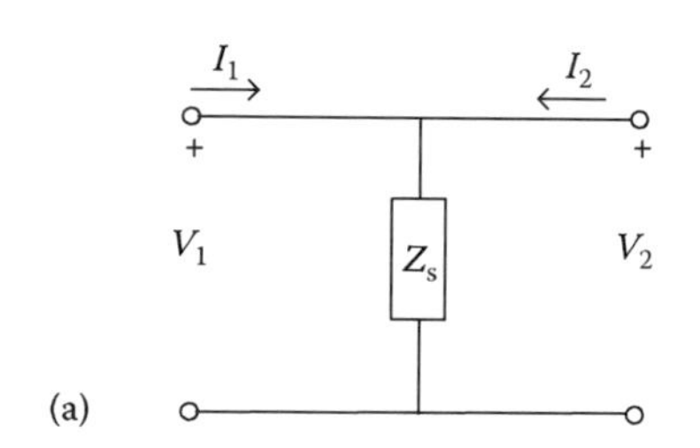

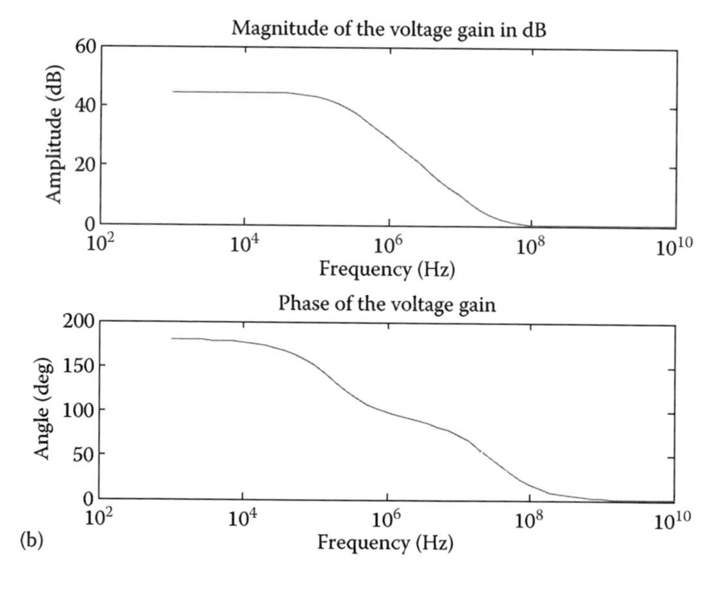

FIGURE 3.21 (a) The equivalent circuit for network 4. (b) Voltage gain and phase responses.

So,

$$[Z]_{N_4} = \begin{bmatrix} Z_S & Z_S \\ Z_S & Z_S \end{bmatrix}$$

- The port parameters for series-connected networks N_4 and N_{23} are found first by converting Y parameters for N_{23} to Z parameters by

$$[Z]_{N_{23}} = ([Y]_{N_{23}})^{-1} = \begin{bmatrix} \dfrac{Y_{C_{gd}} + 1/Z'}{\Delta Y_{N_{23}}} & \dfrac{Y_{C_{gd}}}{\Delta Y_{N_{23}}} \\ \dfrac{Y_{C_{gd}} - g_m}{\Delta Y_{N_{23}}} & \dfrac{Y_{C_{gd}} + Y_{GS}}{\Delta Y_{N_{23}}} \end{bmatrix}$$

As a result, the two-port parameters for series-connected networks are found from

$$[Z]_{N_{234}} = [Z]_{N_{23}} + [Z]_{N_4} = \begin{bmatrix} \dfrac{Y_{C_{gd}} + 1/Z'}{\Delta Y_{N_{23}}} & \dfrac{Y_{C_{gd}}}{\Delta Y_{N_{23}}} \\ \dfrac{Y_{C_{gd}} - g_m}{\Delta Y_{N_{23}}} & \dfrac{Y_{C_{gd}} + Y_{GS}}{\Delta Y_{N_{23}}} \end{bmatrix} + \begin{bmatrix} Z_S & Z_S \\ Z_S & Z_S \end{bmatrix}$$

Hence,

$$[Z]_{N_{234}} = \begin{bmatrix} \dfrac{Y_{C_{gd}} + 1/Z'}{\Delta Y_{N_{23}}} + Z_S & \dfrac{Y_{C_{gd}}}{\Delta Y_{N_{23}}} + Z_S \\ \dfrac{Y_{C_{gd}} - g_m}{\Delta Y_{N_{23}}} + Z_S & \dfrac{Y_{C_{gd}} + Y_{GS}}{\Delta Y_{N_{23}}} + Z_S \end{bmatrix}$$

- We need to convert the resultant central Z parameters to $ABCD$ parameters using

$$[ABCD]_{N_{234}} = \begin{bmatrix} \dfrac{(Z_{11})_{[Z]_{N_{234}}}}{(Z_{21})_{[Z]_{N_{234}}}} & \dfrac{\left|[Z]_{N_{234}}\right|}{(Z_{21})_{[Z]_{N_{234}}}} \\ \dfrac{1}{(Z_{21})_{[Z]_{N_{234}}}} & \dfrac{(Z_{22})_{[Z]_{N_{234}}}}{(Z_{21})_{[Z]_{N_{234}}}} \end{bmatrix}$$

- The complete network parameters of the small-signal model of MOSFETs are found from $ABCD$ parameters since now, N_1, N_{234}, and N_5 are all cascaded as

$$[ABCD] = [ABCD]_{N_1}[ABCD]_{N_{234}}[ABCD]_{N_5}$$

or

$$[ABCD] = \begin{bmatrix} 1 & Z_g \\ 0 & 1 \end{bmatrix} \begin{bmatrix} \dfrac{(Z_{11})_{[Z]_{N234}}}{(Z_{21})_{[Z]_{N234}}} & \dfrac{\left|[Z]_{N_{234}}\right|}{(Z_{21})_{[Z]_{N234}}} \\ \dfrac{1}{(Z_{21})_{[Z]_{N234}}} & \dfrac{(Z_{22})_{[Z]_{N234}}}{(Z_{21})_{[Z]_{N234}}} \end{bmatrix} \begin{bmatrix} 1 & 0 \\ 1/Z_L & 1 \end{bmatrix}$$

b. The MATLAB script is written to obtain the characteristics of the network by finding the voltage gain and the phase of the voltage gain using the MATLAB functions developed and given in Section 3.4, as shown below.

```
%%%%%%%%%%%%%%%%%%%%%%%%%%%%%%%%%%%%%%%%%%%%%%%%%%%%%%%%%%%%%%%%%%%%%%%%%%%%
This m-file is developed to calculate the response of small signal
MOSFET model
%%%%%%%%%%%%%%%%%%%%%%%%%%%%%%%%%%%%%%%%%%%%%%%%%%%%%%%%%%%%%%%%%%%%%%%%%%%%

clear
Rg=5;
Cgs=10e-12;
rgs=0.5;
Cgd=100e-12;
Cds=2e-12;
Gm=20e-3;
Rds=70e3;
RL=10e3;
RS=3;
Rhigh=10e3;
f=logspace(3,10);
omega=2*pi.*f;
Z3=Rds;
Z4=1./(1i.*omega.*Cds)+Rhigh;
Zp=(Z3.*Z4)./(Z3+Z4);

%++++++++++++++++++++++++VCCS (y-parameters)++++++++++++++++++++++++
VCCS_11=1./(rgs.*ones(size(f))+1./(1i.*omega.*Cgs));
VCCS_12=zeros(size(f));
VCCS_21=Gm.*ones(size(f));
VCCS_22=1./(Zp);
%++++++++++++++++++++++++CGD (y-parameters)+++++++++++++++++++++++++
CGD_11= 1i.*omega.*Cgd;
CGD_12=-1i.*omega.*Cgd;
CGD_21=-1i.*omega.*Cgd;
CGD_22= 1i.*omega.*Cgd;
%+++++++++++++++++++++++++RS (z-parameters)+++++++++++++++++++++++++
RS_11=RS.*ones(size(f));
RS_12=RS.*ones(size(f));
RS_21=RS.*ones(size(f));
RS_22=RS.*ones(size(f));
%++++++++++++++++++++++++RL (ABCD-parameters)++++++++++++++++++++++++
RL_11=1.*ones(size(f));
RL_12=0.*ones(size(f));
RL_21=1./(RL);
RL_22=1.*ones(size(f));
```

```
%++++++++++++++++++++++++Ci (ABCD-parameters)++++++++++++++++++++++++
Ci_11=1.*ones(size(f));
Ci_12=(Rg.*ones(size(f)));
Ci_21=0.*ones(size(f));
Ci_22=1.*ones(size(f));
%++++++++++++++++++++++++++++Xa = VCCS and CGD+++++++++++++++++++++++
[Xa_11,Xa_12,Xa_21,Xa_22]=PARALLEL(VCCS_11,VCCS_12,VCCS_21,VCCS_22,
   CGD_11,CGD_12,CGD_21,CGD_22);
%+++++++++++++++++++Convert Admittance to Impedance++++++++++++++++++
[z11,z12,z21,z22]=Y2Z(Xa_11,Xa_12,Xa_21,Xa_22);
%++++++++++++++++++++++Xb = VCCS, CGD and RS+++++++++++++++++++++++++
[Xb_11,Xb_12,Xb_21,Xb_22]=SERIES(z11,z12,z21,z22,RS_11,RS_12,RS_21,
   RS_22);
%+++++++++++++++++++Convert Impedance to ABCD++++++++++++++++++++++
[a11,a12,a21,a22]=Z2A(Xb_11,Xb_12,Xb_21,Xb_22);
%+++++++++++++++++++++Xc = VCCS, CGD, RS and RL+++++++++++++++++++++
[Xc_11,Xc_12,Xc_21,Xc_22]=CASCADE(a11,a12,a21,a22,RL_11,RL_12,RL_21,
   RL_22);
%+++++++++++++++++++Xd = VCCS, CGD, RS, RL and Ci++++++++++++++++++++
[Xd_11,Xd_12,Xd_21,Xd_22]=CASCADE(Xc_11,Xc_12,Xc_21,Xc_22,Ci_11,Ci_12,
   Ci_21,Ci_22);
subplot(211)
semilogx(f,20*log10(abs(1./Xd_11)))
xlabel('Frequency (Hz)')
ylabel('Amplitude (dB)')
title('Magnitude of the voltage gain in dB')
subplot(212)
semilogx(f,((angle(1./Xd_11)).*180)./pi)
xlabel('Frequency (Hz)')
ylabel('Angle (deg)')
title('Phase of the voltage gain')
%====================================================================
```

The voltage gain and phase responses when the program is run are given in Figure 3.21b.

3.4 *S*-SCATTERING PARAMETERS

Scattering parameters are used to characterize RF/microwave devices and components at high frequencies [2]. Specifically, they are used to define the return loss and insertion loss of a component or a device.

3.4.1 ONE-PORT NETWORK

Consider the circuit given in Figure 3.22. The relationship between current and voltage can be written as

$$I = \frac{V_g}{Z_g + Z_L} \tag{3.62}$$

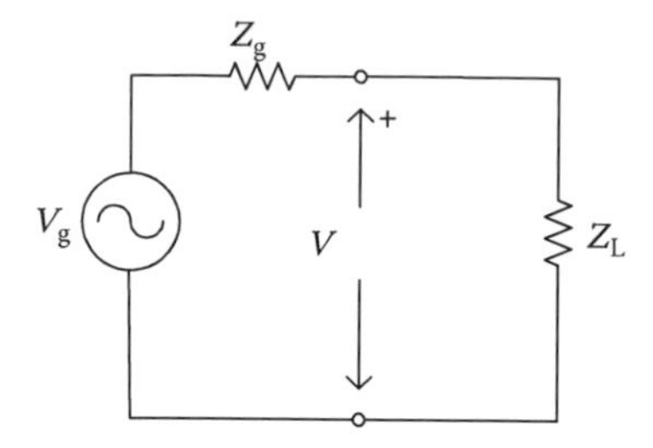

FIGURE 3.22 One-port network for scattering parameter analysis.

and

$$V = \frac{V_g Z_L}{Z_g + Z_L} \tag{3.63}$$

where Z_g is the generator impedance. The incident waves for voltage and current can be obtained when the generator is matched as

$$I_i = \frac{V_g}{Z_g + Z_g^*} = \frac{V_g}{2\,\mathrm{Re}\{Z_g\}} \tag{3.64}$$

and

$$V_i = \frac{V_g Z_g^*}{Z_g + Z_g^*} = \frac{V_g Z_g^*}{2\,\mathrm{Re}\{Z_g\}} \tag{3.65}$$

Then, the reflected waves are found from

$$I = I_i - I_r \tag{3.66}$$

and

$$V = V_i - V_r \tag{3.67}$$

Substituting Equations 3.62 and 3.64 into Equation 3.66 gives the reflected wave as

$$I_r = I_i - I = \left(\frac{Z_L - Z_g^*}{Z_L + Z_g^*}\right) I_i \tag{3.68}$$

or

$$I_r = S^I I_i \tag{3.69}$$

where

$$S^{\mathrm{I}} = \left(\frac{Z_{\mathrm{L}} - Z_{\mathrm{g}}^*}{Z_{\mathrm{L}} + Z_{\mathrm{g}}^*} \right) \tag{3.70}$$

is the scattering matrix for current. A similar analysis can be done to find the reflected voltage wave by substituting Equations 3.63 and 3.65 into Equation 3.67 as

$$V_{\mathrm{r}} = V_{\mathrm{i}} - V = \frac{Z_{\mathrm{g}}}{Z_{\mathrm{g}}^*} \left(\frac{Z_{\mathrm{L}} - Z_{\mathrm{g}}^*}{Z_{\mathrm{L}} + Z_{\mathrm{g}}^*} \right) V_{\mathrm{i}} \tag{3.71}$$

or

$$V_{\mathrm{r}} = \frac{Z_{\mathrm{g}}}{Z_{\mathrm{g}}^*} S^{\mathrm{I}} V_{\mathrm{i}} = S^{\mathrm{V}} V_{\mathrm{i}} \tag{3.72}$$

where

$$S^{\mathrm{V}} = \frac{Z_{\mathrm{g}}}{Z_{\mathrm{g}}^*} S^{\mathrm{I}} \tag{3.73}$$

is the scattering matrix for voltage. It can also be shown that

$$V_{\mathrm{i}} = Z_{\mathrm{g}}^* I_{\mathrm{i}} \tag{3.74}$$

$$V_{\mathrm{r}} = Z_{\mathrm{g}} I_{\mathrm{r}} \tag{3.75}$$

When the generator impedance is purely real, $Z_{\mathrm{g}} = R_{\mathrm{g}}$, then

$$S^{\mathrm{I}} = S^{\mathrm{V}} = \left(\frac{Z_{\mathrm{L}} - R_{\mathrm{g}}}{Z_{\mathrm{L}} + R_{\mathrm{g}}} \right)_{\mathrm{i}} \tag{3.76}$$

3.4.2 *N*-Port Network

The analysis described in Section 3.4.1 can be extended to the *N*-port network shown in Figure 3.23. The analysis is based on the assumption that generators are independent of each other. Hence, the *Z*-generator matrix has no cross-coupling terms, and it can be expressed as a diagonal matrix:

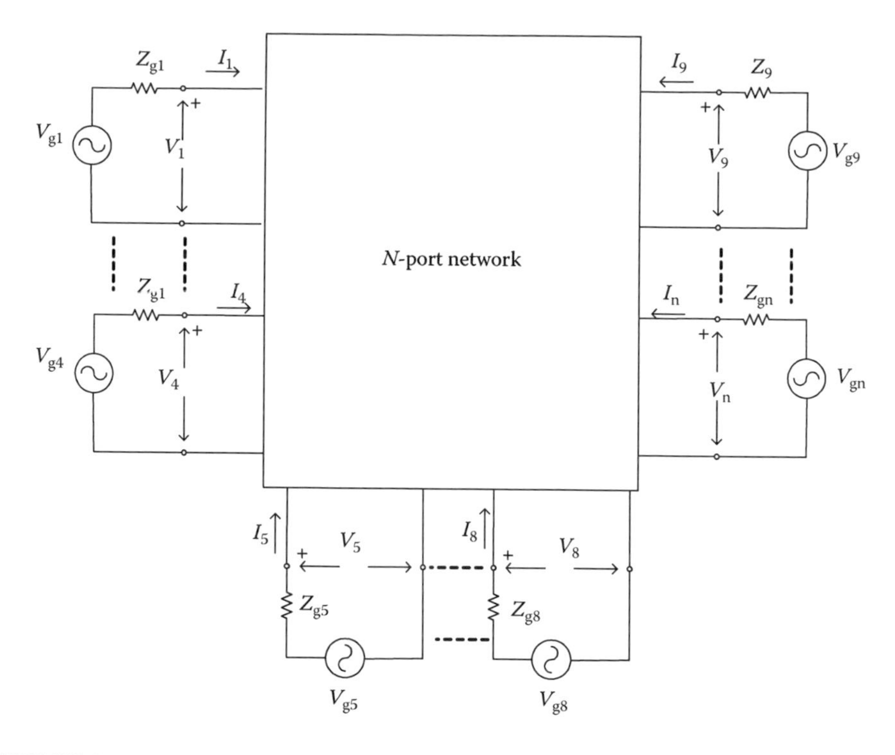

FIGURE 3.23 *N*-port network for scattering analysis.

$$[Z_g] = \begin{bmatrix} Z_{g1} & 0 & \cdots & 0 \\ 0 & Z_{g2} & \cdots & 0 \\ \vdots & \vdots & & \vdots \\ 0 & 0 & \cdots & Z_{gn} \end{bmatrix} \tag{3.77}$$

From Equations 3.66 and 3.67, the incident and reflected waves are related to the actual voltage and current values as

$$[I] = [I_i] - [I_r] \tag{3.78}$$

$$[V] = [V_i] + [V_r] \tag{3.79}$$

From Equations 3.74 and 3.75, the incident and reflected components can be related through

$$[V_i] = [Z_g^*][I_i] \tag{3.80}$$

$$[V_r] = [Z_g][I_r] \tag{3.81}$$

similar to the one-port case as derived before. For the N-port network, Z parameters can be obtained as

$$[V] = [Z][I] \tag{3.82}$$

Using Equations 3.77 through 3.82, we can obtain

$$[V_r] = [V] - [V_i] = [Z][I] - [Z_g^*][I_i] \tag{3.83}$$

Equation 3.83 can also be expressed as

$$[Z_g][I_r] = [Z][I] - [Z_g^*][I_i] = [Z]([I_i] - [I_r]) - [Z_g^*][I_i] \tag{3.84}$$

and simplified to

$$([Z] + [Z_g])[I_r] = ([Z] - [Z_g^*])[I_i] \tag{3.85}$$

Equation 3.85 can be put in the following form:

$$[I_r] = ([Z] + [Z_g])^{-1}([Z] - [Z_g^*])[I_i] \tag{3.86}$$

From Equation 3.70, the scattering matrix for the current for the N-port network is equal to

$$S^I = ([Z] + [Z_g])^{-1}([Z] - [Z_g^*]) \tag{3.87}$$

Then, Equation 3.86 can be expressed as

$$[I_r] = [S^I][I_i] \tag{3.88}$$

For the N-port network, the Y parameters for the short-circuit case can be obtained similarly as

$$[I] = [Y][V] \tag{3.89}$$

It can also be shown that

$$[V_r] = -([Y] + [Y_g])^{-1}([Y] - [Y_g^*])[V_i] \tag{3.90}$$

or

$$[V_r] = [S^V][V_i] \tag{3.91}$$

where

$$S^{V} = -([Y]+[Y_g])^{-1}([Y]-[Y_g^*]) \tag{3.92}$$

Example

Consider a transistor network that is represented as a two port network and connected between the source and the load. It is assumed that the generator or source and load impedances are equal and given to be R_g. The transistor is represented by the following Z parameters, as shown in Figure 3.24. Find the current scattering matrix, S^I.

$$[Z] = \begin{bmatrix} Z_i & Z_r \\ Z_f & Z_o \end{bmatrix}$$

Solution

From Equation 3.87, the scattering matrix for current is

$$S^{I} = ([Z]+[Z_g])^{-1}([Z]-[Z_g^*]) \tag{3.93}$$

The generator Z_g-matrix is

$$[Z_g] = [Z_g^*] = \begin{bmatrix} R_g & 0 \\ 0 & R_g \end{bmatrix} \tag{3.94}$$

Then,

$$[Z]+[Z_g^*] = \begin{bmatrix} Z_i & Z_r \\ Z_f & Z_o \end{bmatrix} + \begin{bmatrix} R_g & 0 \\ 0 & R_g \end{bmatrix} = \begin{bmatrix} Z_i + R_g & 0 \\ 0 & Z_o + R_g \end{bmatrix} \tag{3.95}$$

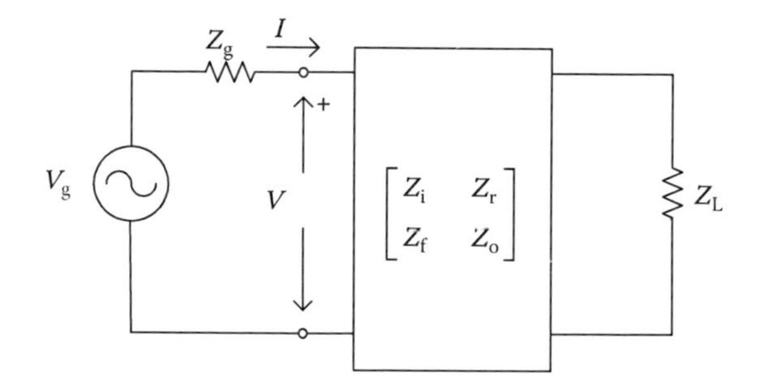

FIGURE 3.24 Two-port transistor network.

The inverse of the matrix in Equation 3.95 is

$$[([Z]+[Z_g^*])]^{-1} = \frac{1}{\left|[Z]+[Z_g^*]\right|}([[Z]+[Z_g^*]]^C)^T \tag{3.96}$$

$\left|[Z]+[Z_g^*]\right|$ is the determinant of $[Z]+[Z_g^*]$ and is calculated as

$$\left|[Z]+[Z_g^*]\right| = (Z_i + R_g)(Z_o + R_g) - Z_r Z_f \tag{3.97}$$

$[[Z]+[Z_g^*]]^C$ is the cofactor matrix for $[Z]+[Z_g^*]$ and is calculated as

$$([Z]+[Z_g^*])^C = \begin{bmatrix} Z_o + R_g & -Z_f \\ -Z_r & Z_i + R_g \end{bmatrix} \tag{3.98}$$

Then,

$$[([Z]+[Z_g^*])^C]^T = \begin{bmatrix} Z_o + R_g & -Z_r \\ -Z_f & Z_i + R_g \end{bmatrix} \tag{3.99}$$

Hence, the inverse of the matrix from Equations 3.97 through 3.99 is equal to

$$[([Z]+[Z_g^*])]^{-1} = \frac{1}{((Z_i + R_g)(Z_o + R_g) - Z_r Z_f)} \begin{bmatrix} Z_o + R_g & -Z_r \\ -Z_f & Z_i + R_g \end{bmatrix} \tag{3.100}$$

Then, from Equation 3.93,

$$S^I = ([Z]+[Z_g])^{-1}([Z]-[Z_g]) = \left(\left[\frac{1}{((Z_i + R_g)(Z_o + R_g) - Z_r Z_f)} \begin{bmatrix} Z_o + R_g & -Z_r \\ -Z_f & Z_i + R_g \end{bmatrix}\right] - \begin{bmatrix} Z_i - R_g & Z_r \\ Z_f & Z_o - R_g \end{bmatrix}\right) \tag{3.101}$$

which can be simplified to

$$S^I = ([Z]+[Z_g])^{-1}([Z]-[Z_g]) = \begin{pmatrix} (Z_o + R_g)(Z_i - R_g) - Z_r Z_f & 2Z_r R_g \\ 2Z_f R_g & (Z_i + R_g)(Z_o - R_g) - Z_r Z_f \end{pmatrix} \tag{3.102}$$

3.4.3 Normalized Scattering Parameters

Normalized scattering parameters can be introduced by Equations 3.103 and 3.104 for the incident and reflected waves as follows:

$$[a] = \frac{1}{\sqrt{2}}\sqrt{([Z_g] + [Z_g^*])}[I_i] \tag{3.103}$$

$$[b] = \frac{1}{\sqrt{2}}\sqrt{([Z_g] + [Z_g^*])}[I_r] \tag{3.104}$$

where

$$\frac{1}{\sqrt{2}}\sqrt{([Z_g] + [Z_g^*])} = \sqrt{\mathrm{Re}\{Z_g\}} = \begin{bmatrix} \sqrt{\mathrm{Re}\{Z_{g1}\}} & 0 & \cdots & 0 \\ 0 & \sqrt{\mathrm{Re}\{Z_{g2}\}} & \cdots & 0 \\ \vdots & \vdots & & \vdots \\ 0 & 0 & \cdots & \sqrt{\mathrm{Re}\{Z_{gn}\}} \end{bmatrix} \tag{3.105}$$

Substituting Equation 3.86 into Equations 3.103 and 3.104 gives

$$\frac{[b]}{\sqrt{\mathrm{Re}\{Z_g\}}} = [I_r] = [S^I][I_i] \tag{3.106}$$

or

$$\frac{[b]}{\sqrt{\mathrm{Re}\{Z_g\}}} = [S^I]\frac{[a]}{\sqrt{\mathrm{Re}\{Z_g\}}} \tag{3.107}$$

Then, from Equations 3.106 and 3.107,

$$[b] = \sqrt{\mathrm{Re}\{Z_g\}}[S^I][\mathrm{Re}\{Z_g\}]^{-1/2}[a] \tag{3.108}$$

Equation 3.108 can be simplified to

$$[b] = [S][a] \tag{3.109}$$

where

$$[S] = \sqrt{\mathrm{Re}\{Z_g\}}[S^I][\mathrm{Re}\{Z_g\}]^{-1/2} \tag{3.110}$$

and

$$[\mathrm{Re}\{Z_g\}]^{-1/2} = \begin{bmatrix} \frac{1}{\sqrt{\mathrm{Re}\{Z_{g1}\}}} & 0 & \cdots & 0 \\ 0 & \frac{1}{\sqrt{\mathrm{Re}\{Z_{g2}\}}} & \cdots & 0 \\ \vdots & \vdots & & \vdots \\ 0 & 0 & \cdots & \frac{1}{\sqrt{\mathrm{Re}\{Z_{gn}\}}} \end{bmatrix} \quad (3.111)$$

The S-matrix in Equation 3.109 is called a normalized scattering matrix. It can be proven that

$$[S^I] = [Z_g]^{-1}[S^V][Z_g^*] \quad (3.112)$$

When the generator or source impedance is real, $Z_g = R_g$, then from Equation 3.112, we obtain

$$[S^I] = [S^V] \quad (3.113)$$

In addition, Equations 3.87 and 3.110 take the following form:

$$S^I = ([Z] + [R_g])^{-1}([Z] - [R_g]) \quad (3.114)$$

$$[S] = \sqrt{R_g}[S^I][Z_g]^{-1/2} \quad (3.115)$$

From Equations 3.113 and 3.115, we obtain

$$[S] = [S^I] = [S^V] \quad (3.116)$$

S parameters can be calculated using the two-port network shown in Figure 3.25. In Figure 3.25, the source or generator impedances are given as R_{g1} and R_{g2}. When Equation 3.109 is expanded,

$$\begin{bmatrix} b_1 \\ b_2 \end{bmatrix} = \begin{bmatrix} S_{11} & S_{12} \\ S_{21} & S_{22} \end{bmatrix} \begin{bmatrix} a_1 \\ a_2 \end{bmatrix} \quad (3.117)$$

From Equation 3.117,

$$b_1 = S_{11}a_1 + S_{12}a_2 \quad (3.118)$$

$$b_2 = S_{21}a_1 + S_{22}a_2 \quad (3.119)$$

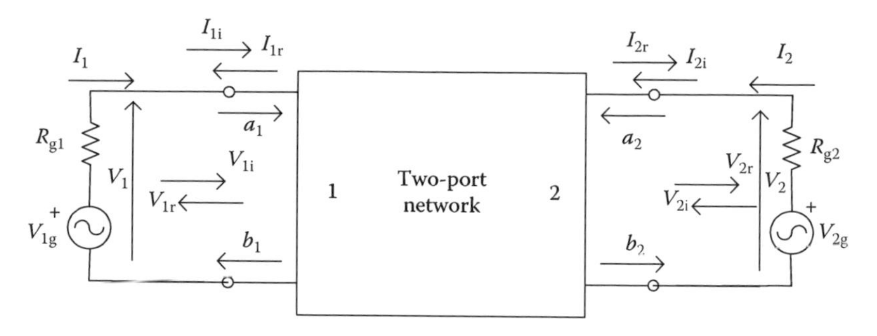

FIGURE 3.25 *S*-parameters for two-port networks.

Hence, S parameters can be defined from Equations 3.118 and 3.119 as

$$S_{11} = \frac{b_1}{a_1}\bigg|_{a_2=0} \qquad S_{12} = \frac{b_1}{a_2}\bigg|_{a_1=0}$$
$$S_{21} = \frac{b_2}{a_1}\bigg|_{a_2=0} \qquad S_{22} = \frac{b_2}{a_2}\bigg|_{a_1=0} \tag{3.120}$$

From Equation 3.120, the scattering parameters are calculated when $a_1 = 0$ or $a_2 = 0$. a represents the incident waves. If Equation 3.103 is reviewed again,

$$[a] = \frac{1}{\sqrt{2}}\sqrt{([Z_g]+[Z_g^*])}[I_i] \tag{3.121}$$

a_2 becomes zero when $I_{2i} = 0$. This can be obtained when there is no source connected to port 2, i.e., $V_{2g} = 0$ with the existence of source impedance, R_{2g}. From KLV for the second port, we obtain

$$V_2 = -I_2R_{g2} \text{ or } V_2 + I_2R_{g2} = 0 \tag{3.122}$$

Substituting Equations 3.78 and 3.79 into Equation 3.122 gives

$$V_2 + I_2R_{g2} = V_{2i} + V_{2r} + R_{g2}(I_{2i} - I_{2r}) \tag{3.123}$$

which leads to

$$V_2 + I_2R_{g2} = I_{2i}R_{g2} + R_{g2}I_{2r} + R_{g2}I_{2i} - I_{2r}R_{g2} \tag{3.124}$$

or

$$V_2 + I_2R_{g2} = 2R_{g2}I_{2i} \tag{3.125}$$

From Equation 3.121, when $Z_g = R_g$,

$$[a] = \sqrt{R_g}\,[I_i] \tag{3.126}$$

Substituting Equation 3.124 into Equation 3.125 gives

$$V_2 + I_2 R_{g2} = 2\sqrt{R_{g2}}\,a_2 \tag{3.127}$$

Then,

$$a_2 = \frac{V_2 + I_2 R_{g2}}{2\sqrt{R_{g2}}} \tag{3.128}$$

It is then proven that when Equation 3.122 is substituted into Equation 3.128, $a_2 = 0$ as expected. This also requires $I_i = 0$ from Equation 3.127. Then, this shows that there is no reflected current, which is the incident current, I_{2i}, at port 2 due to the source generator incident wave from port 1.

A similar analysis can be done at port 1 when $a_1 = 0$. The same steps can be followed, and it can be shown that

$$a_1 = \frac{V_1 + I_1 R_{g1}}{2\sqrt{R_{g1}}} \tag{3.129}$$

Reflected waves b_1 and b_2 can be analyzed the same way using the analysis just presented for the incident waves a_1 and a_2. When there is no source voltage connected at port 1, $a_1 = 0$ in the existence of source voltage R_{g1}, we can write

$$V_1 = -I_1 R_{g1} \text{ or } V_1 + I_1 R_{g1} = 0 \tag{3.130}$$

In terms of the reflected and incident voltage and current, we get

$$V_1 - I_1 R_{g1} = V_{1i} + V_{1r} - R_{g1}(I_{1i} - I_{1r}) \tag{3.131}$$

which leads to

$$V_1 - I_1 R_{g1} = I_{1i} R_{g1} + I_{1r} R_{g1} - I_{1i} R_{g1} + I_{1r} R_{g1} \tag{3.132}$$

or

$$V_1 - I_1 R_{g1} = 2R_{g1} I_{1r} \tag{3.133}$$

From Equation 3.104, when $Z_g = R_g$,

$$[b] = \sqrt{R_g}[I_r] \tag{3.134}$$

Hence, Equation 3.133 can be written as

$$V_1 - I_1 R_{g1} = 2\sqrt{R_{g1}} b_1 \tag{3.135}$$

Then,

$$b_1 = \frac{V_1 - I_1 R_{g1}}{2\sqrt{R_{g1}}} \tag{3.136}$$

It can also be shown that when $a_2 = 0$,

$$b_2 = \frac{V_2 - I_2 R_{g2}}{2\sqrt{R_{g2}}} \tag{3.137}$$

The incident and reflected parameters a and b for the N-port network can be written using the results given in Equations 3.128, 3.129, 3.136, and 3.137 for real generator impedance, R_g, as

$$[a] = \frac{1}{2}[R_g]^{-1/2}([V] + [R_g][I]) \tag{3.138}$$

$$[b] = \frac{1}{2}[R_g]^{-1/2}([V] - [R_g][I]) \tag{3.139}$$

For an arbitrary impedance, Equations 3.138 and 3.139 can be written as

$$[a] = \frac{1}{2}[\mathrm{Re}\{Z_g\}]^{-1/2}([V] + [Z_g][I]) \tag{3.140}$$

$$[b] = \frac{1}{2}[\mathrm{Re}\{Z_g\}]^{-1/2}([V] - [Z_g^*][I]) \tag{3.141}$$

Now, since the conditions when a_1 and a_2 are zero were derived, the equations given by Equation 3.120 can be expanded. When $a_2 = 0$, S_{11} and S_{21} can be calculated. From Equations 3.124, 3.126, 3.129, and 3.136, S_{11} can be expressed as

$$S_{11} = \left.\frac{b_1}{a_1}\right|_{a_2=0} = \left.\frac{\left(\dfrac{V_1 - I_1 R_{g1}}{2\sqrt{R_{g1}}}\right)}{\left(\dfrac{V_1 + I_1 R_{g1}}{2\sqrt{R_{g1}}}\right)}\right|_{I_{2i}=0} = \frac{V_1 - I_1 R_{g1}}{V_1 + I_1 R_{g1}} = \frac{V_{1r}}{V_{1i}} = \frac{\sqrt{R_{1g}} I_{1r}}{\sqrt{R_{1g}} I_{1i}} = \frac{I_{1r}}{I_{1i}} \quad (3.142)$$

or

$$S_{11} = \frac{Z_{11} - R_{g1}}{Z_{11} + R_{g1}} \quad (3.143)$$

In Equation 3.143, S_{11} is the reflection coefficient at port 1 when port 2 is terminated with generator impedance R_{g2}. S_{21} is expressed using Equations 3.126, 3.129, 3.134, and 3.137 as

$$S_{21} = \left.\frac{b_2}{a_1}\right|_{a_2=0} = \left.\frac{\left(\dfrac{V_2 - I_2 R_{g2}}{2\sqrt{R_{g2}}}\right)}{\left(\dfrac{V_1 + I_1 R_{g1}}{2\sqrt{R_{g1}}}\right)}\right|_{I_{2i}=0} = \frac{(V_2 - I_2 R_{g2})\sqrt{R_{g1}}}{(V_1 + I_1 R_{g1})\sqrt{R_{g2}}} = \frac{\sqrt{R_{2g}} I_{2r}}{\sqrt{R_{1g}} I_{1i}} \quad (3.144)$$

When $a_2 = 0$, $V_{2g} = 0$ and that results in $V_2 = -I_2 R_{2g}$ and $V_{1g} = 2I_{1i}R_{1g}$; then Equation 3.144 can be written as

$$S_{21} = \left.\frac{b_2}{a_1}\right|_{a_2=0} = -\frac{\sqrt{R_{2g}} I_2}{\sqrt{R_{1g}}\,(V_{1g}/2R_{1g})} = -2\sqrt{R_{1g}}\sqrt{R_{2g}}\,\frac{I_2}{V_{1g}} = 2\sqrt{\frac{R_{1g}}{R_{2g}}}\,\frac{V_2}{V_{1g}} = \frac{\left(V_2/\sqrt{R_{2g}}\right)}{\left(\dfrac{1}{2} V_{1g}/\sqrt{R_{1g}}\right)}$$

(3.145)

As shown from Equation 3.145, S_{21} is the forward transmission gain of the network from port 1 to port 2. A similar procedure can be repeated to derive S_{22} and S_{12} when $a_1 = 0$. Hence, it can be shown that

$$S_{22} = \left.\frac{b_2}{a_2}\right|_{a_1=0} = \left.\frac{\left(\dfrac{V_2 - I_2 R_{g2}}{2\sqrt{R_{g2}}}\right)}{\left(\dfrac{V_2 + I_2 R_{g2}}{2\sqrt{R_{g2}}}\right)}\right|_{I_{1i}=0} = \frac{V_2 - I_2 R_{g2}}{V_2 + I_2 R_{g2}} = \frac{V_{2r}}{V_{2i}} = \frac{\sqrt{R_{2g}} I_{2r}}{\sqrt{R_{2g}} I_{2i}} = \frac{I_{2r}}{I_{2i}} \quad (3.146)$$

or

$$S_{22} = \frac{Z_{22} - R_{g2}}{Z_{22} + R_{g2}} \tag{3.147}$$

S_{22} is the reflection coefficient of the output. S_{12} can be obtained as

$$S_{12} = \left.\frac{b_1}{a_2}\right|_{a_1=0} = \left.\frac{\left(\dfrac{V_1 - I_1 R_{g1}}{2\sqrt{R_{g1}}}\right)}{\left(\dfrac{V_2 + I_2 R_{g2}}{2\sqrt{R_{g2}}}\right)}\right|_{I_{1i}=0} = \frac{(V_1 - I_1 R_{g1})\sqrt{R_{g2}}}{(V_2 + I_2 R_{g2})\sqrt{R_{g1}}} = \frac{\sqrt{R_{1g}} I_{1r}}{\sqrt{R_{2g}} I_{2i}} \tag{3.148}$$

which can be put in the following form:

$$S_{12} = \left.\frac{b_1}{a_2}\right|_{a_1=0} = -2\sqrt{R_{1g}}\sqrt{R_{2g}}\frac{I_1}{V_{2g}} = 2\sqrt{\frac{R_{2g}}{R_{1g}}}\frac{V_1}{V_{2g}} = \frac{\left(V_1/\sqrt{R_{1g}}\right)}{\left(\dfrac{1}{2}V_{2g}/\sqrt{R_{2g}}\right)} \tag{3.149}$$

S_{12} is the reverse transmission gain of the network from port 2 to port 1. Overall, S parameters are found when $a_n = 0$, which means that there is no reflection at that port. This is only possible by matching all the ports except the measurement port. Insertion loss and return loss in terms of S parameters are defined as

$$\text{Insertion loss(dB)} = \text{IL(dB)} = 20\log(|S_{ij}|),\ i \neq j \tag{3.150}$$

$$\text{Return loss(dB)} = \text{RL(dB)} = 20\log(|S_{ii}|) \tag{3.151}$$

Another important parameter that can be defined using S parameters is the voltage standing wave ratio, VSWR. For instance, VSWR at port 1 is found from

$$\text{VSWR} = \frac{1-|S_{11}|}{1+|S_{11}|} \tag{3.152}$$

The two-port network is reciprocal if

$$S_{21} = S_{12} \tag{3.153}$$

It can be shown that a network is reciprocal if it is equal to its transpose. This is represented for the two-port network as

$$[S] = [S]^t \tag{3.154}$$

or

$$\begin{bmatrix} S_{11} & S_{12} \\ S_{21} & S_{22} \end{bmatrix}^{t} = \begin{bmatrix} S_{11} & S_{21} \\ S_{12} & S_{22} \end{bmatrix} \quad (3.155)$$

When a network is lossless, S parameters can be used to characterize this feature as

$$[S]^{t}[S]^{*} = [U] \quad (3.156)$$

where * defines the complex conjugate of a matrix, and U is the unitary matrix and defined by

$$[U] = \begin{bmatrix} 1 & 0 \\ 0 & 1 \end{bmatrix} \quad (3.157)$$

Equation 3.156 can be applied for a two-port network as

$$[S]^{t}[S]^{*} = \begin{bmatrix} \left(|S_{11}|^{2} + |S_{21}|^{2}\right) & (S_{11}S_{12}^{*} + S_{21}S_{22}^{*}) \\ (S_{12}S_{11}^{*} + S_{22}S_{21}^{*}) & \left(|S_{12}|^{2} + |S_{22}|^{2}\right) \end{bmatrix} = \begin{bmatrix} 1 & 0 \\ 0 & 1 \end{bmatrix} \quad (3.158)$$

It can be further shown that if a network is lossless and reciprocal, it satisfies

$$|S_{11}|^{2} + |S_{21}|^{2} = 1 \quad (3.159)$$

$$S_{11}S_{12}^{*} + S_{21}S_{22}^{*} = 0 \quad (3.160)$$

Example

Find the characteristic impedance of the T-network given in Figure 3.26 to have no return loss at the input port.

Solution

The scattering parameters for T-network are found from Equation 3.120. From Equation 3.120, S_{11} is equal to

$$S_{11} = \left.\frac{b_1}{a_1}\right|_{a_2=0} = \frac{Z_{in} - Z_o}{Z_{in} + Z_o} \quad (3.161)$$

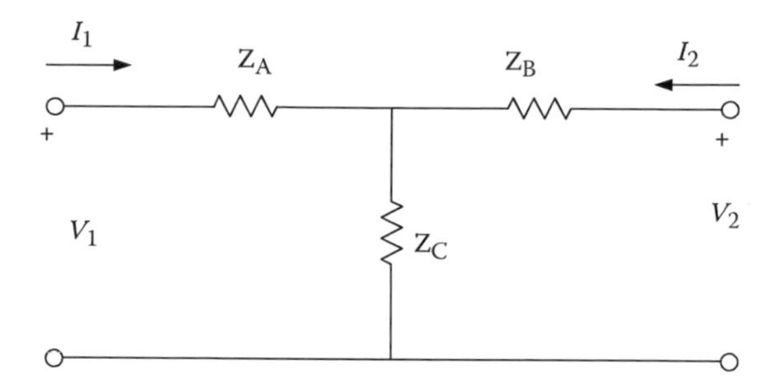

FIGURE 3.26 *T*-network configuration.

where

$$Z_{in} = Z_A + \left[\frac{Z_C(Z_B + Z_o)}{Z_C + (Z_B + Z_o)} \right] \tag{3.162}$$

No return loss is possible when $S_{11} = 0$. This can be satisfied from Equations 3.161 and 3.162 when

$$Z_o = Z_{in} = Z_A + \left[\frac{Z_C(Z_B + Z_o)}{Z_C + (Z_B + Z_o)} \right] \tag{3.163}$$

Example

Consider the typical transformer coupling circuit used for RF power amplifiers given in Figure 3.27. The primary and secondary sides of the transformer circuit become resonant at the frequency of operation. The coupling factor M is also set for maximum power transfer. Derive the S parameters of the circuit.

Solution

The scattering parameters for S_{11} and S_{21} are found by connecting the source and generator impedance only at port 1 and generator impedance at port 2, as shown in Figure 3.28.

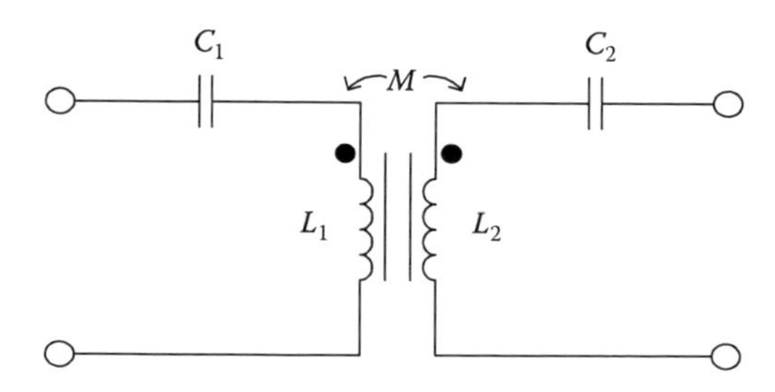

FIGURE 3.27 Transformer circuit.

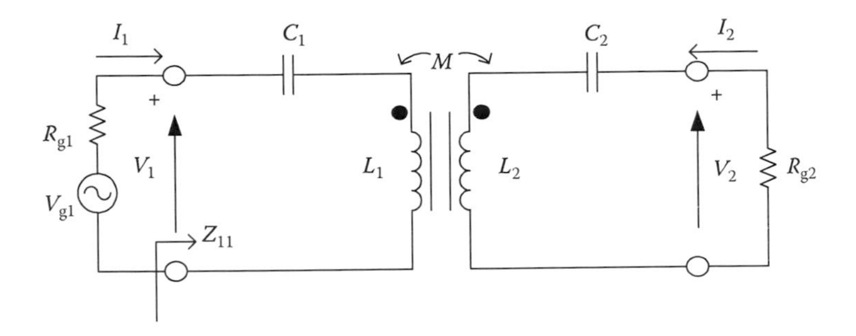

FIGURE 3.28 Transformer coupling circuit for S_{11} and S_{21}.

Application of KVL on the left and right sides of the circuit gives

$$V_{1g} - I_1R_{g1} + jI_1X_{C1} - jI_1X_{L1} - j\omega I_2M = 0 \tag{3.164}$$

$$I_2R_{g2} - jI_2X_{C2} + jI_2X_{L2} + j\omega I_1M = 0 \tag{3.165}$$

From Equations 3.164 and 3.165, we obtain

$$V_{g1} = I_1\left(R_{g1} - jX_{C1} + jX_{L1} + \frac{\omega^2 M}{R_{g2} - jX_{C2} + jX_{L2}}\right) \tag{3.166}$$

$\dfrac{\omega^2 M}{R_{g2} - jX_{C2} + jX_{L2}}$ represented the secondary impedance referred to the primary side. It is communicated in the problem that the circuit is at resonance with the operational frequency. Then, Equation 3.166 is simplified to

$$V_{g1} = I_1\left(R_{g1} + \frac{\omega^2 M}{R_{g2}}\right) \tag{3.167}$$

From Equation 3.167, it is seen that the maximum power transfer occurs when

$$R_{g1} = \frac{\omega^2 M}{R_{g2}} \tag{3.168}$$

Hence, the coupling factor, M, is found from Equation 3.168 as

$$M = \omega\sqrt{R_{g1}R_{g2}} \tag{3.169}$$

As a result of Equations 3.168 and 3.169, Z_{11} is equal to

$$Z_{11} = R_{g1} \tag{3.170}$$

From Equation 3.143, S_{11} is found as

$$S_{11} = 0 \tag{3.171}$$

From Equation 3.145,

$$S_{21} = -2\sqrt{R_{1g}}\sqrt{R_{2g}}\frac{I_2}{V_{1g}} \tag{3.172}$$

When Equation 3.169 is substituted into Equations 3.164 and 3.165, they are simplified to

$$V_{1g} - I_1 R_{g1} - jI_2\sqrt{R_{g1}R_{g2}} = 0 \tag{3.173}$$

$$I_2 R_{g2} + jI_1\sqrt{R_{g1}R_{g2}} = 0 \tag{3.174}$$

Solving Equations 3.173 and 3.174 for the ratio of (I_2/V_{1g}) gives

$$\frac{I_2}{V_{1g}} = \frac{-j}{2\sqrt{R_{1g}R_{2g}}} \tag{3.175}$$

Substitution of Equation 3.175 into Equation 3.172 gives S_{21} as

$$S_{21} = -2\sqrt{R_{1g}}\sqrt{R_{2g}}\left(\frac{-j}{2\sqrt{R_{1g}R_{2g}}}\right) \tag{3.176}$$

Then,

$$S_{21} = j = 1\ \angle 90^\circ \tag{3.177}$$

The scattering parameters for S_{22} and S_{12} are found by connecting the source and generator impedance only at port 2 and the generator impedance at port 1, as shown in Figure 3.29.

We can apply KVL for both sides of the circuit with Equation 3.169 and obtain

$$I_1 R_{g1} + jI_2\sqrt{R_{g1}R_{g2}} = 0 \tag{3.178}$$

$$V_{g2} - I_2 R_{g2} - jI_1\sqrt{R_{g1}R_{g2}} = 0 \tag{3.179}$$

Solving Equations 3.178 and 3.179 for V_{g2} gives

$$V_{g2} = I_2(R_{g1} + R_{g2}) \tag{3.180}$$

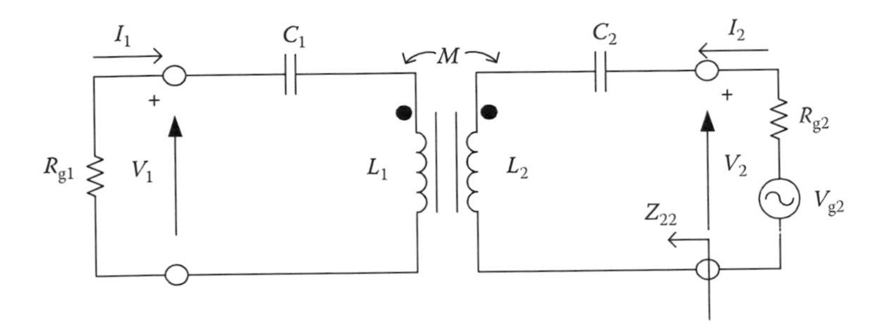

FIGURE 3.29 Transformer coupling circuit for S_{22} and S_{12}.

Since from Equations 3.168 and 3.169,

$$R_{g2} = \frac{\omega^2 M}{R_1} \quad (3.181)$$

Then,

$$Z_{22} = R_{g2} \quad (3.182)$$

Substitution of Equation 3.182 into Equation 3.147 leads to

$$S_{22} = 0 \quad (3.183)$$

S_{12} is calculated by finding the ratio of (I_1/V_{g2}) from Equations 3.178 and 3.179 as

$$\frac{I_1}{V_{2g}} = \frac{-j}{2\sqrt{R_{1g}R_{2g}}} \quad (3.184)$$

From Equation 3.149,

$$S_{12} = -2\sqrt{R_{1g}}\sqrt{R_{2g}}\frac{I_1}{V_{2g}} = -2\sqrt{R_{1g}}\sqrt{R_{2g}}\left(\frac{-j}{2\sqrt{R_{1g}R_{2g}}}\right) \quad (3.185)$$

Hence, S_{12} is equal to

$$S_{12} = j = 1\ \angle 90° \quad (3.186)$$

Then, the S-matrix for a coupling transformer can be written as

$$S = \begin{bmatrix} 0 & j \\ j & 0 \end{bmatrix} \quad (3.187)$$

TABLE 3.2
ABCD and S Parameters of Basic Network Configurations

	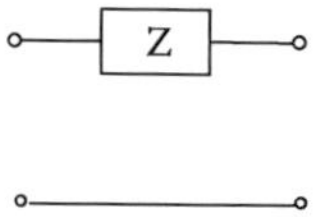 Series Z: $z = \frac{Z}{Z_o}$	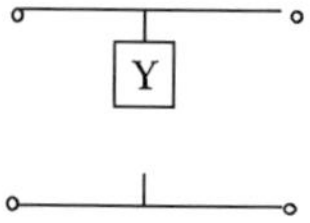 Shunt Y: $y = \frac{Y}{Y_o}$		Transformer $N_1 : N_2$, $n = N_1/N_2$	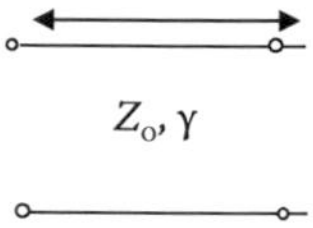 Transmission line ℓ, Z_o, γ, $\gamma = \alpha + j\beta$
A	1	1	**A**	$n = N_1/N_2$	$\cosh(\gamma\ell)$
B	Z	0	**B**	0	$Z_o \sinh(\gamma\ell)$
C	0	Y	**C**	0	$\frac{\sinh(\gamma\ell)}{Z_o}$
D	1	1	**D**	$\frac{1}{n} = \frac{N_2}{N_1}$	$\cosh(\gamma\ell)$
S_{11}	$\frac{z}{z+2}$	$\frac{-y}{y+2}$	S_{11}	$\frac{n^2-1}{n^2+1}$	0
S_{12}	$\frac{2}{z+2}$	$\frac{2}{y+2}$	S_{12}	$\frac{2n}{n^2+1}$	$e^{-\gamma\ell}$
S_{21}	$\frac{2}{z+2}$	$\frac{2}{y+2}$	S_{21}	$\frac{2n}{n^2+1}$	$e^{-\gamma\ell}$
S_{22}	$\frac{z}{z+2}$	$\frac{-y}{y+2}$	S_{22}	$-\frac{n^2-1}{n^2+1}$	0

or

$$S = \begin{bmatrix} 0 & 1\angle 90^\circ \\ 1\angle 90^\circ & 0 \end{bmatrix} \quad (3.188)$$

The *ABCD* and *S* parameters for some of the basic RF components given in the following are shown in Table 3.2.

3.5 MEASUREMENT OF *S* PARAMETERS

In this section, the measurement of scattering parameters for two-port and three-port networks will be discussed. In addition, the design of test fixture to measure scattering parameters will also be given.

3.5.1 Measurement of *S* Parameters for a Two-Port Network

Two-port scattering parameters can be measured by expressing the incident and reflected waves in terms of circuit parameters. From Equations 3.118 and 3.119,

$$b_1 = S_{11}a_1 + S_{12}a_2 \quad (3.189)$$

$$b_2 = S_{21}a_1 + S_{22}a_2 \quad (3.190)$$

It was given by Equations 3.80, 3.81, 3.126, and 3.134 that

$$[V_i] = [Z_g^*][I_i] \quad (3.191)$$

$$[V_r] = [Z_g][I_r] \quad (3.192)$$

$$[a] = \sqrt{R_g}[I_i] \quad (3.193)$$

$$[b] = \sqrt{R_g}[I_r] \quad (3.194)$$

Then, when there are real generator impedances, the following equations can be written:

$$V_{i1} = R_{g1}I_{i1} \quad (3.195)$$

$$V_{r1} = R_{g1}I_{r1} \quad (3.196)$$

$$V_{i2} = R_{g2}I_{i2} \quad (3.197)$$

$$V_{r2} = R_{g2}I_{r2} \quad (3.198)$$

and

$$a_1 = \sqrt{R_{g1}}\, I_{i1} \tag{3.199}$$

$$b_1 = \sqrt{R_{g1}}\, I_{r1} \tag{3.200}$$

$$a_2 = \sqrt{R_{g2}}\, I_{i2} \tag{3.201}$$

$$b_2 = \sqrt{R_{g2}}\, I_{r2} \tag{3.202}$$

When Equations 3.199 through 3.202 are substituted into Equations 3.189 and 3.190, we obtain

$$\sqrt{R_{g1}}\, I_{r1} = S_{11}\sqrt{R_{g1}}\, I_{i1} + S_{12}\sqrt{R_{g2}}\, I_{i2} \tag{3.203}$$

$$\sqrt{R_{g2}}\, I_{r2} = S_{21}\sqrt{R_{g2}}\, I_{i1} + S_{22}\sqrt{R_{g2}}\, I_{i2} \tag{3.204}$$

When the generator impedances at ports 1 and 2 are equal, i.e., $R_{g1} = R_{g2} = R$, then Equations 3.203 and 3.204 simplify to

$$I_{r1} = S_{11} I_{i1} + S_{12} I_{i2} \tag{3.205}$$

$$I_{r2} = S_{21} I_{i1} + S_{22} I_{i2} \tag{3.206}$$

Similarly,

$$V_{r1} = S_{11} V_{i1} + S_{12} V_{i2} \tag{3.207}$$

$$V_{r2} = S_{21} V_{i1} + S_{22} V_{i2} \tag{3.208}$$

Hence, from Equations 3.207 and 3.208, the scattering parameters can be measured as

$$\begin{aligned} S_{11} &= \left.\frac{V_{r1}}{V_{i1}}\right|_{V_{i2}=0} \qquad S_{12} = \left.\frac{V_{r1}}{V_{i2}}\right|_{V_{i1}=0} \\ S_{21} &= \left.\frac{V_{r2}}{V_{i1}}\right|_{V_{i2}=0} \qquad S_{22} = \left.\frac{V_{r2}}{V_{i2}}\right|_{V_{i1}=0} \end{aligned} \tag{3.209}$$

As seen from Equation 3.209, the measurement of the incident and reflected voltages at each port while the other port is terminated by a matched port will give the scattering parameters in Equation 3.209. The incident and reflected voltage waves can be simply measured by directional couplers in practical applications.

3.5.2 Measurement of S Parameters for a Three-Port Network

The following results are obtained by applying the same analysis for a three-port network. The incident and reflected voltage and current in terms of scattering parameters for a three-port network are obtained as

$$I_{r1} = S_{11}I_{i1} + S_{12}I_{i2} + S_{13}I_{i3} \tag{3.210}$$

$$I_{r2} = S_{21}I_{i1} + S_{22}I_{i2} + S_{23}I_{i3} \tag{3.211}$$

$$I_{r3} = S_{31}I_{i1} + S_{32}I_{i2} + S_{33}I_{i3} \tag{3.212}$$

Similarly,

$$V_{r1} = S_{11}V_{i1} + S_{12}V_{i2} + S_{13}V_{i3} \tag{3.213}$$

$$V_{r2} = S_{21}V_{i1} + S_{22}V_{i2} + S_{23}V_{i3} \tag{3.214}$$

$$V_{r3} = S_{31}V_{i1} + S_{32}V_{i2} + S_{33}V_{i3} \tag{3.215}$$

Hence, the scattering parameters from Equations 3.210 through 3.215 are found as

$$\begin{aligned} S_{11} &= \left.\frac{V_{r1}}{V_{i1}}\right|_{V_{i2}=0,V_{i3}=0} \qquad & S_{12} &= \left.\frac{V_{r1}}{V_{i2}}\right|_{V_{i1}=0,V_{i3}=0} \\ S_{31} &= \left.\frac{V_{r3}}{V_{i1}}\right|_{V_{i2}=0,V_{i3}=0} \qquad & S_{33} &= \left.\frac{V_{r3}}{V_{i3}}\right|_{V_{i1}=0,V_{i3}=0} \end{aligned} \tag{3.216}$$

The complete conversion chart between S parameters and two-port parameters is given in Table 3.3.

TABLE 3.3
Conversion Chart between *S* Parameters and Two-Port Parameters

S		*Z*	*Y*	*ABCD*
S_{11}	S_{11}	$\frac{(Z_{11}-Z_0)(Z_{22}+Z_0)-Z_{12}Z_{21}}{\Delta Z}$	$\frac{(Y_0-Y_{11})(Y_0+Y_{22})+Y_{12}Y_{21}}{\Delta Y}$	$\frac{A+B/Z_0-CZ_0-D}{A+B/Z_0+CZ_0+D}$
S_{12}	S_{12}	$\frac{2Z_{12}Z_0}{\Delta Z}$	$\frac{-2Y_{12}Y_0}{\Delta Y}$	$\frac{2(AD-BC)}{A+B/Z_0+CZ_0+D}$
S_{21}	S_{21}	$\frac{2Z_{21}Z_0}{\Delta Z}$	$\frac{-2Y_{21}Y_0}{\Delta Y}$	$\frac{2}{A+B/Z_0+CZ_0+D}$
S_{22}	S_{22}	$\frac{(Z_{11}+Z_0)(Z_{22}-Z_0)-Z_{12}Z_{21}}{\Delta Z}$	$\frac{(Y_0+Y_{11})(Y_0-Y_{22})+Y_{12}Y_{21}}{\Delta Y}$	$\frac{-A+B/Z_0-CZ_0+D}{A+B/Z_0+CZ_0+D}$
Z_{11}	$Z_0\frac{(1+S_{11})(1-S_{22})+S_{12}S_{21}}{(1-S_{11})(1-S_{22})-S_{12}S_{21}}$	Z_{11}	$\frac{Y_{22}}{\lvert Y\rvert}$	$\frac{A}{C}$
Z_{12}	$Z_0\frac{2S_{12}}{(1-S_{11})(1-S_{22})-S_{12}S_{21}}$	Z_{12}	$\frac{-Y_{12}}{\lvert Y\rvert}$	$\frac{AD-BC}{C}$
Z_{21}	$Z_0\frac{2S_{21}}{(1-S_{11})(1-S_{22})-S_{12}S_{21}}$	Z_{21}	$\frac{-Y_{21}}{\lvert Y\rvert}$	$\frac{1}{C}$
Z_{22}	$Z_0\frac{(1-S_{11})(1+S_{22})+S_{12}S_{21}}{(1-S_{11})(1-S_{22})-S_{12}S_{21}}$	Z_{22}	$\frac{Y_{11}}{\lvert Y\rvert}$	$\frac{D}{C}$

(Continued)

TABLE 3.3 (CONTINUED)
Conversion Chart between *S* Parameters and Two-Port Parameters

S		***Z***	***Y***	***ABCD***
Y_{11}	$Y_0 \dfrac{(1-S_{11})(1+S_{22})+S_{12}S_{21}}{(1+S_{11})(1+S_{22})-S_{12}S_{21}}$	$\dfrac{Z_{22}}{\lvert Z\rvert}$	Y_{11}	$\dfrac{D}{B}$
Y_{12}	$Y_0 \dfrac{-2S_{12}}{(1+S_{11})(1+S_{22})-S_{12}S_{21}}$	$\dfrac{-Z_{12}}{\lvert Z\rvert}$	Y_{12}	$\dfrac{BC-AD}{B}$
Y_{21}	$Y_0 \dfrac{-2S_{21}}{(1+S_{11})(1+S_{22})-S_{12}S_{21}}$	$\dfrac{-Z_{21}}{\lvert Z\rvert}$	Y_{21}	$\dfrac{-1}{B}$
Y_{22}	$Y_0 \dfrac{(1+S_{11})(1-S_{22})+S_{12}S_{21}}{(1+S_{11})(1+S_{22})-S_{12}S_{21}}$	$\dfrac{Z_{11}}{\lvert Z\rvert}$	Y_{22}	$\dfrac{A}{B}$
A	$\dfrac{(1+S_{11})(1-S_{22})+S_{12}S_{21}}{2S_{21}}$	$\dfrac{Z_{11}}{Z_{21}}$	$\dfrac{-Y_{22}}{Y_{21}}$	A
B	$Z_0 \dfrac{(1+S_{11})(1+S_{22})-S_{12}S_{21}}{2S_{21}}$	$\dfrac{\lvert Z\rvert}{Z_{21}}$	$\dfrac{-1}{Y_{21}}$	B
C	$\dfrac{1}{Z_0} \dfrac{(1+S_{11})(1+S_{22})-S_{12}S_{21}}{2S_{21}}$	$\dfrac{1}{Z_{21}}$	$\dfrac{-\lvert Y\rvert}{Y_{21}}$	C
D	$\dfrac{(1-S_{11})(1+S_{22})-S_{12}S_{21}}{2S_{21}}$	$\dfrac{Z_{22}}{Z_{21}}$	$\dfrac{-Y_{11}}{Y_{21}}$	D

The MATLAB conversion codes to convert S parameters to Z parameters and $ABCD$ parameters to S parameters are given below.

```
%%%%%%%%%%%%%%%%%%%%%%%%%%%%%%%%%%%%%%%%%%%%%%%%%%%%%%%%%%%%%
This m-file is function program to convert S Parameters to Z Parameters
%%%%%%%%%%%%%%%%%%%%%%%%%%%%%%%%%%%%%%%%%%%%%%%%%%%%%%%%%%%%%

function [z11,z12,z21,z22]=S2Z(s11,s12,s21,s22,Zo)
zhi=(1-s11)*(1-s22)-s12*s21;
z11=Zo*(((1+s11)*(1-s22)+s12*s21)/zhi);
z12=Zo*(2*s12/zhi);
z21=Zo*(2*s21/zhi);
z22=Zo*(((1-s11)*(1+s22)+s12*s21)/zhi);
end
```

$$|Z| = Z_{11}Z_{22} - Z_{12}Z_{21};\ |Y| = Y_{11}Y_{22} - Y_{12}Y_{21};\ \Delta Y = (Y_{11} + Y_0)(Y_{22} + Y_0) - Y_{12}Y_{21};\ \Delta Z = (Z_{11} + Z_0)(Z_{22} + Z_0) - Z_{12}Z_{21};\ Y_0 = 1/Z_0$$

and

```
%%%%%%%%%%%%%%%%%%%%%%%%%%%%%%%%%%%%%%%%%%%%%%%%%%%%%%%%%%%%%
This m-file is function program to convert A Parameters to S Parameters%
%%%%%%%%%%%%%%%%%%%%%%%%%%%%%%%%%%%%%%%%%%%%%%%%%%%%%%%%%%%%%

function [s11,s12,s21,s22]=A2S(a11,a12,a21,a22)
%Assume Zo=50
Zo=50;
DET=(a11+a12/Zo+a21*Zo+a22);
s11=(a11+a12/Zo-a21*Zo-a22)/DET;
s12=2*(a11*a22-a12*a21)/DET;
s21=2/DET;
s22=(-a11+a12/Zo-a21*Zo+a22)/DET;
end
```

3.5.3 Design and Calibration Methods for Measurement of Transistor Scattering Parameters

Vector network analyzers (VNAs) are used to measure and characterize RF and microwave components and devices via scattering (S) parameters. VNAs have coaxial ports and need an interface called test fixture to measure the characteristics of the noncoaxial devices under test (DUTs). When VNA is used to measure the characteristics of the DUT, the S parameters of the fixture and the accompanying cables that are used for interface are also introduced as an error into the measured S parameters. Hence, an accurate characterization of the devices requires the test fixture characteristics with cable effects to be removed from the measured results via a certain calibration technique.

There has been an extensive study on various calibration methods and algorithms in the literature for characterization of RF transistors [3–10]. The network analyzer with test fixture calibration is based on the error models, which are used for correction. Two-port error correction gives more accurate results because it accounts for all of the important sources of systematic error. Systematic errors include VNA measurement errors due to impedance mismatch and leakage terms in the test setup, isolation characteristics between the reference and test signal paths, and system frequency response. The most commonly used error model for two-port calibration with automatic network analyzer is the 12-term error model [11], though modern network analyzers also use 8- or 16-term error models [12]. The 12-term error model requires measurement of a complete *S* parameter set of the two-port device. It has two six-term error models: one for the forward measurement direction and one for the reverse measurement direction. The error terms that are due to fixturing effects can be removed by modeling, de-embedding, or direct measurement of the test fixture. The advantage in using the direct measurement method is due to the fact that the precise characteristics of the fixture do not need to be known beforehand. The characteristics of the fixture are measured during the calibration process. The commonly used calibration methods based on different error models are short–open–load–thru (SOLT) and thru–reflect–line (TRL) techniques. The TRL method is a multiline calibration method using the zero-length through, reflection (short or open), and line standards and was first introduced by Engen and Hoer [13] and then studied extensively by other researchers [14–16]. One of the disadvantages of the TRL calibration method is that the line standards can become physically too long for practical use at low frequencies. The SOLT calibration method is attractive for RF fixtures due to simpler and less-expensive fixtures and standards, which gives more accuracy at lower frequencies in comparison to the TRL method [17–20].

Typical network analyzer measurement setup for direct measurement using two-port calibration to characterize active devices is illustrated in Figure 3.30.

The effects of the test fixture including loss, length, and mismatch from the measurement can be removed if in-fixture calibration standards are available. Full two-port calibration with the conventional SOLT method based on the 12-term error model can be used to remove the effects of fixture in the measurement. Figure 3.31 illustrates the representation of the 12-term error model using forward and reverse models with a signal flow graph. There are six error terms in each model, which must be solved to accurately measure the DUT. The SOLT calibration method allows the error terms to be solved accurately through the measurement of a set of known calibration standards. Hence, when the SOLT method is used, the calibration standards must be well known in terms of their scattering parameters in order to achieve the accurate DUT measurements. This requires accurate modeling and characterization of calibration standards, which will be detailed in Section 3.5.3.1.

The DUT in Figure 3.31 is considered to be a two-port device. When the DUT is a bipolar junction transistor (BJT), it can be treated as a two-port device if it is operated in a common emitter configuration.

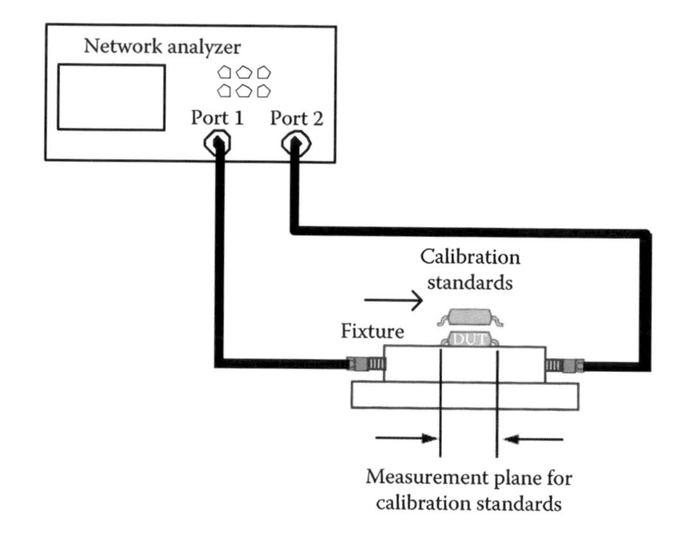

FIGURE 3.30 Network analyzer direct measurement setup for transistors.

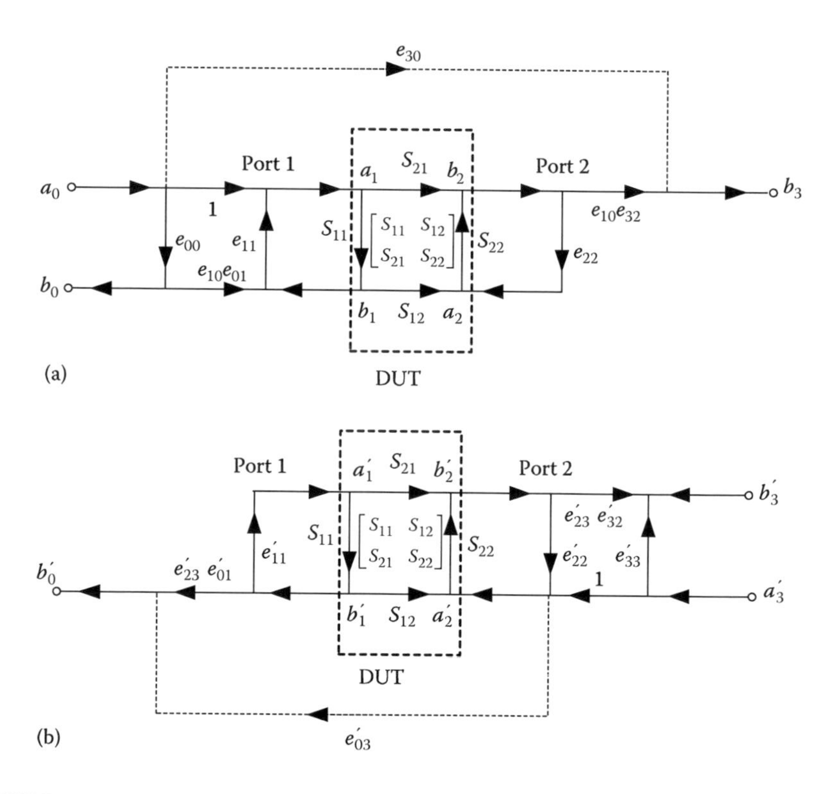

FIGURE 3.31 12-term error models. (a) Forward model. (b) Reverse model.

3.5.3.1 Design of SOLT Test Fixtures Using Grounded Coplanar Waveguide Structure

Prior to characterizing a component with a given network analyzer, it is first necessary to calibrate the instrument for a given test setup. This is done in order to remove the effects of test fixture in the measurement. Most of the modern network analyzers have integrated mathematical algorithms that can be utilized to calibrate out these effects seen by each port using standard network analyzer error models and thus allowing the user to more easily obtain accurate measurements. Better accuracy in measurement of the device characteristics can be obtained using full two-port calibration as discussed with the SOLT method. In SOLT calibration, the analyzer is subjected to a series of known configuration setups, as shown in Figure 3.32. During these measurements, the network analyzer obtains the *S* parameters of the fixture used. Once these are known, the network analyzer can easily remove the effects of fixturing through the utilization of an error matrix generated during calibration.

Test fixtures for the SOLT calibration method are designed using the GCPW structure, as illustrated in Figure 3.33. A grounded coplanar waveguide (GCPW structure consists of a center conductor of width *W*, with a gap of width *s* on either side, separating it from a ground plane. Due to the presence of the center conductor,

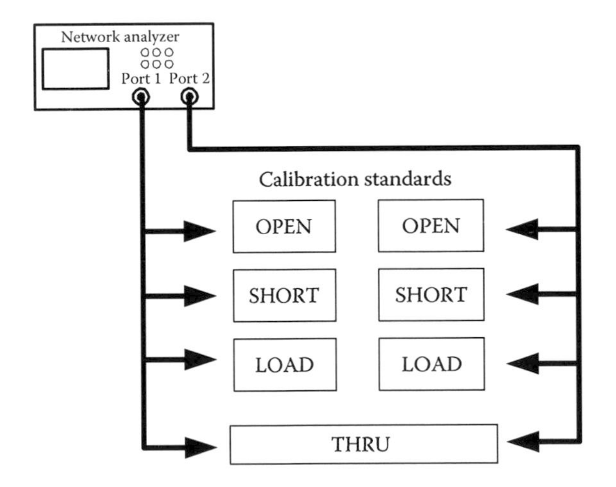

FIGURE 3.32 Implementation of SOLT calibration for network analyzer.

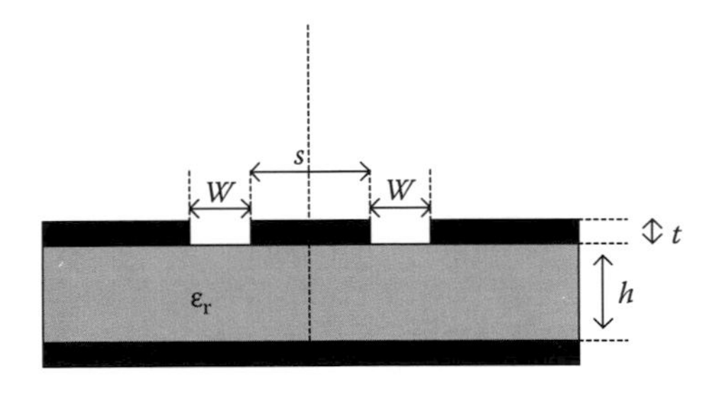

FIGURE 3.33 GCPW for SOLT calibration fixture implementation.

the transmission line can support both even and odd quasi-TEM modes, which is dependent upon the E-fields in the tow gaps that are in either the opposite directions or the same direction. This type of transmission line is therefore considered to be a good design choice for active devices due to both the center conductor and the close proximity of surrounding ground planes. The GCPW shown in Figure 3.33 with finite thickness dielectric h, a finite trace thickness t, a center conductor of width W, a gap of width s, and an infinite ground plane is analyzed using the quasi-static approach given in Refs. [21,22]. The effective permittivity constant in this configuration is obtained from

$$\varepsilon_{re} = 1 + q \cdot (\varepsilon_r - 1) \tag{3.217}$$

In Equation 3.217, q is the filling factor and is defined by

$$q = \frac{\dfrac{K(k_3)}{K'(k_3)}}{\dfrac{K(k_1)}{K'(k_1)} + \dfrac{K(k_3)}{K'(k_3)}} \tag{3.218}$$

where

$$k_1 = \frac{W}{W + 2s} \tag{3.219}$$

$$k_3 = \frac{\tanh\left(\dfrac{\pi W}{4h}\right)}{\tanh\left(\dfrac{\pi \cdot (W + 2s)}{4h}\right)} \tag{3.220}$$

$K(k_1)$ represents the complete elliptic integral of the first kind, $K'(k)$ represents its complement, and $K(k_3)$ represents the complete elliptic integral of the third kind. They are found from

$$\frac{K(k)}{K'(k)} = \frac{\pi}{\ln\left(2\dfrac{1+\sqrt{k'}}{1-\sqrt{k'}}\right)} \quad \text{for} \quad 0 \le k \le \frac{1}{\sqrt{2}} \tag{3.221}$$

$$\frac{K(k)}{K'(k)} = \frac{\ln\left(2\dfrac{1+\sqrt{k}}{1-\sqrt{k}}\right)}{\pi} \quad \text{for} \quad \frac{1}{\sqrt{2}} \le k \le 1 \tag{3.222}$$

k' is the complementary modulus of k and is obtained from $k' = \sqrt{1-k^2}$. The resulting impedance of the GCPW is then obtained as

$$Z = \frac{60\pi}{\sqrt{\varepsilon_{re}}} \cdot \frac{1}{\frac{K(k_1)}{K'(k_1)} + \frac{K(k_3)}{K'(k_3)}} \tag{3.223}$$

The effects of the trace thickness are included using a first-order correction factor on the conductor width and gap as

$$s_e = s - \Delta \tag{3.224}$$

$$W_e = W + \Delta \tag{3.225}$$

The correction factor, Δ, is given by

$$\Delta = \frac{1.25t}{\pi} \cdot \left(1 + \ln\left(\frac{4\pi W}{t}\right)\right) \tag{3.226}$$

The trace thickness can be included in the impedance calculation given by Equation 3.223 using effective modulus k_e, which is given by

$$k_e = \frac{W_e}{W_e + 2s_e} \approx k_1 + \left(1 - k_1^2\right) \cdot \frac{\Delta}{2s} \tag{3.227}$$

and effective modulus ε_{re}^{t} given by

$$\varepsilon_{re}^{t} = \varepsilon_{re} - \frac{0.7 \cdot (\varepsilon_{re} - 1) \cdot \frac{t}{s}}{\frac{K(k_1)}{K'(k_1)} + 0.7 \cdot \frac{t}{s}} \tag{3.228}$$

In Equation 3.228, t represents the trace thickness, and s represents the noncorrected gap of the coplanar waveguide (CPW). Substituting the modulus values of k_e and ε_{re}^{t} back into the impedance formula gives the final impedance of the GCPW as

$$Z = \frac{60\pi}{\sqrt{\varepsilon_{re}^{t}}} \cdot \frac{1}{\frac{K(k_e)}{K'(k_e)} + \frac{K(k_3)}{K'(k_3)}} \tag{3.229}$$

Since the impedance of GCPW is obtained and given by Equation 3.229, the next design parameter to consider is the physical length of the transmission line. The physical length of the transmission line is obtained from

$$l = \frac{e_1 \cdot \lambda}{360} \tag{3.230}$$

where

$$\lambda = \frac{v_p}{f} \tag{3.231}$$

$$v_p = \frac{c}{\sqrt{\varepsilon_{re}^t}} \tag{3.232}$$

In Equations 3.230 through 3.232, λ is the wavelength, v_p is the phase velocity, and c is the speed of light. e_l represents the desired electrical length in degrees of the line with respect to the wavelength of the structure. The following MATLAB script can be used to design the GCPW structure for the test fixture.

```
%%%%%%%%%%%%%%%%%%%%%%%%%%%%%%%%%%%%%%%%%%%%%%%%%%%%%%%%%%%%%%%%
 This m-file implements a coplanar waveguide calculator.
 It is used to calculate the dimensions of a coplanar waveguide for
 a given characteristic impedance and frequency. The algorithm does
 an approximation using the elliptic integral in order to do the
 calculations. NOTE all input parameters must be in mils.

  Inputs:
      W - Trace Width
      er - Relative Permittivity of Dielectric
      s - Trace gap
      h - Dielectric Thickness
      f - Target Frequency
      l - Length of Line

  Outputs:
      Zo - Characteristic Impedance of Line
      el - Electrical Length of Line
%%%%%%%%%%%%%%%%%%%%%%%%%%%%%%%%%%%%%%%%%%%%%%%%%%%%%%%%%%%%%%%%
clear
%%%%%%%%%%%%%%%%%%%%%%%%%%%%%%%%%%%%%%%%%%%%%%%%%%%%%%%%%%%%%%%%
 COPLANAR WAVEGUIDE DESIGN PARAMETERS
%%%%%%%%%%%%%%%%%%%%%%%%%%%%%%%%%%%%%%%%%%%%%%%%%%%%%%%%%%%%%%%%
W = 49.25;              % Trace width (mils)
er = 3.38;              % Relative permittivity
s = 10;                 % Trace gap (mils)
h = 32;                 % Dielectric thickness (mils)
t = 1.4;                % Trace thickness (mils)
c = 3e8*1000/0.0254;    % Speed of light (mils/s)
f = 700e6;              % Frequency
l = 2816;               % Length (mils)

% Effects of the trace thickness
dlt = ((1.25*t)/pi)*(1+log((4*pi*W)/(t)));
se = s-dlt;
We = W+dlt;

% Elliptic integral input
k1 = W/(W+2*s);
```

```
% Elliptic integral input
k3num = tanh((pi*W)/(4*h));
k3den = tanh((pi*(W+2*s))/(4*h));
k3   = k3num/k3den;

% Elliptic integral input
ke = We/(We+2*se);
ke = k1+(1-k1^2)*(dlt/(2*s));
%%%%%%%%%%%%%%%%%%%%%%%%%%%%%%%%%%%%%%%%%%%%%%%%%%%%%%%%%%%%%%%%%%%
 ELLIPTIC INTEGRAL APPROXIMATIONS
%%%%%%%%%%%%%%%%%%%%%%%%%%%%%%%%%%%%%%%%%%%%%%%%%%%%%%%%%%%%%%%%%%%

k1u=sqrt(1-k1^2);
if k1>=0 & k1<=(1/sqrt(2))
 kk1=pi/log((2*(1+sqrt(k1u))/(1-sqrt(k1u))));
elseif k1>(1/sqrt(2)) & k1<=1
 kk1=log((2*(1+sqrt(k1)))/(1-sqrt(k1)))/pi;
end;
k3u=sqrt(1-k3^2);
if k3>=0 & k3<=(1/sqrt(2))
 kk3=pi/log((2*(1+sqrt(k3u))/(1-sqrt(k3u))));
elseif k3>(1/sqrt(2)) & k3<=1
 kk3=log((2*(1+sqrt(k3)))/(1-sqrt(k3)))/pi;
end;

keu=sqrt(1-ke^2);
if ke>=0 & ke<=(1/sqrt(2))
 kke=pi/log((2*(1+sqrt(keu))/(1-sqrt(keu))));
elseif ke>(1/sqrt(2)) & ke<=1
 kke=log((2*(1+sqrt(ke)))/(1-sqrt(ke)))/pi;
end;

% Filling factor
q = (kk3)/(kk1+kk3);

%%%%%%%%%%%%%%%%%%%%%%%%%%%%%%%%%%%%%%%%%%%%%%%%%%%%%%%%%%%%%%%%%%%
 EFFECTIVE PERMITTIVITY
%%%%%%%%%%%%%%%%%%%%%%%%%%%%%%%%%%%%%%%%%%%%%%%%%%%%%%%%%%%%%%%%%%%
ere3  = 1+q*(er-1);
ere4  = ere3-((0.7*(ere3-1)*(t/s))/(kk1+0.7*(t/s)));

%%%%%%%%%%%%%%%%%%%%%%%%%%%%%%%%%%%%%%%%%%%%%%%%%%%%%%%%%%%%%%%%%%%
 ELECTRICAL LENGTH (degrees)
%%%%%%%%%%%%%%%%%%%%%%%%%%%%%%%%%%%%%%%%%%%%%%%%%%%%%%%%%%%%%%%%%%%
vp     = c/sqrt(ere4);              % Phase velocity
lambda = vp/f;                      % Lambda
round = floor(l/lambda)*lambda;     % Nearest whole wavelength
l     = l-round;                    % 360 degree constrained
el     = (l/lambda)*(360)           % Electrical length (degrees)

%%%%%%%%%%%%%%%%%%%%%%%%%%%%%%%%%%%%%%%%%%%%%%%%%%%%%%%%%%%%%%%%%%%
 IMPEDANCE OF THE LINE
%%%%%%%%%%%%%%%%%%%%%%%%%%%%%%%%%%%%%%%%%%%%%%%%%%%%%%%%%%%%%%%%%%%
Zo = ((60*pi)/(sqrt(ere4)))*(1/(kke+kk3))
```

Design Example

Design and build a test fixture, and then measure the S parameters of the BFR92 transistor at 700 MHz, and compare the results with published data.

Solution

The design details and measurement results are given next.

3.5.3.1.1 CPW Design

In order to provide an efficient, repeatable solution, the design equations were modeled in the MATLAB code given above. The model functions as a CPW calculator. It allows the user to enter specific parameters and requirements into the model such as width, gap, relative permittivity of the dielectric, trace thickness, frequency, and length. The model then generates the resulting characteristic impedance and electrical length based on the input parameters. The user may continue to tune the input parameters until the characteristic impedance and electrical length meet the defined requirements. The analysis of the transistor is requested to be done using a CPW with a characteristic impedance of 50 Ω at 700 MHz. In order to easily verify the solution, a quarter wave (90°) was targeted as the electrical length of the line. Using the MATLAB model, the values detailed in Table 3.4 were calculated for the width, gap, and length of the CPW.

After using the model to determine the dimensions, Ansoft Designer's TRL calculator was used to verify the calculations. The TRL calculator in Ansoft Designer has more accurate modeling properties when it comes to the board and substrate materials. Plugging in the resulting values from the MATLAB model revealed the values detailed in Table 3.5.

When Tables 3.4 and 3.5 are compared, the MATLAB model results match the Ansoft Designer TRL calculator results.

TABLE 3.4
MATLAB CPW Calculations

MATLAB Model Calculations			
Input Parameters		**Resulting Calculations**	
Width (W)	49.25 mils		
Gap (s)	10 mils		
Length (l)	2816 mils		
Relative permittivity (ε_r)	3.38		
Trace thickness (t)	1.4 mils		
Dielectric thickness (h)	32 mils		
Frequency (f)	700 MHz		
		Impedance (Z_0)	50.0105 Ω
		Electrical length (e_l)	90.0535°

TABLE 3.5
Ansoft Designer CPW Calculations

Ansoft Designer TRL Calculations			
Input Parameters		**Resulting Calculations**	
Width (W)	49.25 mils		
Gap (s)	10 mils		
Length (l)	2816 mils		
Relative permittivity (ε_r)	3.38		
Trace thickness (t)	1.4 mils		
Dielectric thickness (h)	32 mils		
Frequency (f)	700 MHz		
		Impedance (Z_0)	50.1467 Ω
		Electrical length (e_l)	90.5856°

3.5.3.1.2 Quarter-Wave Stub Design

The stubs were to be designed using a parallel coaxial line and attached to both the input and output ports. The stubs are "invisible" at the design frequency of 700 MHz. The target frequency chosen was 100 MHz for the quarter-wave stub design. Ansoft Designer's coaxial line TRL calculator was used to design the quarter-wave 50 Ω coaxial stub lines at 100 MHz. From the calculator, a length of 29,507 mils is calculated for the stub.

3.5.3.1.3 SOLT Test Fixtures

The SOLT test fixtures are designed repeating the method described above; they are implemented as shown in Figures 3.34 through 3.37. Rogers 4003 with relative permittivity of 3.38 is used as a dielectric substrate. The board material has substrate thickness of 0.060 in. with 1-oz. copper plating on both sides. The gap size of 10 mils is used as discussed and illustrated in Figure 3.33. In addition, several biases were

FIGURE 3.34 "Thru" calibration test fixture—top view.

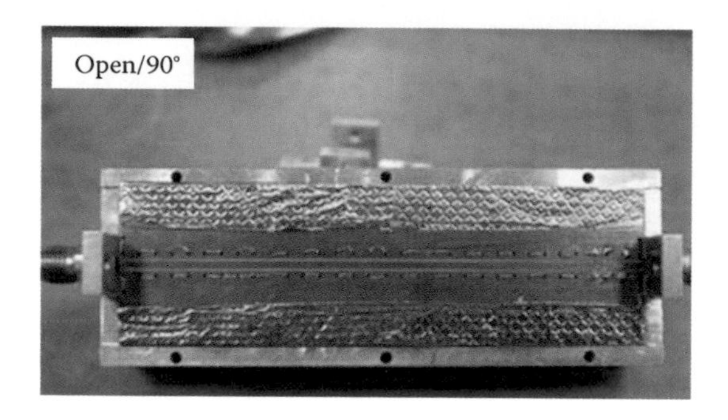

FIGURE 3.35 "Open" calibration test fixture.

FIGURE 3.36 "Load" calibration test fixture.

FIGURE 3.37 "Short" calibration test fixture.

placed through the board in order to provide a good electrical connection between the top and bottom layer ground planes to further improve the grounding of the fixture. Each board layout for the various fixtures incorporated a 0.25-in. section of transmission line at the output of each connector. This consistent length of transmission line is important in order to ensure that the ports of the network analyzer are subjected to the same additional line impedance regardless of the fixture connected. By doing so, the error matrix generated by the analyzer will be appropriate for this

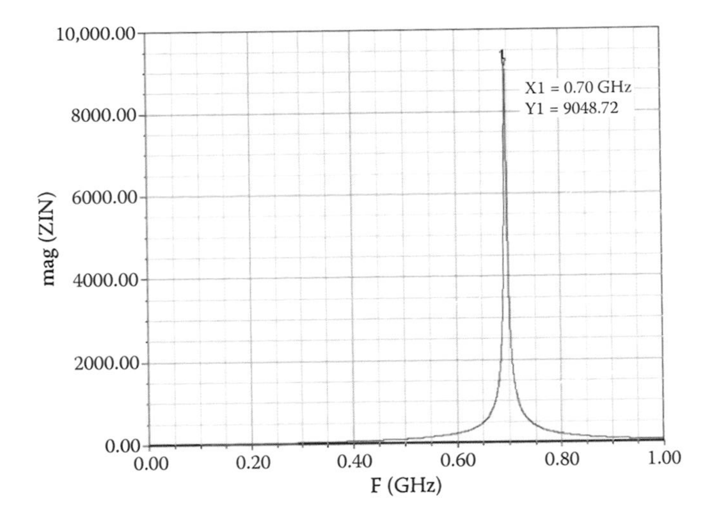

FIGURE 3.42 Quarter-wave short-circuit simulation.

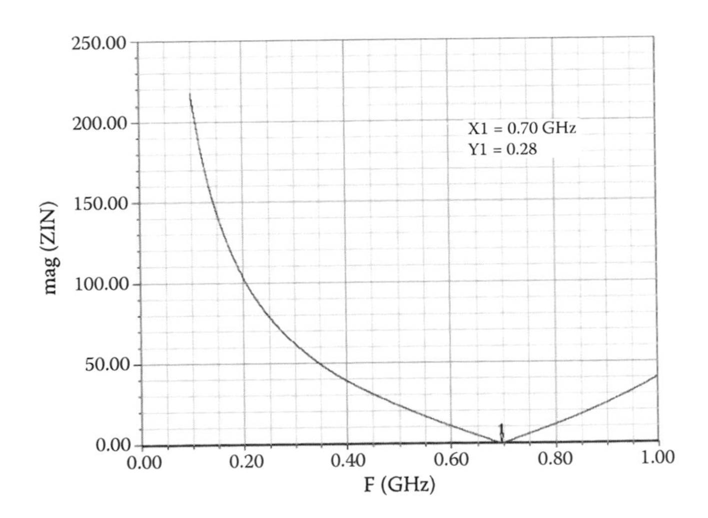

FIGURE 3.43 Quarter-wave open-circuit simulation.

in Ansoft Designer. Before attaching the stubs, the S parameters of the CPW transmission line were captured to serve as a reference when comparing the results with the stubs attached. Figure 3.44 provides the plot of the S parameters with no stubs attached. It is shown in Figure 3.35 that $S_{11} = S_{22}$ and $S_{12} = S_{21}$. This is consistent with what would be expected. There should be no forward (S_{21}) or reverse gain (S_{12}) and very little reflection at both ends (S_{11} and S_{22}) due to the 50 Ω transmission line, which provides a good impedance match to the ports.

Figure 3.45 shows the circuit in Ansoft Designer used for the analysis to verify that short-circuit stub attached to the input port is "invisible" at the design frequency of 700 MHz. Ansoft Designer is also used to do the same analysis, only replacing the

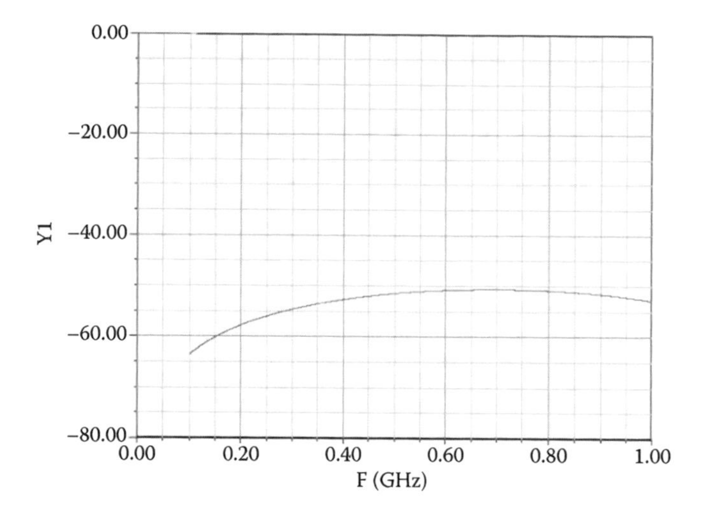

FIGURE 3.44 *S* parameters of CPW transmission line without stubs attached.

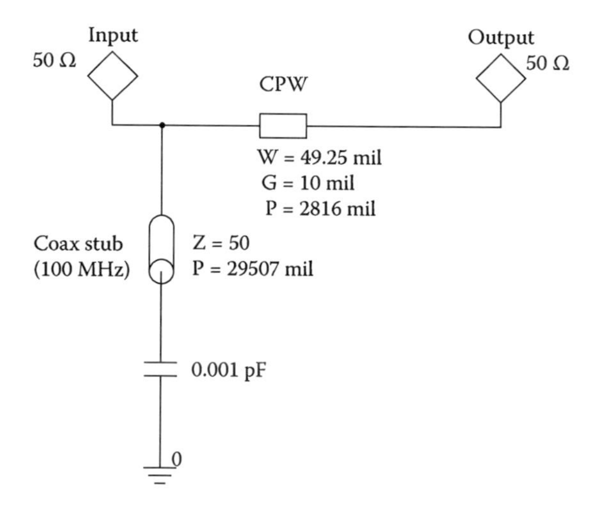

FIGURE 3.45 Quarter-wave stub attached to input port.

short end of the stub by a small capacitor, which results in an open-circuit stub at high frequencies. Simulations were run, and the *S* parameters were captured and plotted. Figures 3.46 and 3.47 provide the plots of the *S* parameters for the short-circuit and capacitor-attached cases, respectively. Note that the capacitor was removed for the short-circuit stub analysis.

From the results in Figures 3.46 and 3.47, it is clear that neither the short- nor open-circuit quarter-wave stubs had any negative effects on the circuit's performance at the design frequency of 700 MHz. In both cases, the forward and reverse gains as well as the reflection coefficients at both ports were unaffected. Ansoft Designer

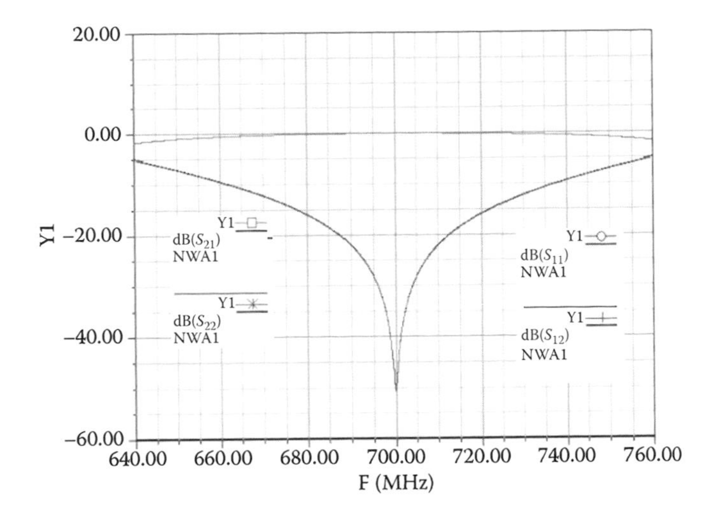

FIGURE 3.46 Quarter-wave short-circuit stub attached at input port.

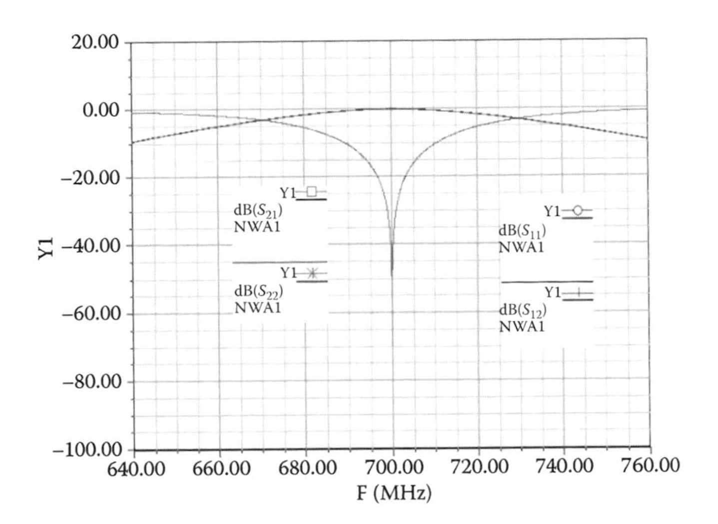

FIGURE 3.47 Quarter-wave stub with capacitor attached at input port.

is also used to attach the same short and open stub to the output port and verify that the stub is again "invisible" at the design frequency of 700 MHz, as shown in Figure 3.48. Simulations were run, and the *S* parameters were captured and plotted. Figures 3.49 and 3.50 provide the plots of the *S* parameters for the short-circuit and capacitor-attached cases, respectively. Note that the capacitor was removed for the short-circuit stub analysis.

From the results in Figures 3.44 and 3.45, it is clear that neither the short- nor open-circuit quarter-wave stubs had any negative effects on the circuit's performance

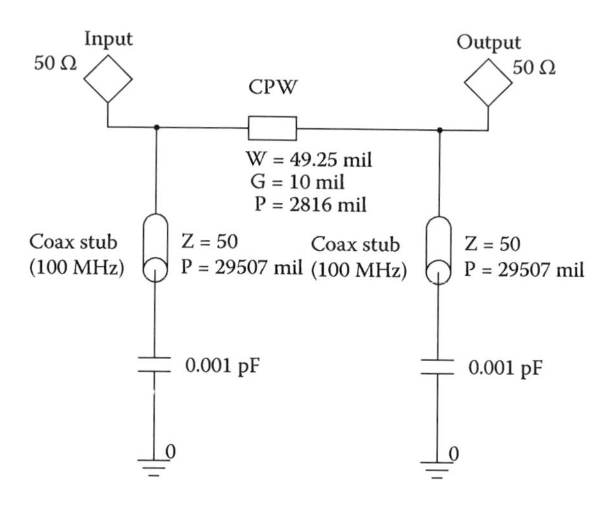

FIGURE 3.48 Quarter-wave stub attached to both input and output ports.

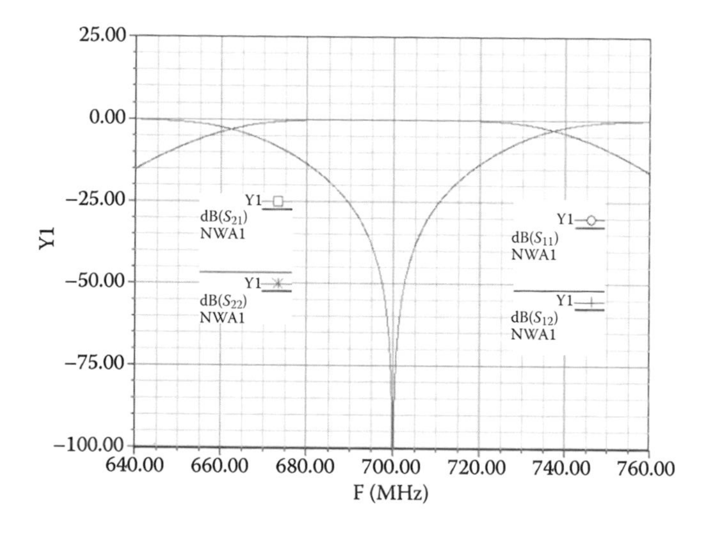

FIGURE 3.49 Quarter-wave short-circuit stub attached to both input and output ports.

at the design frequency of 700 MHz. Both the forward and reverse gains were unaffected in both cases. However, the reflection coefficients in the short-circuit case actually provided better performance (20–30 dB) than the original circuit.

3.5.3.1.5 Measurement Results

S parameter characterization data are obtained with the manufactured calibration set shown in Figure 3.40 and test fixtures shown in Figures 3.34 through 3.37. The BFR92 transistor was measured under the six different bias conditions detailed in Table 3.1. The measured DC gain under each biasing condition is summarized in the

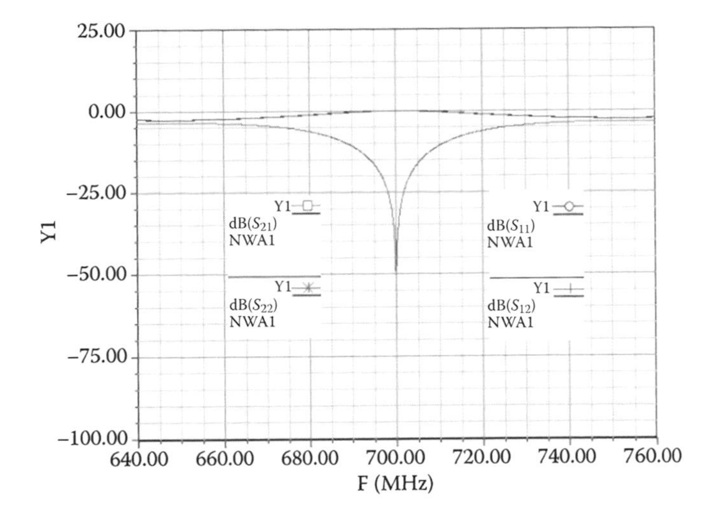

FIGURE 3.50 Quarter-wave stub with capacitor attached at both input and output ports.

TABLE 3.6
DC Bias Conditions

V_{CE}	I_C	V_{BE}	V_{base}	$V_{collector}$	I_B	β
10 V	5 mA	0.769 V	0.823 V	11.04 V	59 μa	85
10 V	10 mA	0.765 V	0.865 V	12.11 V	109 μa	91
10 V	15 mA	0.751 V	0.889 V	13.16 V	150 μa	100
5 V	5 mA	0.785 V	0.845 V	6.05 V	65 μa	77
5 V	10 mA	0.794 V	0.909 V	7.10 V	125 μa	80
5 V	15 mA	0.797 V	0.966 V	8.15 V	184 μa	82

column titled "β" in Table 3.6. The measured *S* parameters were then plotted with the vendor-supplied spice model under the same biasing conditions, as well as the catalog vendor data for the given biases. These results are detailed in Figures 3.51 through 3.54 when $V_{CE} = 10[\text{V}]$ and $I_c = 5[\text{mA}]$. As detailed in the *S* parameter plots, the data measured using the BFR92 test fixture aligned themselves closely with both the vendor-supplied spice model and the vendor catalog data. The comparison of the measured and simulated data has also been done for all other conditions shown in Table 3.6. The agreement again was seen on all of them.

3.6 CHAIN SCATTERING PARAMETERS

When amplifier networks are cascaded, as shown in Figure 3.55, and the relation between the incident and reflected waves using scattering parameters is requested to be established, mathematically, it is more efficient to use a direct matrix multiplication

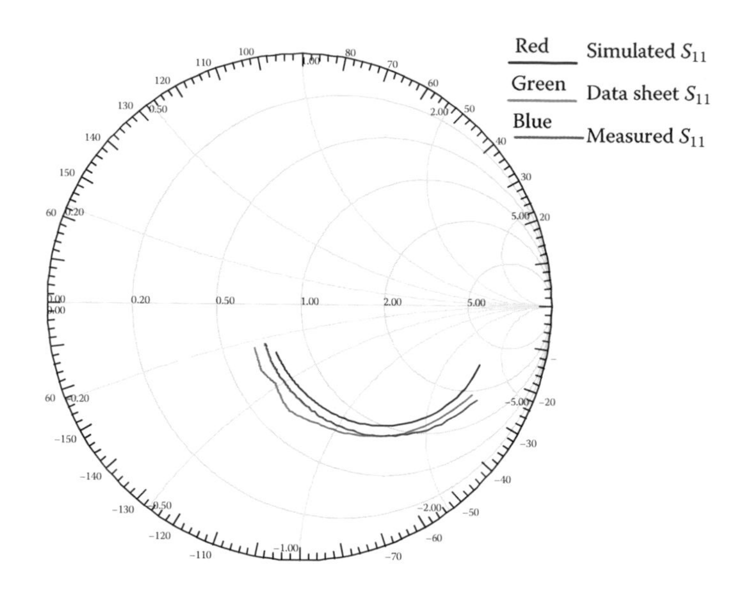

FIGURE 3.51 Input return loss comparison (S_{11}); V_{CE} = 10 V, I_C = 5 mA.

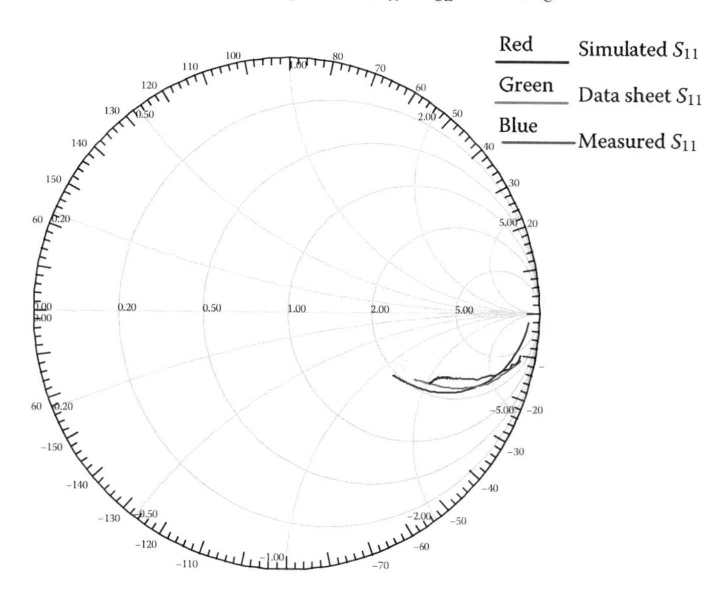

FIGURE 3.52 Output return loss comparison (S_{22}); V_{CE} = 10 V, I_C = 5 mA.

similar to *ABCD* matrices. The chain scattering matrix, *T*, is introduced to fulfill this requirement. The chain scattering matrix, *T*, can be expressed in terms of incident and reflected waves as

$$\begin{bmatrix} a_1 \\ b_1 \end{bmatrix} = \begin{bmatrix} T_{11} & T_{12} \\ T_{21} & T_{22} \end{bmatrix} \begin{bmatrix} b_2 \\ a_2 \end{bmatrix} \tag{3.233}$$

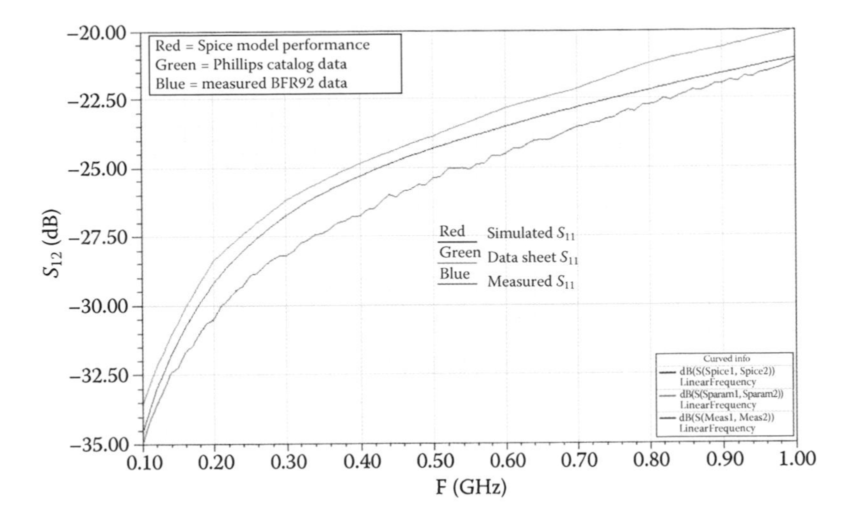

FIGURE 3.53 Reverse isolation comparison (S_{12}); $V_{CE} = 10$ V, $I_C = 5$ mA.

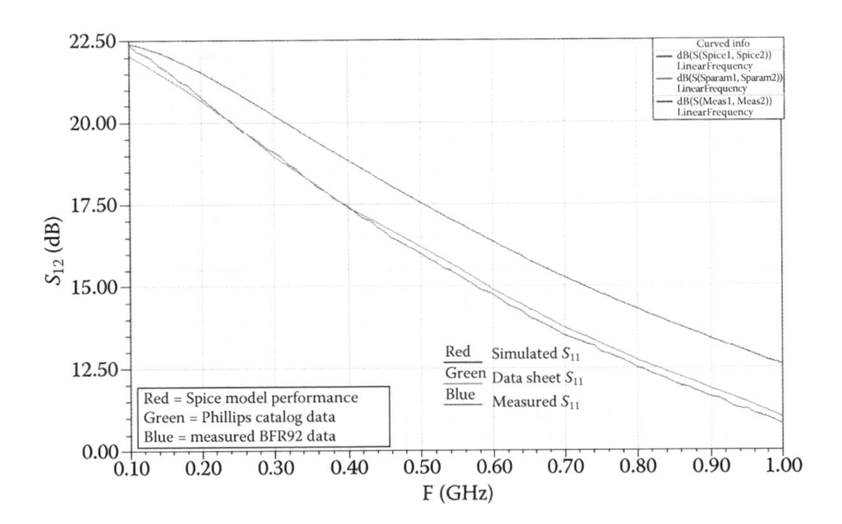

FIGURE 3.54 Forward gain comparison (S_{21}); $V_{CE} = 10$ V, $I_C = 5$ mA.

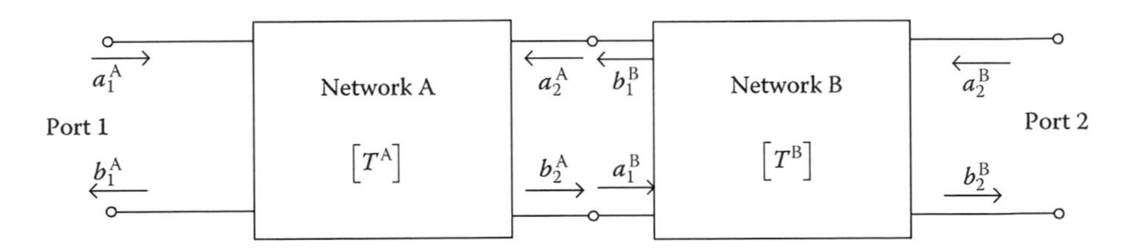

FIGURE 3.55 Illustration of chain scattering matrix for cascaded networks.

Hence, the chain scattering matrices for networks A and B can be written as

$$\begin{bmatrix} a_1^A \\ b_1^A \end{bmatrix} = \begin{bmatrix} T_{11}^A & T_{12}^A \\ T_{21}^A & T_{22}^A \end{bmatrix} \begin{bmatrix} b_2^A \\ a_2^A \end{bmatrix} \tag{3.234}$$

$$\begin{bmatrix} a_1^B \\ b_1^B \end{bmatrix} = \begin{bmatrix} T_{11}^B & T_{12}^B \\ T_{21}^B & T_{22}^B \end{bmatrix} \begin{bmatrix} b_2^B \\ a_2^B \end{bmatrix} \tag{3.235}$$

It is also seen from Figure 3.55 that

$$\begin{bmatrix} a_2^A \\ b_2^A \end{bmatrix} = \begin{bmatrix} b_1^B \\ a_1^B \end{bmatrix} \tag{3.236}$$

Then, using Equation 3.236, we can turn Equation 3.235 into Equation 3.234 and obtain

$$\begin{bmatrix} a_1^A \\ b_1^A \end{bmatrix} = \begin{bmatrix} T_{11}^A & T_{12}^A \\ T_{21}^A & T_{22}^A \end{bmatrix} \begin{bmatrix} T_{11}^B & T_{12}^B \\ T_{21}^B & T_{22}^B \end{bmatrix} \begin{bmatrix} b_2^B \\ a_2^B \end{bmatrix} \tag{3.237}$$

or

$$\begin{bmatrix} a_1^A \\ b_1^A \end{bmatrix} = [T] \begin{bmatrix} b_2^B \\ a_2^B \end{bmatrix} \tag{3.238}$$

where

$$[T] = [T^A][T^B] \tag{3.239}$$

and

$$[T^A] = \begin{bmatrix} T_{11}^A & T_{12}^A \\ T_{21}^A & T_{22}^A \end{bmatrix} \tag{3.240}$$

and

$$[T^B] = \begin{bmatrix} T_{11}^B & T_{12}^B \\ T_{21}^B & T_{22}^B \end{bmatrix} \tag{3.241}$$

The chain scattering parameters are found from scattering parameters and are defined by

$$T_{11} = \frac{1}{S_{21}}$$
$$T_{21} = \frac{S_{11}}{S_{21}} \tag{3.242}$$

$$T_{12} = -\frac{S_{22}}{S_{21}}$$
$$T_{22} = \frac{-(S_{11}S_{22} - S_{12}S_{21})}{S_{21}} = -\frac{\Delta S}{S_{21}} \tag{3.243}$$

So,

$$\begin{bmatrix} T_{11} & T_{12} \\ T_{21} & T_{22} \end{bmatrix} = \begin{bmatrix} \dfrac{1}{S_{21}} & -\dfrac{S_{22}}{S_{21}} \\ \dfrac{S_{11}}{S_{21}} & S_{12} - \dfrac{S_{11}S_{22}}{S_{21}} \end{bmatrix} \tag{3.244}$$

3.7 SYSTEMATIZING RF AMPLIFIER DESIGN BY NETWORK ANALYSIS

Network techniques greatly facilitate the analysis of amplifiers using network parameters as illustrated before. This approach can be systematized by following the steps outlined in the following for the given amplifier circuit shown in Figure 3.56.

Step 1. In this step, the transistor network parameters in the amplifier will first be represented in matrix form:

- Record the operational frequency of the amplifier shown in Figure 3.56.
- Represent the transistor scattering parameters, S_T, in matrix form, as shown in Figure 3.57.

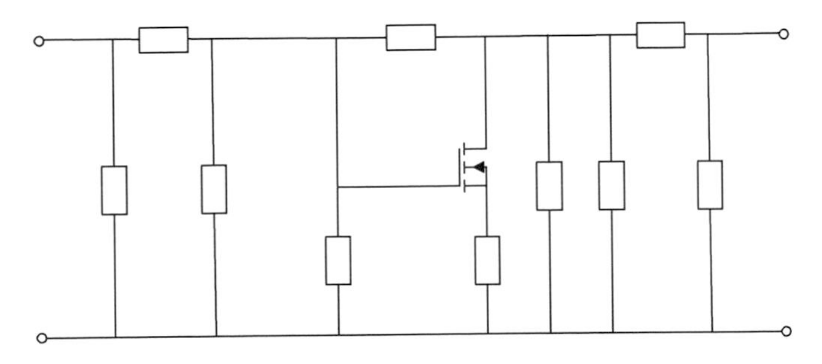

FIGURE 3.56 Transistor amplifier representation as a two-port network.

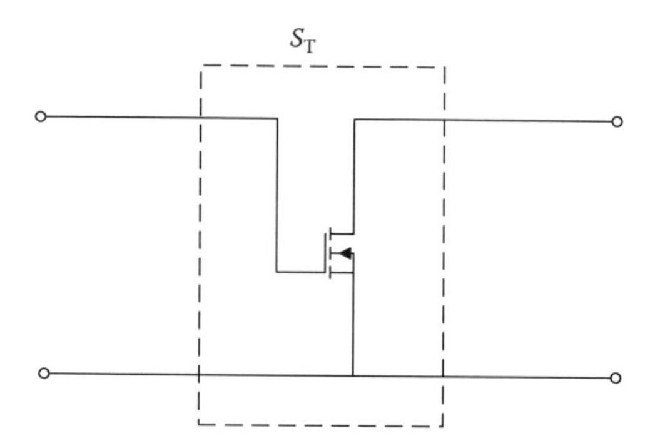

FIGURE 3.57 Transistor scattering parameter representation.

Step 2. In this stage, the network parameters of the shunt feedback network shown in Figure 3.58 will be included. Shunt feedback network and transistor are connected in parallel, as shown in Figure 3.50. Hence, Y parameters will be used to find the network parameters of the parallel-connected two networks. For this reason,

- Convert the transistor scattering parameters S_T into Y parameters, Y_T.
- Represent the shunt feedback network in Y parameters, Y_{feed}.
- Add the Y parameters of these two networks as

$$Y_{t1} = Y_T + Y_{feed} \tag{3.245}$$

Step 3. In this stage, the source feedback network shown in Figure 3.59 will be included to the network given in Figure 3.60. The source feedback network, Z_S, is connected in series to the network given as shown in Figure 3.61. To find the network parameters of the overall network that consists of two parallel-connected network and one series-connected network, we proceed as follows:

- Convert the Y parameter from Step 2, Y_{t1}, into Z parameters.
- Represent the series feedback network, source impedance in Z parameters, Z_s.
- Add the Z parameters of the networks as

$$Z_{tot1} = Z_{t1} + Z_{feed} \tag{3.246}$$

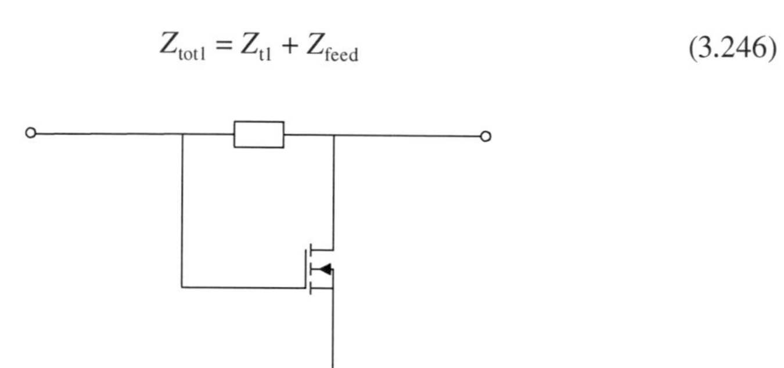

FIGURE 3.58 Transistor with shunt feedback network.

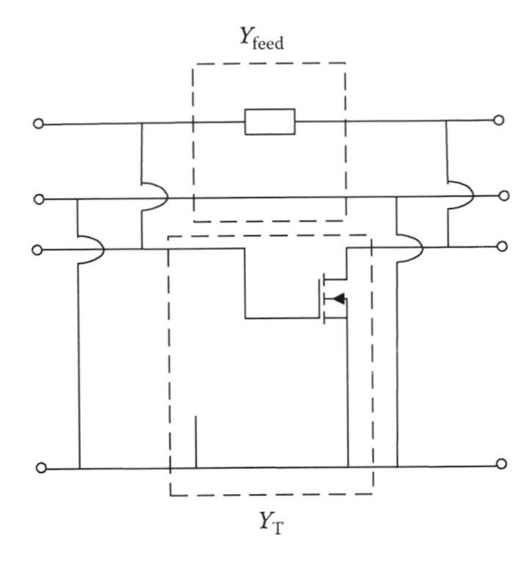

FIGURE 3.59 Illustration of amplifier with source impedance.

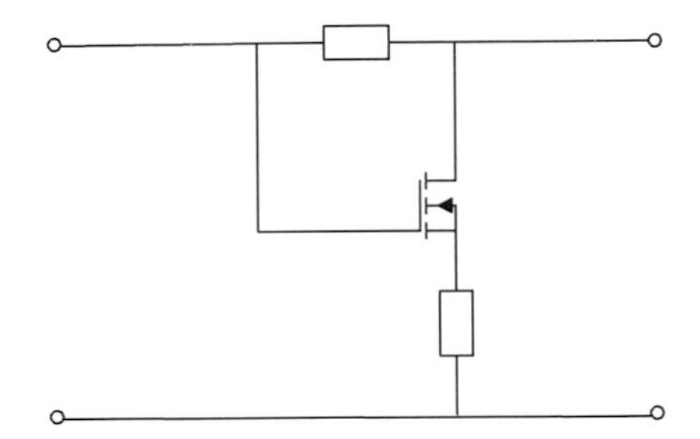

FIGURE 3.60 Parallel network connection of transistor with shunt feedback network.

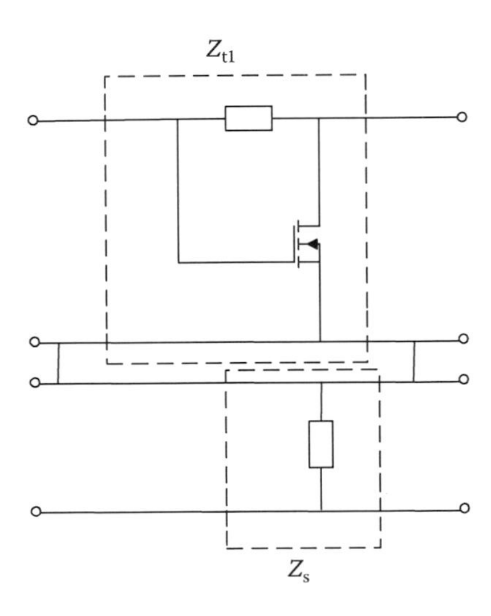

FIGURE 3.61 Network connection with source impedance.

Step 4. In this stage, the gate and drain bias elements will be added to the transistor amplifier consisting of source and feedback networks, as shown in Figure 3.62. These two bias networks are connected in cascade to the network shown in Figure 3.60, as illustrated in Figure 3.63.

To find the network parameters of the overall network that consists of three cascaded networks, we need to

- Convert the Z parameters from Step 3, Z_{totl}, into $ABCD$ parameters, $(ABCD)_{\text{totl}}$.
- Represent the gate and drain bias networks in $ABCD$ matrix forms, $(ABCD)_{\text{g}}$ and $(ABCD)_{\text{d}}$.
- Multiply the $ABCD$ matrices of the three networks to find the $ABCD$ matrix of the equivalent network.

$$(ABCD)_{\text{Ampl}} = (ABCD)_{\text{g}}(ABCD)_{\text{totl}}(ABCD)_{\text{d}} \tag{3.247}$$

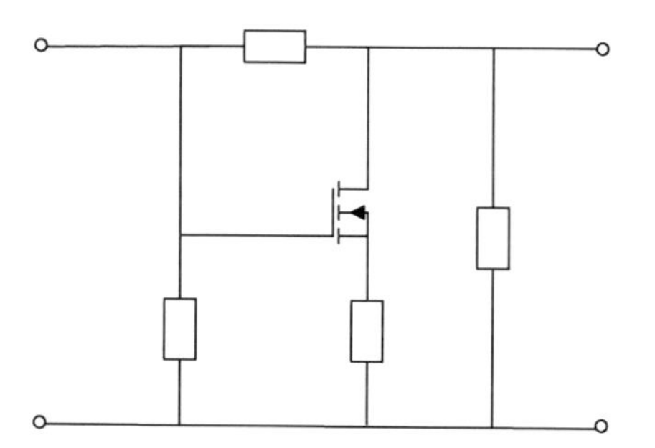

FIGURE 3.62 Transistor amplifier with gate and drain bias networks.

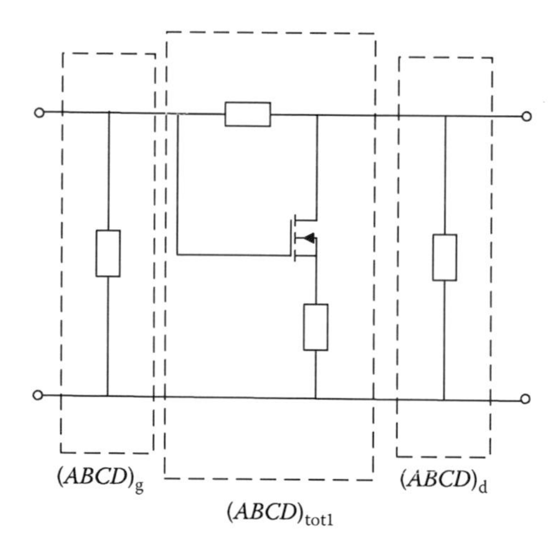

FIGURE 3.63 Network representation of transistor amplifier with gate and drain bias networks.

Step 5. The input and output matching networks are now included to the amplifier network shown in Figure 3.62, as illustrated in Figure 3.64.

The input and output matching networks are connected in cascade to the network shown in Step 4, as illustrated in Figure 3.65.

The transistor amplifier network representation with all the networks can be found by

- Representing the input and output matching networks in *ABCD* matrix forms, $(ABCD)_{IM}$ and $(ABCD)_{OM}$.
- Multiplying the *ABCD* matrices of the three networks to find the *ABCD* matrix of the equivalent network.

$$(ABCD)_{AMPtot} = (ABCD)_{IM}(ABCD)_{Ampl}(ABCD)_{OM} \tag{3.248}$$

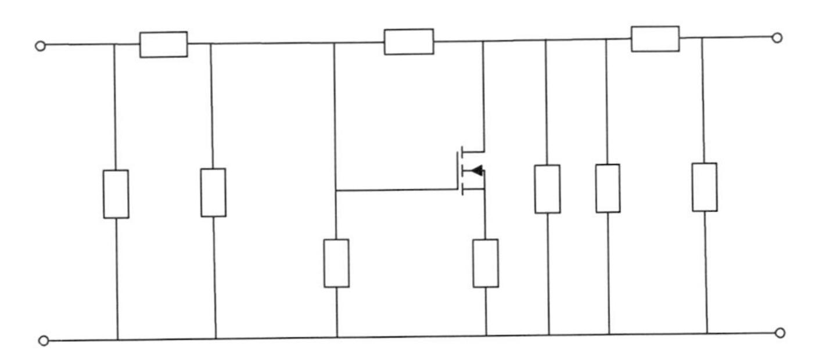

FIGURE 3.64 Transistor amplifier with all the networks including matching networks.

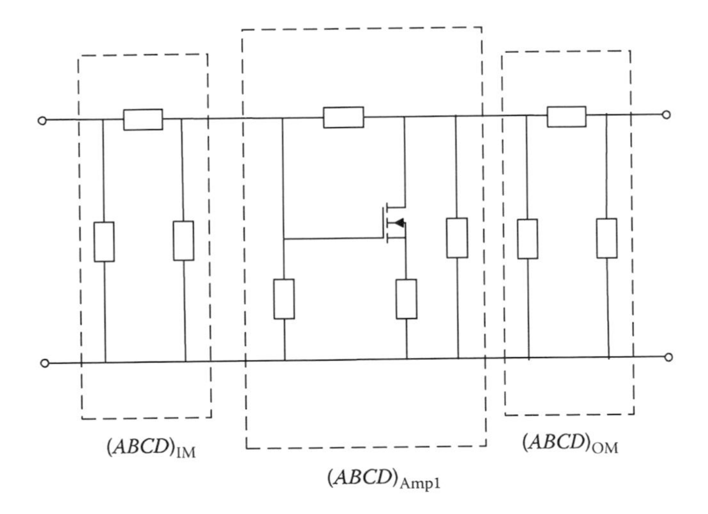

FIGURE 3.65 Network representation of transistor amplifier with all the networks including matching networks.

Step 6. In this final stage, the transistor amplifier response is found by converting the *ABCD* matrix in Step 5, $(ABCD)_{AMPtot}$, to a scattering matrix to obtain the complete amplifier response using the parameters including

- Maximum available gain
- Transducer power gain
- Mismatch losses
- Stability factor
- Unilateral figure of merit

3.8 EXTRACTION OF PARASITICS FOR MOSFET DEVICES

The intrinsic and extrinsic parasitics of MOSFETs are critical to the device and, as a result, amplifier performance because extrinsic effects such as parasitic capacitances and inductances cannot be ignored at high frequencies. Hence, the extraction of these parameters is important so that they can be included in the design of RF amplifiers. An example of a widely used RF power transistor package, TO-247, is illustrated in Figure 3.66. The packaged device is treated as a two-port RF network with a peripheral circuit of lumped parasitic elements.

The simplified model showing extrinsic and intrinsic parameters of this device is shown in Figure 3.67. The illustration of the MOSFET equivalent with intrinsic parameters only is given in Figure 3.68. The complete model with package and intrinsic parasitics for MOSFETs is illustrated in Figure 3.69. It can be shown that under zero bias conditions, $V_{GS} = V_{DS} = 0$, the device amplifier properties, $g_m = 0$, $R_{ds} = \infty$, and inductance effects can then be ignored.

Hence, at zero bias, the network-measured *S* parameters are dominated by capacitances, and hence, the small-signal two-port network reduces to the circuit in Figure 3.70. The package parasitic resistances can be found by converting *S* parameters to *Z* parameters. The real party of the *Z* parameters is equal to the package inductance resistance values as given by

$$\text{Re}\{Z_{11}\} = R_g + R_s \tag{3.249}$$

$$\text{Re}\{Z_{22}\} = R_d + R_s \tag{3.250}$$

$$\text{Re}\{Z_{21}\} = \text{Re}\{Z_{21}\} = R_s \tag{3.251}$$

FIGURE 3.66 MOSFET TO-247 transistor package.

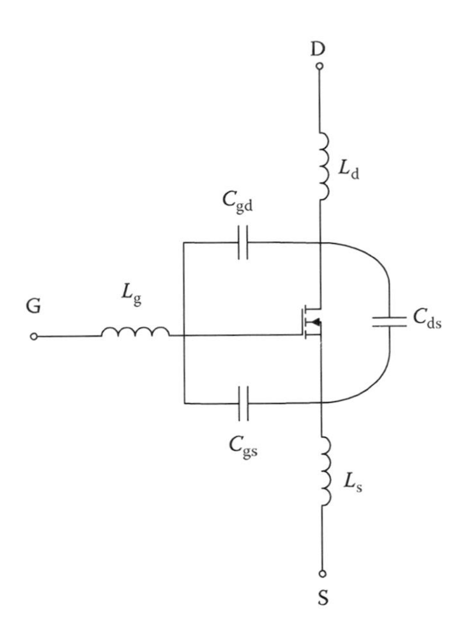

FIGURE 3.67 Illustration of extrinsic and intrinsic parameters of TO-247 transistor package.

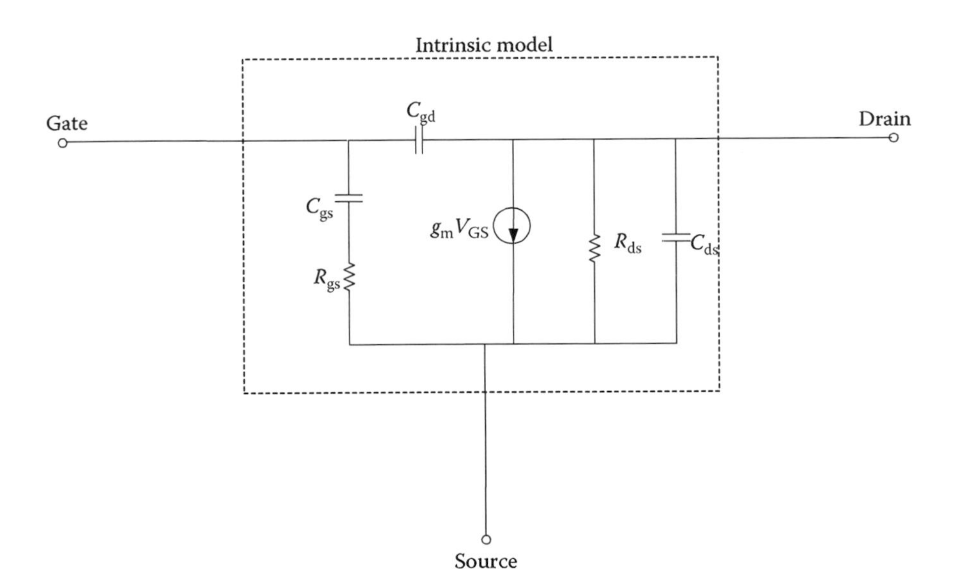

FIGURE 3.68 MOSFET package parasitics with intrinsic parameters.

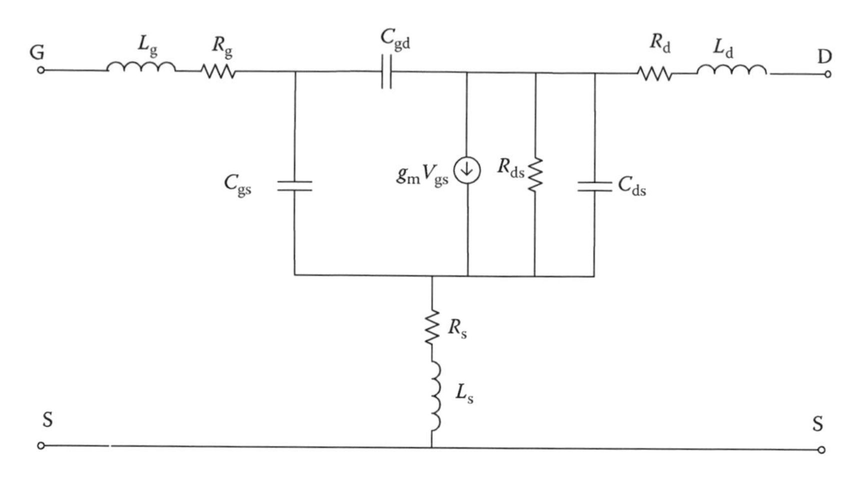

FIGURE 3.69 Small-signal two-port network representation of MOSFETs with parasitics.

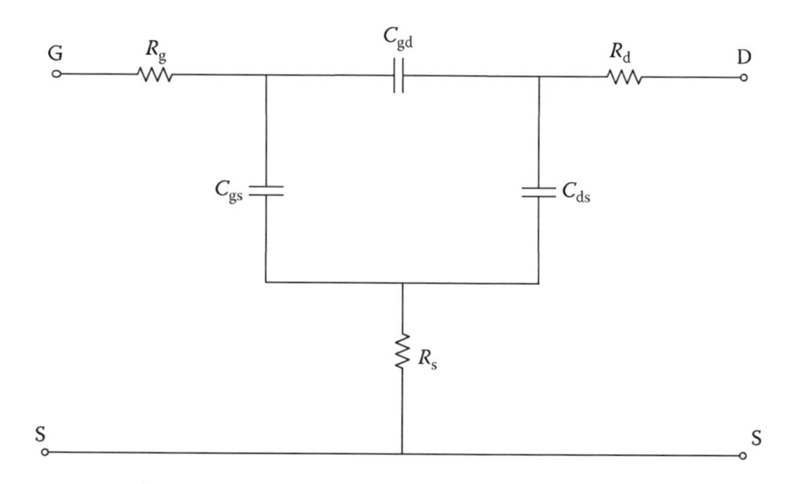

FIGURE 3.70 Zero-biased, small-signal two-port network at low frequencies.

Frequency change does not affect the parasitic resistance values obtained in Equations 3.249 through 3.257. Z parameters, Z^{DUT}, are the parameters measured with zero bias as described before. After parasitic resistances are identified using Equations 3.249 through 3.251, the device-intrinsic parameter Z^i is found from

$$Z_{11}^{i} = Z_{11}^{DUT} - (R_g + R_s) \tag{3.252}$$

$$Z_{22}^{i} = Z_{22}^{DUT} - (R_d + R_s) \tag{3.253}$$

$$Z_{12}^{i} = Z_{12}^{DUT} - R_s \tag{3.254}$$

$$Z_{21}^{i} = Z_{21}^{DUT} - R_s \tag{3.255}$$

The Y parameters of the intrinsic MOSFET components, Y^{i}, can be obtained from

$$Y^{\mathrm{i}} = \frac{1}{Z^{\mathrm{i}}} \tag{3.256}$$

or using the conversion parameters given previously. Y parameters can be obtained as

$$Y_{11}^{\mathrm{i}} = j\omega(C_{\mathrm{gs}} + C_{\mathrm{gd}}) \tag{3.257}$$

$$Y_{22}^{\mathrm{i}} = j\omega(C_{\mathrm{gd}} + C_{\mathrm{ds}}) + \frac{1}{R_{\mathrm{ds}}} \tag{3.258}$$

$$Y_{12}^{\mathrm{i}} = -j\omega(C_{\mathrm{gd}}) \tag{3.259}$$

$$Y_{21}^{\mathrm{i}} = g_{\mathrm{m}} - j\omega C_{\mathrm{gd}} \tag{3.260}$$

Hence, the MOSFET intrinsic parameters are

$$C_{\mathrm{gd}} = \frac{-\mathrm{Im}\left(Y_{12}^{\mathrm{i}}\right)}{2\pi f} \tag{3.261}$$

$$C_{\mathrm{gs}} = \frac{\mathrm{Im}\left(Y_{11}^{\mathrm{i}}\right) + \mathrm{Im}\left(Y_{12}^{\mathrm{i}}\right)}{2\pi f} \tag{3.262}$$

$$C_{\mathrm{ds}} = \frac{\mathrm{Im}\left(Y_{22}^{\mathrm{i}}\right) + \mathrm{Im}\left(Y_{12}^{\mathrm{i}}\right)}{2\pi f} \tag{3.263}$$

$$R_{\mathrm{ds}} = \frac{1}{\mathrm{Re}\left(Y_{22}^{\mathrm{i}}\right)} \tag{3.264}$$

$$g_{\mathrm{m}} e^{-j\omega\tau} = Y_{21}^{\mathrm{i}} - Y_{12}^{\mathrm{i}} \tag{3.265}$$

where

$$g_{\mathrm{m}} = \left|Y_{21}^{\mathrm{i}} - Y_{12}^{\mathrm{i}}\right| \tag{3.266}$$

$$\tau = \frac{\tan^{-1}\left(\dfrac{\mathrm{Im}\left\{Y_{21}^{\mathrm{i}} - Y_{12}^{\mathrm{i}}\right\}}{\mathrm{Re}\left\{Y_{21}^{\mathrm{i}} - Y_{12}^{\mathrm{i}}\right\}}\right)}{2\pi f} \tag{3.267}$$

Using Equations 3.257 through 3.260, Z^{i} parameters can be obtained as

$$Z_{11}^{\mathrm{i}} = \frac{Y_{22}^{\mathrm{i}}}{|Y^{\mathrm{i}}|} = \frac{R_{\mathrm{ds}} + j\omega(C_{\mathrm{gd}} + C_{\mathrm{ds}})}{Y_{11}^{\mathrm{i}}Y_{22}^{\mathrm{i}} - Y_{12}^{\mathrm{i}}Y_{21}^{\mathrm{i}}} \tag{3.268}$$

$$Z_{12}^{\mathrm{i}} = -\frac{Y_{12}^{\mathrm{i}}}{|Y^{\mathrm{i}}|} = \frac{j\omega C_{\mathrm{gd}}}{Y_{11}^{\mathrm{i}}Y_{22}^{\mathrm{i}} - Y_{12}^{\mathrm{i}}Y_{21}^{\mathrm{i}}} \tag{3.269}$$

$$Z_{21}^{\mathrm{i}} = -\frac{Y_{21}^{\mathrm{i}}}{|Y^{\mathrm{i}}|} = \frac{-g_{\mathrm{m}} + j\omega C_{\mathrm{gd}}}{Y_{11}^{\mathrm{i}}Y_{22}^{\mathrm{i}} - Y_{12}^{\mathrm{i}}Y_{21}^{\mathrm{i}}} \tag{3.270}$$

$$Z_{22}^{\mathrm{i}} = \frac{Y_{11}^{\mathrm{i}}}{|Y^{\mathrm{i}}|} = \frac{j\omega(C_{\mathrm{gd}} + C_{\mathrm{gs}})}{Y_{11}^{\mathrm{i}}Y_{22}^{\mathrm{i}} - Y_{12}^{\mathrm{i}}Y_{21}^{\mathrm{i}}} \tag{3.271}$$

When the intrinsic and extrinsic parameters of the device are combined, we then obtain the Z parameters of the device Z^{DUT} from Equations 3.252 through 3.255 as

$$Z_{11}^{\mathrm{DUT}} = [(R_{\mathrm{g}} + R_{\mathrm{s}}) + j\omega(L_{\mathrm{g}} + L_{\mathrm{s}})] + Z_{11}^{\mathrm{i}} = [(R_{\mathrm{g}} + R_{\mathrm{s}}) + j\omega(L_{\mathrm{g}} + L_{\mathrm{s}})] + \frac{g_{\mathrm{ds}} + j\omega(C_{\mathrm{gd}} + C_{\mathrm{ds}})}{Y_{11}^{\mathrm{i}}Y_{22}^{\mathrm{i}} - Y_{12}^{\mathrm{i}}Y_{21}^{\mathrm{i}}} \tag{3.272}$$

$$Z_{12}^{\mathrm{DUT}} = [(R_{\mathrm{s}}) + j\omega(L_{\mathrm{s}})] + Z_{12}^{\mathrm{i}} = [R_{\mathrm{s}} + j\omega L_{\mathrm{s}}] + \frac{j\omega C_{\mathrm{gd}}}{Y_{11}^{\mathrm{i}}Y_{22}^{\mathrm{i}} - Y_{12}^{\mathrm{i}}Y_{21}^{\mathrm{i}}} \tag{3.273}$$

$$Z_{21}^{\mathrm{DUT}} = [(R_{\mathrm{s}}) + j\omega(L_{\mathrm{s}})] + Z_{21}^{\mathrm{i}} = [R_{\mathrm{s}} + j\omega L_{\mathrm{s}}] + \frac{-g_{\mathrm{m}} + j\omega C_{\mathrm{gd}}}{Y_{11}^{\mathrm{i}}Y_{22}^{\mathrm{i}} - Y_{12}^{\mathrm{i}}Y_{21}^{\mathrm{i}}} \tag{3.274}$$

$$Z_{22}^{\mathrm{DUT}} = [(R_{\mathrm{d}} + R_{\mathrm{s}}) + j\omega(L_{\mathrm{d}} + L_{\mathrm{s}})] + Z_{22}^{\mathrm{i}} = [(R_{\mathrm{d}} + R_{\mathrm{s}}) + j\omega(L_{\mathrm{d}} + L_{\mathrm{s}})] + \frac{j\omega(C_{\mathrm{gd}} + C_{\mathrm{gs}})}{Y_{11}^{\mathrm{i}}Y_{22}^{\mathrm{i}} - Y_{12}^{\mathrm{i}}Y_{21}^{\mathrm{i}}} \tag{3.275}$$

In practice, the device parasitics are measured using a test fixture that interfaces the device with the equipment. The DUT can be characterized accurately by removing the test fixture characteristics from the measured results. VNA is commonly used as the measurement equipment to characterize the RF and microwave components. To characterize the device parasitics, the S parameters for the DUT must first be de-embedded from the total measured S parameters. The input and output sides of the test fixture also have some reactance caused by the coaxial-to-CPW transition.

Design Example

A manufacturer gives the following measured S parameters for the high-power TO-247 MOSFET. It is communicated that

At 3 MHz, low-frequency measurement

$$S_{11} = 0.09 - j0.31; \quad S_{12} = 0.89 + j0.01$$

$$S_{21} = 0.89 + j0.01; \quad S_{22} = 0.02 - j0.34$$

At 300 MHz, high-frequency measurement

$$S_{11} = -0.35 + j0.76; \quad S_{12} = 0.2 + j0.1$$

$$S_{21} = 0.2 + j0.1; \quad S_{22} = -0.4 + j0.85$$

Calculate the extrinsic and intrinsic parameters of this device by ignoring the test fixture effects. Compare your results with the exact given high-power TO-247 MOSFET extrinsic and intrinsic values, which are

- $C_{iss} = C_{gs} + C_{gd} = 2700$ pF, $C_{gs} = 2625$ pF
- $C_{rss} = C_{gd} = 75$ pF
- $C_{oss} = C_{ds} + C_{gd} = 350$ pF, $C_{ds} = 275$ pF
- $L_g = 13$ nH, $L_s = 13$ nH, $L_d = 5$ nH
- $R_s = 0.95\ \Omega$, $R_d = 0.5\ \Omega$, $R_g = 5\ \Omega$

Solution

The zero bias MOSFET model given in Figure 3.70 will be used. The following MATLAB script is written to extract the intrinsic and extrinsic values of the MOSFET using the formulation given by Equations 3.249 through 3.275.

```
% This program extracts extrinsic and intrinsic parameters of MOSFET
% It uses zero bias network with no test fixturing effects.
clear;
Zo=50;
        % Enter low and high frequency S parameters
       fl = input('Enter Low frequency ');
       fh = input('Enter High frequency ');
               wl=2*pi*fl;
       wh=2*pi*fh;
 s11_mes = input('Enter S1_11 (Measured S11 in rectangular for Low
    Frequency: ');
 s12_mes = input('Enter S1_12 (Measured S12 in rectangular for Low
    Frequency: ');
 s21_mes = input('Enter S1_21 (Measured S21 in rectangular for Low
    Frequency: ');
 s22_mes = input('Enter S1_22 (Measured S22 in rectangular for Low
    Frequency: ');
```

```
sc11_mes = input('Enter S1_11 (Measured S11 in rectangular for High
   Frequency: ');
sc12_mes = input('Enter S1_12 (Measured S12 in rectangular for High
   Frequency: ');
sc21_mes = input('Enter S1_21 (Measured S21 in rectangular for High
   Frequency: ');
sc22_mes = input('Enter S1_22 (Measured S22 in rectangular for High
   Frequency: ');

        %Convert Zero Bias S parameters to Z parameters
        [z11,z12,z21,z22]=S2Z(s11_mes,s12_mes,s21_mes,s22_mes,Zo);
        [y11,y12,y21,y22]=Z2Y(z11,z12,z21,z22);

        %Extract Extrinsic Resistances using the formulation
        %Use Equations 3.232 through 3.234
        rd=real(z21)
        rs=real(z22)-rd
        rg=real(z11)-rd

        %Extract Intrinsic Capacitances using the formulation
        %Use Equations 3.240 through 3.243
        cgs=-imag(y12)/wl
        cgd=(imag(y11+y12))/wl
        cds=(imag(y22+y12))/wl

        %Convert High Frequency S parameters to Z parameters
        [z11,z12,z21,z22]=S2Z(sc11_mes,sc12_mes,sc21_mes,
            sc22_mes,Zo);

        %Extract Extrinsic Inductances using the formulation

        Ld=imag(z12)/wh
        Lg=imag(z11)/wh-Ld
        Ls=imag(z22)/wh-Ld

        %Enter Frequency range
        f=[1e6:500000:500e6];
        w=2*pi*f;
        %Using the formulation in the book
        D=-w.^2.*(cgs+cgd).*(cgs+cds)+w.^2.*cgs.^2;
Z11=rg+rd+w.*(Lg+Ld)*1i+1i*w.*(cds+cgs)./D;
Z12=rd+w.*(Ld)*1i+1i*w.*(cgs)./D;
Z21=rd+w.*(Ld)*1i+1i*w.*(cgs)./D;
Z22=rs+rd+w.*(Ls+Ld)*1i+1i*w.*(cgd+cgs)./D;

[s11x,s12x,s21x,s22x]=Z2S(Z11,Z12,Z21,Z22,Zo);

smith_chart(2)

for k=1:999

rd1(k)=abs(s11x(k));
alpha1(k)=angle(s11x(k));
rd2(k)=abs(s12x(k));
alpha2(k)=angle(s12x(k));
rd3(k)=abs(s22x(k));
alpha3(k)=angle(s22x(k));
```

```
hold on
plot(rd1(k)*cos(alpha1(k)),rd1(k)*sin(alpha1(k)),'r','linewidth',10)
hold on
plot(rd2(k)*cos(alpha2(k)),rd2(k)*sin(alpha2(k)),'k','linewidth',5)
hold on
plot(rd3(k)*cos(alpha3(k)),rd3(k)*sin(alpha3(k)),'linewidth',15)

end
```

When the program is run, the extracted parameters are found as

- $C_{iss} = C_{gs} + C_{gd} = 2526.478$ pF, $C_{gs} = 2461.7$ pF
- $C_{rss} = C_{gd} = 64.778$ pF
- $C_{oss} = C_{ds} + C_{gd} = 347.678$ pF, $C_{ds} = 282.9$ pF
- $L_g = 12.555$ nH, $L_s = 12.529$ nH, $L_d = 4.6354$ nH
- $R_s = 0.3502\ \Omega$, $R_d = 0.2736\ \Omega$, $R_g = 5.2779\ \Omega$

The Smith chart plot obtained using the MATLAB script vs. frequency is given in Figure 3.71.

Now, the extracted values of the MOSFET were compared with its exact values given in the question vs. frequency using Ansoft Designer with *S* parameters. The zero bias equivalent circuit of the MOSFET with the exact parameters is simulated with Ansoft Designer and is given in Figure 3.72. The simulated Smith chart plot using the exact values with Ansoft Designer is given in Figure 3.73. As shown, the difference between extracted and exact values is acceptably close with the use of zero bias equivalent circuit with low-frequency and high-frequency *S* parameters.

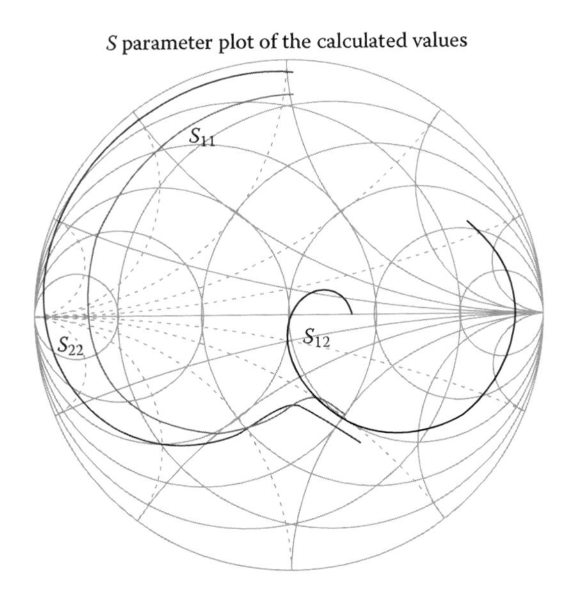

FIGURE 3.71 *S* parameter plot using extracted values of TO-247 MOSFET.

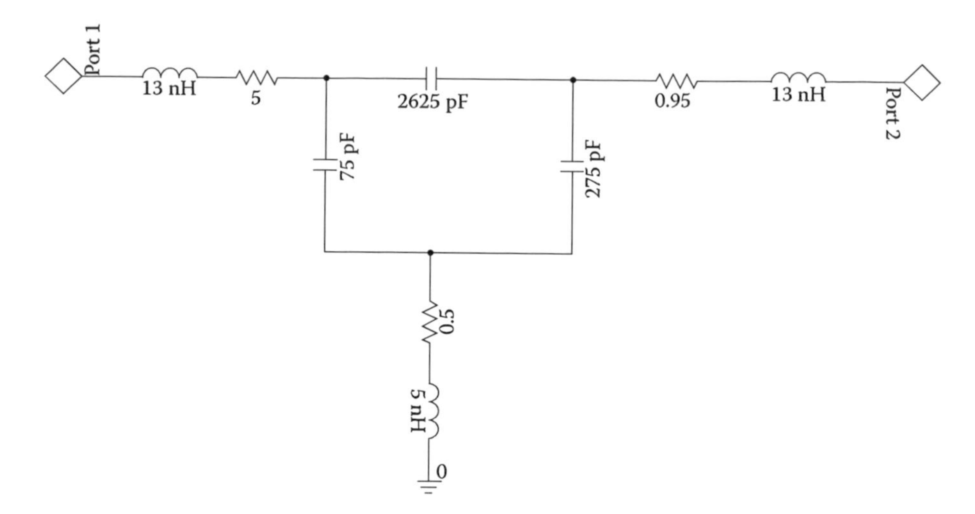

FIGURE 3.72 TO-247 MOSFET zero bias equivalent circuit.

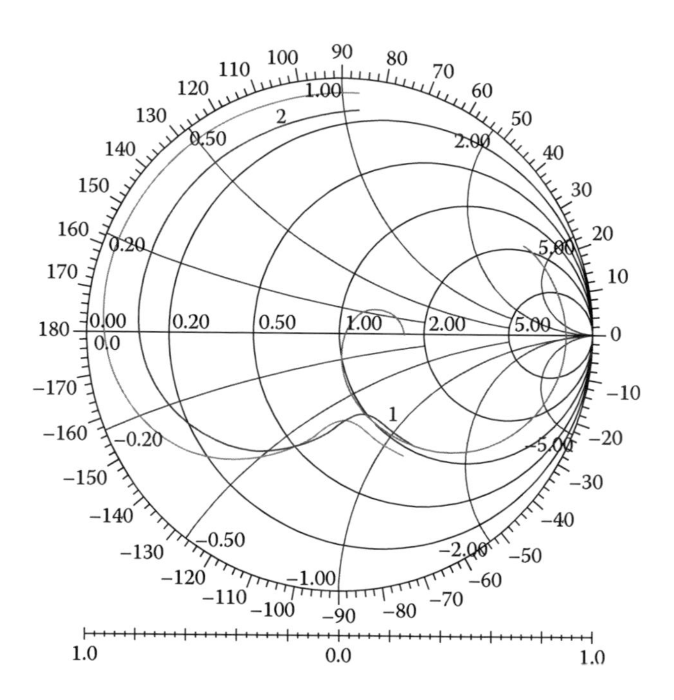

FIGURE 3.73 Simulation of S parameter plot using exact values of TO-247 MOSFET.

3.8.1 De-Embedding Techniques

In order to accurately measure the package parasitics, a method called de-embedding exists to remove the test fixture capacitance and inductance from the component measurements [23]. De-embedding is a mathematical process that removes the effects of unwanted portions of the measurement structure that are embedded in the measured data by subtracting their contributions. This can be shown by the relation

$$[S\text{ parameters}]_{\text{DUT}} = [S\text{ parameters}]_{\text{DUT with fixture}} - [S\text{ parameters}]_{\text{fixture}} \quad (3.276)$$

De-embedding uses a model of the test fixture and mathematically removes the fixture characteristics from the overall measurement. The process of de-embedding a test fixture from the DUT measurement can be performed using chain scattering parameters.

Accurate modeling of the fixture is needed to obtain S parameters of the DUT as described in Section 3.5. Accurate modeling of the test fixture can be obtained from empirical measured data or simulation-based models using Equations 3.249 through 3.275. The typical test fixture shown in Figure 3.74 itself has capacitance and inductance values that will embed themselves into the measurements of the component parasitics.

Signal flow graph can be used to illustrate the test fixture and the DUT as separate two-port networks, as shown in Figure 3.75. The discussion on signal flow graphs is given in Chapter 5. Fixtures A and B in the signal flow (Figure 3.75) show each side of the test fixture where the coaxial to non-coaxial interface exists.

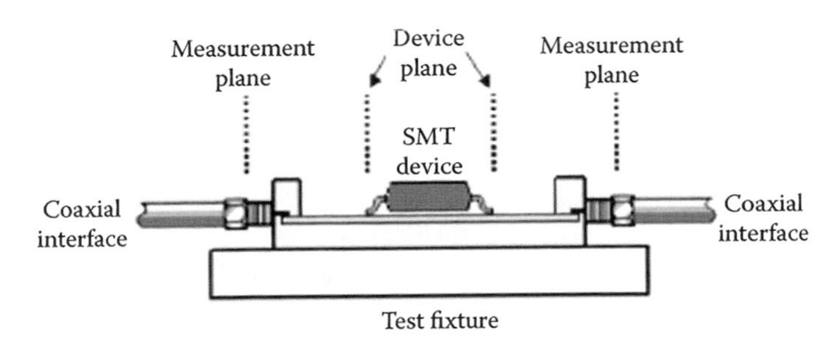

FIGURE 3.74 Test fixture to measure device parasitics.

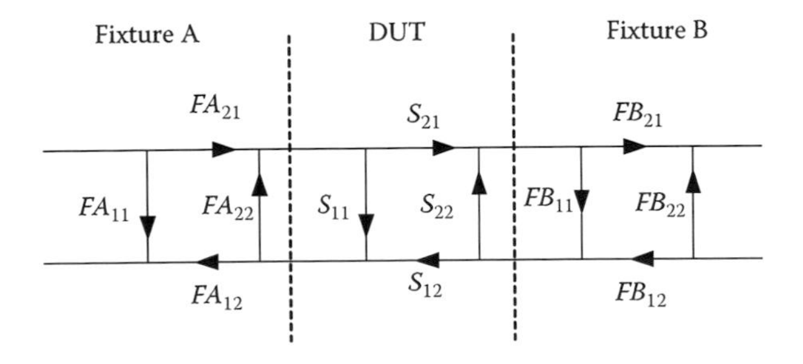

FIGURE 3.75 Signal flow graph representing test fixture and DUT.

The FA_{xy} and FB_{xy} designators represent the S parameters of the test fixture on each side. The effect of the test fixture on the measurement of the device parasitics of the DUT and why de-embedding needs to be performed for it can be better visualized with the illustration given in Figure 3.76.

Typically, the de-embedding process is performed after the measurements have been taken, but often, it is preferable to display the de-embedded measurements on the VNA in real time. This can be done by modifying the error coefficients using the calibration process. A calibration procedure is used to characterize the test fixtures before the measurement of the DUT. The S or T parameter network for each half of the test fixture needs to be modeled before the process of de-embedding of the test fixture parameters can mathematically begin.

Conventional de-embedding is performed using open/short test element group. The implementation of the conventional de-embedding method can be described by considering one side of the test fixtures shown in Figure 3.77.

The analysis of the fixture in Figure 3.77 begins with calculating

$$\Gamma_L = \frac{Z_L - Z_0}{Z_L + Z_0} \tag{3.277}$$

$$b_1 = S_{11}a_1 + S_{12}a_2 = S_{11}a_1 + S_{12}b_2\Gamma_L \tag{3.278}$$

$$b_2 = S_{21}a_1 + S_{22}a_2 = S_{21}a_1 + S_{22}b_2\Gamma_L \tag{3.279}$$

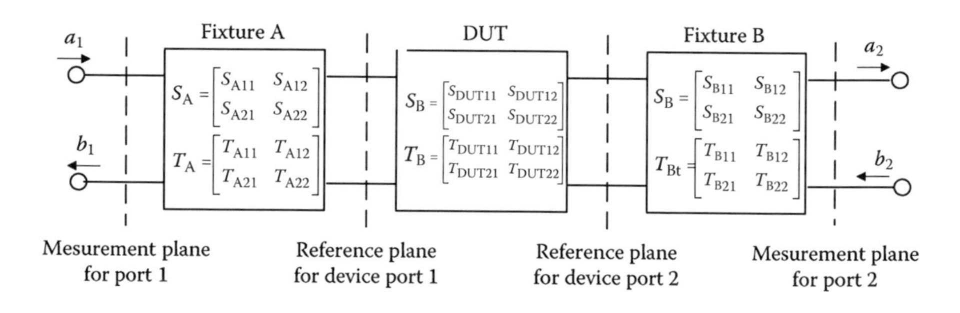

FIGURE 3.76 Effect of test fixture in DUT measurement.

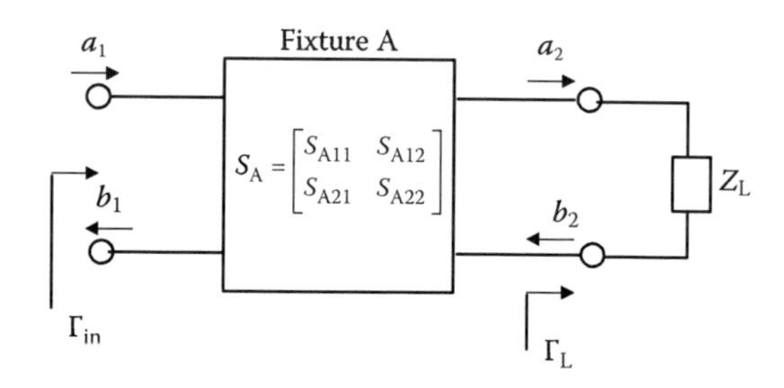

FIGURE 3.77 Fixture characterization.

$$b_2(1 - S_{22}\Gamma_L) = S_{21}a_1 \tag{3.280}$$

$$b_2 = \frac{S_{21}a_1}{(1 - S_{22}\Gamma_L)} \tag{3.281}$$

$$b_1 = S_{11}a_1 + S_{12}\frac{S_{21}a_1\Gamma_L}{(1 - S_{22}\Gamma_L)} \tag{3.282}$$

$$\Gamma_{in} = \frac{b_1}{a_1} = S_{11} + \frac{S_{12}S_{21}\Gamma_L}{1 - S_{22}\Gamma_L} \tag{3.283}$$

Assume that the test fixture is reciprocal, i.e., $Z_{12} = Z_{21}$. Then,

$$\Gamma_{in} = S_{11} + \frac{S_{12}^2\Gamma_L}{1 - S_{22}\Gamma_L} \tag{3.284}$$

To calibrate the test fixture means to find scattering parameters S_{ij} of fixture A. The following three situations are considered:

1. The load is short circuited; $\Gamma_L = -1$

$$\Gamma_{in,s} = S_{11} - \frac{S_{12}^2}{1 + S_{22}} \tag{3.285}$$

2. The load is open circuited; $\Gamma_L = 1$

$$\Gamma_{in,o} = S_{11} + \frac{S_{12}^2}{1 - S_{22}} \tag{3.286}$$

3. The load is matched; $\Gamma_L = 0$

$$\Gamma_{in} = S_{11} \tag{3.287}$$

By solving Equations 3.285 through 3.287, we obtain S_{11}, S_{12}, and S_{22}. We then get the scattering parameters S_{ij} of the test fixture. Working with a cascade of three two-port networks, we convert the S parameters of fixtures A and B to the corresponding T parameters. As a result, the overall measurement transmission matrix is

$$\begin{bmatrix} T_{Am11} & T_{Am12} \\ T_{Am21} & T_{Am22} \end{bmatrix} = \begin{bmatrix} T_{A11} & T_{A12} \\ T_{A21} & T_{A22} \end{bmatrix} \begin{bmatrix} T_{DUT11} & T_{DUT12} \\ T_{DUT21} & T_{DUT22} \end{bmatrix} \begin{bmatrix} T_{Bm11} & T_{Bm12} \\ T_{Bm21} & T_{Bm22} \end{bmatrix} \tag{3.288}$$

The T parameters of the DUT are

$$\begin{bmatrix} T_{\mathrm{DUT11}} & T_{\mathrm{DUT12}} \\ T_{\mathrm{DUT21}} & T_{\mathrm{DUT22}} \end{bmatrix} = \begin{bmatrix} T_{\mathrm{A11}} & T_{\mathrm{A12}} \\ T_{\mathrm{A21}} & T_{\mathrm{A22}} \end{bmatrix}^{-1} \begin{bmatrix} T_{\mathrm{Am11}} & T_{\mathrm{Am12}} \\ T_{\mathrm{Am21}} & T_{\mathrm{Am22}} \end{bmatrix} \begin{bmatrix} T_{\mathrm{Bm11}} & T_{\mathrm{Bm12}} \\ T_{\mathrm{Bm21}} & T_{\mathrm{Bm22}} \end{bmatrix}^{-1} \quad (3.289)$$

Finally, the T parameters of the DUT are converted into S parameters.

As described, the de-embedding process of the test fixture using the conventional method [24] is quite involved. This method suffers at high frequencies, mainly due to incompleteness of open and short pattern and approximation of a parasitic circuit by an equivalent circuit topology.

One other alternative is accurate modeling of the test fixtures using electromagnetic (EM) simulators [25,26]. Ansoft High-Frequency Structure Simulator (HFSS) is a 3D EM simulator tool that can be used to model the test fixture that is used to measure the DUT accurately. The test fixture circuit is characterized by the EM simulator and has no approximation and gives accurate results.

The de-embedding process can be done using two techniques: static approach and real-time approach. The details of these methods will be given next.

3.8.2 De-Embedding Technique with Static Approach

The static approach uses measured data from the VNA, and the de-embedding is performed by processing the data using the T parameter matrix calculations [23]. Once the measurements are de-embedded, the data are displayed statically on a computer screen or can be downloaded into the analyzer's memory for display. The procedure for de-embedding of the S parameters for the DUT using the static de-embedding method can be outlined as follows:

- Simulate the fixture that will be used to measure the scattering parameters of the DUT and obtain an accurate model.
- Obtain the S parameters of the fixture on the input and output sides from simulation. Convert the S parameters to T parameters.
- Calibrate the VNA with a standard coaxial calibration kit. Measure the combined S parameters of the device and fixture.
- Convert the measured S parameters to T parameters.
- Apply the de-embedding equation from

$$[T_{\mathrm{meas}}] = [T_{\mathrm{A}}][T_{\mathrm{DUT}}][T_{\mathrm{A}}] \quad (3.290)$$

from

$$[T_{\mathrm{DUT}}] = [T_{\mathrm{A}}]^{-1}[T_{\mathrm{meas}}][T_{\mathrm{B}}]^{-1} \quad (3.291)$$

Convert the T parameters of the DUT back to S parameters. This represents the S parameters of the device only; test fixture effects have been removed.

3.8.3 De-Embedding Technique with Real-Time Approach

The second method of de-embedding is the real-time approach, which uses the VNA to directly perform the de-embedding calculations allowing the de-embedded response to be viewed in real time. Real-time analysis can be performed using two methods that will allow the de-embedded calculation to be performed directly on a network analyzer. One method accounts for simple corrections for fixture effects by modifying the calibration offsets (offset delay, offset loss, and offset impedance Z_o) to take into account the offsets from the fixture. There is also the method of modifying the 12-term error model. Using the 12-term model allows for better results (given the accuracy of the model) and is what will be used in this effort. Modifying the 12-term error model requires creating a detailed model of the test fixture. The accuracy of this model directly affects the accuracy of the measurements of the DUT. The model is used to generate the *S* parameters of the test fixture on both sides through analysis in an EM design suite such as HFSS. These *S* parameters derived from the test fixture analysis will be combined with the analyzer's error correction values to derive the error values that will be used in real-time de-embedding VNA measurements.

In order to derive the 12-term error model, the VNA must first be calibrated to correct for any measurement error that would be the result of the VNA. This is done during the VNA calibration procedure where the VNA measures the magnitude and phase responses of known devices such as open, short, and load adapters. After the calibration of the VNA, the measurement errors that would have resulted from the VNA have been accounted for (de-embedding the VNA system errors from the measurements).

There are six error terms shown in Figure 3.78 including the forward directivity error term resulting from signal leakage through the directional coupler on port 1 (E_{df}), the forward reflection tracking term resulting from the path differences between the test and reference paths (E_{rf}), the forward source match term resulting from the VNA's test port impedance not being perfectly matched to the source impedance (E_{sf}), as well as the forward transmission error (E_{tf}), the forward load match error (E_{lf}), and the forward crosstalk error (E_{xf}). These errors also exist in the reverse direction, which results in the 12 error terms.

If the forward error model shown in Figure 3.78 was modified to include the test fixture before and after the DUT, it would look like the signal flow graph shown in Figure 3.79. This diagram shows the original calibration terms being cascaded with the *S* parameters from the test fixture.

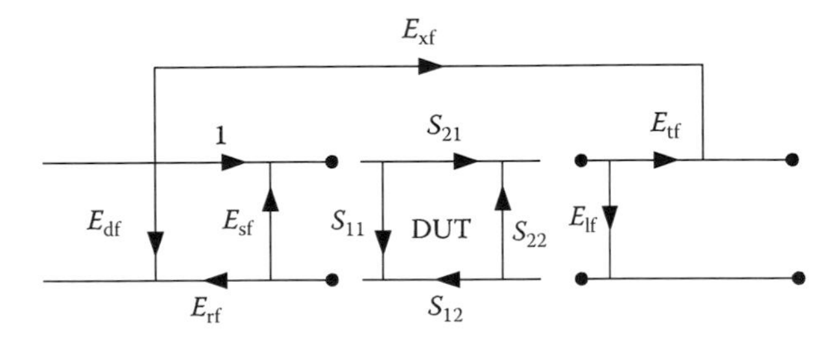

FIGURE 3.78 Forward model for six-error term.

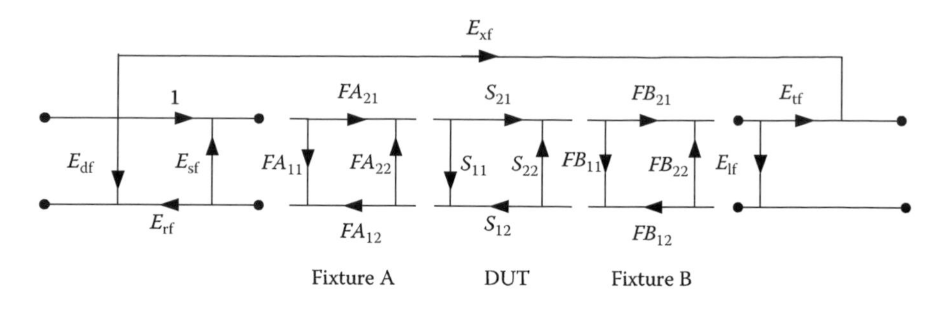

FIGURE 3.79 Forward model for six-error term with test fixture.

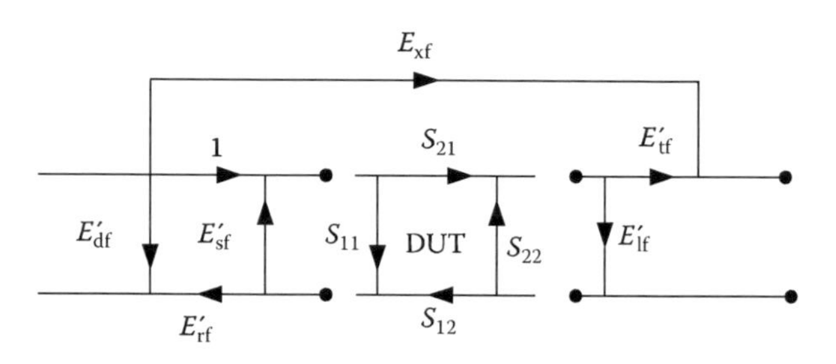

FIGURE 3.80 Modified forward model for six-error term with test fixture.

The cascading of the calibration error terms with the S parameters from the test fixture allows a new signal flow graph to be derived where new error terms exist that include the test fixture S parameters, as shown in Figure 3.80.

Using this new signal flow graph, the error coefficients can be derived and given by Equations 3.292 through 3.302 [23].

$$E'_{df} = E_{df} + \frac{(E_{rf}FA_{11})}{(1 - E_{sf}FA_{11})} \tag{3.292}$$

$$E'_{sf} = FA_{22} + \frac{(E_{sf}FA_{12}FA_{21})}{(1 - E_{sf}FA_{11})} \tag{3.293}$$

$$E'_{rf} = \frac{(E_{rf}FA_{12}FA_{21})}{(1 - E_{sf}FA_{11})^2} \tag{3.294}$$

$$E'_{lf} = FB_{11} + \frac{(E_{lf}FB_{12}FB_{21})}{(1 - E_{lf}FB_{22})} \tag{3.295}$$

$$E'_{tf} = \frac{(E_{tf}FA_{21}FB_{21})}{((1 - E_{lf}FB_{22})(1 - E_{sf}FA_{11}))} \tag{3.296}$$

$$E'_{dr} = E_{dr} + \frac{(E_{rr}FB_{22})}{(1 - E_{sr}FB_{22})} \tag{3.297}$$

$$E'_{sr} = FB_{11} + \frac{(E_{sr}FB_{12}FB_{21})}{(1 - E_{sr}FB_{22})} \tag{3.298}$$

$$E'_{lr} = FA_{22} + \frac{(E_{lr}FA_{12}FA_{21})}{(1 - E_{lr}FA_{11})} \tag{3.299}$$

$$E'_{tr} = \frac{(E_{tr}FA_{12}FB_{12})}{((1 - E_{lr}FA_{11})(1 - E_{sr}FB_{22}))} \tag{3.300}$$

$$E'_{xf} = E_{xf} \tag{3.301}$$

$$E'_{xr} = E_{xr} \tag{3.302}$$

In Equations 3.292 through 3.302,

E_{df} = forward (port 1) directivity
E_{sf} = forward (port 1) source match
E_{rf} = forward (port 1) reflection tracking
E_{xf} = forward (port 1) isolation
E_{lf} = forward (port 1) load match
E_{tf} = forward (port 1) transmission tracking

E_{dr} = reverse (port 2) directivity
E_{sr} = reverse (port 2) source match
E_{rr} = reverse (port 2) reflection tracking
E_{xr} = reverse (port 2) isolation
E_{lr} = reverse (port 2) load match
E_{rr} = reverse (port 2) transmission tracking

Using these derived equations along with the S parameters obtained using modeling via an EM simulator, we can obtain accurate test results from the VNA.

Design Example

The setup including the test fixture shown in Figure 3.81 is used to measure the MOSFET with TO-247 package considered in the previous example. It is communicated that the whole setup with the fixture gives the following measured S parameters:

- At 3 MHz, low-frequency measurement

$$S_{11} = 0.09 - j0.31; \quad S_{12} = 0.89 - j0.01$$

$$S_{21} = 0.89 + j0.01; \quad S_{22} = 0.02 - j0.34$$

- At 300 MHz, high-frequency measurement

$$S_{11} = -0.31 + j0.78; \quad S_{12} = 0.2 + j0.1$$

$$S_{21} = 0.2 + j0.1; \quad S_{22} = -0.35 + j0.87$$

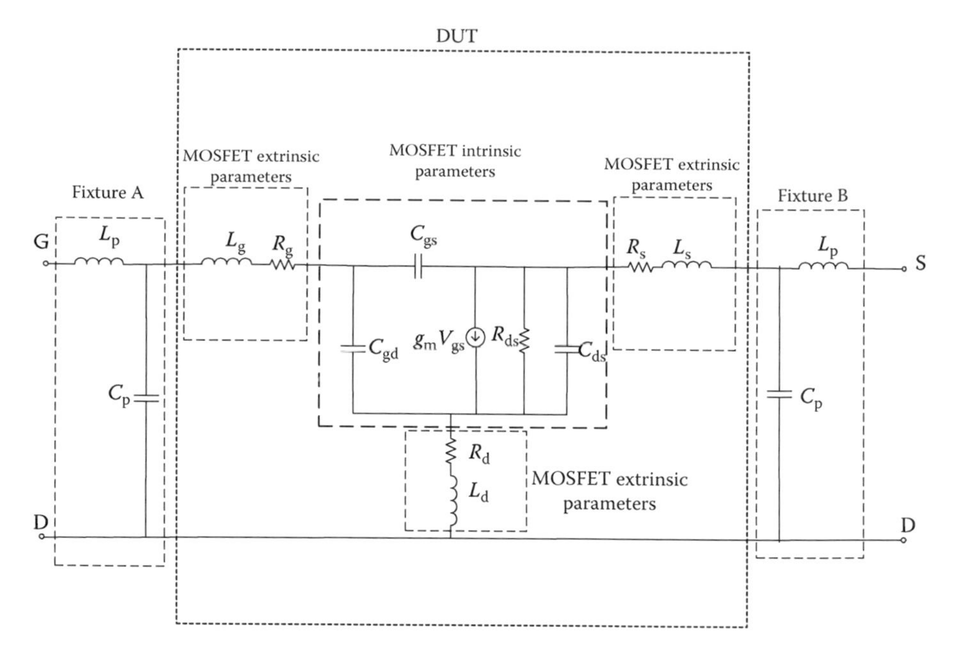

FIGURE 3.81 TO-247 MOSFET package measurement setup.

Calculate the extrinsic and intrinsic parameters of this device by ignoring the test fixture effects. The test fixture is symmetric and can be represented by the LC network, which has $C_p = 0.1$ pF and $L_p = 1$ nH.

Compare again your results with typical high-power TO-247 MOSFET extrinsic and intrinsic values, which are

- $C_{iss} = C_{gs} + C_{gd} = 2700$ pF, $C_{gs} = 2625$ pF
- $C_{rss} = C_{gd} = 75$ pF
- $C_{oss} = C_{ds} + C_{gd} = 350$ pF, $C_{ds} = 275$ pF
- $L_g = 13$ nH, $L_s = 13$ nH, $L_d = 5$ nH
- $R_s = 13$ nH
- $L_d = 5$ nH

Solution

The steps given in the static de-embedding method are followed. The only difference in this example is that the *S* parameters of the fixture are obtained using ideal calculated component values. Then, the de-embedding equation given by Equation 3.291 is used to obtain the measurement values for only the DUT. The MATLAB script calculates the extrinsic and intrinsic parameters of the MOSFET with TO-247 package. When the program is run, the extracted parameters are found as

- $C_{iss} = C_{gs} + C_{gd} = 2526.478$ pF, $C_{gs} = 2465.9$ pF
- $C_{rss} = C_{gd} = 64.682$ pF

- $C_{oss} = C_{ds} + C_{gd} = 347.678$ pF, $C_{ds} = 283.2$ pF
- $L_g = 14.43$ nH, $L_s = 14.49$ nH, $L_d = 4.863$ nH
- $R_s = 0.3516\ \Omega$, $R_d = 0.2717\ \Omega$, $R_g = 5.2762\ \Omega$

The MATLAB plot showing the sweep of S parameters on a Smith chart is shown in Figure 3.82.

The MOSFET with TO-247 package is simulated with Ansoft Designer with the exact values including test fixture, as shown in Figure 3.83. The plot of scattering parameters on a Smith chart in Figure 3.84 shows that the results are in agreement with the extracted results shown in Figure 3.82.

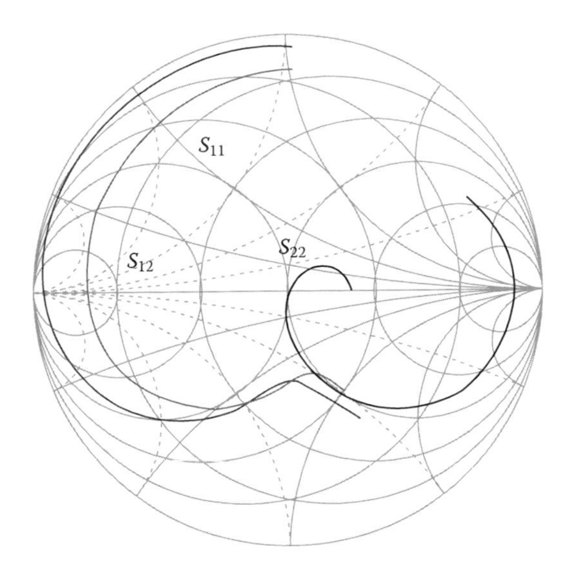

FIGURE 3.82 S parameter plot using extracted values of TO-247 MOSFET with fixture effects.

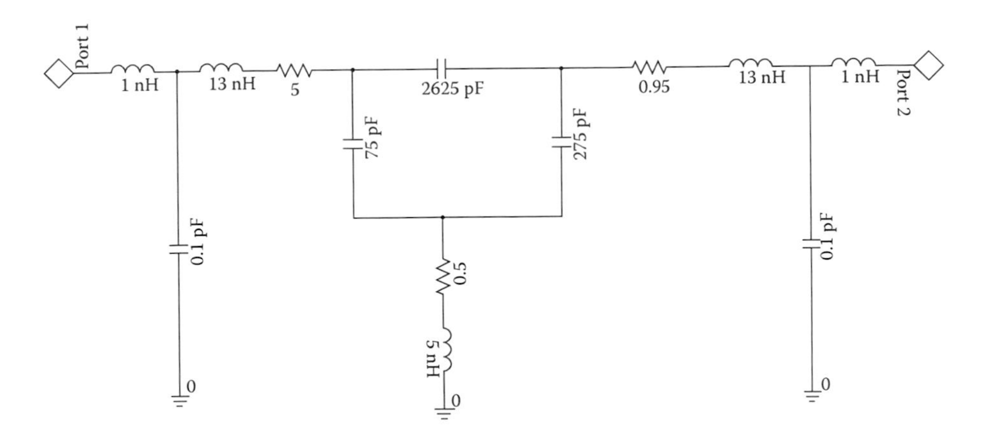

FIGURE 3.83 Simulated TO-247 MOSFET zero bias equivalent circuit with test fixture.

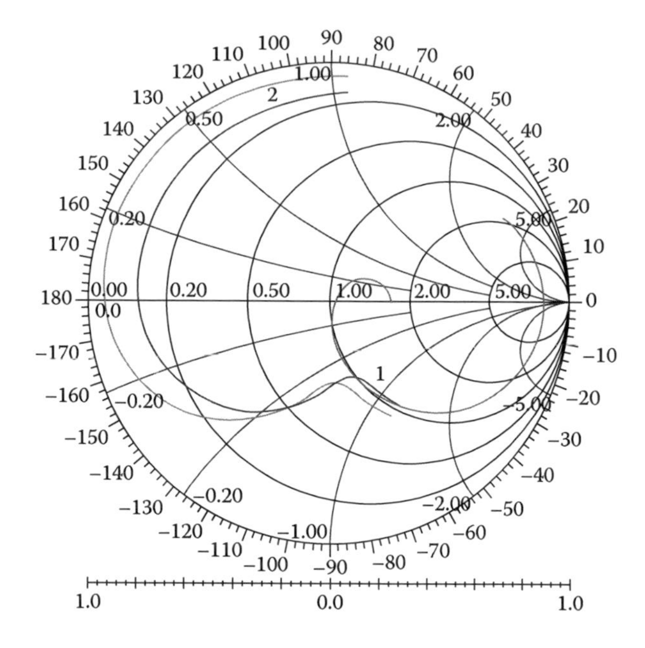

FIGURE 3.84 Simulation of S parameter plot using exact values of TO-247 MOSFET with fixture.

PROBLEMS

1. Obtain the Z and Y parameters of the circuits in Figure 3.85a and b.
2. Find the $ABCD$ and h parameters of the transformer given in Figure 3.86.
3. Consider the AC-coupled amplifier circuit shown in Figure 3.87.

   The amplifier small-signal model is shown in Figure 3.88.

   Amplifier parameters are given as $g_m = 50$ mA/V, $R_s = 2$ kΩ, $R_i = 8$ kΩ, $R_o = 15$ kΩ, and $R_L = 10$ kΩ, $C_1 = 5$ [pF], $C_o = 1$ [pF], $C_1 = 0.01$ [μF], and $C_4 = 0.01$ [μF]m. Use the two-port parameter method to plot and calculate the amplifier gain at 500 kHz.

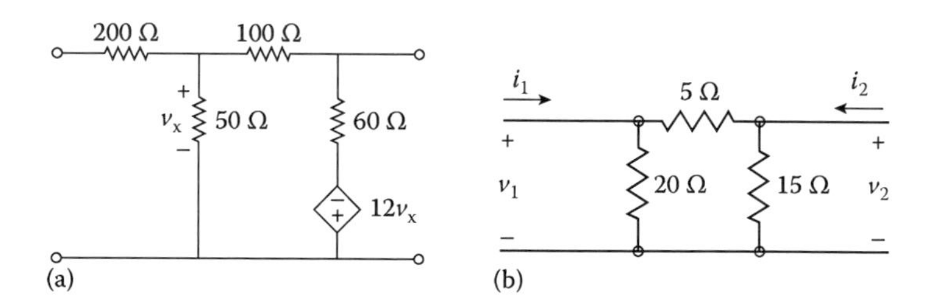

FIGURE 3.85 Z and Y parameters of the network (a) PI network with dependent source and (b) resistive PI network.

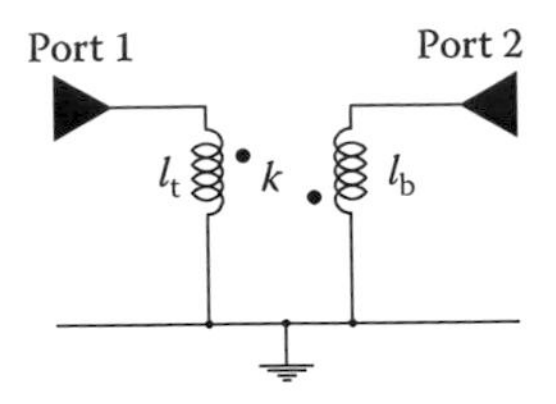

FIGURE 3.86 *ABCD* and *h* parameters of the transformer.

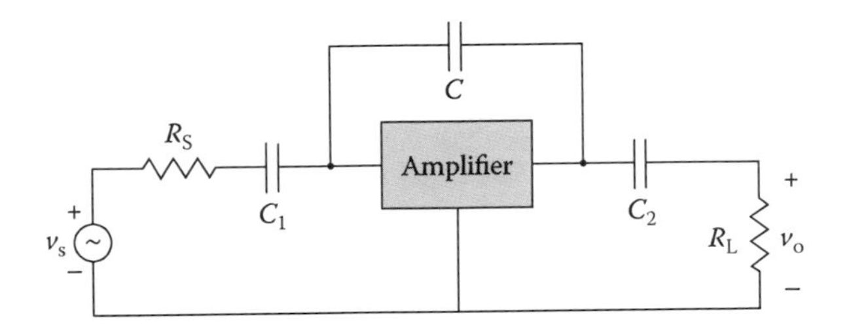

FIGURE 3.87 AC-coupled amplifier.

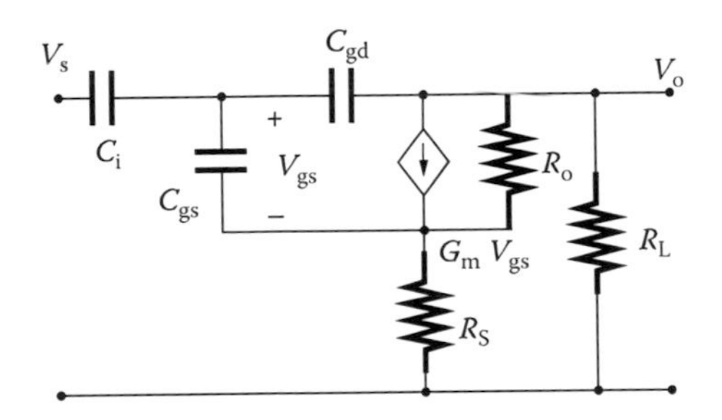

FIGURE 3.88 Small-signal model of an amplifier.

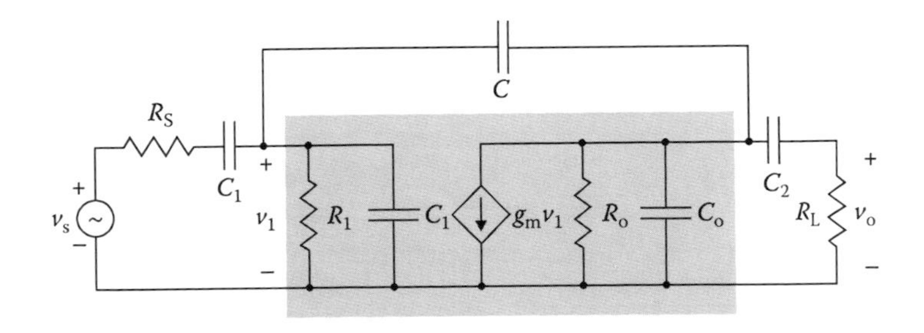

FIGURE 3.89 High-frequency model of an amplifier.

4. Derive and obtain the voltage gain and phase of the voltage gain of the model shown in Figure 3.89 vs. frequency between 10 MHz and 1 GHz when $C_i = 1e - 6$ F, $C_{gs} = 10e - 12$, $C_{gd} = 1e - 12$ F, $C_{ds} = 2e - 12$ F, $G_m = 20e - 3$ S, $R_S = 100\ \Omega$, and $R_L = 70 \times 10^3\ \Omega$.

REFERENCES

1. A. Eroglu. 2013. *RF Circuit Design Techniques for MF-UHF Applications*. CRC Press, Boca Raton, FL.
2. G. Matthaei, E.M.T. Jones, and L. Young. 1980. *Microwave Filters, Impedance-Matching Networks, and Coupling Structures*. Artech House, Norwood, MA.
3. A. Ferrero, V. Teppati, M. Garelli, and A. Neri. 2008. A novel calibration algorithm for a special class of multi-port vector network analyzers. *IEEE Transactions on Microwave Theory and Techniques*, Vol. 56, No. 3, pp. 693–699.
4. J. Hu, K.G. Gard, and M.B. Steer. 2011. Calibrated non-linear vector network measurement without using a multi-harmonic generator. *IET Microwaves, Antennas & Propagation*, Vol. 5, No. 5, pp. 616–624, May.
5. W.S. El-Deeb, M.S. Hashmi, S. Bensmida, N. Boulejfen, and F.M. Ghannouchi. 2010. Thru-less calibration algorithm and measurement system for on-wafer large-signal characterization of microwave devices. *IET Microwaves, Antennas & Propagation*, Vol. 4, No. 11, pp. 1773–1781, November.
6. U. Stumper. 2005. Uncertainty of VNA S-parameter measurement due to non-ideal TRL calibration items. *IEEE Transactions on Instrumentation and Measurement*, Vol. 54, No. 2, pp. 676–679.
7. W. Zhao, H.-B. Qin, and L. Qiang. 2012. A calibration procedure for two port VNA with three measurement channels based on T-matrix. *Progress in Electromagnetics Research Letters*, Vol. 29, pp. 35–42.
8. I.M. Kang, S.-J. Jung, T.-H. Choi, J.-H. Jung, C. Chung, H.-S. Kim, H. Oh, H. W. Lee, G. Jo, Y.-K. Kim, H.-G. Kim, and K.-M. Choi. 2009. Five-step (pad–pad short–pad open–short–open) de-embedding method and its verification. *IEEE Electron Device Letters*, Vol. 30, No. 4, pp. 398–400, April.
9. S. Padmanabhan, L. Dunleavy, and J.E. Daniel. 2006. Broadband space conservative on-wafer network analyzer calibrations with more complex load and thru models. *IEEE Transactions on Microwave Theory and Techniques*, Vol. 54, No. 9, pp. 3583–3593.
10. W.M. Okamura, M.M. DuFault, and A.K. Sharma. 2000. A comprehensive millimeter-wave calibration development and verification approach. *2000 IEEE MTT-S International Microwave Symposium Digest*, pp. 1477–1480, June 11–16.
11. S. Rehnmark. 1974. On the calibration process of automatic network analyzer systems. *IEEE Transactions on Microwave Theory and Techniques*, Vol. MTT-22, pp. 457–458, April.
12. J.V. Butler, D.K. Rytting, M.F. Iskander, R.D. Pollard, and M.V. Bossche. 1991. 16-term error model and calibration procedure for on-wafer network analysis measurements. *IEEE Transactions on Microwave Theory and Techniques*, Vol. 39, No. 12, pp. 2211–2217, December.
13. G.F. Engen, and C.A. Hoer. 1979. Thru-reflect-line: An improved technique for calibrating the dual six-port automatic network analyzer. *IEEE Transactions on Microwave Theory and Techniques*, Vol. 27, No. 12, pp. 987–993, December.
14. D.C. DeGroot, J.A. Jargon, and R.B. Marks. 2002. Multiline TRL revealed. *60th ARFTG Conference Digest*, pp. 131–155.

15. C. Shih. 1998. Advanced TRL (through-reflect-line) fixture design and error analyses for RF high power transistor characterization and automatic load pull measurement. *51st ARFTG Conference Digest*, pp. 72–76.
16. R.B. Marks. 1991. A multiline method of network analyzer calibration. *IEEE Transactions on Microwave Theory and Techniques*, Vol. 39, No. 7, pp. 1205–1215.
17. W. Kruppa, and K.F. Sodomsky. 1971. An explicit solution for the scattering parameters of a linear two-port measured with an imperfect test set (correspondence). *IEEE Transactions on Microwave Theory and Techniques*, Vol. 19, No. 1, pp. 122–123, January.
18. J. Fitzpatrick. 1978. Error models for system measurements. *Microwave Journal*, Vol. 21, No. 5, pp. 63–66, May.
19. M. Imparato, T. Weller, and L. Dunleavy. 1999. On-wafer calibration using space conservative (SOLT) standards. *1999 IEEE MTT-S Int'l Microwave Symposium*, Vol. 4, pp. 1643–1646, June.
20. S. Padmanabhan, P. Kirby, J. Daniel, and L. Dunleavy. 2003. Accurate broadband on wafer SOLT calibrations with complex load and thru models. *61st ARFTG Conference Digest*, June.
21. B.C. Wadell. 1991. *Transmission Line Design Handbook*. Artech House, Norwood, MA.
22. I. Wolf. 2006. *Coplanar Microwave Integrated Circuits*. Wiley, New York.
23. Agilent Technologies. 2004. Agilent de-embedding and embedding S-parameter networks using a vector network analyzer. Application Note 1364-1.
24. M.C.A.M. Koolen, J.A.M. Geelen, and M.P.J.G. Versleijen. 1991. An improved de-embedding technique for on-wafer high-frequency characterization. *Proceedings of the IEEE, Bipolar/BiCMOS Circuits and Technology Meeting*, pp. 188–191, September.
25. T. Hirano, J. Hirokawa, M. Ando, H. Nakano, and Y. Hirachi. 2008. De-embedding of lumped element characteristics with the aid of EM analysis. *Digest of IEEE AP-S International Symposium and USNC/URSI National Radio Science Meeting*, Session: 436.3, July 5–12.
26. T. Hirano, H. Nakano, Y. Hirachi, J. Hirokawa, and M. Ando. 2010. De-embedding method using an electromagnetic simulator for characterization of transistors in the millimeter-wave band. *IEEE Transactions on Microwave Theory and Techniques*, Vol. 58, No. 10, pp. 2663–2672.

4 Resonator Networks for Amplifiers

4.1 INTRODUCTION

Resonators have frequency characteristics that give them the ability to present specific impedance, quality factor, and bandwidth. They can eliminate the reactive component effects and introduce only the resistive portion of the impedance at a frequency called resonance frequency. The circuit that is capable of producing these effects is called a resonant circuit. An ideal resonant circuit acts like a filter and eliminates the unwanted signal content out of the frequency of interest, as shown in Figure 4.1. Resonant circuits can also be used as part of the impedance matching networks to transform one impedance at one point to another impedance. In RF amplifier circuits, it is a commonly used technique to present a matched impedance at one frequency and introduce high impedance levels at others. When the amplifier is matched at the input and output for maximum gain, it is possible to deliver the highest amount of power by keeping the circuit stable. The stability of the circuit is accomplished most of the time by using filters and resonators to eliminate the spurious contents and oscillations.

In this chapter, a discussion on resonant networks, transmission lines, Smith chart, and impedance matching networks is given.

4.2 PARALLEL AND SERIES RESONANT NETWORKS

4.2.1 Parallel Resonance

Consider the parallel resonant circuit given in Figure 4.2. The response of the circuit can be obtained by finding the voltage with an application of Kirchhoff's current law (KCL). Application of KCL gives the first-order differential equation for voltage v as

$$\frac{v}{R}+\frac{1}{L}\int_0^t v\,\mathrm{d}\tau+I_o+C\frac{\mathrm{d}v}{\mathrm{d}t}=0 \tag{4.1}$$

where I_o is the initial charged current on the inductor. Equation 4.1 can be written as

$$\frac{\mathrm{d}^2v}{\mathrm{d}t^2}+\frac{1}{RC}\frac{\mathrm{d}v}{\mathrm{d}t}+\frac{v}{LC}=0 \tag{4.2}$$

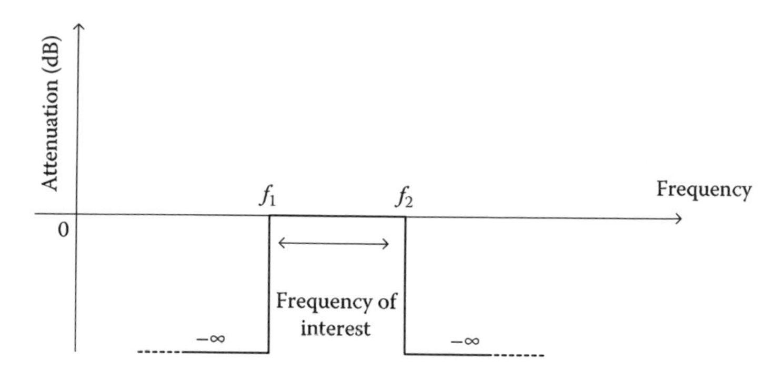

FIGURE 4.1 Ideal resonant network response.

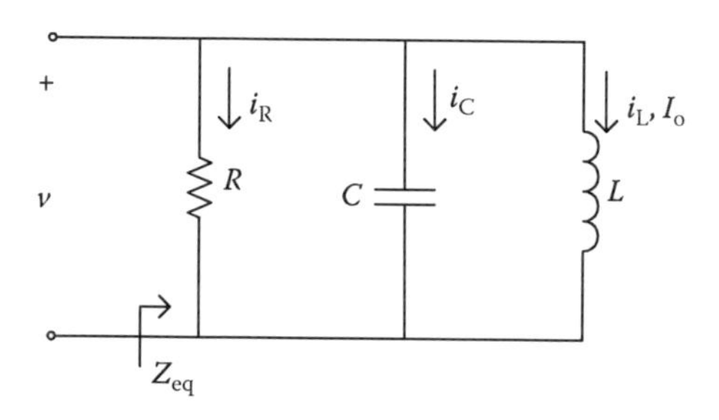

FIGURE 4.2 Parallel resonant circuit.

The solution for the voltage in Equation 4.2 will be in the following form:

$$v = Ae^{st} \tag{4.3}$$

where A is constant, and $s = j\omega$. Substitution of Equation 4.3 into Equation 4.2 gives

$$Ae^{st}\left(s^2 + \frac{s}{RC} + \frac{1}{LC}\right) = 0 \tag{4.4}$$

which can be simplified to

$$s^2 + \frac{s}{RC} + \frac{1}{LC} = 0 \tag{4.5}$$

Equation 4.5 is called the characteristic equation. The roots of the equation are

$$s_1 = -\frac{1}{2RC} + \sqrt{\left(\frac{1}{2RC}\right)^2 - \left(\frac{1}{LC}\right)} \tag{4.6}$$

$$s_2 = -\frac{1}{2RC} - \sqrt{\left(\frac{1}{2RC}\right)^2 - \left(\frac{1}{LC}\right)} \tag{4.7}$$

The complete solution for the voltage v is then obtained as

$$v = v_1 + v_2 = A_1 e^{s_1 t} + A_2 e^{s_2 t} \tag{4.8}$$

The roots given by Equations 4.6 and 4.7 can be expressed as

$$s_1 = -\alpha + \sqrt{\alpha^2 - \omega_0^2} \tag{4.9}$$

$$s_2 = -\alpha - \sqrt{\alpha^2 - \omega_0^2} \tag{4.10}$$

In Equation 4.10, α is the damping coefficient, and ω_0 is the resonant frequency. At resonant frequency, the reactive components cancel each other. The damping coefficient and the resonant frequency are given by the following equations:

$$\alpha = \frac{1}{2RC} \tag{4.11}$$

and

$$\omega_0 = \frac{1}{\sqrt{LC}} \tag{4.12}$$

When

$$\begin{aligned} &\omega_0^2 < \alpha^2, \quad s_1 \text{ and } s_2 \quad \text{are real and distinct, voltage is overdamped} \\ &\omega_0^2 > \alpha^2, \quad s_1 \text{ and } s_2 \quad \text{are complex, voltage is underdamped} \\ &\omega_0^2 = \alpha^2, \quad s_1 \text{ and } s_2 \quad \text{are real and equal, voltage is critically damped} \end{aligned} \tag{4.13}$$

The time domain representation of a parallel resonant network voltage response to illustrate underdamped and overdamped cases is illustrated in Figure 4.3a and b, respectively. The L and C values are taken to be 0.1 H and 0.001 F for an underdamped case, whereas for an overdamped case, the L and C values are taken to be 50 mH and 0.2 μF. R values are varied to see their effect on voltage response for damping.

The quality factor and the bandwidth of the parallel resonant network are

$$Q = \frac{R}{\omega_0 L} = \omega_0 RC \tag{4.14}$$

$$BW = \frac{\omega_0}{Q} = \frac{1}{RC} \tag{4.15}$$

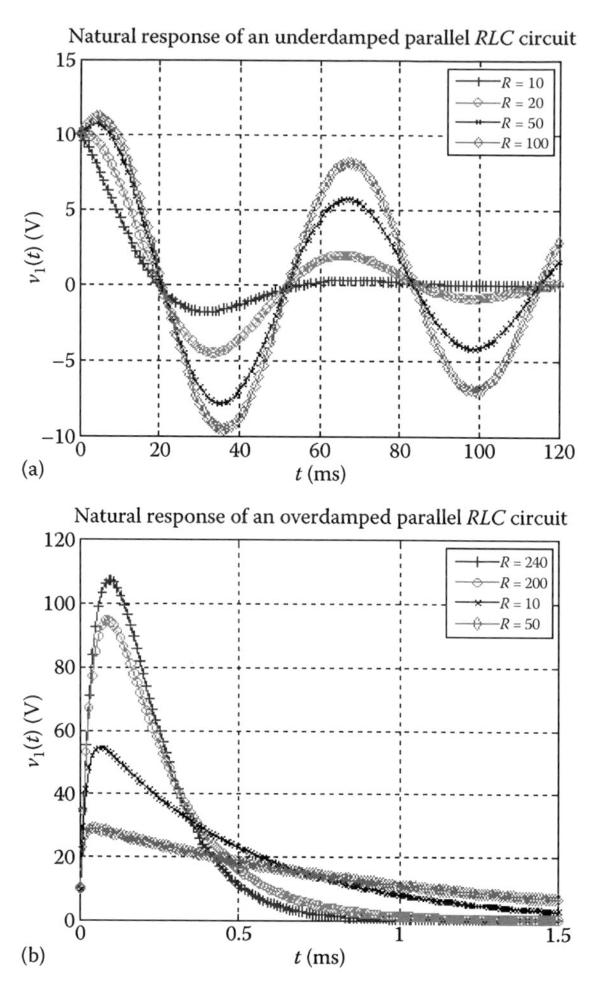

FIGURE 4.3 (a) Parallel resonant network response for an underdamped case. (b) Parallel resonant network response for an overdamped case.

In terms of the quality factor, the roots given by Equations 4.6 and 4.7 can be expressed as

$$s_1 = \omega_0 \left[-\frac{1}{2Q} + \sqrt{\left(\frac{1}{2Q}\right)^2 - 1} \right] \tag{4.16}$$

$$s_2 = \omega_0 \left[-\frac{1}{2Q} - \sqrt{\left(\frac{1}{2Q}\right)^2 - 1} \right] \tag{4.17}$$

Now, assume that there is a source current connected to the parallel resonant network in Figure 4.2, as illustrated in Figure 4.4.

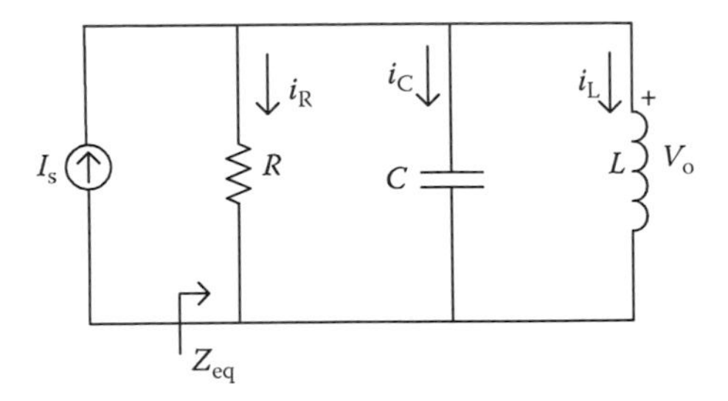

FIGURE 4.4 Parallel resonant circuit with source current.

The equivalent impedance of the parallel resonant network is found from Figure 4.4 as

$$Z_{eq}(s) = \frac{V_o(s)}{I_s(s)} = \frac{s/C}{s^2 + s(1/RC) + (1/LC)} = \frac{s/C}{(s - s_1)(s + s_2)} \tag{4.18}$$

which can be written as

$$Z_{eq}(j\omega) = \frac{1}{R} + j\frac{\omega L}{(1 - \omega^2 LC)} \tag{4.19}$$

In Equation 4.14, s_1 and s_2 are now the poles of the impedance. When

$$\left(\frac{1}{2RC}\right)^2 \geq \left(\frac{1}{LC}\right) \quad \text{or} \quad R \leq \left(\frac{\omega_0 L}{2}\right) = \left(\frac{1}{2\omega_0 C}\right) \tag{4.20}$$

the poles of the impedance lie on the negative real axis. Hence, the value of R is small in comparison to the values of the reactances, and as a result, the resonant network has a broadband response. When

$$\left(\frac{1}{2RC}\right)^2 < \left(\frac{1}{LC}\right) \text{ or } R > \left(\frac{\omega_0 L}{2}\right) \tag{4.21}$$

the poles become complex, and they take the following form:

$$s_1 = -\alpha + j\sqrt{\omega_0^2 - \alpha^2} = -\alpha + j\beta \tag{4.22}$$

$$s_2 = -\alpha - j\sqrt{\omega_0^2 - \alpha^2} = -\alpha - j\beta \tag{4.23}$$

This can be illustrated on the pole-zero diagram, as shown in Figure 4.5.

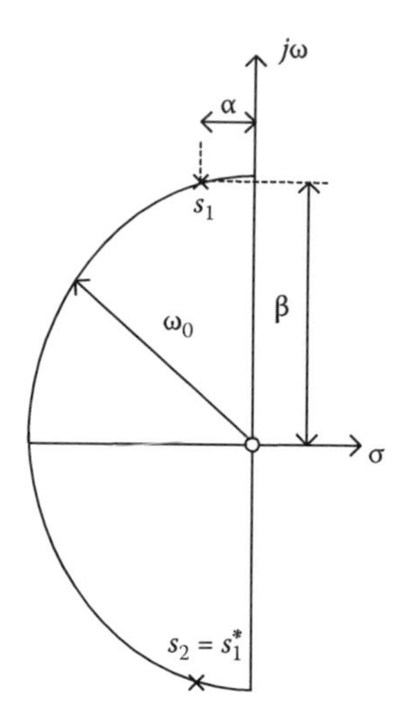

FIGURE 4.5 Pole-zero diagram for complex conjugate roots.

The transfer function for the parallel resonant network is found using Figure 4.2 as

$$|H(\omega)| = \left|\frac{I_R}{I_S}\right| = \frac{\omega(L/R)}{\sqrt{(1-\omega^2 LC)^2 + (\omega L/R)^2}} \tag{4.24}$$

At the resonant frequency, the transfer function will be real and be equal to its maximum value as

$$|H(\omega=\omega_0)| = \frac{\omega_0(L/R)}{\sqrt{\left(1-\omega_0^2 LC\right)^2 + (\omega_0 L/R)^2}} = 1 = |H(\omega)|_{\max} \tag{4.25}$$

The network response is obtained using the transfer function given by Equation 4.25 for different values of R, as shown in Figure 4.6. The values of the inductance and capacitance are taken to be 1.25 μH and 400 nF.

This gives the resonant frequency as 0.22508 MHz. The condition for a broadband network is accomplished when $R = 5$, as shown in Figure 4.6. As R increases, the quality factor of the network also increases, which agrees with Equation 4.14.

Quality factor, Q, is an important parameter in resonant network response as it can be used as a measure for the loss and bandwidth of the circuit. The quality factor of the circuit defines the ratio of the peak energy stored to the energy dissipated per cycle, as given by

$$Q = \frac{2\pi(\text{Peak energy stored})}{(\text{Energy dissipated per cycle stored})} = \frac{2\pi\left(\frac{1}{2}CV^2\right)}{(2\pi/\omega_0)(V^2/2R)} = \omega_0 CR \tag{4.26}$$

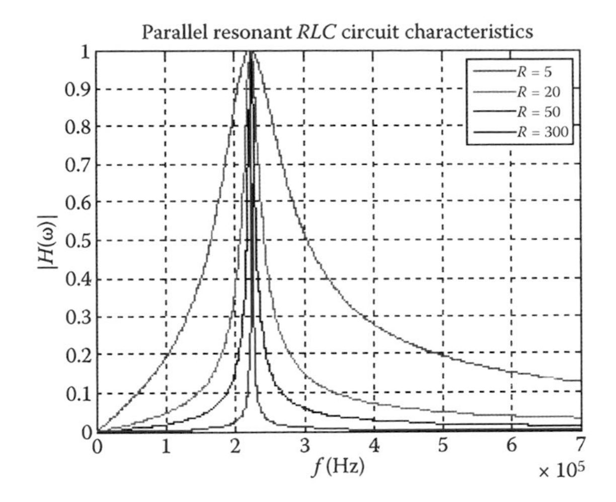

FIGURE 4.6 Parallel resonant circuit transfer function characteristics.

Example

A parallel resonant circuit has a source resistance of 50 Ω and a load resistance of 25 Ω. The loaded Q must be equal to 12 at the resonant frequency of 60 MHz.

a. Design the resonant circuit.
b. Calculate the 3-dB bandwidth of the resonant circuit.
c. Use MATLAB® to obtain the frequency response of this circuit versus frequency, i.e., plot 20log (V_o/V_{in}) vs. frequency.

Solution

a. The effective parallel resistance across a parallel resonance circuit is

$$R_p = \frac{(50)25}{50+25} = 16.67\,[\Omega]$$

Then,

$$X_p = \frac{R_p}{Q} = \frac{16.67}{12} = 1.4$$

Since,

$$X_p = \omega L = \frac{1}{\omega C}$$

then the resonance element values are

$$L = \frac{X_p}{\omega} = \frac{1.4}{2\pi(60 \times 10^6)} = 3.71[\text{nH}]$$

and

$$C = \frac{1}{\omega X_p} = \frac{1}{2\pi(60 \times 10^6)(1.4)} = 1894.7\,[\text{pF}]$$

b. The BW is found from

$$Q = \frac{f_c}{BW_{3dB}} \rightarrow BW_{3dB} = \frac{f_c}{Q} = \frac{60 \times 10^6}{12} = 5 \times 10^6\,[\text{Hz}]$$

c. The MATLAB script to obtain the attenuation profile is given below.

```
clear
f = linspace(1,100*10^6);
RL = 25;
RS = 50;
RP = 50*25/(50+25);
Q = 12;
XP = RP/Q;

fc = 60*10^6;
wc = 2*pi()*fc;
L = XP/(wc);
C = 1/(wc*XP);

w = 2*pi.*f;
XL=1j.*w.*L;
XC=-1j./(w.*C);
Xeq=(XL.*XC)./(XL+XC);
Zeq=(RL.*Xeq)./(RL+Xeq);
S21=20.*log10(abs(Zeq./(Zeq+RS)));

plot(f,S21);

grid on
title('Attenuation Profile')
xlabel('Frequency (Hz)')
ylabel('Attenuation (dB)')
```

The plot of the attenuation profile is given in Figure 4.7.

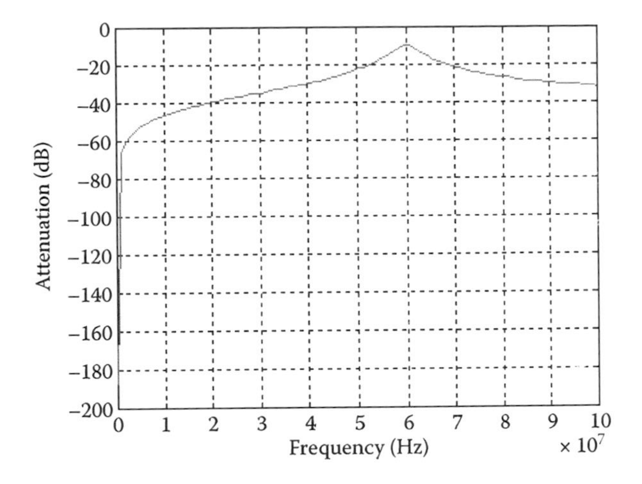

FIGURE 4.7 Attenuation profile vs. frequency.

4.2.2 Series Resonance

Consider the series resonant circuit given in Figure 4.8. The response of the circuit can be obtained by application of Kirchhoff's voltage law (KVL), which gives the first-order differential equation for current i as

$$Ri + L\frac{\mathrm{d}i}{\mathrm{d}t} + \frac{1}{C}\int_0^t i\,\mathrm{d}\tau + V_o = 0 \tag{4.27}$$

where V_o is the initial charged voltage on a capacitor. Equation 4.25 can be written as

$$\frac{\mathrm{d}^2 i}{\mathrm{d}t^2} + \frac{R}{L}\frac{\mathrm{d}i}{\mathrm{d}t} + \frac{i}{LC} = 0 \tag{4.28}$$

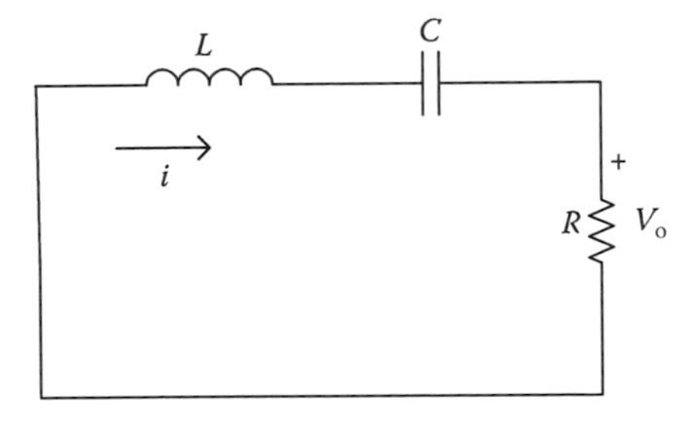

FIGURE 4.8 Series resonant network.

Following the same solution technique for a parallel resonant network leads to the following equation in the frequency domain:

$$s^2 + \frac{R}{L}s + \frac{1}{LC} = 0 \tag{4.29}$$

The roots of the equation are

$$s_1 = -\frac{R}{2L} + \sqrt{\left(\frac{R}{2L}\right)^2 - \left(\frac{1}{LC}\right)} \tag{4.30}$$

$$s_1 = -\frac{R}{2L} - \sqrt{\left(\frac{R}{2L}\right)^2 - \left(\frac{1}{LC}\right)} \tag{4.31}$$

Equations 4.30 and 4.31 can be expressed as

$$s_1 = -\alpha + \sqrt{\alpha^2 - \omega_0^2} \tag{4.32}$$

$$s_2 = -\alpha - \sqrt{\alpha^2 - \omega_0^2} \tag{4.33}$$

where α and ω are defined for a series resonant network as

$$\alpha = \frac{R}{2L} \tag{4.34}$$

and

$$\omega_0 = \frac{1}{\sqrt{LC}} \tag{4.35}$$

The quality factor and the bandwidth of the series resonant network are

$$Q = \frac{\omega_0 L}{R} = \frac{1}{\omega_0 RC} \tag{4.36}$$

$$BW = \frac{\omega_0}{Q} = \frac{R}{L} \tag{4.37}$$

In terms of quality factor, the roots given by Equations 4.30 and 4.31 can be obtained as

$$s_1 = \omega_0 \left[-\frac{1}{2Q} + \sqrt{\left(\frac{1}{2Q}\right)^2 - 1} \right] \tag{4.38}$$

$$s_2 = \omega_0 \left[-\frac{1}{2Q} - \sqrt{\left(\frac{1}{2Q}\right)^2 - 1} \right] \tag{4.39}$$

Equations 4.38 and 4.39 are identical to the ones obtained for a parallel resonant circuit. The damping characteristics of the series resonant network follow the conditions listed in Equation 4.13. The time domain representation of a series resonant network current response illustrating underdamped and overdamped cases is illustrated in Figures 4.9 and 4.10, respectively. L and C values are taken to be 100 mH and 10 μF for an underdamped case, whereas for an overdamped case, L and C values are taken to be 200 mH and 10 μF. R values are varied to see their effect on current response for damping.

The transfer function of this network can be found by connecting a source voltage, as shown in Figure 4.11 and obtained as

$$|H(j\omega)| = \left|\frac{V_o(s)}{V_s(s)}\right| = \frac{\omega \frac{R}{L}}{\sqrt{\left(\frac{1}{LC} - \omega^2\right)^2 + \left(\omega \frac{R}{L}\right)^2}} \tag{4.40}$$

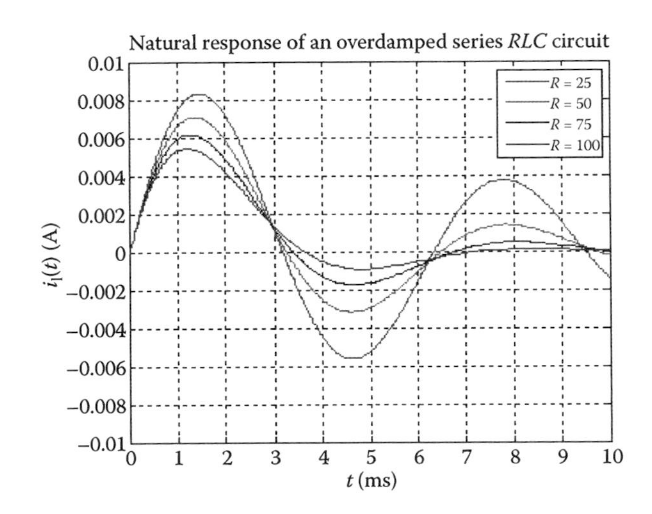

FIGURE 4.9 Series resonant network response for an underdamped case.

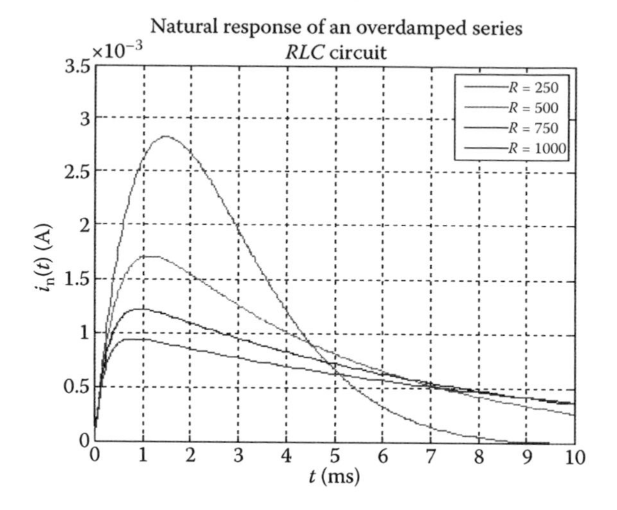

FIGURE 4.10 Series resonant network response for an overdamped case.

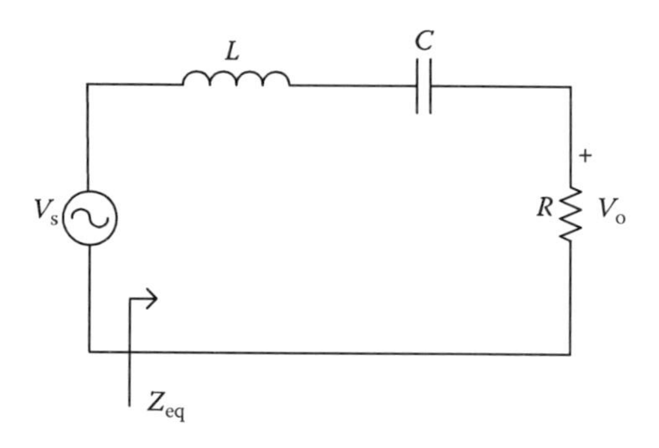

FIGURE 4.11 Series resonant network with source voltage.

The phase of the transfer function is found from

$$\theta(j\omega) = 90° - \tan^{-1}\left(\frac{\omega\dfrac{R}{L}}{\dfrac{1}{LC} - \omega^2}\right) \tag{4.41}$$

At resonant frequency, the transfer function is maximum and will be equal to

$$|H(j\omega)| = \frac{\sqrt{\dfrac{1}{LC}}\dfrac{R}{L}}{\sqrt{\left(\dfrac{1}{LC} - \dfrac{1}{LC}\right)^2 + \left(\sqrt{\dfrac{1}{LC}}\dfrac{R}{L}\right)^2}} = 1 = |H(j\omega)|_{max} \tag{4.42}$$

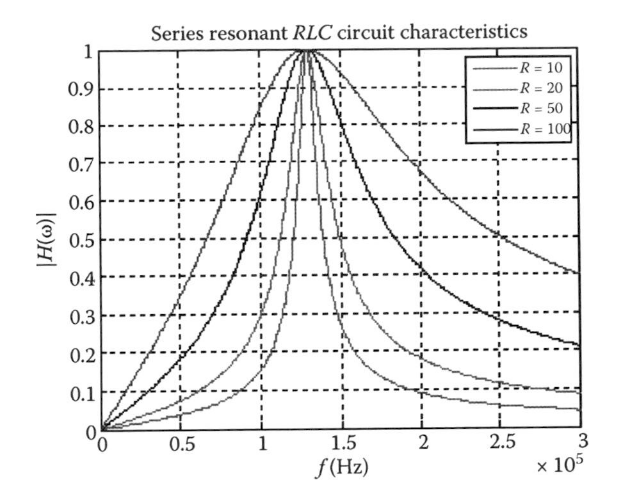

FIGURE 4.12 Series resonant circuit transfer function characteristics.

The resonant characteristics of the network can be obtained by plotting the transfer function given by Equation 4.40 vs. different values of R, as shown in Figure 4.12. The values of the inductance and capacitance are taken to be 150 μH and 10 nF.

4.3 PRACTICAL RESONANCES WITH LOSS, LOADING, AND COUPLING EFFECTS

4.3.1 Component Resonances

RF components such as resistors, inductors, and capacitors in practice exhibit resonances at high frequencies due to their high-frequency characteristics. The high-frequency representation of an inductor and a capacitor is given in Figure 4.13.

As seen from the equivalent circuit in Figure 4.13, an inductor will act as such until it reaches resonant frequency; then it gets into resonance and exhibits capacitive effects after. The expression that gives these characteristics for an inductor can be obtained as

$$Z = \frac{(j\omega L + R)\dfrac{1}{j\omega C_s}}{(j\omega L + R) + \dfrac{1}{j\omega C_s}} = \frac{R}{(1-\omega^2 LC_s)^2 + (\omega RC_s)^2} + j\frac{\omega(L - R^2 C_s) - \omega^3 L^2 C_s}{(1-\omega^2 LC_s)^2 + (\omega RC_s)^2} \tag{4.43}$$

Equation 4.43 can also be written as

$$Z = R_s + jX_s \tag{4.44}$$

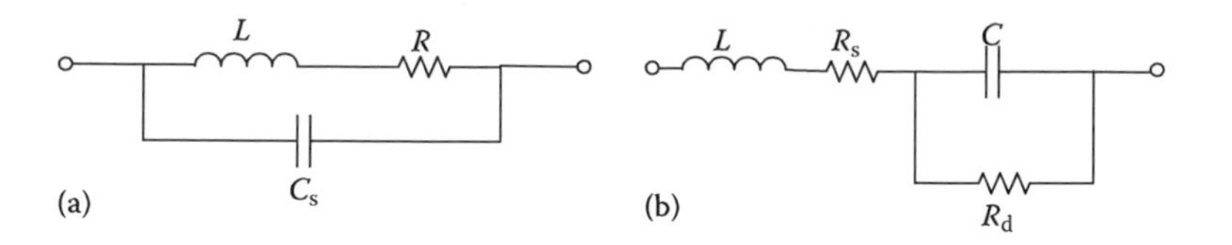

FIGURE 4.13 High-frequency representation of (a) an inductor and (b) a capacitor.

where

$$R_s = \frac{R}{(1-\omega^2 LC_s)^2 + (\omega RC_s)^2} \tag{4.45}$$

and

$$X_s = \frac{\omega(L - R^2C_s) - \omega^3 L^2 C_s}{(1-\omega^2 LC_s)^2 + (\omega RC_s)^2} \tag{4.46}$$

We can now use Equations 4.45 and 4.46 and represent the circuit given in Figure 4.13a with an equivalent series circuit shown in Figure 4.14.

The resonance frequency is found when $X_s = 0$ as

$$f_r = \frac{1}{2\pi}\sqrt{\frac{L - R^2C_s}{L^2C_s}} \tag{4.47}$$

The quality factor is obtained from

$$Q = \frac{|X_s|}{R_s} \tag{4.48}$$

R_s in Figure 4.14 is the series resistance of the inductor and includes the distributed resistance effect of the wire. It is calculated from

$$R = \frac{l_W}{\sigma A} \tag{4.49}$$

FIGURE 4.14 Equivalent series circuit.

C_s in Figure 4.13a is the capacitance including the effects of distributed capacitance of the inductor and is given by

$$C_s = \frac{2\pi\varepsilon_0 daN^2}{l_W} \tag{4.50}$$

For an air core inductor, the value of L shown in Figure 4.13a and used in Equation 4.46 is found from

$$L = \frac{d^2N^2}{18d + 40l}[\mu\text{H}] \tag{4.51}$$

In this equation, L is given as inductance in [μH], d is the coil inner diameter in inches, l is the coil length in inches, and N is the number of turns in the coil. The formula given in Equation 4.51 can be extended to include the spacing between each turn of the air coil inductor. Then, Equation 4.51 can be modified as

$$L = \frac{d^2N^2}{18d + 40(Na + (N-1)s)}[\mu\text{H}] \tag{4.52}$$

In Equation 4.52, a represents the wire diameter in inches, and s represents the spacing in inches between each turn.

When air is replaced with a magnetic material such as a toroidal core, thc inductance of the formed inductor can be calculated using

$$L = \frac{4\pi N^2\mu_i A_{Tc}}{l_e}[\text{nH}] \tag{4.53}$$

In Equation 4.53, L is the inductance in nanohenries, N is the number of turns, μ_i is the initial permeability, A_{Tc} is the total cross-sectional area of the core in square centimeters, and l_e is the effective length of the core in centimeter. The details of the derivation and implementation of inductor design and design tables are given in Ref. [1].

From Figure 4.13b, it is clear that a non-ideal capacitor also has resonances due to its high-frequency characteristics. The high-frequency model of the capacitor has parasitic components such as lead inductance, L, conductor loss, R_s, and dielectric loss, R_d, which only become relevant at high frequencies. The characteristics of the capacitor can be obtained by finding the equivalent impedance as

$$Z = (j\omega L_s + R_s) + \left(\frac{1}{G_d + j\omega C}\right) = \frac{R_s G_d^2 + (\omega C)^2 + G_d}{G_d^2 + (\omega C)^2} + j\frac{\omega L G_d^2 + L\omega(\omega C)^2 - \omega C}{G_d^2 + (\omega C)^2} \tag{4.54}$$

Then, Equation 4.54 can be expressed as

$$Z = R_s + jX_s \tag{4.55}$$

where

$$R_s = \frac{R_s G_d^2 + (\omega C)^2 + G_d}{G_d^2 + (\omega C)^2} \tag{4.56}$$

$$X_s = \frac{\omega L G_d^2 + L\omega(\omega C)^2 - \omega C}{G_d^2 + (\omega C)^2} \tag{4.57}$$

The impedance given in Equation 4.55 can be converted to admittance as

$$Y = Z^{-1} = \frac{R_s}{R_s^2 + X_s^2} + j\frac{-X_s}{R_s^2 + X_s^2} = G + jB \tag{4.58}$$

or

$$Y = \frac{\left(R_s G_d^2 + (\omega C)^2 + G_d\right)\left(G_d^2 + (\omega C)^2\right)}{\left(R_s G_d^2 + (\omega C)^2 + G_d\right)^2 + \left(\omega L G_d^2 + L\omega(\omega C)^2 - \omega C\right)^2} + j\frac{\left(\omega C - \omega L G_d^2 - L\omega(\omega C)^2\right)\left(G_d^2 + (\omega C)^2\right)}{\left(R_s G_d^2 + (\omega C)^2 + G_d\right)^2 + \left(\omega L G_d^2 + L\omega(\omega C)^2 - \omega C\right)^2} \tag{4.59}$$

Then, the capacitor can be represented by a parallel equivalent circuit, as shown in Figure 4.15, where

$$G = \frac{\left(R_s G_d^2 + (\omega C)^2 + G_d\right)\left(G_d^2 + (\omega C)^2\right)}{\left(R_s G_d^2 + (\omega C)^2 + G_d\right)^2 + \left(\omega L G_d^2 + L\omega(\omega C)^2 - \omega C\right)^2} \tag{4.60}$$

$$B = \frac{\left(\omega C - \omega L G_d^2 - L\omega(\omega C)^2\right)\left(G_d^2 + (\omega C)^2\right)}{\left(R_s G_d^2 + (\omega C)^2 + G_d\right)^2 + \left(\omega L G_d^2 + L\omega(\omega C)^2 - \omega C\right)^2} \tag{4.61}$$

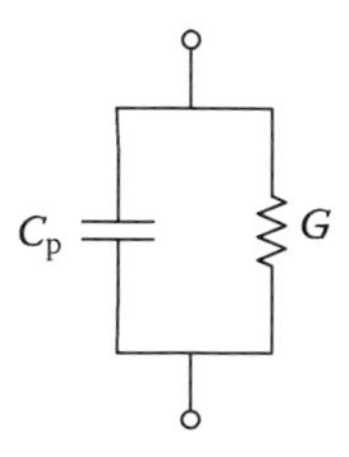

FIGURE 4.15 Equivalent parallel circuit.

The resonance frequency for the circuit shown in Figure 4.14 is found when $B = 0$ as

$$f_r = \frac{1}{2\pi}\sqrt{\frac{R^2C - L}{R^2C^2L}} \tag{4.62}$$

The quality factor for the parallel network is then obtained from

$$Q = \frac{|B|}{G} = \frac{R_p}{|X_p|} \tag{4.63}$$

Design Example

Develop a MATLAB graphic user interface (GUI) to design an air core inductor with a user-specified inductance value and with air core diameter. The program should also be able to identify the correct wire gauge for the user-entered current amount. Assume that the operational voltage is 50 [V_{rms}] and the frequency is 27.12 [MHz]. With your program,

a. Calculate the number of turns.
b. Determine the minimum gauge wire that needs to be used.
c. Obtain the high-frequency characteristic of the inductor.
d. Identify its resonant frequency.
e. Find its quality factor.
f. Find the length of the wire that will be used.
g. Find the length of the inductor.

Solution

The following is the MATLAB GUI that is developed.

```
clear
%Gui Prompt
prompt = {'Inductance [uH]:','Inductor Inner Diameter [in]: ',...
    'Current [A]:'};
dlg_title = 'Air Core Inductor Parameters';
num_lines = 1;
def = {'0.180','0.25','5'};
answer = inputdlg(prompt,dlg_title,num_lines,def, 'on');

%convert the strings received from the GUI to numbers
valuearray = str2double(answer);

%Give variable names to the received numbers
Lnominal=valuearray(1);
d=valuearray(2);
I=valuearray(3);

%Convert inner diameter to meters
dm=d*0.0254;
```

```
%Establish the tables of wire gauge, ampacity, and diameter as vectors
awgvector=[40 38 36 34 32 30 28 26 24 22 20 18 16 14 12 10 8 6 4 2 0];
ampacityvector=[0.226134423 0.314649454 0.437811623 0.609182742...
    0.847633078 1.179419221 1.641075289 2.283435826 3.177233372 ...
    4.420887061 6.151339898 8.559138024 11.90941241 16.57107334...
    23.05743241 32.08272502 44.64075732 62.11433765 86.42754229 ...
    120.2575822 167.3296];
wirediametervector=[0.0799 0.101 0.127 0.16 0.202 0.255...
    0.321 0.405 0.511 0.644 0.812 1.024 1.291 1.628 ...
    2.053 2.588 3.264 4.115 5.189 6.544 8.251];

%Initialize variables for use in the wire size search loop
Irating=0;
awgselected=0;
k=1;
flag=0;
wirediametermm=0;
errorflag=0;

%Start with the smallest wire and see if it meets the required current,
%if it does not, go to next size up and repeat, stop when wire is
  large enough
while flag==0
    Irating=ampacityvector(k);
    if I>Irating
        k=k+1;
        if k>21
            msgbox( sprintf(['The required current exceeds '...
                 'the rating of all available wire stock']));
            errorflag=1;
            break
         end
    elseif I<Irating
        awgselected=awgvector(k);
        wirediametermm=wirediametervector(k);
        flag=1;
    end
end

%Convert wire diameter to inches
wirediameterin=wirediametermm*0.0393701;
a=wirediameterin;

%Calculate the nominal number of turns
Nnominal=max(roots([d^2 -40*a*Lnominal -18*d*Lnominal]));

%Convert turns to an integer
N=round(Nnominal);

%Recalculate inductance value based on integer N
L=(d^2*N^2)/(18*d+40*N*a)/1e6;

%Calculate length of wire and wire cross-sectional area
lw=2*pi*(dm/2)*N;
A=pi*(wirediametermm/2000)^2;
```

```
%Total wire resistance assuming copper wire
R=lw/((59.6e6)*A);

%Total capacitance calculation
C=(2*pi*8.84194128e-12*dm*a*0.0254*N^2)/lw;

%Axial inductor length
inductorlength=N*wirediametermm;

%Calculate resonant frequency
resonant=1/(2*pi)*sqrt((L-R^2*C)/(L^2*C))/1e6;

%Generate high frequency charateristic plot and Q plot, plot f out to
  2x
%resonant frequency to make each plot consistent
fvector=linspace(1,resonant*2,200);
Zvector=1:200;
Qvector=1:200;
j=sqrt(-1);

%Index through vector of frequencies and generate Z and Q at each
  frequency
for p=1:200
    f=fvector(p);
    w=2*pi*f*1e6;
    Zvector(p)=((j*w*L+R)/(j*w*C))/((j*w*L+R)+(1/(j*w*C)));
    Rs=R/((1-w^2*L*C)^2+(w*R*C)^2);
    Xs=(w*(L-R^2*C)-(w^3*L^2*C))/((1-w^2*L*C)^2+(w*R*C)^2);
    Qvector(p)=abs(Xs)/Rs;
end

%Convert values to proper unit magnitude for display
L=L*1e6;
lw=lw*1000;
lw_in=lw/25.4;
inductorlength_in=N*wirediametermm/25.4;
R=R*1000;
dmm=d*25.4;

%Calculate Inductor value difference
Ldiff=(abs(L-Lnominal)/Lnominal)*100;

%If Inductor value difference is greater than 10 percent or the
  inductor length
%is greater than 10", display error if one hasnt already been
  displayed
if (Ldiff>10 && errorflag==0)
    msgbox( sprintf([
          'The given design parameters result in an Inductance\n'...
         'value that is greater than 10 percent from the desired value.
            \n\n'...
         'Try selecting a different Inductor inner
            diameter.'],Lnominal,d,dmm,I));
    errorflag=1;
elseif (inductorlength_in>10 && errorflag==0)
    msgbox( sprintf([
```

```
        'The given design parameters result in an Inductor\n'...
        ' with an axial length greater than 10 inches.\n\n'...
        'Try selecting a different Inductor inner
           diameter.'],Lnominal,d,dmm,I));
    errorflag=1;
end

%If there are no design errors, display results and plots
if errorflag==0;
    subplot(1,2,1)
    plot(fvector,abs(Zvector))
    xlabel('f (MHz)')
    ylabel('Impedance Magnitude Z')
    title('High Frequency Inductor Response')

    subplot(1,2,2)
    plot(fvector,Qvector)
    xlabel('f (MHz)')
    ylabel('Quality Factor Z')
    title('High Frequency Q Response')
msgbox( sprintf([...
        'Given design parameters: \n'...
        '    Nominal Inductance: %4.4f uH \n'...
        '    Inductor Inner Diameter: %4.3f in, %4.3f mm\n'...
        '    Rated Current: %4.2f A \n'...
        '\n'...
        'The specifications of the air-core inductor are as
follows: \n'...
        '    Actual Inductance: %6.4f uH\n'...
        '    Inductance value difference: %4.2f percent\n'...
        '    Turns: %0.0f \n'...
        '    Wire Gauge: %0.0f AWG\n'...
        '    Length of wire: %4.2f in, %4.2f mm\n'...
        '        (Add extra for connection leads)\n'...
        '    Axial Length of Inductor: %4.2f in, %4.2f mm\n'...
        '    Resonant Frequency: %6.2f MHz\n'...
        '    Series Resistance: %4.3f mohm\n'...
        '\n'...
        'High frequency response and Quality '...
        'factor plots are shown in Figure 1\n\n\n']...
         ,Lnominal,d,dmm,I,L,Ldiff,N,awgselected,...
        lw_in,lw,inductorlength_in,inductorlength,resonant,R));
end
```

The program output for an air core inductor with an inductance of 180 [nH] with 0.25-in. diameter, which can handle 5 [A_{rms}], is shown in Figures 4.16 and 4.17.

4.3.2 Parallel LC Networks

4.3.2.1 Parallel LC Networks with Ideal Components

When the ideal components are used, the typical circuit would be represented as the one shown in Figure 4.18. At resonance, the magnitudes of the reactances of the L and C elements are equal. The reactances of the two components have opposite signs

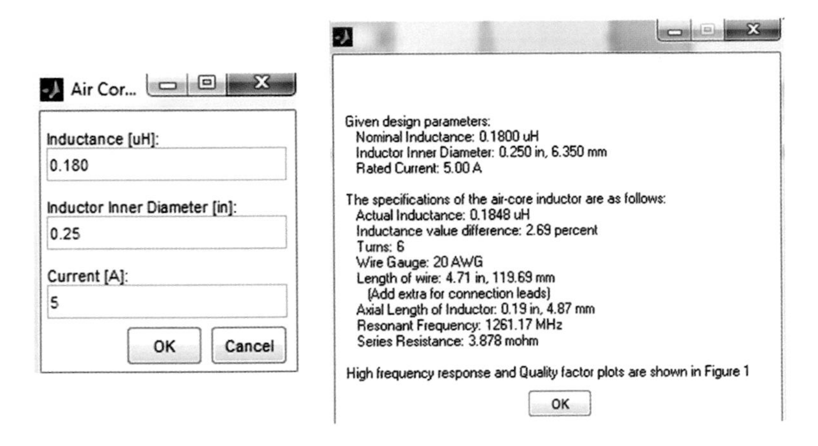

FIGURE 4.16 MATLAB GUI window for air core inductor design.

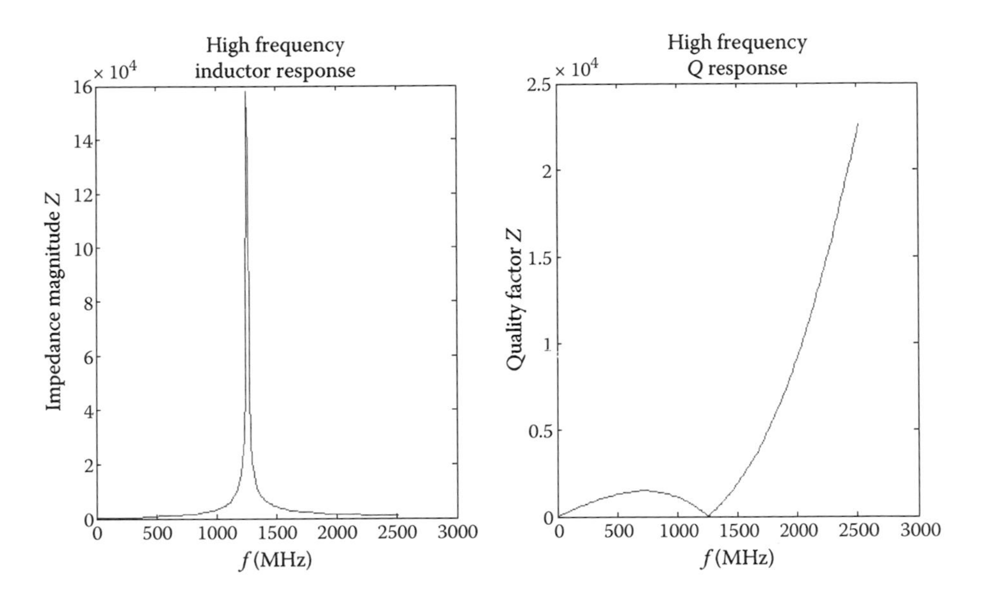

FIGURE 4.17 MATLAB GUI plots for resonance and quality factor.

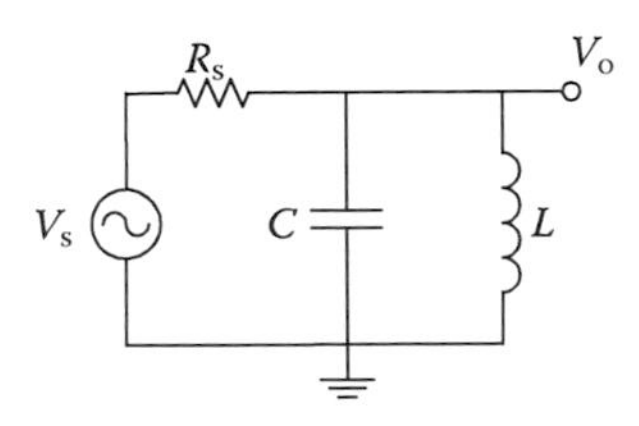

FIGURE 4.18 LC resonant network with ideal components and source.

so the net reactance is zero for a series circuit or infinity for a parallel circuit. Hence, the resonance frequency can be obtained from

$$\omega_0 L = \frac{1}{\omega_0 C} \rightarrow \omega_0 = \frac{1}{\sqrt{LC}} \rightarrow f_0 = \frac{1}{2\pi\sqrt{LC}} \tag{4.64}$$

The transfer function for the LC resonant network in Figure 4.18 is found from

$$|H(\omega)|_{\text{dB}} = \left|\frac{V_{\text{out}}}{V_{\text{in}}}\right|_{\text{dB}} = 20\log\left(\frac{X_{\text{total}}}{R_{\text{s}} + X_{\text{total}}}\right) \tag{4.65}$$

where

$$X_{\text{total}} = \frac{X_{\text{C}} X_{\text{L}}}{X_{\text{C}} + X_{\text{L}}} \tag{4.66}$$

Substitution of Equation 4.65 into Equation 4.64 gives

$$|H(\omega)|_{\text{dB}} = \left|\frac{V_{\text{out}}}{V_{\text{in}}}\right|_{\text{dB}} = 20\log\left|\frac{j\omega L/R_{\text{s}}}{(1-\omega^2 LC) + j\omega L/R_{\text{s}}}\right| \tag{4.67}$$

The frequency characteristics of the circuit are plotted in Figure 4.19 when $L = 0.5$ μH, $C = 2500$ pF, and $R = 50$ Ω.

The quality factor of the circuit is found from its equivalent impedance as

$$Z_{\text{eq}} = R_{\text{s}} + \frac{X_{\text{C}} X_{\text{L}}}{X_{\text{C}} + X_{\text{L}}} = \frac{R_{\text{s}}}{(1-\omega^2 LC)} + j\frac{\omega L}{(1-\omega^2 LC)} \tag{4.68}$$

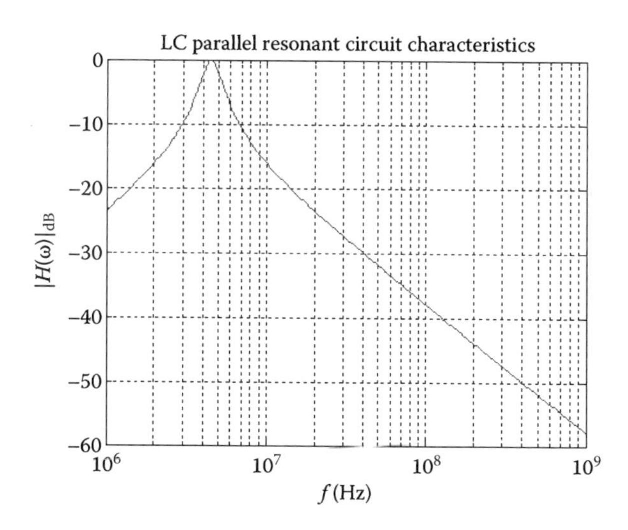

FIGURE 4.19 The frequency characteristics of LC network with source.

Then, the quality factor of the circuit is

$$Q = \frac{|X|}{R} = \frac{\omega L}{R_s} \tag{4.69}$$

4.3.2.2 Parallel LC Networks with Non-Ideal Components

Now, assume that we have some additional loss for inductor for the LC network, as shown in Figure 4.20.

The equivalent impedance of the network in Figure 4.20 takes the following form with the addition of loss component, r:

$$Z_{eq} = R_s + \frac{r + j\omega L}{(j\omega rC - \omega^2 LC + 1)} = R_s + \frac{(r + j\omega L)(1 - \omega^2 LC - j\omega rC)}{(1 - \omega^2 LC)^2 + (\omega rC)^2} \tag{4.70}$$

which can be simplified to

$$Z_{eq} = \frac{R_s[(1 - \omega^2 LC)^2 + (\omega rC)^2] + r}{(1 - \omega^2 LC)^2 + (\omega rC)^2} + j\frac{\omega[(L - Cr^2) - \omega^2 L^2 C]}{(1 - \omega^2 LC)^2 + (\omega rC)^2} \tag{4.71}$$

The resonant frequency of the network is now equal to

$$\omega_0 = \sqrt{\frac{L - Cr^2}{L^2 C}} \rightarrow f_0 = \frac{1}{2\pi}\sqrt{\frac{L - Cr^2}{L^2 C}} \tag{4.72}$$

The loaded quality factor of the circuit is obtained as

$$Q_L = \frac{\omega[(L - Cr^2) - \omega^2 L^2 C]}{R_s[(1 - \omega^2 LC)^2 + (\omega rC)^2]} \tag{4.73}$$

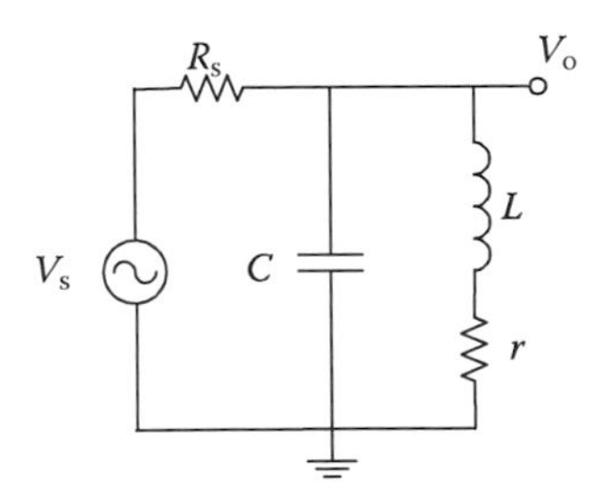

FIGURE 4.20 Addition of loss to parallel LC network.

When $r = 0$ for the lossless case, Equations 4.72 and 4.73 reduce to the ones given in Equations 4.64 and 4.69. The transfer function with the loss resistance changes to

$$\left|H(\omega)\right|_{\text{dB}} = \left|\frac{V_{\text{out}}}{V_{\text{in}}}\right|_{\text{dB}} = 20\log\left|\frac{r + j\omega[(L - Cr^2) - \omega^2 L^2 C]}{R_s[(1-\omega^2 LC)^2 + (\omega r C)^2] + j\omega[(L - Cr^2) - \omega^2 L^2 C]}\right| \tag{4.74}$$

The transfer function showing the attenuation profile with different loss resistance values when $L = 0.5$ μH, $C = 2500$ pF, and $R = 50$ Ω is illustrated in Figure 4.21. The network bandwidth broadens as the value of r increases, as expected.

The quality factor of the network with the addition of loss resistance significantly differs from the original LC parallel resonant network. The original network has an ideally infinite value of quality factor. In agreement with this, a very large value of the quality factor for the original network is obtained at resonance frequency when $r = 0$ is also seen from Figure 4.22.

4.3.2.3 Loading Effects on Parallel LC Networks

The resonant circuit becomes loaded when it is connected to a load or when fed by a source. The Q of the circuit under these conditions is called loaded Q or simply Q_L. The loaded Q of the circuit then depends on source resistance, load resistance, and individual Q of the reactive components.

When the reactive components are lossy, they impact the Q factor of the overall circuit. For instance, consider the resonant circuit with source resistance in Figure 4.18. The quality factor of the circuit vs. various source resistance values when $L = 0.5$ μH and $C = 2500$ pF have been illustrated in Figure 4.23. For the same frequency, the quality factor increases as the value of source resistance, R_s, increases. As a

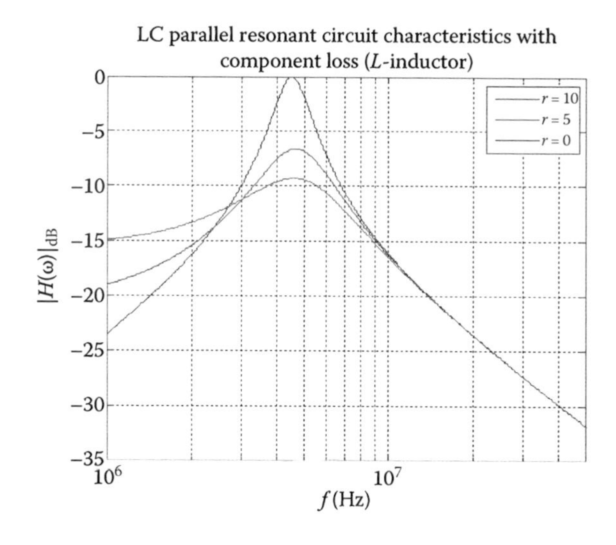

FIGURE 4.21 Attenuation profile of LC network with loss resistor.

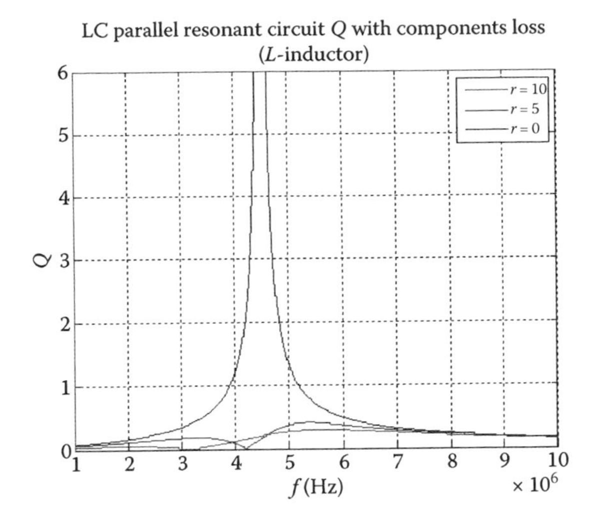

FIGURE 4.22 Quality factor of LC network with loss resistor.

result, the selectivity of the network can be adjusted by setting the value of source resistance. The attenuation profile showing the response when the source resistance is changed is given in Figure 4.24.

The circuit shown in Figure 4.25 has both source and load resistances. The impedance of the loaded resonant network can be obtained as

$$Z_{eq} = R_s + \frac{j\omega R_L L}{\omega L + j\omega R_L(\omega^2 LC - 1)} = R_s + \frac{\omega R_L L[\omega L - j\omega R_L(\omega^2 LC - 1)]}{(\omega L)^2 + [\omega R_L(\omega^2 LC - 1)]^2} \quad (4.75)$$

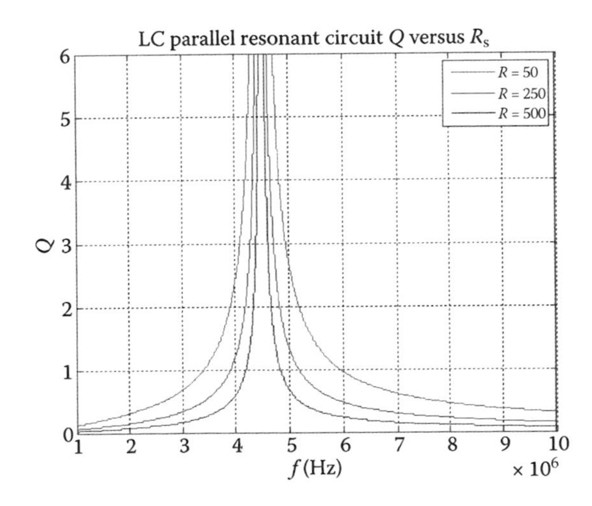

FIGURE 4.23 Quality factor of LC network for different source resistance values.

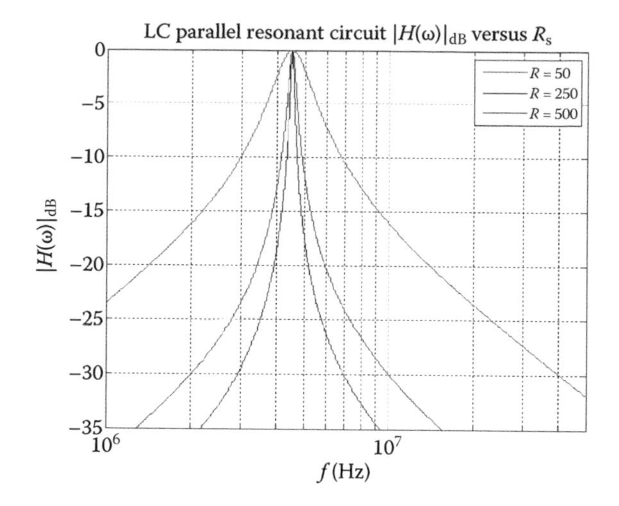

FIGURE 4.24 Attenuation profile of LC network for different source resistance values.

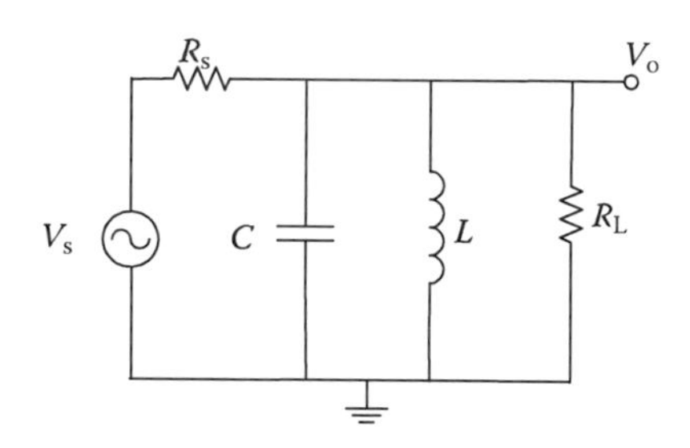

FIGURE 4.25 Loaded LC resonant circuit.

which can be simplified to

$$Z_{eq} = \frac{R_s\left((\omega L)^2 + [\omega R_L(\omega^2 LC - 1)]^2\right) + (\omega L)^2 R_L}{(\omega L)^2 + [\omega R_L(\omega^2 LC - 1)]^2} + j\frac{\omega R_L^2 L(1 - \omega^2 LC)}{(\omega L)^2 + [\omega R_L(\omega^2 LC - 1)]^2} \tag{4.76}$$

The loaded quality factor of the circuit is obtained as

$$Q_L = \frac{\omega R_L^2 L(1 - \omega^2 LC)}{R_s\left((\omega L)^2 + [\omega R_L(\omega^2 LC - 1)]^2\right) + (\omega L)^2 R_L} \tag{4.77}$$

At resonant frequency, this circuit simplifies to the one in Figure 4.26.

The equivalent impedance from Equation 4.76 is equal to

$$Z_{eq} = R_s + R_L \tag{4.78}$$

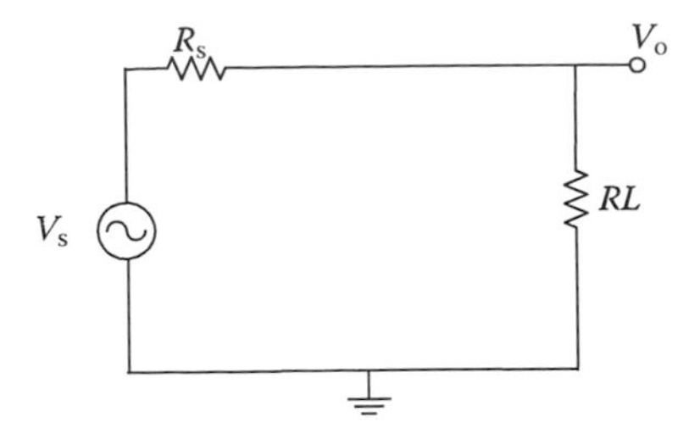

FIGURE 4.26 Equivalent loaded LC resonant circuit at resonance.

It is important to note that at resonant frequency,

$$\omega_0 L = \frac{1}{\omega_0 C} \tag{4.79}$$

4.3.2.4 LC Network Transformations

4.3.2.4.1 RL Networks

When any one of the reactive components is lossy, impedance transformation from parallel to series or series to parallel will greatly facilitate the analysis of the problem.

The equivalent impedances of the series and parallel RL networks shown in Figure 4.27 are

$$Z_s = R_s + j\omega L_s \tag{4.80}$$

$$Z_p = \frac{\omega^2 L_p^2 R_p}{R_p^2 + \omega^2 L_p^2} + j\frac{\omega L_p R_p^2}{R_p^2 + \omega^2 L_p^2} \tag{4.81}$$

To transform the parallel network to a series network in Figure 4.27, it is assumed that their quality factors are the same, $Q_p = Q_s$, and the impedance equality is used as

$$Z_s = Z_p \tag{4.82}$$

This can be accomplished by equating the real and imaginary parts as

$$R_s = R_p \quad \text{or} \quad R_s = \frac{\omega^2 L_p^2 R_p}{R_p^2 + \omega^2 L_p^2} \tag{4.83}$$

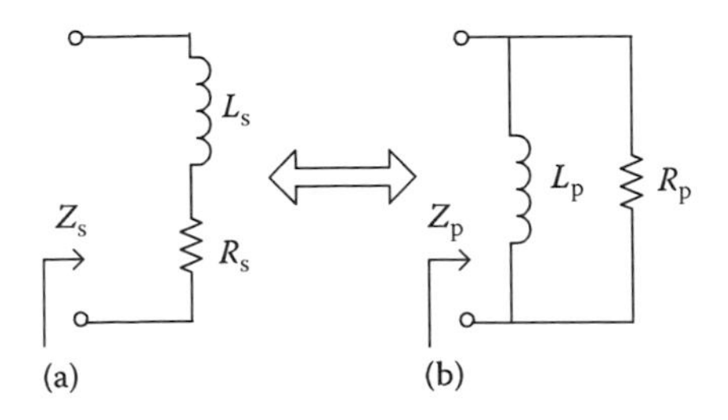

FIGURE 4.27 (a) RL series network and (b) equivalent RL parallel network.

which can be rewritten as

$$R_{s} = \frac{R_{p}}{(R_{p}/\omega L_{p})^{2} + 1} \tag{4.84}$$

Since the quality factor of the parallel network, Q_p, in Figure 4.27b is

$$Q = \frac{R_{p}}{\omega L_{p}} \tag{4.85}$$

substitution of Equation 4.85 into Equation 4.84 gives the equation that relates the resistances of two networks via the quality factor as

$$R_{s} = \frac{R_{p}}{Q_{p}^{2} + 1} \tag{4.86}$$

When the imaginary parts are equated,

$$\omega L_{s} = \frac{\omega L_{p} R_{p}^{2}}{R_{p}^{2} + \omega^{2} L_{p}^{2}} \quad \text{or} \quad L_{s} = \frac{L_{p}}{1 + (\omega L_{p}/R_{p})^{2}} \tag{4.87}$$

which can be expressed as

$$L_{s} = L_{p}\left(\frac{Q_{p}^{2}}{1 + Q_{p}^{2}}\right) \tag{4.88}$$

The same procedure can be applied to transform a series RC network to a parallel RC network. The results are summarized in Table 4.1.

TABLE 4.1
Series RL Network Transformation

(a) Series Network		(b) Parallel Network	
$Q_{s} = \frac{\omega L_{s}}{R_{s}}$	(4.89)	$Q_{p} = \frac{R_{p}}{\omega L_{p}}$	(4.92)
$L_{s} = L_{p}\left(\frac{Q_{p}^{2}}{1 + Q_{p}^{2}}\right)$	(4.90)	$L_{p} = L_{s}\left(\frac{1 + Q_{s}^{2}}{Q_{s}^{2}}\right)$	(4.93)
$R_{s} = \frac{R_{p}}{1 + Q_{p}^{2}}$	(4.91)	$R_{p} = R_{s}\left(1 + Q_{s}^{2}\right)$	(4.94)

Example

A parallel resonant circuit has a 3-dB bandwidth of 5 MHz and a center frequency of 40 MHz. It is given that the resonant circuit has source and load impedances of 100 Ω. The Q of the inductor is given to be 120. The capacitor is assumed to be an ideal capacitor.

a. Design the resonant circuit.
b. What is the loaded Q of the resonant circuit?
c. What is the insertion loss of the network?
d. Obtain the frequency response of this circuit vs. frequency, i.e., plot 20log (V_o/V_{in}) vs. frequency.

Solution

a. Since the inductor is lossy, we need to make a conversion from series to parallel circuit, as shown in Figure 4.28.

   For this,

$$X_p = \frac{R_p}{Q_p} \rightarrow R_p = Q_p X_p = 120X_p$$

The loaded Q of the resonant circuit is

$$Q = \frac{R_{total}}{X_p} \rightarrow 8 = \frac{R_{total}}{X_p}$$

R_{total} is found from

$$8 = \frac{R_{total}}{X_p} = \frac{\dfrac{R_p(50)}{R_p + (50)}}{X_p} \rightarrow 8 = \frac{120X_p(50)}{(120X_p + 50)X_p}$$

So,

$$X_p = \frac{5600}{960} = 5.83\,[\Omega]$$

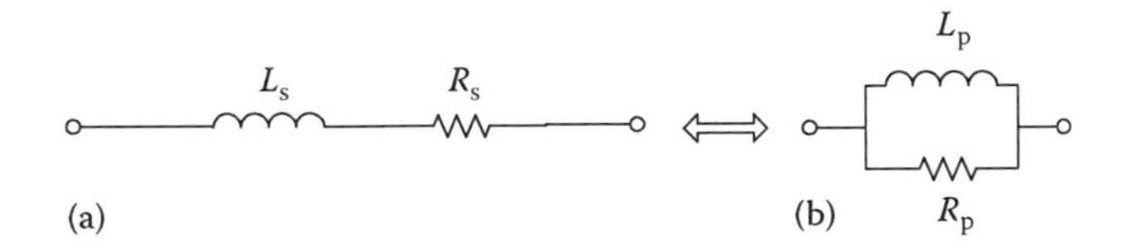

FIGURE 4.28 (a) Series RL network and (b) equivalent parallel RL network.

Then, $R_p = 120X_p = 699.6$ [Ω]. The values of L and C are found from

$$L = \frac{X_p}{\omega} = \frac{5.83}{2\pi(40\times10^6)} = 23.2\,[\text{nH}]$$

and

$$C = \frac{1}{\omega X_p} = \frac{1}{2\pi(40\times10^6)(5.83)} = 682.5\,[\text{pF}]$$

b. $$Q = \frac{f_c}{f_2 - f_1} = \frac{40}{5} = 8$$

c. The load voltage with the resonant circuit is found from

$$V_L = \frac{84.1}{84.1+100}V_s = 0.457V_s$$

The insertion loss is then found from

$$IL = 20\log\left(\frac{0.457V_s}{0.5V_s}\right) = 0.78\text{ dB}$$

d. The attenuation profile is obtained using MATLAB. The MATLAB script and the attenuation profile that is obtained with the program are given by Figure 4.29.

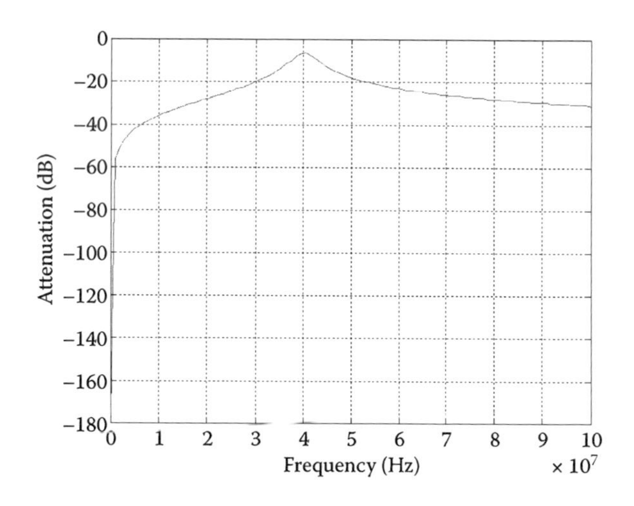

FIGURE 4.29 Attenuation profile for a parallel resonant network.

```
clear
f = linspace(1,100*10^6);
RL = 100;
RS = 100;
RP = 699.6;
Q = 12;
XP = 5.83;
fc = 40*10^6;
wc = 2*pi()*fc;
L = XP/(wc);
C = 1/(wc*XP);

w = 2*pi.*f;
XL=1j.*w.*L;
XC=-1j./(w.*C);
Xeq=(XL.*XC)./(XL+XC);
Req=(RP*RL)./(RP+RL);
Zeq=(Req.*Xeq)./(Req+Xeq);
S21=20.*log10(abs(Zeq./(Zeq+RS)));

plot(f,S21);

grid on
title('Attenuation Profile')
xlabel('Frequency (Hz)')
ylabel('Attenuation (dB)')
```

4.3.2.4.2 RC Networks

Consider the RC parallel circuits shown in Figure 4.30. To transform the parallel RC network to a series RC network so that both circuits would have the same quality factors, the same approach in Section 4.3.2.4.1 is used, and the impedances are equated as

$$Z_s = Z_p \tag{4.95}$$

where

$$Z_p = \frac{R_p}{1+(\omega R_p C_p)^2} - \frac{\omega R_p^2 C_p}{1+(\omega R_p C_p)^2} \tag{4.96}$$

$$Z_s = R_s - \frac{j}{\omega C_s} \tag{4.97}$$

Equating the real and imaginary parts gives

$$R_s = \frac{R_p}{1+(\omega R_p C_p)^2} = \frac{R_p}{1+Q_p^2} \tag{4.98}$$

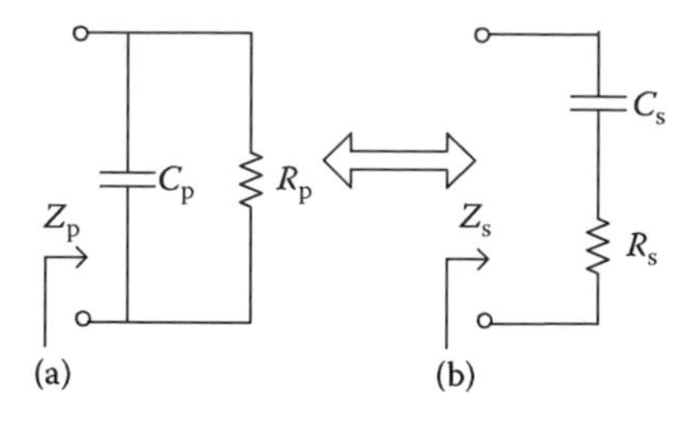

FIGURE 4.30 (a) Parallel RC network and (b) equivalent series RC network.

TABLE 4.2
Parallel RL Network Transformation

(a) Series Network		(b) Parallel Network	
$Q_s = \dfrac{1}{\omega R_s C_s}$	(4.99)	$Q_p = \omega R_p C_p$	(4.102)
$C_s = C_p\left(\dfrac{1+Q_p^2}{Q_p^2}\right)$	(4.100)	$C_p = C_s\left(\dfrac{Q_s^2}{1+Q_s^2}\right)$	(4.103)
$R_s = \dfrac{R_p}{1+Q_p^2}$	(4.101)	$R_p = R_s\left(1+Q_s^2\right)$	(4.104)

and

$$\frac{1}{\omega C_s} = \frac{\omega R_p^2 C_p}{1+(\omega R_p C_p)^2} \tag{4.105}$$

which leads to

$$C_s = C_p\left(\frac{1+(\omega R_p C_p)^2}{(\omega R_p C_p)^2}\right) = C_p\left(\frac{1+Q_p^2}{Q_p^2}\right) \tag{4.106}$$

The same procedure can be applied to transform the series network to a parallel RL network. The results are summarized in Table 4.2.

4.3.2.5 LC Network with Series Loss

When a parallel LC network has a component with series loss, we can then use the transformations discussed. Sections 4.3.2.4.1 and 4.3.2.4.2 can be used to simplify the analysis of the circuit. Consider the parallel LC network with series loss shown in Figure 4.31.

The transformation from Figure 4.31a to b can be done using the relations given by

$$R_p = R_s\left(1+Q_s^2\right) \tag{4.107}$$

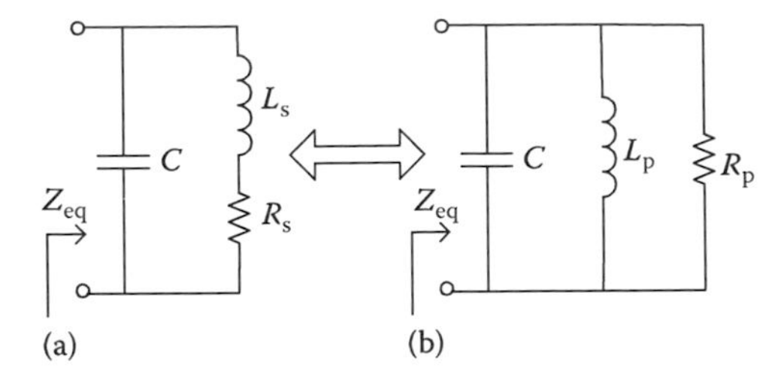

FIGURE 4.31 (a) Series loss added for inductor and (b) equivalent parallel network.

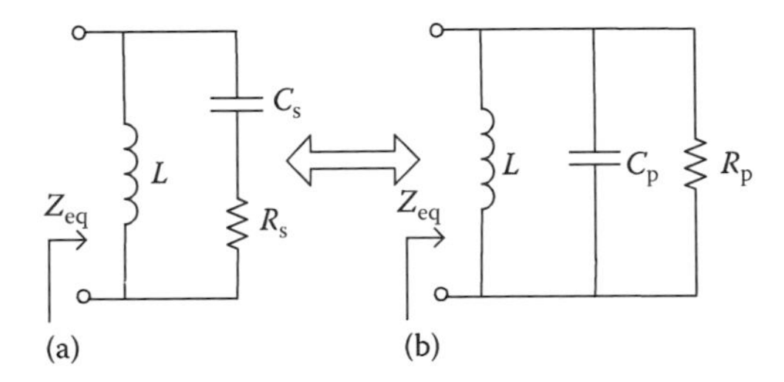

FIGURE 4.32 (a) Series loss added for capacitor and (b) equivalent parallel network.

$$L_p = L_s\left(\frac{1+Q_s^2}{Q_s^2}\right) \tag{4.108}$$

$$Q_s = \frac{\omega L_s}{R_s} \tag{4.109}$$

The same procedure can be applied if the capacitor has a series loss, as shown in Figure 4.32a. The transformation from Figure 4.32a to b can then be performed using Equations 4.107, 4.109, and

$$C_p = C_s\left(\frac{Q_s^2}{1+Q_s^2}\right) \tag{4.110}$$

4.4 COUPLING OF RESONATORS

A single resonator can be coupled via capacitors or inductors to produce a wide, flat passband and steeper skirts. The conventional way of coupling single identical resonators via inductor is shown in Figure 4.33. The circuit shown in Figure 4.33 will be exhibited as a single shunt tapped inductance with a 6-dB/octave slope below resonance since each reactive element presents this amount for the slope. The circuit will be behaving like a three-element, low-pass filter above resonance, as shown in Figure 4.34 with an 18-dB/octave slope.

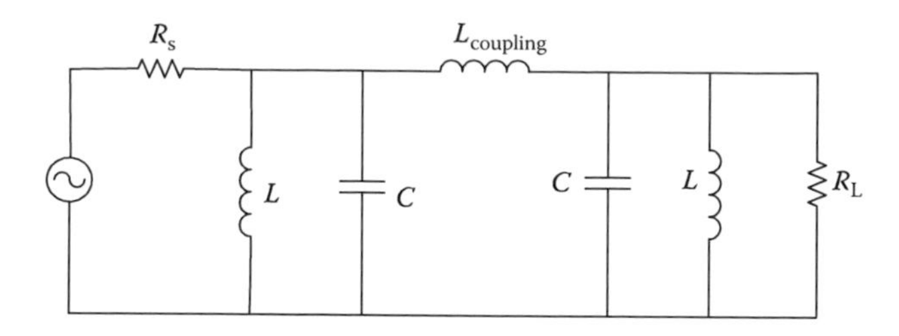

FIGURE 4.33 Inductively coupled resonators.

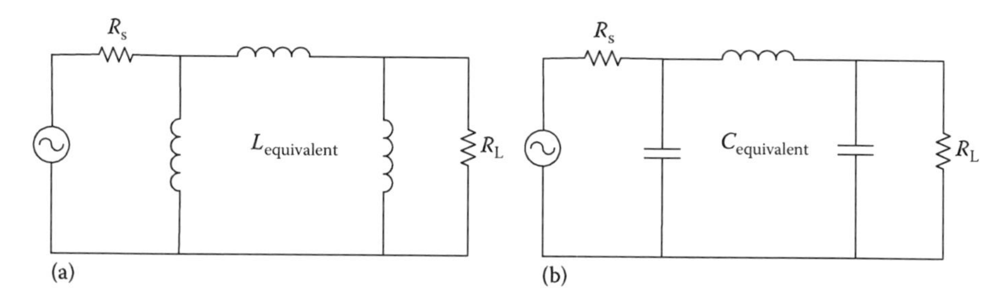

FIGURE 4.34 Inductively coupled resonators: (a) below resonance and (b) above resonance.

The value of the inductor used to couple two identical resonant circuits is found from

$$L_{coupling} = Q_R L \tag{4.111}$$

where Q_R is the loaded quality factor of the single resonator. Coupling of single identical resonators via a capacitor is shown in Figure 4.35. The circuit shown in Figure 4.35 presents a three-element, low-pass filter with an 18-dB/octave slope below resonance and effective single shunt tapped capacitance with 6-dB/octave slope above resonance, as shown in Figure 4.36.

The value of the capacitor used to couple two identical resonant circuits is found from

$$C_{coupling} = \frac{C}{Q_R} \tag{4.112}$$

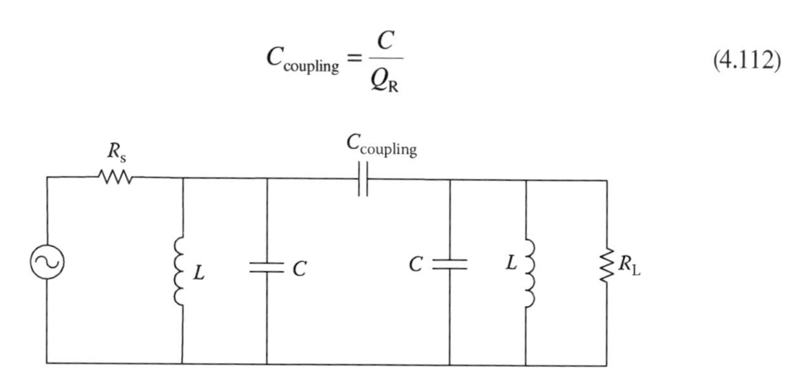

FIGURE 4.35 Capacitively coupled resonators.

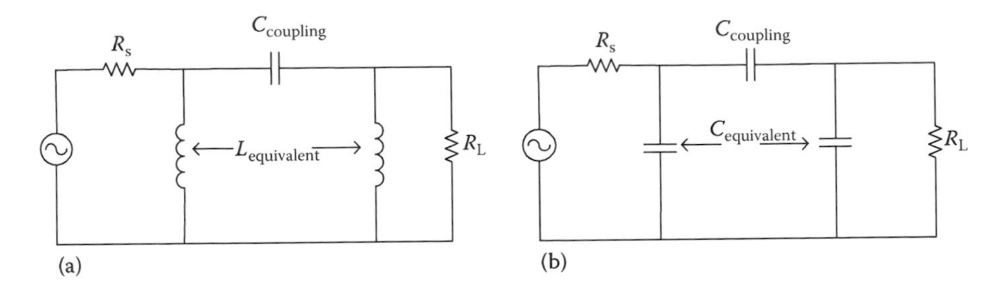

FIGURE 4.36 Inductively coupled resonators: (a) below resonance and (b) above resonance.

The relation between the loaded quality factor of the single resonator and the total loaded Q_T of the entire resonator circuit is obtained from

$$Q_R = \frac{Q_T}{0.707} \tag{4.113}$$

Example

Design a two-resonator tuned circuit at a resonant frequency of 75 MHz, a 3-dB bandwidth of 3.75 MHz, and source and load impedances of 100 and 1000 Ω, respectively, using inductively coupled and capacitively coupled circuits shown in Figures 4.33 and 4.35, respectively. Assume that inductors Q of 85 are at the frequency of interest. Use CAD to obtain the frequency response of the coupled resonator circuits.

Solution

The loaded quality factor of the overall resonant circuit is

$$Q_{total} = \frac{f_o}{BW} = \frac{75}{3.75} = 20 \tag{4.114}$$

Then, the quality factor of the single resonator is found from

$$Q_R = \frac{Q_{total}}{0.707} = \frac{20}{0.707} = 28.3 \tag{4.115}$$

The inductor is lossy; its quality factor is found from

$$Q_p = \frac{R_p}{X_p} = 85 \quad \text{or} \quad R_p = 85X_p \tag{4.116}$$

The loaded quality factor of a single resonator found in Equation 4.115 can also be found from

$$Q_R = \frac{R_{total}}{X_p} \tag{4.117}$$

where

$$R_{\text{total}} = \frac{R_s' R_p}{R_s' + R_p} \tag{4.118}$$

$$R_s' = 1000\,[\Omega] \tag{4.119}$$

Combining Equations 4.117 through 4.119 leads to

$$Q_R = \frac{R_s' R_p}{(R_s' + R_p) X_p} = 28.3 \tag{4.120}$$

So,

$$X_p = \frac{R_s' R_p}{(R_s' + R_p) Q_R} = \frac{1000 R_p}{(1000 + R_p) 28.3} \tag{4.121}$$

or

$$X_p = \frac{1000(85) X_p}{(1000 + (85) X_p) 28.3} = 23.57\,\Omega \tag{4.122}$$

Then,

$$R_p = 85 X_p = 2003\ \Omega \tag{4.123}$$

The component values are

$$L_1 = L_2 = \frac{X_p}{\omega} = 50\,\text{nH} \tag{4.124}$$

$$C_s = \frac{1}{X_p \omega} = 90\,\text{pF} \tag{4.125}$$

The coupling inductance is found from

$$L_{12} = Q_R L = (28.3) 50\ \text{nH} = 1.415\ \mu\text{H} \tag{4.126}$$

and the coupling capacitance is found from

$$C_{12} = \frac{C}{Q_R} = \frac{90 \times 10^{-12}}{28.3} = 3.18 \times 10^{-12} \tag{4.127}$$

The attenuation profile is obtained using Ansoft Designer with the circuits shown in Figures 4.37 and 4.38. The response for capacitive coupling is given in Figure 4.39, and inductive coupling is given in Figure 4.40.

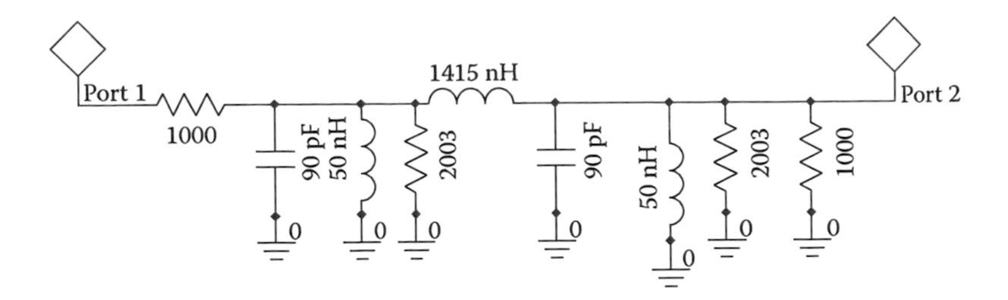

FIGURE 4.37 Inductively coupled resonators.

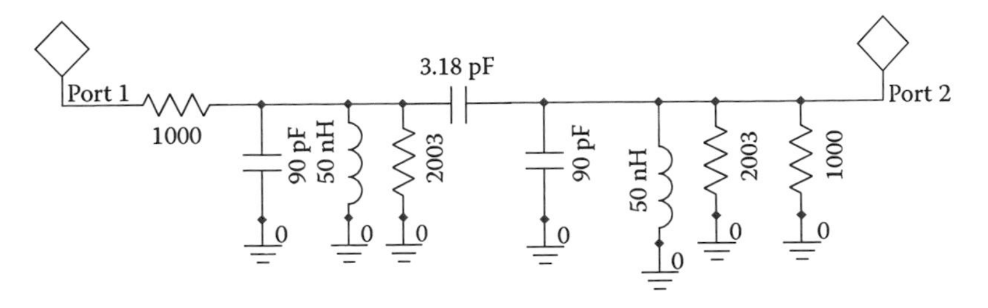

FIGURE 4.38 Capacitively coupled resonators.

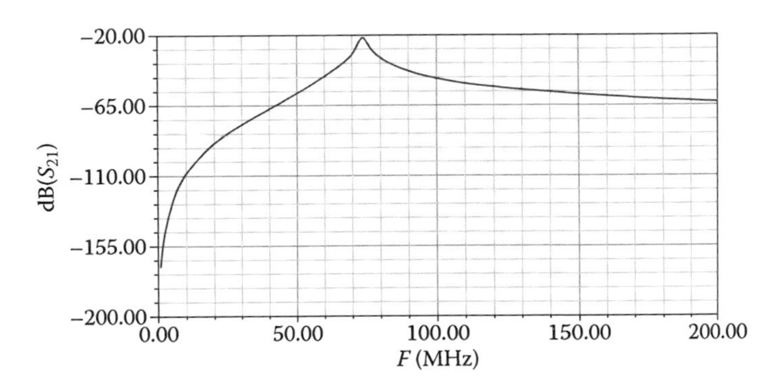

FIGURE 4.39 Attenuation profile for capacitively coupled resonators.

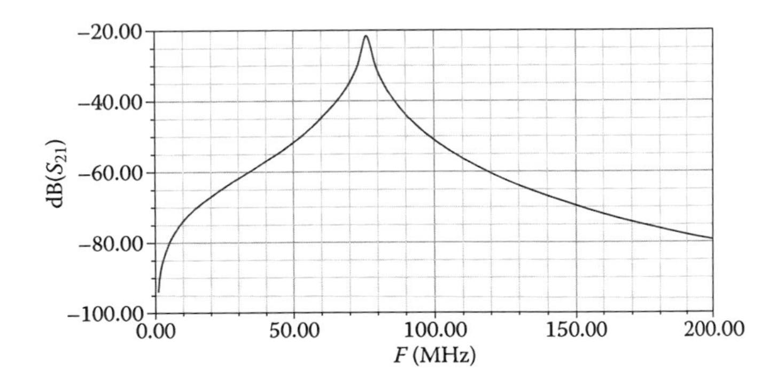

FIGURE 4.40 Attenuation profile for inductively coupled resonators.

4.5 LC RESONATORS AS IMPEDANCE TRANSFORMERS

4.5.1 INDUCTIVE LOAD

Consider the LC parallel network shown in Figure 4.20 by ignoring the source resistance. This time, assume that the loss resistor is part of the load resistance, *R*. The new circuit can be illustrated in Figure 4.41.

The equivalent impedance at the input for the circuit in Figure 4.41 can be written as

$$Z_{eq} = \frac{R}{(1-\omega^2 LC)^2 + (\omega RC)^2} + j\frac{\omega[(L - CR^2) - \omega^2 L^2 C]}{(1-\omega^2 LC)^2 + (\omega RC)^2} \tag{4.128}$$

The resonant frequency of the network is now equal to

$$\omega_0 = \sqrt{\frac{L - CR^2}{L^2 C}} \rightarrow f_0 = \frac{1}{2\pi}\sqrt{\frac{L - CR^2}{L^2 C}} \tag{4.129}$$

At resonance frequency, the equivalent impedance will be purely resistive, $Z_{eq} = R_{eq}$, and equal to

$$R_{eq} = \frac{R}{(1-\omega^2 LC)^2 + (\omega RC)^2} = \frac{L}{RC} \tag{4.130}$$

Hence, the network at resonance converts the inductive load impedance to a resistive impedance. Since the quality factor, Q_{load}, of the load at resonance is

$$Q_{load} = \frac{\omega_0 L}{R} \tag{4.131}$$

the following relation can be written between the load quality factor and the equivalent impedance at resonance as

$$R_{eq} = \frac{L}{RC} = \left(Q_{load}^2 + 1\right)R \tag{4.132}$$

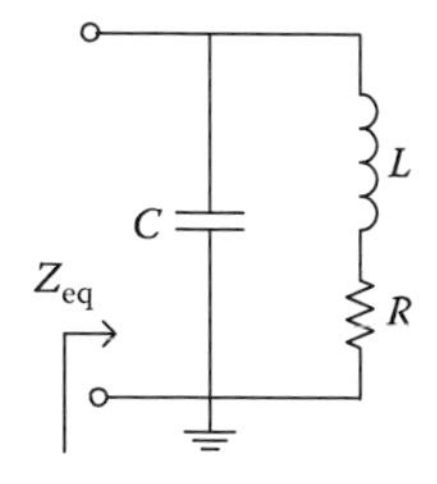

FIGURE 4.41 LC impedance transformer for inductive load.

4.5.2 Capacitive Load

The same principle for inductive load can be applied to convert the capacitive load to a resistive load at the resonant frequency using the LC resonant circuit shown in Figure 4.42.

The equivalent impedance at the input for the circuit in Figure 4.42 can be written as

$$Z_{eq} = \frac{\omega^4 RL^2C^2}{(1-\omega^2 LC)^2 + (\omega RC)^2} + j\frac{\omega L\left[\omega^2(R^2C^2 - LC)+1\right]}{(1-\omega^2 LC)^2 + (\omega RC)^2} \quad (4.133)$$

The resonant frequency of the network is now equal to

$$\omega_0 = \sqrt{\frac{1}{LC - R^2C^2}} \rightarrow f_0 = \frac{1}{2\pi}\sqrt{\frac{1}{LC - R^2C^2}} \quad (4.134)$$

At resonance frequency, $Z_{eq} = R_{eq}$, and it can be expressed as

$$R_{eq} = \frac{L}{RC} \quad (4.135)$$

Hence, the network at resonance converts the capacitive load impedance to a resistive input impedance. Since the quality factor, Q_{load}, of the load at resonance is

$$Q_{load} = \frac{1}{\omega_0 RC} \quad (4.136)$$

then the following relation can be established:

$$R_{eq} = \frac{L}{RC} = \omega_0 L Q_{load} \quad (4.137)$$

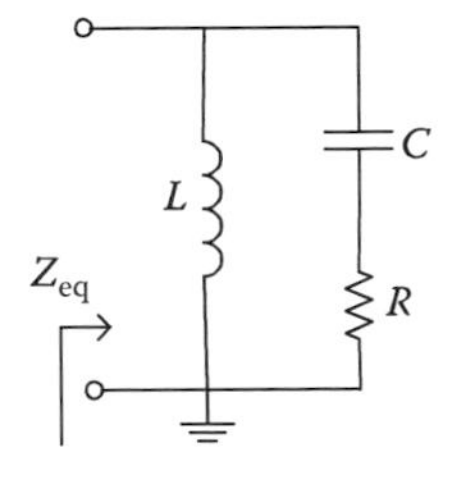

FIGURE 4.42 LC impedance transformer for capacitive load.

Example

An amplifier output needs to be terminated with a load line resistance of 2000 Ω at 1.6 MHz. It is given in the data sheet that the transistor has 20 pF at 1.6 MHz. There is an inductive load connected to the output of the load line circuit of the amplifier with $R_L = 5\ \Omega$. The configuration of this circuit is given in Figure 4.43.

a. Calculate the values of L and C by assuming that the load inductor has a negligible loss, i.e., $r = 0$.
b. The inductor is changed to a magnetic core inductor, which has the quality factor of 50. Calculate the loss resistance, r, for the reactive component values obtained in (a). What is the value of new load line resistance?
c. If the quality factor is 50, the inductor is 50, and the load line resistor is required to be 2000 Ω as set in the problem, what are the values of L and C with $R_L = 5\ \Omega$?

Solution

It is given that $R_{eq} = 2000\ \Omega$, $R_L = R = 5\ \Omega$, $C_{tran} = 20$ pF, $f = 1.6$ MHz, $\omega_0 = 10^7$ rad/s, and $C_T = C_{tran} + C$.

a. When $r = 0\ \Omega$, Equation 4.132 can be used as

$$R_{eq} = \left(Q_{load}^2 + 1\right)R \rightarrow \frac{R_{eq}}{R} - 1 = Q_{load}^2 \rightarrow Q_{load} = 19.98 \tag{4.138}$$

From Equation 4.131,

$$Q_{load} = \frac{\omega_0 L}{R} \rightarrow L = \frac{Q_{load} R}{\omega_0} \rightarrow L = 10\,[\mu\text{H}] \tag{4.139}$$

Now, using Equation 4.130,

$$R_{eq} = \frac{L}{RC_T} \rightarrow C_T = \frac{L}{R_{eq} R} \rightarrow C_T = 1\,[\text{nF}] \tag{4.140}$$

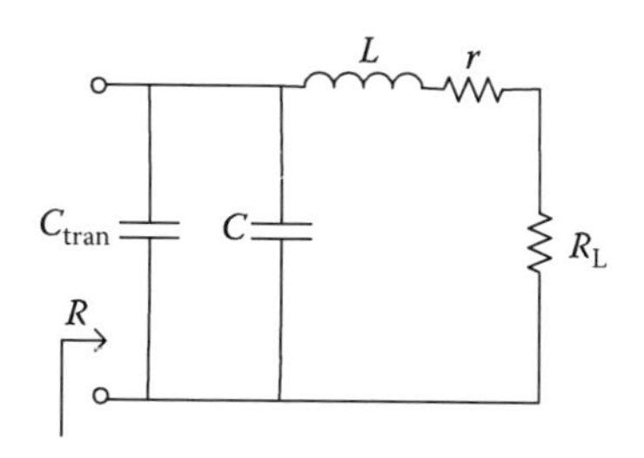

FIGURE 4.43 Amplifier output load line circuit.

Since

$$C_T = C_{tran} + C \rightarrow C = C_T - C_{tran} \rightarrow C = 980 \text{ [pF]} \tag{4.141}$$

b. The Q of the inductor is given to be equal to 50. Then,

$$Q_{inductor} = \frac{\omega_0 L}{r} \rightarrow r = \frac{\omega_0 L}{Q_{inductor}} \rightarrow r = 2 \ [\Omega] \tag{4.142}$$

So, the new load resistance, R_{eq}, from Equation 4.120 is

$$R_{eq} = \frac{L}{(R+r)C_T} = \frac{10\times 10^{-6}}{(7)(1\times 10^{-9})} = 1428.6 \ [\Omega] \tag{4.143}$$

c. The Q of the inductor is given to be equal to 50. Then,

$$Q_{inductor} = \frac{\omega_0 L}{r} \rightarrow r = 2\times 10^5 L = aL \tag{4.144}$$

Since

$$R_{eq} = R\left(Q_{load}^2 + 1\right) \rightarrow R_{eq} = (R+r)\left(Q_{load}^2 + 1\right) = \frac{\omega_0^2 L^2 + (R+aL)^2}{R+aL} \tag{4.145}$$

which leads to the solution for L as

$$L^2 - L\frac{a(R_{eq} - 2R)}{\omega_0^2 + a^2} - \frac{R(R_{eq} - R)}{\omega_0^2 + a^2} = 0 \tag{4.146}$$

From Equation 4.146, the inductance value is found as L = 12.2 [μH]. Substituting the value of L into Equation 4.144 gives the value of r as

$$r = 2 \times 10^5 L = 2.44 \ [\Omega]$$

where

$$L = 12.2 \ [\mu\text{H}]. \tag{4.147}$$

Using Equation 4.140,

$$R_{eq} = \frac{L}{RC_T} \rightarrow C_T = \frac{L}{R_{eq}R} \rightarrow C_T = 820 \text{ [pF]} \tag{4.148}$$

Since

$$C_T = C_{tran} + C \rightarrow C = 820 - 20 \rightarrow C = 800 \text{ [pF]} \tag{4.149}$$

Design Example

Develop a MATLAB program for the amplifier network given in Figure 4.44 to interface 10-Ω differential output of the amplifier 1 (Amp 1) to 100-Ω input impedance of the second amplifier (Amp 2) using an unbalanced *L–C* network at 100 MHz.

Solution

The LC unbalanced matching network will be implemented, as shown in Figure 4.45. The LC matching network can be simplified and shown in Figure 4.46.

The following generic MATLAB script designs an unbalanced LC network to match Amplifier 1 output differential impedance to Amplifier 2 input differential impedance.

```
%Script takes in user defined inputs for output and input impedance
%of differential amplifier circuit, current in, voltage in, and operational
%frequency.  Based off these inputs, an unbalanced LC matching network is
%designed.  A toroidal inductor with powdered iron core is then designed
%to satisfy inductor requirements.

%User-defined input
R1 = input('Enter output impedance of amplifier 1 (ohms): ');
R2 = input('Enter input impedance of amplifier 2 (ohms): ');
fop = input('Enter operational frequency (Hz): ');

%Calculates unbalanced LC network
Ra = .5*R1;
Rb = .5*R2;
Qlc = sqrt(max(Ra,Rb)/min(Ra,Rb)-1);     %Determines Q of network
fopG = fop*1e-9;    %Expresses operational frequency in Ghz
L = 0.159*Qlc*min(Ra,Rb)/fopG;
Ca = 159*Qlc/(fopG*max(Ra,Rb));
L = L*10^-9;    %Inductor value for LC network
Ca = Ca*10^-12; %Capacitor value for LC network

disp('You will need 2 inductors of inductance (H): '); disp(L);
disp('You will need 2 capacitors of capacitance (F): '); disp(Ca);
```

When the program is executed, the following calculated values are displayed via MATLAB Command Window.

```
Enter output impedance of amplifier 1 (ohms): 10
Enter input impedance of amplifier 2 (ohms): 100
Enter operational frequency (Hz): 100e6
You will need 2 inductors of inductance (H):
  2.3850e-008
You will need 2 capacitors of capacitance (F):
  9.5400e-011
```

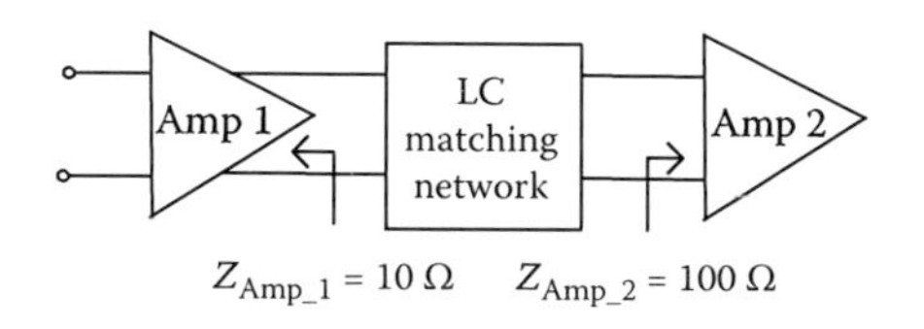

FIGURE 4.44 Amplifier impedance transformer design.

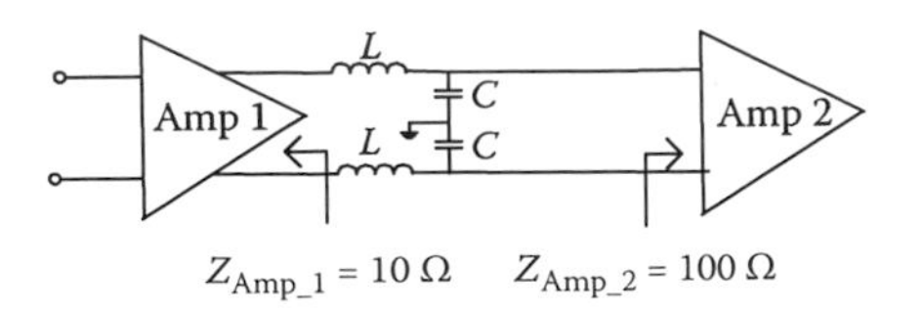

FIGURE 4.45 Implementation of unbalanced LC network.

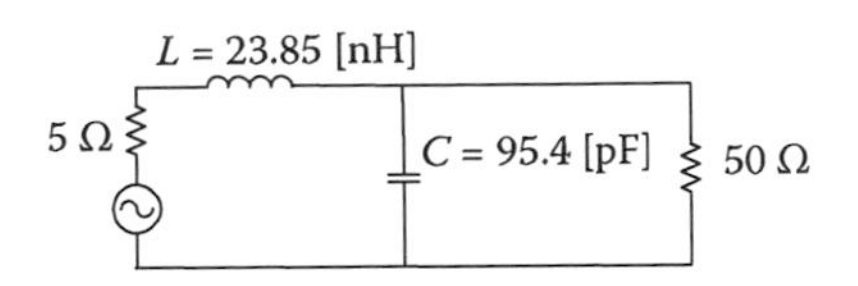

FIGURE 4.46 Simplified illustration of LC matching network for amplifier network.

4.6 TAPPED RESONATORS AS IMPEDANCE TRANSFORMERS

4.6.1 Tapped-*C* Impedance Transformer

To understand the operation of a tapped-*C* impedance transformer, consider the capacitive voltage divider circuit shown in Figure 4.47. The output voltage can be found from

$$v_o = v_i \frac{1/(j\omega C_2)}{1/(j\omega C_2)+1/(j\omega C_1)} = v_i \frac{C_1}{C_1+C_2} \tag{4.150}$$

which can be expressed as

$$v_o = v_i n$$

where

$$n = \frac{C_1}{C_1+C_2} \tag{4.151}$$

Now, assume that there is a load resistor connected to the output of the capacitor and a resonator inductor connected to the input of the divider circuit, as shown in Figure 4.48.

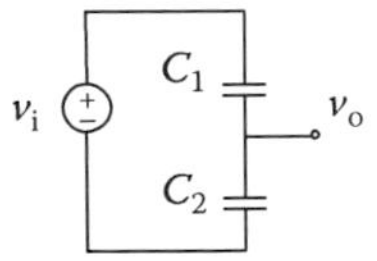

FIGURE 4.47 Capacitive voltage divider.

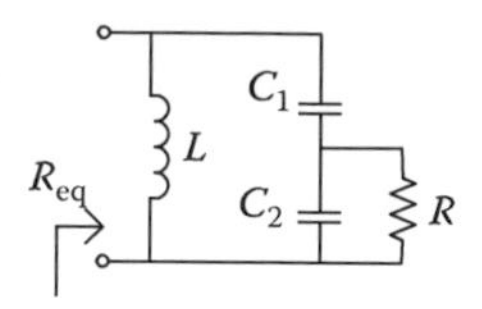

FIGURE 4.48 Capacitive voltage divider with load resistor.

The output shunt connected circuit is then converted to series connection by using parallel-to-series conversion introduced before, as shown in Figure 4.49.

The relation of the components in Figurcs 4.48 and 4.49 is

$$C_s = C_2\left(\frac{Q_p^2+1}{Q_p^2}\right) \tag{4.152}$$

$$R_s = \frac{R}{Q_p^2+1} \tag{4.153}$$

$$R_s = \frac{R_{eq}}{Q_r^2+1} \tag{4.154}$$

where

$$Q_p = \frac{R}{X_{C_2}} = \omega_0 R C_2 \tag{4.155}$$

$$Q_r = \frac{R_{eq}}{\omega_0 L} = \frac{1}{\omega_0 R_s C} \tag{4.156}$$

The equivalent capacitance can then be written as

$$C = \frac{C_1 C_s}{C_1 + C_s} \tag{4.157}$$

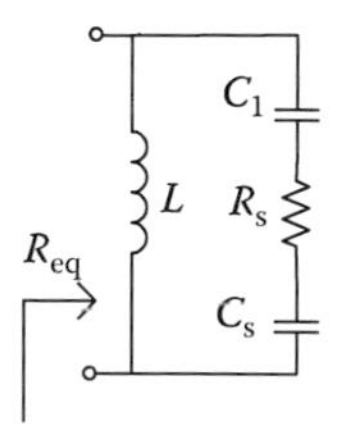

FIGURE 4.49 Capacitive voltage divider with parallel-to-series transformation.

Equating Equations 4.153 through 4.156 gives

$$Q_{\mathrm{p}} = \sqrt{\left[\left(Q_{\mathrm{r}}^{2}+1\right)\frac{R}{R_{\mathrm{eq}}}-1\right]} \tag{4.158}$$

Overall, using the transformations given, the tapped-C circuit in Figure 4.48 can be simplified and transformed to the one in Figure 4.50 with the following relations as

$$R_{\mathrm{s}}' = R_{\mathrm{s}}\left(1+\frac{C_1}{C_2}\right)^2 \tag{4.159}$$

and

$$C_{\mathrm{T}} = \frac{C_1 C_2}{C_1 + C_2} \tag{4.160}$$

At resonance, the circuit can be simplified to the impedance transformer circuit shown in Figure 4.51 with

$$N^2 = \frac{R}{R_{\mathrm{eq}}} \tag{4.161}$$

Substitution of Equations 4.159 and 4.160 into Equation 4.158 gives

$$Q_{\mathrm{p}} = \sqrt{\left[\frac{\left(Q_{\mathrm{r}}^{2}+1\right)}{N^2}-1\right]} \tag{4.162}$$

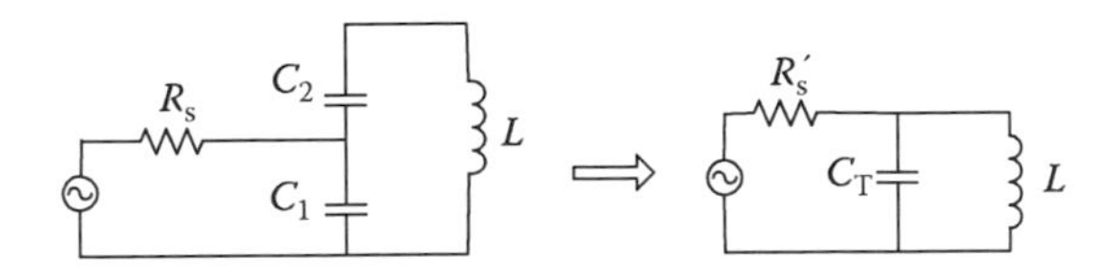

FIGURE 4.50 Tapped equivalent circuit.

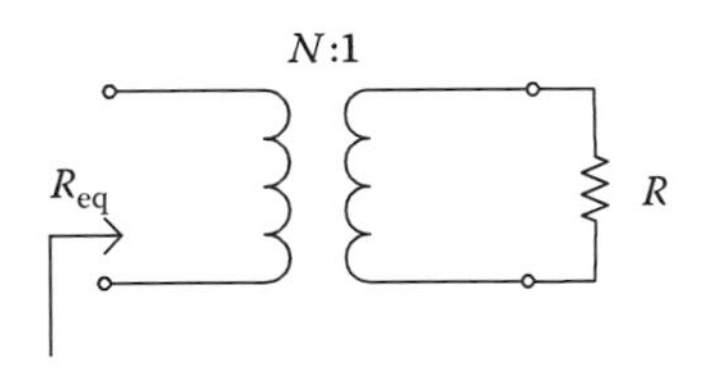

FIGURE 4.51 Equivalent tapped-C circuit representation using transformer.

Example

Design a parallel resonant circuit with the tapped-C approach where the 3-dB bandwidth is 3 MHz and the center frequency is 27.12 MHz. The resonant circuit will operate between a source resistance of 50 Ω and a load resistance of 100 Ω. Assume that the Q of the inductor is 150 at 27.12 MHz.

a. Obtain the element values of the circuit shown in Figure 4.52a.
b. Obtain the element values of the equivalent circuit shown in Figure 4.52b.
c. Use MATLAB to obtain the frequency response of the circuits shown in Figure 4.52a and b.

Solution

The equivalent circuit shown in Figure 4.52b has

$$R_s' = 100\ \Omega$$

Since

$$R_s' = R_s\left(1+\frac{C_1}{C_2}\right)^2 \rightarrow \left(1+\frac{C_1}{C_2}\right)^2 = 2 \rightarrow C_1 = 0.414C_2$$

Since the inductor is lossy,

$$X_p = \frac{R_p}{Q_p} \rightarrow R_p = Q_p X_p = 150X_p$$

The loaded Q of the resonant circuit is found from

$$Q = \frac{f_c}{f_2 - f_1} = \frac{27.12}{3} = 9.04$$

Since

$$Q = \frac{R_{\text{total}}}{X_p} \rightarrow 9.04 = \frac{R_{\text{total}}}{X_p} = \frac{50R_p}{(50+R_p)X_p} \rightarrow 9.04 = \frac{(150X_p)50}{(50+150X_p)X_p}$$

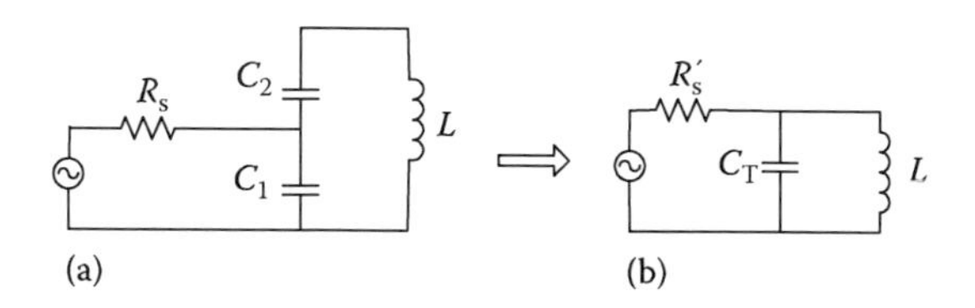

FIGURE 4.52 (a) Parallel resonant circuit with tapped-C and (b) equivalent tapped-C network.

then

$$X_p = \frac{7048}{1356} = 5.2\ [\Omega]$$

So,

$$R_p = 150X_p = 780\ [\Omega]$$

The values of L and C are found from

$$L = \frac{X_p}{\omega} = \frac{5.2}{2\pi(27.12\times10^6)} = 30.5\ [\text{nH}]$$

$$C_T = \frac{1}{\omega X_p} = \frac{1}{2\pi(27.12\times10^6)(5.2)} = 1128\ [\text{pF}]$$

The capacitor values for the circuit in Figure 4.52a are found from

$$C_T = \frac{C_1C_2}{C_1+C_2} \rightarrow 1128 = \frac{0.414C_2}{1.414} \rightarrow C_2 = 3852.6\ [\text{pF}] \quad \text{and} \quad C_1 = 1595\ [\text{pF}]$$

The attenuation profile is obtained with the MATLAB script given below and plotted in Figure 4.53.

```
clear
f = linspace(1,100*10^6);
RL = 100;
RS = 50;
RP = 780;
XP = 5.2;
fc = 27.12*10^6;
wc = 2*pi()*fc;
L = XP/(wc);
C = 1/(wc*XP);

w = 2*pi.*f;
XL=1j.*w.*L;
XC=-1j./(w.*C);
Xeq=(XL.*XC)./(XL+XC);
Req=(RP*RL)./(RP+RL);
Zeq=(Req.*Xeq)./(Req+Xeq);
S21=20.*log10(abs(Zeq./(Zeq+RS)));

plot(f,S21);

grid on
title('Attenuation Profile')
xlabel('Frequency (Hz)')
ylabel('Attenuation (dB)')
```

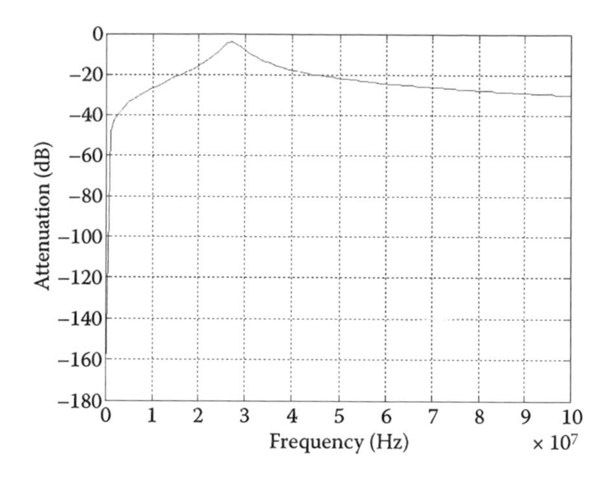

FIGURE 4.53 Attenuation profile for parallel resonant circuit with tapped-C.

4.6.2 Tapped-L Impedance Transformer

The typical tapped-L impedance transformer circuit is shown in Figure 4.54a. The same procedure outlined in Section 4.6.1 can be followed, and the circuit can be converted to its equivalent circuit shown in Figure 4.54b by using parallel-to-series transformation relations as previously done.

Overall, the tapped-L circuit in Figure 4.54a can be simplified to the one shown in Figure 4.55 using the transformations with the following relation as

$$R_s' = R_s \left(\frac{n}{n_1} \right)^2 \tag{4.163}$$

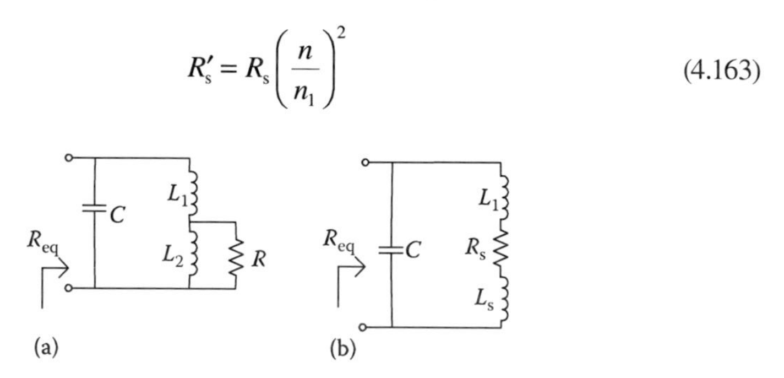

FIGURE 4.54 (a) Tapped-L impedance transformer. (b) Tapped-L impedance transformer with parallel-to-series transformation.

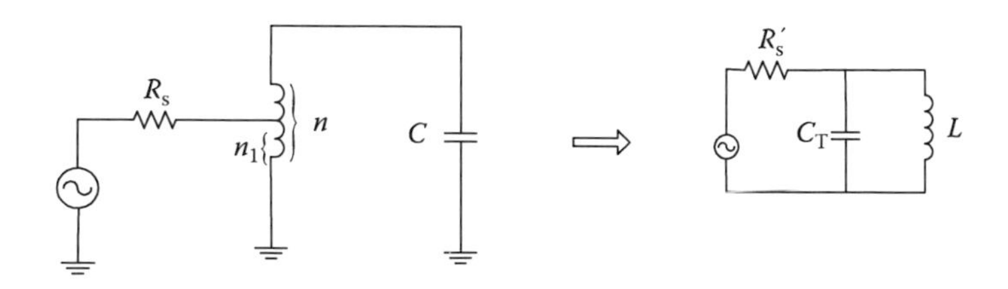

FIGURE 4.55 Tapped-L equivalent circuit.

Design Example

Consider the transistor amplifier circuit given in Figure 4.56. It is required to match the low transistor input impedance 14.1 Ω to 50 Ω using the *C*-tapped circuit at the input and match the output of the transistor impedance 225 Ω down to 50 Ω using the *L*-tapped circuit at 100 MHz. The 3-dB bandwidth of the amplifier circuit is given to be 10 MHz. The loaded *Q* of the input matching network is given to be 5, and the loaded *Q* of the output matching network is given to be equal to 7.5.

a. Calculate C_1, C_2, C_3, L_1, L_2, and the impedance transformer ratio n_1/n_2 for this amplifier circuit.
b. Develop the MATLAB GUI to match the input and output impedances of the transistor to the given source impedance at the input using the *C*-tapped circuit and the given load impedance at the output using the *L*-tapped circuit just like the amplifier circuit shown in Figure 4.56. Your program should take source impedance, load impedance, 3-dB bandwidth, center frequency, and quality factors of the input and output matching networks as input values, and calculate and illustrate C_1, C_2, C_3, L_1, L_2, and the impedance transformer ratio n_1/n_2 as output values. Test the accuracy of your program using the values in part (a).
c. Consider the amplifier circuit given in Figure 4.56 with the calculated values from part (a). Represent the complete circuit with *ABCD* network parameters and calculate the overall *ABCD* parameters of the network and gain. Check your analytical results with the results of your program.

Solution

For part (a), the input side of the network is given in Figure 4.57.

From Equation 4.159,

$$R_s' = R_s\left(1+\frac{C_b}{C_2}\right)^2 \rightarrow \frac{C_b}{C_2} = C_{ratio} = \sqrt{\frac{R_s'}{R_s}} - 1 = \sqrt{\frac{50}{14.1}} - 1 = 0.883 \qquad (4.164)$$

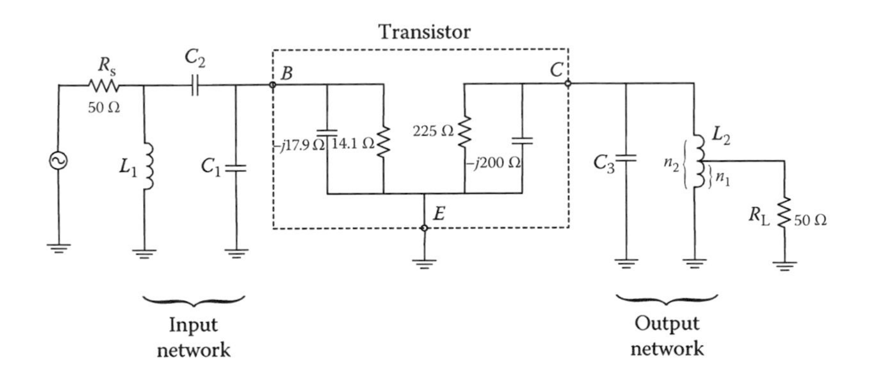

FIGURE 4.56 Tapped-*C* and -*L* implementation for amplifiers.

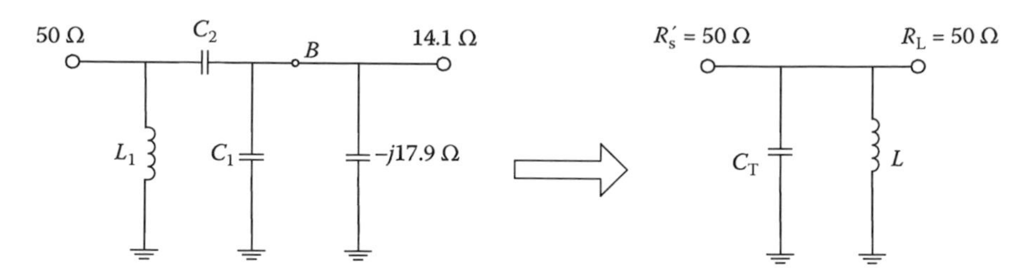

FIGURE 4.57 Input network transformation for tapped-C transformer.

Since the quality factor of the input matching network is 5, $Q_{in} = 5$,

$$X_{C_b} = \frac{R_s}{Q_{in}} = \frac{14.1}{5} = 2.82 \rightarrow C_b = \frac{1}{\omega X_{C_b}} = 564.3\ [\text{pF}] \tag{4.165}$$

From the given information for the transistor,

$$C_{\text{tran\_in}} = \frac{1}{\omega(17.9)} = 88.91\ [\text{pF}] \tag{4.166}$$

Then,

$$C_1 = C_b - C_{\text{trans\_in}} = 564.3 - 88.91 = 475.47\ [\text{pF}] \tag{4.167}$$

Now, C_2 can be found from

$$\frac{C_b}{C_2} = C_{\text{ratio}} = 0.883 \rightarrow C_2 = \frac{C_b}{C_{\text{ratio}}} = \frac{564.3}{0.883} = 639\ [\text{pF}] \tag{4.168}$$

Now, using Equation 4.160,

$$C_T = \frac{C_b C_2}{C_b + C_2} = 299.7\ [\text{pF}] \tag{4.169}$$

Since

$$X_{C_T} = \frac{1}{\omega C_T} = 5.31 \tag{4.170}$$

then inductance L_1 can be found from

$$L_1 = \frac{X_{C_T}}{\omega} = 8.45\ [\text{nH}] \tag{4.171}$$

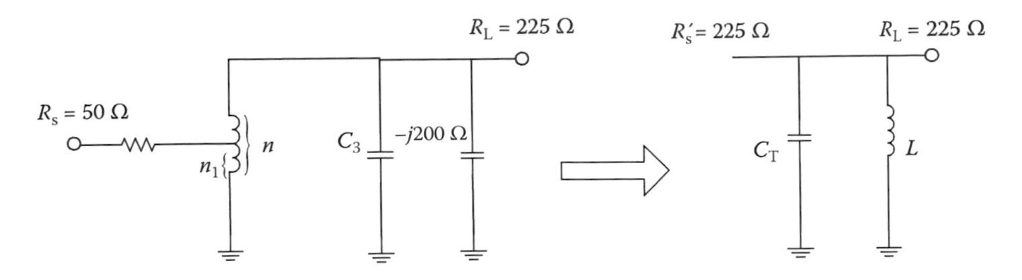

FIGURE 4.58 Output network transformation for tapped-*L* transformer.

The transformation of the output network can be done, as shown in Figure 4.58, with the transformations obtained before.

Using Equation 4.163,

$$R_s' = R_s\left(\frac{n}{n_1}\right)^2 \rightarrow \left(\frac{n}{n_1}\right) = \sqrt{\frac{225}{50}} = 2.12 \tag{4.172}$$

It is given that the quality of the output network is 7.5, $Q_{out} = 7.5$. So,

$$X_p = \frac{R_p}{Q_p} = \frac{(225/2)}{7.5} = 15 \rightarrow C_p = \frac{1}{\omega X_p} = 106.1 \text{ [pF]} \tag{4.173}$$

Since

$$C_p = C_3 + 7.95 \rightarrow C_3 = 106.1 - 7.95 = 98.15 \text{ [pF]} \tag{4.174}$$

where 7.95 pF is obtained from the reactance given in the question, $-j200\ \Omega$, the inductance, L_2, can then be found as

$$L_2 = \frac{X_p}{\omega} = 23.8 \text{ [nH]} \tag{4.175}$$

For parts (b) and (c), the following MATLAB GUI is developed to match any transistor input and output impedances to the desired impedances via tapped-*C* impedance transformer at the input and tapped-*L* impedance transformer at the output.

```
function varargout = AmplifierGUI(varargin)
% AMPLIFIERGUI MATLAB code for AmplifierGUI.fig
%    AMPLIFIERGUI, by itself, creates a new AMPLIFIERGUI or raises the
%
%    AMPLIFIERGUI('CALLBACK',hObject,eventData,handles,...) calls
%    the localfunction named CALLBACK in AMPLIFIERGUI.M with the given
%    input arguments. AMPLIFIERGUI('Property','Value',...) creates a new
%    AMPLIFIERGUI or raises the existing singleton*.  Starting from
%    the left, property value pairs are
%    applied to the GUI before AmplifierGUI_OpeningFcn gets called.
```

```
% Begin initialization code - DO NOT EDIT
gui_Singleton = 1;
gui_State = struct('gui_Name',       mfilename, ...
                   'gui_Singleton',  gui_Singleton, ...
                   'gui_OpeningFcn', @AmplifierGUI_OpeningFcn, ...
                   'gui_OutputFcn',  @AmplifierGUI_OutputFcn, ...
                   'gui_LayoutFcn',  [] , ...
                   'gui_Callback',   []);
if nargin && ischar(varargin{1})
    gui_State.gui_Callback = str2func(varargin{1});
end

if nargout
    [varargout{1:nargout}] = gui_mainfcn(gui_State, varargin{:});
else
    gui_mainfcn(gui_State, varargin{:});
end
% End initialization code - DO NOT EDIT


% --- Executes just before AmplifierGUI is made visible.
function AmplifierGUI_OpeningFcn(hObject, eventdata, handles,
  varargin)

% hObject    handle to figure
% eventdata  reserved - to be defined in a future version of MATLAB
% handles    structure with handles and user data (see GUIDATA)
% varargin   command line arguments to AmplifierGUI (see VARARGIN)

% Choose default command line output for AmplifierGUI
handles.output = hObject;

% Update handles structure
guidata(hObject, handles);

% UIWAIT makes AmplifierGUI wait for user response (see UIRESUME)
% uiwait(handles.figure1);

% --- Outputs from this function are returned to the command line.
function varargout = AmplifierGUI_OutputFcn(hObject, eventdata,
  handles)
% varargout  cell array for returning output args (see VARARGOUT);
% hObject    handle to figure
% eventdata  reserved - to be defined in a future version of MATLAB
% handles    structure with handles and user data (see GUIDATA)

% Get default command line output from handles structure
varargout{1} = handles.output;
% --- Executes on button press in pushbutton1.
function pushbutton1_Callback(hObject, eventdata, handles)
% hObject    handle to pushbutton1 (see GCBO)
% eventdata  reserved - to be defined in a future version of MATLAB
% handles    structure with handles and user data (see GUIDATA)
Qin=str2num(get(handles.Qin,'String'));
Qout=str2num(get(handles.Qout,'String'));
RT1=str2num(get(handles.RT1,'String'));
TR2=str2num(get(handles.TR2,'String'));
```

```
TC1=str2double(get(handles.TC1,'String'));
TC2=str2double(get(handles.TC2,'String'));
Cw=str2num(get(handles.Cw,'String'));
w=str2num(get(handles.w,'String'));
Rsource=str2double(get(handles.Rsource,'String'));
Rload=str2num(get(handles.Rload,'String'));
Qp=w/Cw;
C1P=Qin/(2*pi*w*RT1);
set(handles.C1p,'String',C1P);
XTC1=1/(2*pi*w*1i*TC1);
c1=C1P-XTC1;
set(handles.C1,'String',c1);
c2=C1P/(sqrt(Rsource/RT1)-1);
set(handles.C2,'String',c2);
Ceq=(C1P*c2)/(C1P+c2);
Xp=1/(2*pi*w*Ceq);
L1=Xp/(2*pi*w);
set(handles.L1,'String',L1);
XCT2=1/(2*pi*w*1i*TC2);
N=sqrt(TR2/Rload);
set(handles.N,'String',N);
LRin=(Rload*(Qp*Qp+1))/(Qout*Qout+1);
C3p=Qp/(2*pi*w*LRin);

set(handles.C3p,'String',C3p);
C3=C3p-XCT2;
set(handles.C3,'String',C3);
L21=Rload/(2*pi*w*Qout);
L22=(L21*(Qp*Qout-(Qout*Qout)))/((Qout*Qout)+1);
L2=L22+L21;
set(handles.L2,'String',L2);
set(handles.L21,'String',L21);
set(handles.L22,'String',L22);
A_1=1;B_1=Rsource;C_1=0;D_1=1;
ABCD_1=[A_1 B_1;C_1 D_1];
YC1=1/(2*pi*w*c1*1i);
YC11=1/YC1;
YC2=1/(2*pi*w*c2*1i);
YC22=1/YC2;
YL1=(1i*L1*2*pi*w);
YL11=1/YL1;
A_2=1+(YC22/YC11);
B_2=1/YC11;
C_2=YL11+YC22+((YL11*YC22)/YC11);
D_2=1+(YL11/YC11);
ABCD_2=[A_2 B_2;C_2 D_2];
ABCDtotal_1=ABCD_1*ABCD_2;
set(handles.CA,'String',[real(ABCDtotal_1(1,1)),
  imag(ABCDtotal_1(1,1))].');
set(handles.CB,'String',[real(ABCDtotal_1(1,2)),
  imag(ABCDtotal_1(1,2))].');
set(handles.CC,'String',[real(ABCDtotal_1(2,1)),
  imag(ABCDtotal_1(2,1))].');
set(handles.CD,'String',[real(ABCDtotal_1(2,2)),
  imag(ABCDtotal_1(2,2))].');
TY11=str2double(get(handles.edit4,'String'));
TY12=str2double(get(handles.edit5,'String'));
```

```
TY21=str2double(get(handles.edit6,'String'));
TY22=str2double(get(handles.edit7,'String'));
Ytrans=(1/1000).*[TY11 TY12; TY21 TY22];
Ydet=det(Ytrans);
ABCDtrans=[-Ytrans(2,2)/Ytrans(2,1) -1./Ytrans(2,1);-Ydet/Ytrans(2,1)
  -Ytrans(1,1)/Ytrans(2,1)];
set(handles.TA,'String',[real(ABCDtrans(1,1)), imag(ABCDtrans(1,1))].');
set(handles.TB,'String',[real(ABCDtrans(1,2)), imag(ABCDtrans(1,2))].');
set(handles.TC,'String',[real(ABCDtrans(2,1)), imag(ABCDtrans(2,1))].');
set(handles.TD,'String',[real(ABCDtrans(2,2)), imag(ABCDtrans(2,2))].');
ImpC3=1/(2*pi*w*C3*1i);
YL1=(2*pi*w*L22*1i);
YL2-(2*pi*w*L21*1i);
Series=((Rload*YL2)/(Rload+YL2))+YL1;
Outputsimple=(Series*ImpC3)/((Series+ImpC3));
A=1;B=0;C=1./Outputsimple;D=1;
ABCDOutput=[A B;C D];
set(handles.LA,'String',[real(ABCDOutput(1,1)),
  imag(ABCDOutput(1,1))].');
set(handles.LB,'String',[real(ABCDOutput(1,2)),
  imag(ABCDOutput(1,2))].');
set(handles.LC,'String',[real(ABCDOutput(2,1)),
  imag(ABCDOutput(2,1))].');
set(handles.LD,'String',[real(ABCDOutput(2,2)),
  imag(ABCDOutput(2,2))].');
ABCD=ABCDtotal_1*ABCDtrans*ABCDOutput;
set(handles.A,'String',[real(ABCD(1,1)), imag(ABCD(1,1))].');
set(handles.B,'String',[real(ABCD(1,2)), imag(ABCD(1,2))].');
set(handles.C,'String',[real(ABCD(2,1)), imag(ABCD(2,1))].');
set(handles.D,'String',[real(ABCD(2,2)), imag(ABCD(2,2))].');
Vg=20*log10(abs(1/ABCD(1,1)));
set(handles.Vg,'String',Vg);
%Frequency Response Plotting over range of Frequencies.
range=[1:100000:2*w];
n=1;
V=length(range);
for n=(1:1:V)
freq=n*100000;
Qp=freq/(Cw);
A_1=1;B_1=Rsource;C_1=0;D_1=1;
ABCD_1=[A_1 B_1;C_1 D_1];
YC1=1/(2*pi*freq*c1*1i);
YC11=1/YC1;
YC2=1/(2*pi*freq*c2*1i);
YC22=1/YC2;
YL1=(1i*L1*2*pi*freq);
YL11=1/YL1;
A_2=1+(YC22/YC11);
B_2=1/YC11;
C_2=YL11+YC22+((YL11*YC22)/YC11);
D_2=1+(YL11/YC11);
ABCD_2=[A_2 B_2;C_2 D_2];
ABCDtotal_1=ABCD_1*ABCD_2;
Ytrans=(1/1000).*[TY11 TY12; TY21 TY22];
Ydet=det(Ytrans);
```

```
ABCDtrans=[-Ytrans(2,2)/Ytrans(2,1) -1./Ytrans(2,1);-Ydet/Ytrans(2,1)
   -Ytrans(1,1)/Ytrans(2,1)];
ImpC3=1/(2*pi*freq*C3*1i);
YL1=(2*pi*freq*L22*1i);
YL2=(2*pi*freq*L21*1i);
Series=((Rload*YL2)/(Rload+YL2))+YL1;
Outputsimple=(Series*ImpC3)/((Series+ImpC3));
A=1;B=0;C=1./Outputsimple;D=1;
ABCDOutput=[A B;C D];
ABCD=ABCDtotal_1*ABCDtrans*ABCDOutput;
gain=1/(ABCD(1,1));
VG(n)= 20*log10(abs(gain));
 end
VG;
plot(range,VG);
title('Frequency Response');
xlabel('Frequency (Hz)');
ylabel('Gain (dB)');
```

The last section of the program is not included due to its standard format for GUIs. When the program is executed, the MATLAB GUI window showing the results appears and is illustrated in Figure 4.59.

Design Example

Consider the capacitively coupled amplifier circuit shown in Figure 4.60. Design the resonator-tuned amplifier circuit at a resonant frequency of 100 MHz, 3-dB bandwidth of 5 MHz, and source and load impedances of 50 and 12 Ω, respectively. Assume that the inductor *Q*s are 65 at the frequency of interest. Integrate a tapped-*C* transformer to your circuit to match load impedance to source

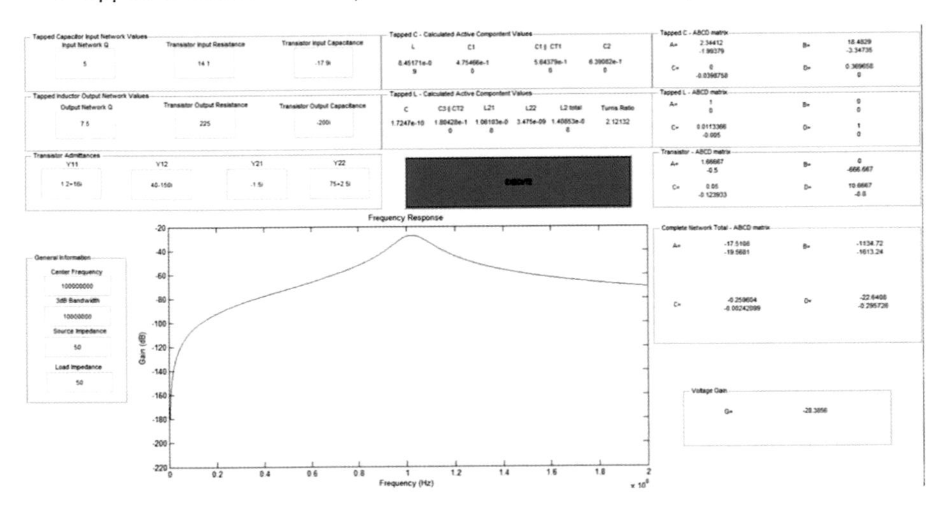

FIGURE 4.59 MATLAB GUI to design tapped-*C* and tapped-*L* impedance transformers for amplifiers.

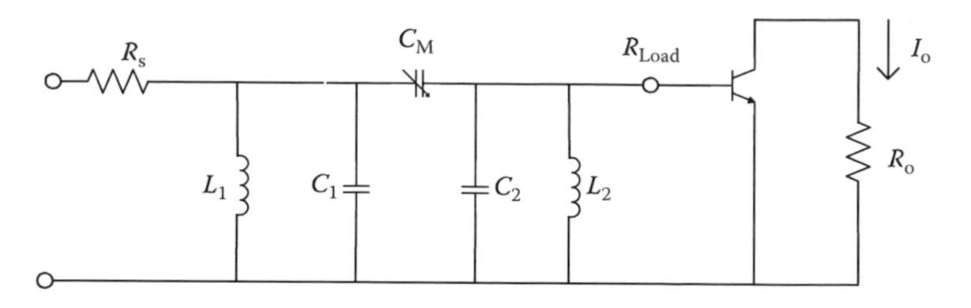

FIGURE 4.60 Capacitively coupled amplifier circuit.

impedance. (a) Calculate resonator and tapped-C component values. (b) Now, using the results you obtained, develop a generic MATLAB program to design a capacitively coupled circuit with a tapped-*C* transformer to match the input and output impedances. Your program should take the input values as source impedance, load impedance, center frequency, 3-dB bandwidth, and inductor quality factor, and calculate and illustrate the resonator and tapped-*C* component values. Test the accuracy of your program using the values in part (a).

Solution

The given parameters are

$$f_c = 100\ [\text{MHz}],\quad f_{3\text{dB}} = 5\ [\text{MHz}],\quad R_s = 50\ [\Omega],\quad R_L = 12\ [\Omega],\quad Q_{\text{ind}} = 65 \tag{4.176}$$

From the given parameters, the Q of the network is found as

$$Q_t = \frac{f_c}{f_{3\text{dB}}} = \frac{100}{5} = 20 \tag{4.177}$$

The quality factor of the resonator can be found from Equation 4.99 as

$$Q_R = \frac{Q_t}{0.707} = \frac{20}{0.707} = 28.3 \tag{4.178}$$

Since the Q of the inductor is given, then,

$$Q_{\text{ind}} = \frac{R_p}{X_p} = 65 \rightarrow R_p = 65X_p \tag{4.179}$$

The Q of the single resonator can also be found from

$$Q_R = \frac{R_{\text{tot}}}{X_p} = 28.3$$

where

$$R_{tot} = R'_s // R_p = R_s // R_p = \frac{50R_p}{50 + R_p} \quad (4.180)$$

Hence,

$$Q_R = \left(\frac{50R_p}{50 + R_p}\right)\frac{1}{X_p} = 28.3 \quad (4.181)$$

From Equations 4.179 and 4.181,

$$X_p = \frac{50(Q_{ind} - Q_R)}{Q_{ind}Q_R} = 1\ [\Omega] \quad (4.182)$$

Now, using Equation 4.179, we find

$$R_p = 65X_p = 65\ [\Omega] \quad (4.183)$$

The values of the inductors are obtained from

$$L_1 = L_2 = \frac{X_p}{\omega} = \frac{1}{2\pi(100\times 10^6)} = 1.59\ [\text{nH}] \quad (4.184)$$

and the capacitor is found from

$$C_T = \frac{1}{\omega X_p} = \frac{1}{2\pi(100\times 10^6)(1)} = 1590\ [\text{pF}] \quad (4.185)$$

Since

$$C_T = \frac{C'_1C'_2}{C'_1 + C'_2} \rightarrow \frac{C'_1}{C'_2} = \sqrt{\frac{R'_s}{R_s}} - 1 \quad (4.186)$$

hence

$$\frac{C'_1}{C'_2} = \sqrt{\frac{R'_s}{R_s}} - 1 = \sqrt{\frac{50}{12}} - 1 = 1.04 \rightarrow C'_1 = 1.04C'_2 \quad (4.187)$$

So,

$$C_T = \frac{C'_1C'_2}{C'_1 + C'_2} = \frac{1.04C_2'^2}{2.04C'_1} = 1590 \rightarrow C'_2 = 3119\ [\text{pF}] \quad (4.188)$$

Using Equation 4.187,

$$C_1' = 1.04C_2' = 3244\ [\text{pF}] \tag{4.189}$$

The coupling capacitor is found from Equation 4.112 as

$$C_M = \frac{C_T}{Q_R} = \frac{1590}{28.3} = 56.2\ [\text{pF}] \tag{4.190}$$

The following MATLAB program generates a user interface to match the input of the amplifier for the given operational frequency, bandwidth, quality factor, and source impedance and input impedance of the transistor.

```
clear
%Gui Prompt
prompt = {'Source Impedance (Ohms):', 'Load Impedance (Ohms):',...
    'Center Frequency (MHz):','-3dB Bandwidth (MHz):',...
    'Inductor Quality Factor:'};
dlg_title = 'Input';
num_lines = 1;
def = {'50','12','100','5','65',};
answer = inputdlg(prompt,dlg_title,num_lines,def);

%convert the strings received from the GUI to numbers
valuearray=str2double(answer);

%give the recieved numbers variable names
Z_in=valuearray(1);
Z_out=valuearray(2);
f=valuearray(3)*1000000;
bw=valuearray(4)*1000000;
Qp=valuearray(5);

%Establish frequency n rad/s
w=2*pi*f;

%Calculate values for total Q and loaded Q
Qtot=f/bw;
Qr=Qtot/0.707;

%Find resonant reactance and Rp
Xp=((Qp*Z_in)/Qr-Z_in)/Qp;
Rp=Qp*Xp;

%Inductance values are equal for the double resonator system
L1=Xp/w;
L2=Xp/w;

%Ctotal and C2 have the same reactance as L
Ctot=1/(w*Xp);
C2=Ctot;
```

```
%Design the tapped C network based on stepping up or down impedance
if(Z_in>Z_out)
  Rsprime=Z_in;
  Rs=Z_out;
else
  Rsprime=Z_out;
  Rs=Z_in;
end

Cratio=sqrt(Rsprime/Rs)-1;
Ca=(Cratio+1)*Ctot/Cratio;
C1=Cratio*Ca;

%Coupling Capacitor
Cm=Ctot/Qr;

%Show the required values in a display window
msgbox( sprintf('The required values of components are: \nC1 = %d F\
  nC2 = %d F\nCm = %d F\nCa = %d F\nL1 = %d H\nL2 = %d H\n',
  C1,C2,Cm,Ca,L1,L2));

%To find the frequency response, we use the ABCD parameters of the
  circuit
%Turn the input into a T network
Za1=Z_in;
Zb1=1/(j*w*Ca);
Zc1=j*w*L1;

%Place into an ABCD matrix
Am1=1+Za1/Zc1;
Bm1=Za1+Zb1+Za1*Zb1/Zc1;
Cm1=1/Zc1;
Dm1=1+Zb1/Zc1;
net1=[Am1 Bm1; Cm1 Dm1];

%The remaining components form a pi network
Za2=1/(j*w*C1);
Zb2=1/(j*w*Cm);

%Put the last three elements in parallel
ZC2=1/(j*w*C2);
ZL2=j*w*L2;
ZRL=Z_out;
Zc2=1/(1/ZC2+1/ZL2+1/ZRL);

%Transform to admittance
Ya2=1/Za2;
Yb2=1/Zb2;
Yc2=1/(Zc2);

%Convert to ABCD parameters
Am2=1+Yb2/Yc2;
Bm2=1/Yc2;
Cm2=Ya2+Yb2+(Ya2*Yb2)/Yc2;
Dm2=1+Ya2/Yc2;
net2=[Am2 Bm2; Cm2 Dm2];
```

```
%Complete ABCD Parameters are the product of the two networks
totalabcd=net1*net2;

%Transfer function of the total system
Vgain=20*log10(abs(1/totalabcd(1,1)));

%Create a vector of frequencies and an empty output vector
fvector=10000000:1000000:191000000;
vgainvector=1:182;

%Iterate through the code, changing the frequency each time
%to give the frequency response across a large range of
  frequencies
for k=1:182
f=fvector(k);
w=2*pi*f;
Za1=Z_in;
Zb1=1/(j*w*Ca);
Zc1=j*w*L1;
Am1=1+Za1/Zc1;
Bm1=Za1+Zb1+Za1*Zb1/Zc1;
Cm1=1/Zc1;
Dm1=1+Zb1/Zc1;
net1=[Am1 Bm1; Cm1 Dm1];
Za2=1/(j*w*C1);
Zb2=1/(j*w*Cm);
ZC2=1/(j*w*C2);
ZL2=j*w*L2;
ZRL=Z_out;
Zc2=1/(1/ZC2+1/ZL2+1/ZRL);
Ya2=1/Za2;
Yb2=1/Zb2;
Yc2=1/Zc2;
Am2=1+Yb2/Yc2;
Bm2=1/Yc2;
Cm2=Ya2+Yb2+(Ya2*Yb2)/Yc2;
Dm2=1+Ya2/Yc2;
net2=[Am2 Bm2; Cm2 Dm2];
totalabcdz=net1*net2;
%Voltage gain of the total system
vgainvector(k)=20*log10(abs(1/totalabcdz(1,1)));
end

%Use the individual V gains calculated at each f to plot the total
%frequency response

plot(fvector,vgainvector)
grid on
xlabel('f (Hz)')
ylabel('|H(w)| (dB)')
title('Frequency Response of Capacitively Coupled Resonant Network')
```

The output of the program when executed gives the design parameters and frequency response of the designed tapped-*C* impedance transformer shown in Figure 4.61.

The frequency response of the network is given in Figure 4.62.

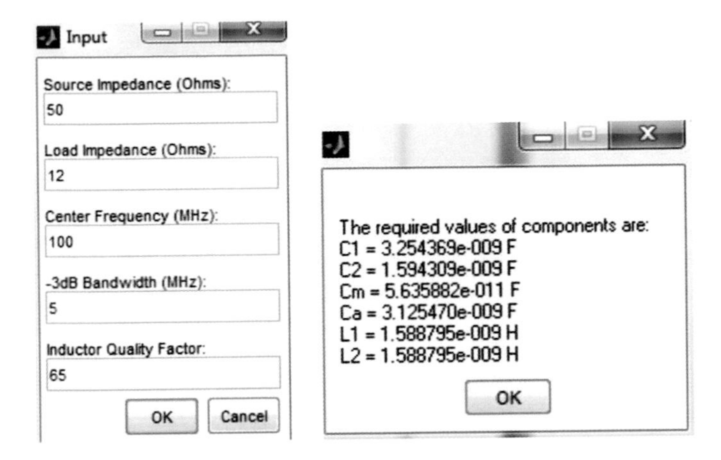

FIGURE 4.61 MATLAB GUI for the user input to design interfacing circuit with tapped capacitor.

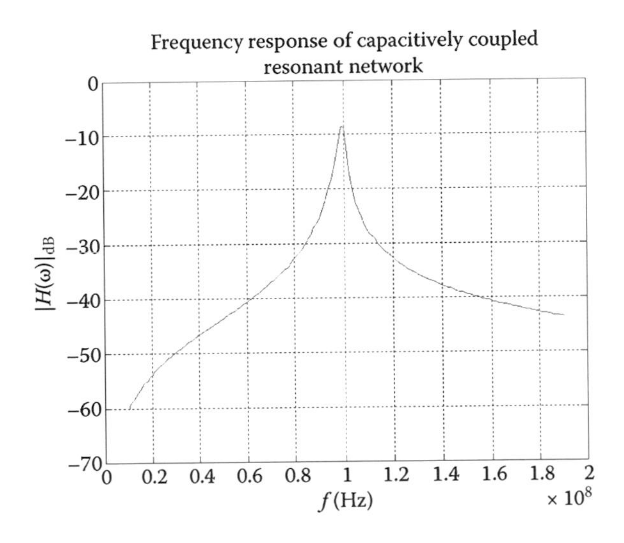

FIGURE 4.62 Frequency response of the input impedance matching network with tapped-C transformer.

PROBLEMS

1. A parallel resonant circuit with a 3-dB bandwidth of 5 MHz and a center frequency of 40 MHz is given. It is also given that the resonant circuit has source and load impedances of 100 Ω. The Q of the inductor is given to be 120. The capacitor is assumed to be an ideal capacitor.
   a. Design the resonant circuit.
   b. What is the loaded Q of the resonant circuit?
   c. What is the insertion loss of the network?
   d. Obtain the frequency response of this circuit vs. frequency.

2. An amplifier output needs to be terminated with a load line resistance of 4000 Ω at 2 MHz. It is given in the data sheet that the transistor has 40 pF at 2 MHz. There is an inductive load connected to the output of the load line circuit of the amplifier with $R_L = 15\ \Omega$. The configuration of this circuit is given in Figure 4.63.
   a. Calculate the values of L and C by assuming that the load inductor has a negligible loss, i.e., $r = 0$.
   b. The inductor is changed to a magnetic core inductor, which has a quality factor of 50. Calculate loss resistance, r, for the reactive component values obtained in (a). What is the value of the new load linc rcsistance?
   c. If the quality factor is 50, the inductor is 50, and the load line resistor is required to be 4000 Ω as set in the problem, what are the values of L and C?
3. Consider the amplifier network given in Figure 4.64. In this network, it is required to interface 10-Ω differential output of the amplifier 1 (Amp 1) to 100-Ω input impedance of the second amplifier (Amp 2) using an unbalanced L–C network at 100-MHz. The current and voltage at the output of Amp 1 are 12 [A_{rms}] and 72 [V_{rms}], respectively. Design the magnetic core inductor with a powdered iron material using stacked core configuration, and
   a. Identify the core material to be used.
   b. Calculate the number of turns.
   c. Determine the minimum gauge wire that needs to be used.

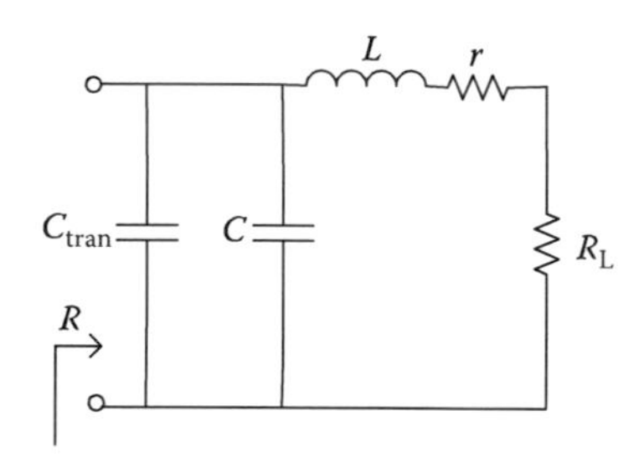

FIGURE 4.63 Termination load for an amplifier.

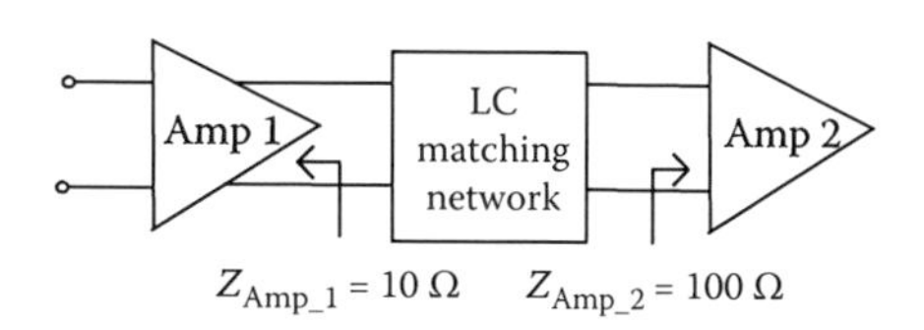

FIGURE 4.64 Amplifier network interface.

d. Obtain the high-frequency characteristic of the inductor.
e. Identify its resonant frequency.
f. Find its quality factor.
g. Determine the length of the wire that will be used.
h. Calculate the total core power loss.
i. Find out the maximum operation flux density.

4. Design a two-resonator tuned circuit at a resonant frequency of 125 MHz, 3-dB bandwidth of 5.75 MHz, and source and load impedances of 250 and 2500 Ω, respectively, using top-C and top-L coupling techniques, as shown in Figure 4.65a and b. Assume that the inductor Qs are 65 at the frequency of interest. Finally, use a tapped-C transformer to present an effective source resistance (R_s) of 1000 Ω to the filter. Use MATLAB to obtain the frequency response of the circuits shown in Figure 4.65a and b.
5. Design a resonant LC circuit driven by a source with a resistance 50 Ω and a load impedance 2 kΩ shown in Figure 4.66. The network Q should be 20, and the center frequency is set to be 100 MHz. Use a capacitive transformer to match the load with the source for maximum power transfer. Assume that lossless capacitors and an inductor with $Q = 20$ are used, and input voltage = 1 mV. Also, calculate the power transferred to the load. Use simulation to verify your results.

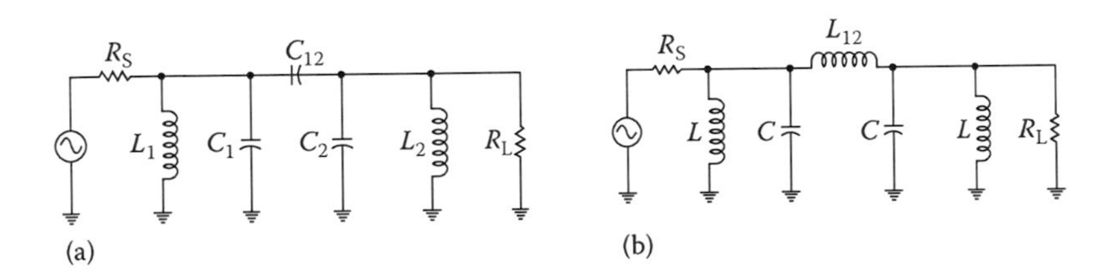

FIGURE 4.65 Resonator networks coupling. (a) Capacitively coupled resonator network and (b) inductively coupled resonator network.

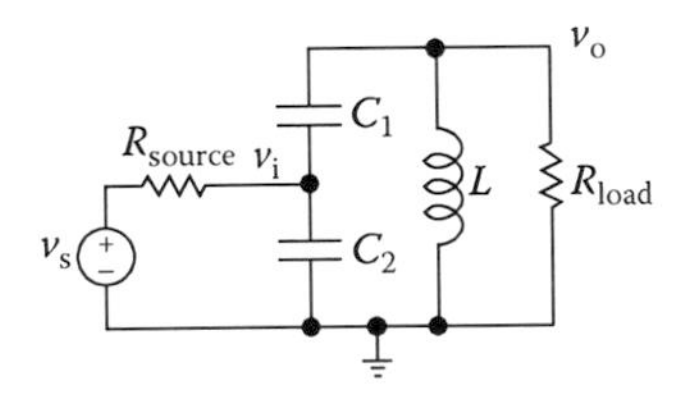

FIGURE 4.66 Capacitive transformer (tapped-C) circuit.

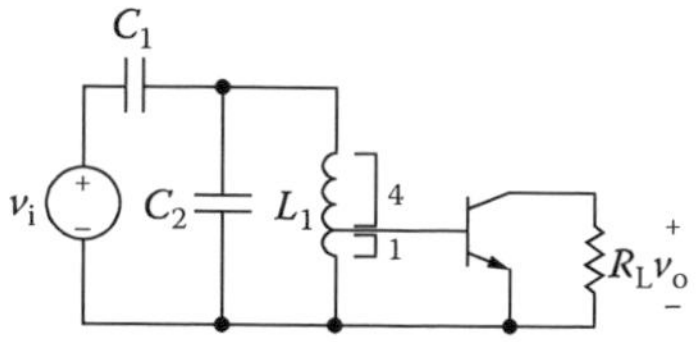

FIGURE 4.67 Tapped-L circuit.

6. Consider the tuned amplifier circuit given in Figure 4.67. What are the center frequency, Q, and midband gain of the amplifier if $L_1 = 5\ \mu\text{H}$, $C_1 = 10$ pF, $I_c = 1$ mA, $C_\pi = 5$ pF, $R_L = 5\ \text{k}\Omega$, $r_\pi = 2.5\ \text{k}\Omega$, and $C_\mu = 1$ pF?

REFERENCE

1. A. Eroglu. 2013. *RF Circuit Design Techniques for MF-UHF Applications*. CRC Press, Boca Raton, FL.

5 Impedance Matching Networks

5.1 INTRODUCTION

Radio frequency (RF) power amplifiers consist of several stages as illustrated in Figure 5.1 [1]. Impedance matching networks are used to provide the optimum power transfer from one stage to another so that the energy transfer is maximized. This can be accomplished by having matching networks between the stages. Matching networks can be implemented using distributed or lumped elements based on the frequency of operation and application. Distributed elements are implemented using transmission lines for high-frequency operation where lumped elements are used for lower frequencies. Matching networks when designed with lumped elements are implemented using a ladder network structure. In the design of matching networks, there are several important parameters such as bandwidth and quality factor of the network. These can be investigated with design tools such as the Smith chart. The Smith chart helps the designer to visualize the performance of the matching network for the operational conditions under consideration. In this chapter, analysis of transmission lines, the Smith chart, and the design of impedance matching networks will be detailed, and several application examples will be given.

5.2 TRANSMISSION LINES

A transmission line is a distributed-parameter network, where voltages and currents can vary in magnitude and phase over the length of the line. Transmission lines usually consist of two parallel conductors that can be represented with a short segment of Δz. This short segment of transmission line can be modeled as a lumped-element circuit, as shown in Figure 5.2.

In Figure 5.2, R is the series resistance per unit length for both conductors, $R(\Omega/\text{m})$, L is the series inductance per unit length for both conductors, $L(\text{H/m})$, G is the shunt conductance per unit length, $G(\text{S/m})$, and C represents the shunt capacitance per unit length, $C(\text{F/m})$, in the transmission line. Application of Kirchhoff's voltage and current laws gives

$$v(z,t) - R\Delta z i(z,t) - L\Delta z \frac{\partial i(z,t)}{\partial t} - v(z+\Delta z,t) = 0 \tag{5.1}$$

$$i(z,t) - G\Delta z v(z+\Delta z,t) - C\Delta z \frac{\partial v(z+\Delta z,t)}{\partial t} - i(z+\Delta z,t) = 0 \tag{5.2}$$

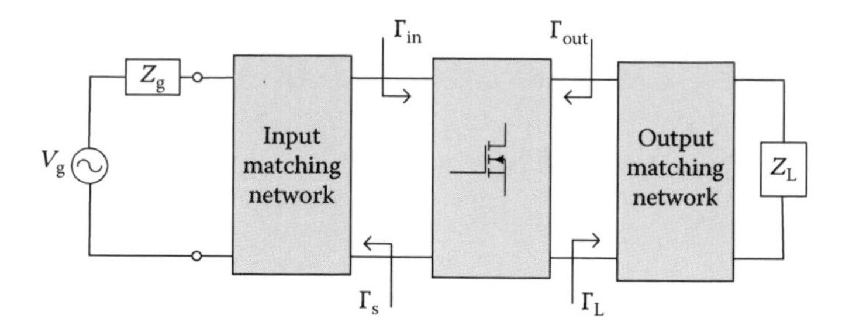

FIGURE 5.1 Matching network implementation for RF power amplifiers.

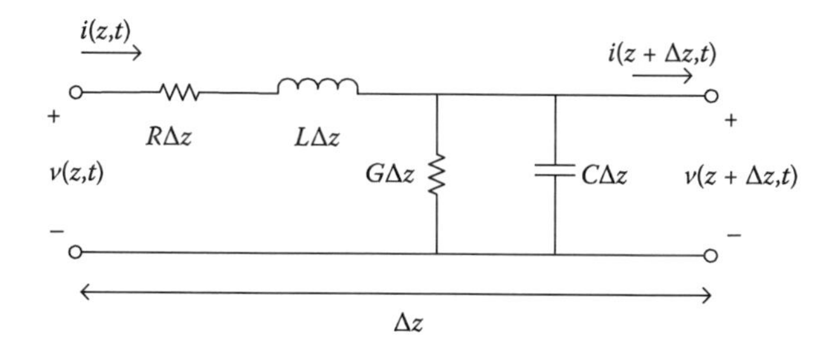

FIGURE 5.2 Short segment of transmission line.

Dividing Equations 5.1 and 5.2 by Δz and assuming that $\Delta z \to 0$, we obtain

$$\frac{\partial v(z,t)}{\partial z} = -Ri(z,t) - L\frac{\partial i(z,t)}{\partial t} \tag{5.3}$$

$$\frac{\partial i(z,t)}{\partial z} = -Gv(z,t) - C\frac{\partial v(z,t)}{\partial t} \tag{5.4}$$

Equations 5.3 and 5.4 are known as the time-domain form of the transmission line, or telegrapher, equations. Assuming the sinusoidal steady-state condition with application cosine-based phasors, Equations 5.3 and 5.4 take the following forms:

$$\frac{\mathrm{d}V(z)}{\mathrm{d}z} = -(R + j\omega L)I(z) \tag{5.5}$$

$$\frac{\mathrm{d}I(z)}{\mathrm{d}z} = -(G + j\omega C)V(z) \tag{5.6}$$

By eliminating either $I(z)$ or $V(z)$ from Equations 5.5 and 5.6, we obtain the wave equations as

$$\frac{d^2V(z)}{dz^2} = -\gamma^2 V(z) \tag{5.7}$$

$$\frac{d^2I(z)}{dz^2} = -\gamma^2 I(z) \tag{5.8}$$

where

$$\gamma = \alpha + j\beta = \sqrt{(R + j\omega L)(G + j\omega C)} \tag{5.9}$$

In Equation 5.9, γ is the complex propagation constant, α is the attenuation constant, and β is known as the phase constant. In transmission lines, phase velocity is defined as

$$v_p = \frac{\omega}{\beta} \tag{5.10}$$

The wavelength can be defined using

$$\lambda = \frac{2\pi}{\beta} \tag{5.11}$$

The traveling wave solutions to the equations obtained in Equations 5.7 and 5.8 are

$$V(z) = V_0^+ e^{-\gamma z} + V_0^- e^{+\gamma z} \tag{5.12}$$

$$I(z) = I_0^+ e^{-\gamma z} + I_0^- e^{+\gamma z} \tag{5.13}$$

Substitution of Equation 5.12 into Equation 5.5 gives

$$I(z) = \frac{\gamma}{R + j\omega L}\left[V_0^+ e^{-\gamma z} + V_0^- e^{+\gamma z}\right] \tag{5.14}$$

From Equation 5.14, the characteristic impedance, Z_0, is defined as

$$Z_0 = \frac{R + j\omega L}{\gamma} = \sqrt{\frac{R + j\omega L}{G + j\omega C}} \tag{5.15}$$

Hence,

$$\frac{V_0^+}{I_0^+} = Z_0 = -\frac{V_0^-}{I_0^-} \tag{5.16}$$

and

$$I(z) = \frac{V_0^+}{Z_0} e^{-\gamma z} - \frac{V_0^-}{Z_0} e^{+\gamma z} \tag{5.17}$$

Using the formulation derived, we can find the voltage and current at any point on the transmission line shown in Figure 5.3. At the load, $z = 0$,

$$V(0) = Z_L I(0) \tag{5.18}$$

$$V_0^+ + V_0^- = \frac{Z_L}{Z_0}\left(V_0^+ - V_0^-\right) \tag{5.19}$$

or

$$V_0^-\left(1 + \frac{Z_L}{Z_0}\right) = V_0^+\left(\frac{Z_L}{Z_0} - 1\right) \tag{5.20}$$

which leads to

$$\frac{V_0^-}{V_0^+} = \left(\frac{Z_L - Z_0}{Z_L + Z_0}\right) \tag{5.21}$$

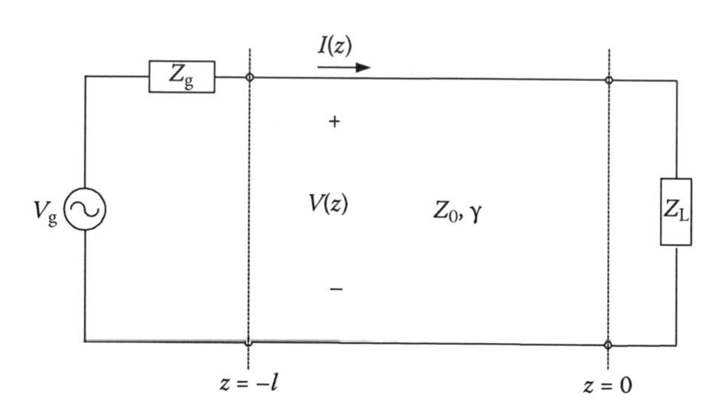

FIGURE 5.3 Finite terminated transmission line.

Equation 5.21 is then defined as the reflection coefficient at the load, and it is expressed as

$$\Gamma_L = \left(\frac{Z_L - Z_0}{Z_L + Z_0}\right) \tag{5.22}$$

Voltage and current can be expressed in terms of reflection coefficient as

$$V(z) = V_0^+\left(e^{-\gamma z} + \Gamma_L e^{+\gamma z}\right) \tag{5.23}$$

$$I(z) = \frac{1}{Z_0} V_0^+\left(e^{-\gamma z} - \Gamma_L e^{+\gamma z}\right) \tag{5.24}$$

The input impedance can be found at any point on the transmission line shown in Figure 5.4 from

$$Z_{in}(z) = \frac{V(z)}{I(z)} \tag{5.25}$$

We then have

$$Z_{in}(z) = Z_0 \frac{\left(e^{-\gamma z} + \Gamma_L e^{+\gamma z}\right)}{\left(e^{-\gamma z} - \Gamma_L e^{+\gamma z}\right)} \tag{5.26}$$

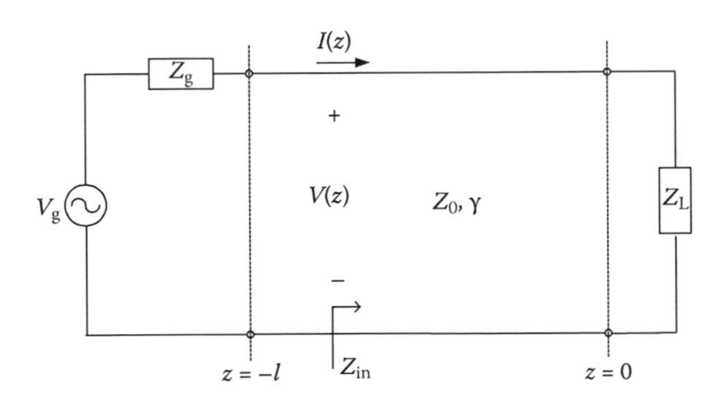

FIGURE 5.4 Input impedance calculation on the transmission line.

which can be expressed as

$$Z_{in}(z) = Z_0\left(\frac{1+\left(\frac{Z_L - Z_0}{Z_L + Z_0}\right)e^{+2\gamma z}}{1-\left(\frac{Z_L - Z_0}{Z_L + Z_0}\right)e^{+2\gamma z}}\right) = Z_0\left(\frac{(Z_L + Z_0)+(Z_L - Z_0)e^{+2\gamma z}}{(Z_L + Z_0)-(Z_L - Z_0)e^{+2\gamma z}}\right)$$

$$= Z_0\left(\frac{(Z_L + Z_0)e^{-\gamma z}+(Z_L - Z_0)e^{+\gamma z}}{(Z_L + Z_0)e^{-\gamma z}-(Z_L - Z_0)e^{+\gamma z}}\right) \quad (5.27)$$

Equation 5.27 can be rewritten as

$$Z_{in}(z) = Z_0\left(\frac{(Z_L + Z_0)e^{-\gamma z}+(Z_L - Z_0)e^{+\gamma z}}{(Z_L + Z_0)e^{-\gamma z}-(Z_L - Z_0)e^{+\gamma z}}\right)$$

$$= Z_0\left(\frac{Z_L(e^{+\gamma z}+e^{-\gamma z})-Z_0(e^{+\gamma z}-e^{-\gamma z})}{-Z_L(e^{+\gamma z}-e^{-\gamma z})+Z_0(e^{+\gamma z}+e^{-\gamma z})}\right) \quad (5.28)$$

which can also be expressed as

$$Z_{in}(z) = Z_0\left(\frac{Z_L - Z_0\tanh(\gamma z)}{Z_0 - Z_L\tanh(\gamma z)}\right) \quad (5.29)$$

At the input when $z = -l$, the impedance can be found from Equation 5.29 as

$$Z_{in}(z) = Z_0\left(\frac{Z_L + Z_0\tanh(\gamma l)}{Z_0 + Z_L\tanh(\gamma l)}\right) \quad (5.30)$$

5.2.1 Limiting Cases for Transmission Lines

There are three cases that can be considered as the limiting case for transmission lines. These are lossless lines, low-loss lines, and distortionless lines.

a. Lossless line ($R = G = 0$)

Transmission lines can be considered as lossless when $R = G = 0$. When $R = G = 0$, the defining equations for the transmission lines can be simplified as

$$\gamma = \alpha + j\beta = j\omega\sqrt{LC} \Rightarrow \alpha = 0 \quad (5.31)$$

$$\beta = \omega\sqrt{LC} \quad (5.32)$$

$$v_p = \frac{\omega}{\beta} = \frac{1}{\sqrt{LC}} \tag{5.33}$$

$$Z_0 = \sqrt{\frac{L}{C}} = R_0 + jX_0 \Rightarrow R_0 = \sqrt{\frac{L}{C}}, \quad X_0 = 0 \tag{5.34}$$

b. Low-loss line ($R \ll \omega L$, $G \ll \omega C$)

For low-loss transmission lines, $R \ll \omega L$, $G \ll \omega C$, and the defining equations simplify to

$$\gamma = \alpha + j\beta = j\omega\sqrt{LC}\left(1 + \frac{R}{j\omega L}\right)^{1/2}\left(1 + \frac{G}{j\omega C}\right)^{1/2} \tag{5.35}$$

$$\alpha \cong \frac{1}{2}\left(R\sqrt{\frac{C}{L}} + G\sqrt{\frac{L}{C}}\right) \tag{5.36}$$

$$\beta \cong \omega\sqrt{LC} \tag{5.37}$$

$$v_p = \frac{\omega}{\beta} \cong \frac{1}{\sqrt{LC}} \tag{5.38}$$

$$Z = R_0 + jX_0 = \sqrt{\frac{L}{C}}\left(1 + \frac{R}{j\omega L}\right)^{1/2}\left(1 + \frac{G}{j\omega C}\right)^{-1/2} \tag{5.39}$$

c. Distortionless line ($R/L = G/C$)

In distortionless transmission lines, $R/L = G/C$, and the defining equations can be simplified as

$$\gamma = \alpha + j\beta = \sqrt{\frac{C}{L}}(R + j\omega L) \tag{5.40}$$

$$\alpha = R\sqrt{\frac{C}{L}} \tag{5.41}$$

$$\beta = \omega\sqrt{LC} \tag{5.42}$$

$$v_p = \frac{1}{\sqrt{LC}} \tag{5.43}$$

$$Z_0 = \sqrt{\frac{L}{C}} \tag{5.44}$$

5.2.2 Terminated Lossless Transmission Lines

Consider the lossless transmission line shown in Figure 5.5. The voltage and current at any point on the line can be written as

$$V(z) = V_0^+ e^{-j\beta z} + V_0^- e^{j\beta z} \tag{5.45}$$

$$I(z) = \frac{V_0^+}{Z_0} e^{-j\beta z} - \frac{V_0^-}{Z_0} e^{+j\beta z} \tag{5.46}$$

The voltage and current at the load, $z = 0$, in terms of the load reflection coefficient, respectively, are

$$V(z) = V_0^+ \left[e^{-j\beta z} + \Gamma e^{j\beta z}\right] \tag{5.47}$$

$$I(z) = \frac{V_0^+}{Z_0} \left[e^{-j\beta z} - \Gamma e^{j\beta z}\right] \tag{5.48}$$

It is seen that the voltage and current on the line consist of a superposition of an incident and reflected wave, which represents standing waves. When $\Gamma = 0$, then it is

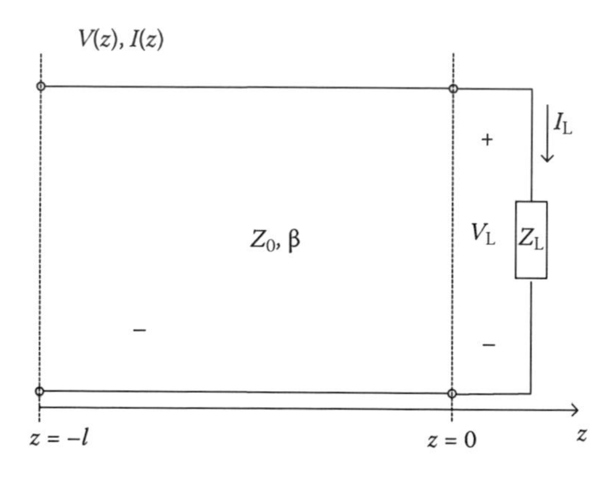

FIGURE 5.5 Lossless transmission line.

a matched condition. The time–average power flow along the line at the point z can be written as

$$P_{avg} = \frac{1}{2}\text{Re}\{V(z)I^*(z)\} = \frac{1}{2}\frac{|V_0^+|^2}{Z_0}\text{Re}\{1 - \Gamma^* e^{-2j\beta z} + \Gamma e^{2j\beta z} - |\Gamma|^2\} \tag{5.49}$$

or

$$P_{avg} = \frac{1}{2}\frac{|V_0^+|^2}{Z_0}\left(1 - |\Gamma|^2\right) \tag{5.50}$$

When the load is mismatched, not all of the available power from the generator is delivered to the load. The power that is lost is known as return loss, RL, and this can be found from

$$\text{RL} = -20\log|\Gamma|\,\text{dB} \tag{5.51}$$

Under a mismatched condition, the voltage on the line can be written as

$$|V(z)| = |V_0^+||1 + \Gamma e^{2j\beta z}| = |V_0^+||1 + \Gamma e^{-2j\beta l}| = |V_0^+||1 + |\Gamma| e^{j(\theta - 2\beta l)}| \tag{5.52}$$

The minimum and maximum values of the voltage from Equation 5.42 are found as

$$V_{max} = |V_0^+|(1 + |\Gamma|) \quad \text{and} \quad V_{min} = |V_0^+|(1 - |\Gamma|) \tag{5.53}$$

A measure of the mismatch of a line called the voltage standing wave ratio (VSWR) can be expressed as the ratio of the maximum voltage to the minimum voltage as

$$\text{VSWR} = \frac{V_{max}}{V_{min}} = \frac{1 + |\Gamma|}{1 - |\Gamma|} \tag{5.54}$$

From Equation 5.52, the distance between two successive voltage maxima (or minima) is $l = 2\pi/2\beta = \lambda/2$ $(2\beta l = 2\pi)$, whereas the distance between a maximum and a minimum is $l = \pi/2\beta = \lambda/4$. From Equation 5.48, with $z = -l$,

$$\Gamma(l) = \frac{V_0^- e^{-j\beta l}}{V_0^+ e^{j\beta l}} = \Gamma(0)e^{-2j\beta l} \tag{5.55}$$

For the current,

$$I(z) = V_0^+ e^{-j\beta z}\left(\frac{1}{Z_0}\right)\left(1 - |\Gamma| e^{+j(\phi - 2\beta l)}\right) \tag{5.56}$$

or

$$|I(z)| = |V_0^+|\left(\frac{1}{Z_0}\right)\left|1 - |\Gamma| e^{+j(\phi - 2\beta l)}\right| \tag{5.57}$$

Hence, the maximum and minimum values of the current on the line can be written as

$$I_{\max} = |I(z)|_{\max} = |V_0^+|\left(\frac{1}{Z_0}\right)(1 + |\Gamma|) \tag{5.58}$$

$$I_{\min} = |I(z)|_{\min} = |V_0^+|\left(\frac{1}{Z_0}\right)(1 - |\Gamma|) \tag{5.59}$$

The current standing wave ratio, ISWR, is

$$\text{ISWR} = \frac{I_{\max}}{I_{\min}} = \frac{1 + |\Gamma|}{1 - |\Gamma|} \tag{5.60}$$

Hence, VSWR = ISWR from Equations 5.54 and 5.60. VSWR will be used throughout the book for analysis. The voltage waveform vs. the length of the transmission line along the axis is plotted in Figure 5.6.

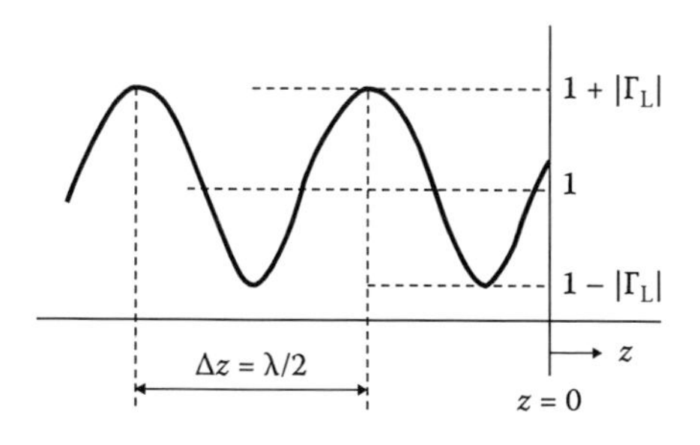

FIGURE 5.6 Voltage vs. transmission line length.

At a distance $l = -z$, the input impedance is then equal to

$$Z_{in} = \frac{V(-l)}{I(-l)} = Z_0 \frac{Z_L + jZ_0 \tan\beta l}{Z_0 + jZ_L \tan\beta l} \tag{5.61}$$

Example

A 2-m lossless, air-spaced transmission line having a characteristic impedance 50 Ω is terminated with an impedance 40 + j30 (Ω) at an operating frequency of 200 MHz. Find the input impedance.

Solution

The phase constant is found from

$$\beta = \frac{\omega}{v_p} = \frac{4}{3}\pi$$

Since it is given that $R_0 = 50\ \Omega$, $Z_L = 40 + j30$, and $\ell = 2$ m, the input impedance is obtained from Equation 5.61 as

$$Z_i = 50 \frac{(40 + j30) + j50 \cdot \tan\left(\frac{4\pi}{3} \cdot 2\right)}{50 + j(40 + j30) \cdot \tan\left(\frac{4\pi}{3} \cdot 2\right)} = 26.3 - j9.87$$

Example

For a transmission, it is given that $Z_L = 17.4 - j30$ [Ω] and $Z_0 = 50$ [Ω]. Calculate Γ_L, SWR, z_{min}, V_{max}, and V_{min} on the transmission line.

Solution

From the given information, we find the load reflection coefficient as

$$\Gamma_L = \frac{Z_L - Z_0}{Z_L + Z_0} = -0.24 - j0.55 = 0.6e^{-j(1.99)}$$

The VSWR is found from

$$\text{SWR} = \frac{V_{max}}{V_{min}} = \frac{1 + |\Gamma_L|}{1 - |\Gamma_L|} = \frac{1 + 0.6}{1 - 0.6} = 4.0$$

This leads to

$$V_{max}/|V^+| = 1 + |\Gamma_L| = 1.6$$

$$V_{min}/|V^+| = 1 - |\Gamma_L| = 0.4$$

Hence, the maximum and minimum values of the voltage are obtained when

$$\begin{aligned} V_{max} \quad &\text{when} \quad \phi + 2\beta z = 0, -2\pi, \ldots \\ V_{min} \quad &\text{when} \quad \phi + 2\beta z = -\pi, -3\pi, \ldots \end{aligned}$$

So, the distance that will give the minimum value of the voltage is found from

$$z_{min} = \frac{-\pi - \phi}{2\beta} = \frac{(-\pi + 1.99)}{2(2\pi/\lambda)} = -0.092\lambda$$

When the voltage waveform is plotted vs. transmission line length, the results agree with the calculated results, as shown in Figure 5.7a.

Example

The SWR on a lossless 50-Ω line terminated in an unknown load impedance is 4. The distance between the successive minimum is 30 cm. And the first minimum is located at 6 cm from the load. Determine Γ, Z_L, and l_m.

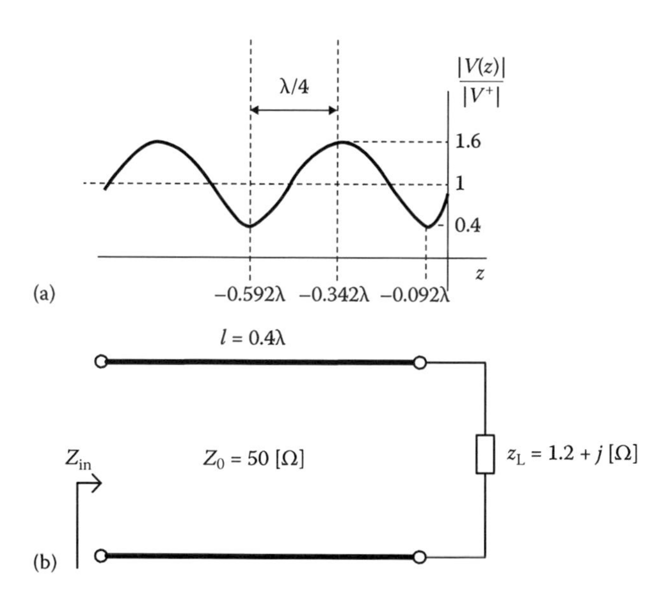

FIGURE 5.7 (a) Voltage vs. transmission length for the example. (b) Transmission line circuit.

Solution

From the given information, the wavelength can be found as

$$\frac{\lambda}{2} = 0.3 \Rightarrow \lambda = 0.6\,\text{m}, \quad \beta = \frac{2\pi}{\lambda} = 3.33\pi$$

The reflection coefficient is equal to

$$|\Gamma| = \frac{4-1}{4+1} = 0.6, \quad z'_{\text{m}} = 0.06\,\text{m} \Rightarrow \ell_{\text{m}} = \frac{\lambda}{2} - z'_{\text{m}} = 0.24\,\text{m}$$

$$\theta_\Gamma = 2\beta z'_m - \pi = -0.6\pi, \quad \Gamma = |\Gamma| e^{j\theta_\Gamma} = 0.6e^{-j0.6\pi} = -0.185 - j0.95$$

The load impedance is then equal to

$$Z_{\text{L}} = Z_0 \frac{1+\Gamma_{\text{L}}}{1-\Gamma_{\text{L}}} = 50 \cdot \frac{1+(-0.185 - j0.95)}{1-(-0.185 - j0.95)} = 1.43 - j41.17$$

Example

Calculate the parameters given below for the transmission line shown in Figure 5.7b when a normalized load of $1.2 + j$ [Ω] is connected.

a. The VSWR on the line
b. Load reflection coefficient
c. Admittance of the load
d. Impedance at the input of the line
e. The distance from the load to the first voltage minimum
f. The distance from the load to the first voltage maximum

Solution

Since the load impedance is already normalized, we can skip the normalization process and start calculations as

a. SWR = 2.5
b. $\Gamma_{\text{L}} = 0.42 \angle 54.5°$
c. $Y_{\text{L}} = \frac{y_{\text{L}}}{Z_0} = \frac{0.5 - j0.42}{50\,\Omega} = (10 - j8.4)\,\text{mS}$
d. $Z_{\text{in}} = z_{\text{in}} \cdot Z_0 = (0.5 + j0.4) \cdot Z_0 = (25 + j20)\ \Omega$
e. $\ell_{\text{min}} = 0.5\lambda - 0.174\lambda = 0.326\lambda$
f. $\ell_{\text{max}} = 0.25\lambda - 0.174\lambda = 0.076\lambda$

5.2.3 Special Cases of Terminated Transmission Lines

a. Short-circuited line

Consider the short-circuited transmission line shown in Figure 5.8. When a transmission line is short circuited, $Z_L = 0 \rightarrow \Gamma = -1$, then the voltage and current can be written as

$$V(z) = V_0^+ \left[e^{-j\beta z} - e^{j\beta z}\right] = -2jV_0^+ \sin\beta z \tag{5.62}$$

$$I(z) = \frac{V_0^+}{Z_0}\left[e^{-j\beta z} + e^{j\beta z}\right] = 2\frac{V_0^+}{Z_0}\cos\beta z \tag{5.63}$$

The input impedance when $z = -l$ is then equal to

$$Z_{in} = jZ_0 \tan \beta l \tag{5.64}$$

The impedance variation of the line along the z is given in Figure 5.9.

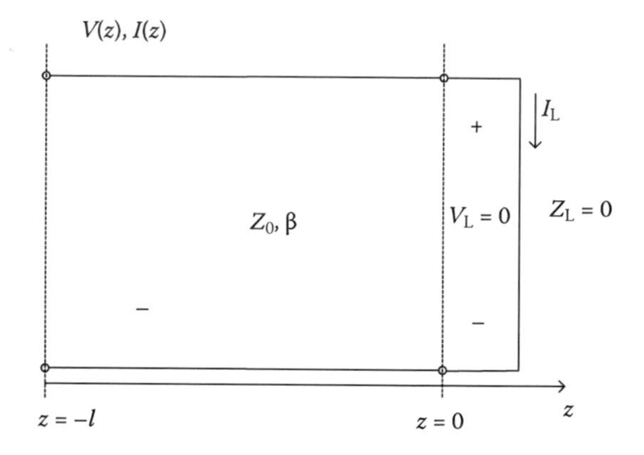

FIGURE 5.8 Short-circuited transmission line.

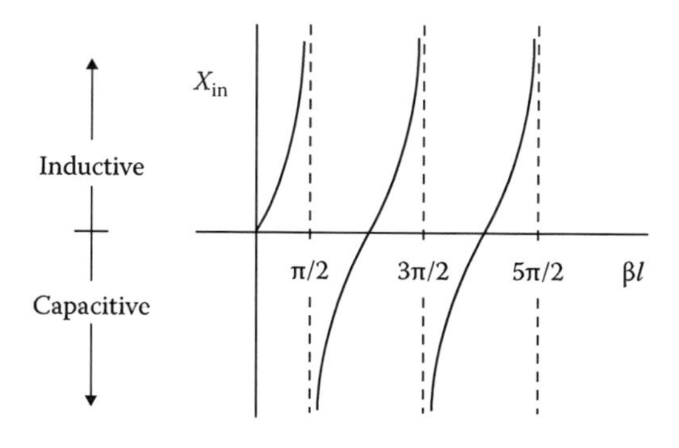

FIGURE 5.9 Impedance variation for a short-circuited transmission line.

At lower frequencies, Equation 5.64 can be written as

$$X_{in} \approx Z_0(\beta l) = \sqrt{\frac{L}{C}}(\omega\sqrt{LC}l) = \omega(Ll) \tag{5.65}$$

Then, the lumped element equivalent model of the transmission line can be represented, as shown in Figure 5.10.

b. Open-circuited line

Consider the open-circuited transmission line shown in Figure 5.11. When the transmission line is short circuited, $Z_L = \infty \rightarrow \Gamma = 1$, then the voltage and current can be written as

$$V(z) = V_0^+\left[e^{-j\beta z} + e^{j\beta z}\right] = 2V_0^+ \cos\beta z \tag{5.66}$$

$$I(z) = \frac{V_0^+}{Z_0}\left[e^{-j\beta z} - e^{j\beta z}\right] = \frac{-2jV_0^+}{Z_0}\sin\beta z \tag{5.67}$$

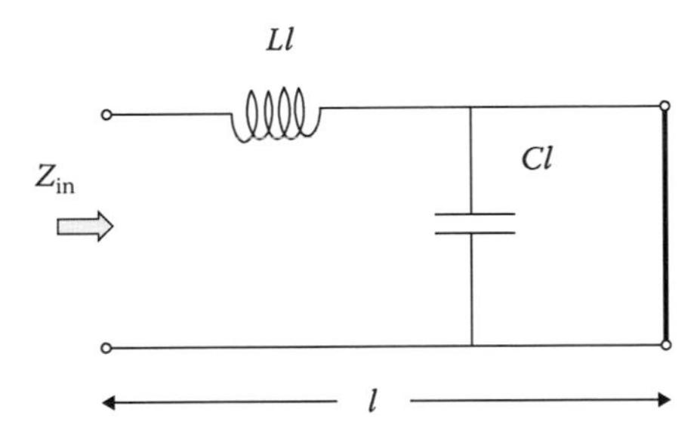

FIGURE 5.10 Low-frequency equivalent circuit of short-circuited transmission line.

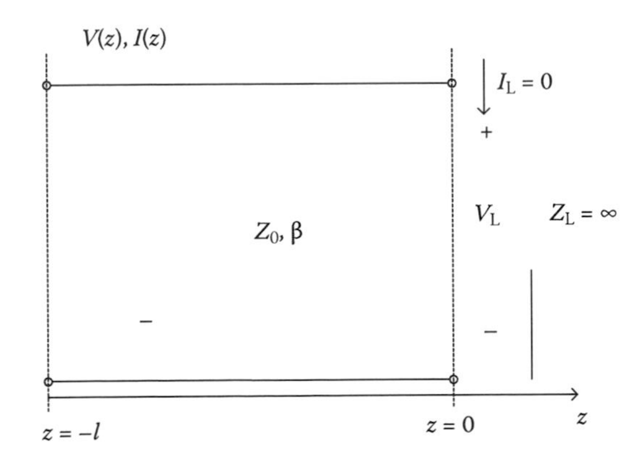

FIGURE 5.11 Open-circuited transmission line.

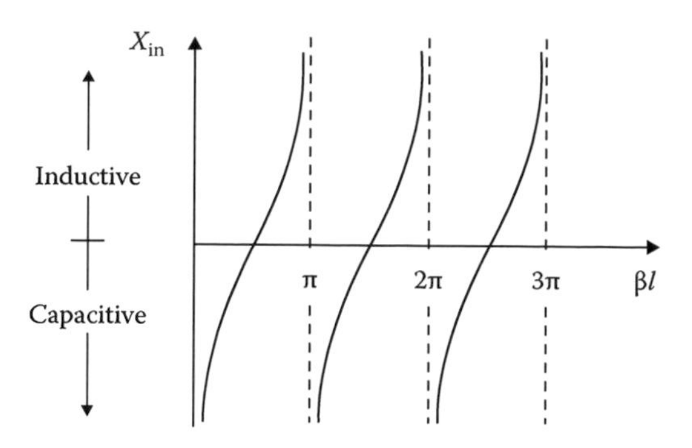

FIGURE 5.12 Impedance variation for an open-circuited transmission line.

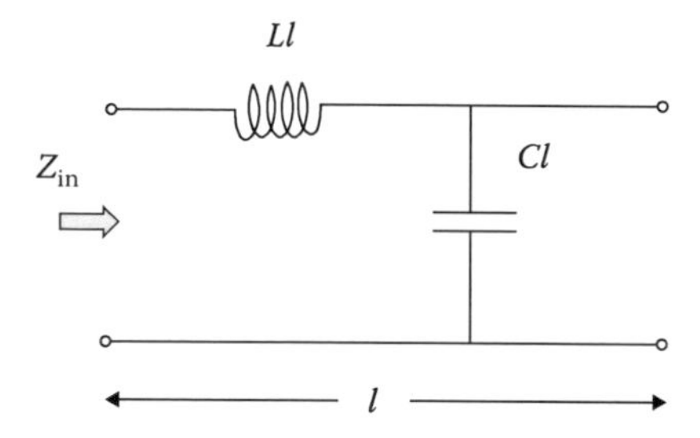

FIGURE 5.13 Low-frequency equivalent circuit of open-circuited transmission line.

The input impedance when $z = -l$ is then equal to

$$Z_{in} = -jZ_0 \cot \beta l \tag{5.68}$$

The impedance variation of the line along the z is given in Figure 5.12.

At lower frequencies, Equation 5.64 can be written as

$$X_{in} \approx -Z_0/(\beta l) = -\sqrt{\frac{L}{C}}\left(\frac{1}{\omega\sqrt{LCl}}\right) = \frac{-1}{\omega(Cl)} \tag{5.69}$$

Then, the lumped element equivalent model of the transmission line can be represented, as shown in Figure 5.13.

5.3 SMITH CHART

The Smith chart is a conformal mapping between the normalized complex impedance plane and the complex reflection coefficient plane. It is a graphical method of displaying impedances and all related parameters using the reflection coefficient. It was invented by Phillip Hagar Smith while he was working at Radio Corporation of

America (RCA). The process of establishing the Smith chart begins with normalizing the impedance, as shown by

$$z_L = \frac{Z_L}{Z_0} = \frac{R_L + jX_L}{Z_0} \tag{5.70}$$

Now, consider the right-hand portion of the normalized complex impedance plane, as illustrated in Figure 5.14. All values of impedance such that $R \geq 0$ are represented by points in the plane. The impedance of all passive devices will be represented by points in the right-half plane.

The complex reflection coefficient may be written as a magnitude and a phase or as real and imaginary parts.

$$\Gamma_L = |\Gamma_L| e^{\angle\Gamma_L} = \Gamma_{Lr} + j\Gamma_{Li} \tag{5.71}$$

The reflection coefficient in terms of the load Z_L terminating line Z_0 is defined as

$$\Gamma_L = \frac{Z_L - Z_0}{Z_L + Z_0} \tag{5.72}$$

The above equation can be rearranged to get

$$Z_L = Z_0 \frac{1 + \Gamma_L}{1 - \Gamma_L} \tag{5.73}$$

In terms of normalized quantities, Equation 5.73 can be written as

$$z_L = r_L + jx_L = \frac{Z_L}{Z_0} = \frac{1 + \Gamma_L}{1 - \Gamma_L} \tag{5.74}$$

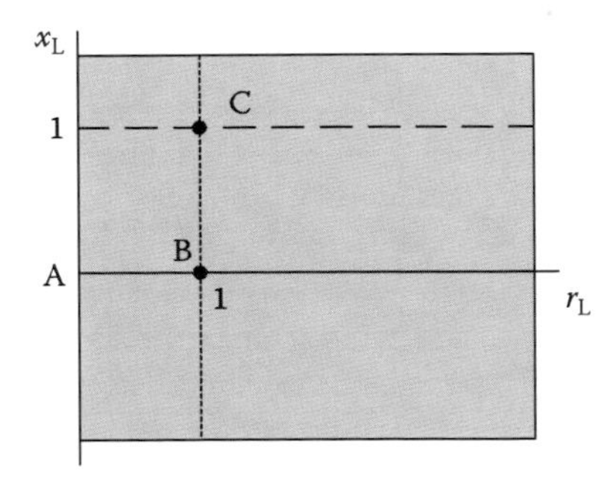

FIGURE 5.14 Right-hand portion of the normalized complex impedance plane.

Substituting in the complex expression for Γ_L and equating real and imaginary parts, we find the two equations that represent circles in the complex reflection coefficient plane as

$$\left(\Gamma_{Lr}-\frac{r_L}{1+r_L}\right)^2+(\Gamma_{Li}-0)^2=\left(\frac{1}{1+r_L}\right)^2 \tag{5.75}$$

$$(\Gamma_{Lr}-1)^2+\left(\Gamma_{Li}-\frac{1}{x_L}\right)^2=\left(\frac{1}{x_L}\right)^2 \tag{5.76}$$

The first circle is centered at

$$\left(\frac{r_L}{1+r_L},0\right) \tag{5.77}$$

and the second circle is centered at

$$\left(1,\frac{1}{x_L}\right) \tag{5.78}$$

The location of the first circle that is always inside the unit circle in the complex reflection coefficient plane with the corresponding radius is

$$\frac{1}{1+r_L} \tag{5.79}$$

Hence, this circle will always be fully contained within the unit circle because the radius can never be greater than unity. This conformal mapping represents the mapping of the real resistance circle and is shown in Figure 5.15 using the mapping equation:

$$\left(\Gamma_r-\frac{r}{1+r}\right)^2+(\Gamma_L)^2=\left(\frac{1}{1+r}\right)^2 \tag{5.80}$$

The location of the second circle that is always outside the unit circle in the complex reflection coefficient plane with the corresponding radius is

$$\left(\frac{1}{x_L}\right) \tag{5.81}$$

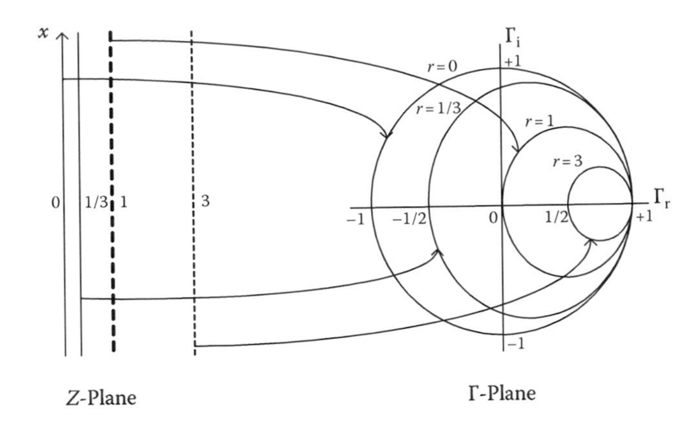

FIGURE 5.15 Conformal mapping of constant resistances.

The value of the radius can vary between 0 and infinity. This conformal mapping represents the mapping of the imaginary reactance circle and is shown in Figure 5.16 using the mapping equation:

$$(\Gamma_r - 1)^2 + \left(\Gamma_i - \frac{1}{x}\right)^2 = \left(\frac{1}{x}\right)^2 \tag{5.82}$$

The circles centered on the real axis represent lines of the constant real part of the load impedance (r_L is constant; x_L varies), and the circles whose centers reside outside the unit circle represent lines of the constant imaginary part of the load impedance (x_L is constant; r_L varies). Combining the results of two mappings

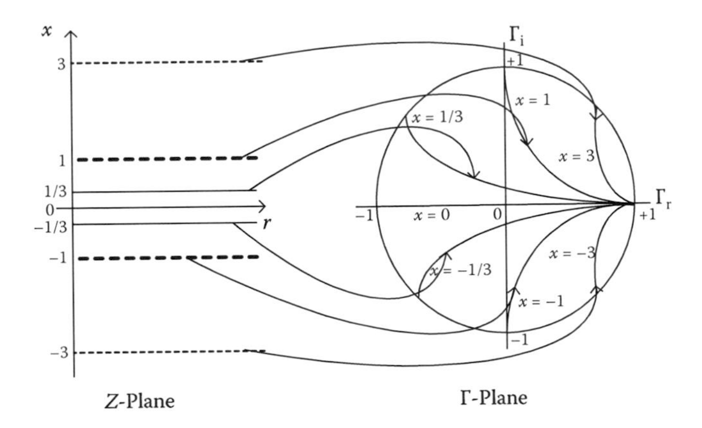

FIGURE 5.16 Conformal mapping of constant reactances.

into a single mapping gives the display of the complete Smith chart, as shown in Figure 5.17.

In summary, the properties of the r-circles are as follows:

- The centers of all r-circles lie on the Γ_r-axis.
- The $r = 0$ circle, having a unity radius and centered at the origin, is the largest.
- The r-circles become progressively smaller as r increases from 0 to ∞, ending at the ($\Gamma_r = 1, \Gamma_i = 0$) point for an open circuit.
- All r circles pass through the ($\Gamma_r = 1, \Gamma_i = 0$) point.

Similarly, the properties of the x-circles are as follows:

- The centers of all x-circles lie on the $\Gamma_r = 1$ line, those for $x > 0$ (inductive reactance) lie above the Γ_r-axis, and those for $x < 0$ (capacitive reactance) lie below the Γ_r-axis.
- The $x = 0$ circle becomes the Γ_r-axis.
- The x-circle becomes progressively smaller as $|x|$ increases from 0 to ∞, ending at the ($\Gamma_r = 1, \Gamma_i = 0$) point for an open circuit.
- All x circles pass through the ($\Gamma_r = 1, \Gamma_i = 0$) point.

Hence, in the combined display of the Smith chart,

- All $|\Gamma|$ circles are centered at the origin, and their radii vary uniformly from 0 to 1.
- The angle, measured from the positive real axis, of the line drawn from the origin through the point representing z_L equals θ_Γ.
- The value of the r circle passing through the intersection of the $|\Gamma|$-circle and the positive-real axis equals the standing-wave ratio SWR.

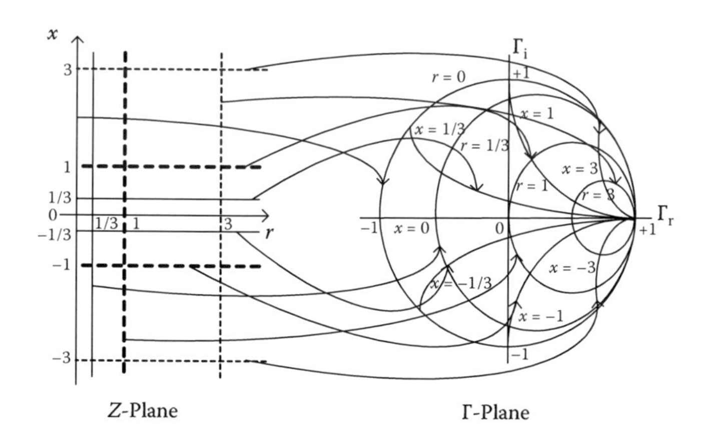

FIGURE 5.17 Combined conformal mapping leading to the display of the Smith chart.

Example

Locate the following normalized impedances on the Smith chart, and calculate the standing wave ratios and reflection coefficients: (a) $z = 0.2 + j0.5$; (b) $z = 0.4 + j0.7$; (c) $z = 0.6 + j0.1$.

Solution

The generic MATLAB® code given below is developed to calculate mark impedance points, draw VSWR circles, and calculate reflection coefficients on the Smith chart at single frequency.

```
%This program marks impedance points, draws VSWR circle, calculates
%reflection coefficients, and marks them on the Smith Chart at single
%frequency

clear all;
close all;
global Z0;
Set_Z0(1); %Set Z0 to 1

%Gui Prompt
 prompt = {'ZL1', 'ZL2: ','ZL3'};
dlg_title = 'Enter Impedance ';
num_lines = 1;
def = { '0.2+j*0.5','0.4+j*0.7','0.6+j*0.1'};
answer = inputdlg(prompt,dlg_title,num_lines,def, 'on');
 %convert the strings received from the GUI to numbers
valuearray=str2double(answer);

%Give variable names to the received numbers
ZL1=valuearray(1);
ZL2=valuearray(2);
ZL3=valuearray(3);

%part a
gamma1=(ZL1-Z0)/(ZL1+Z0);
VSWR1=(1+abs(gamma1))/(1-abs(gamma1));
[th1,rl1]=cart2pol(real(gamma1),imag(gamma1));
smith; %Call Smith Chart Program
s_point(ZL1);
const_SWR_circle(ZL1,'r--');
hold on;
text(real(gamma1)+0.04,imag(gamma1)-0.03,'\bf\Gamma_1');
%part b
gamma2=(ZL2-Z0)/(ZL2+Z0);
VSWR2=(1+abs(gamma2))/(1-abs(gamma2));
[th2,rl2]=cart2pol(real(gamma2),imag(gamma2));
s_point(ZL2);

%part c
gamma3=(ZL3-Z0)/(ZL3+Z0);
VSWR3=(1+abs(gamma3))/(1-abs(gamma3));
[th3,rl3]=cart2pol(real(gamma3),imag(gamma3));
s_point(ZL3);
```

```
const_SWR_circle(ZL3,'r--');
hold on;
text(real(gamma3)+0.04,imag(gamma3)-0.03,'\bf\Gamma_3');

msgbox( sprintf([...
        'Calculated Parameters for Z1 \n'...
        '    Reflection coefficient for Z1:  gamma1 =%f +j(%f)\n'...
        '    Reflection Coefficent for Z1 In Polar form
             :|gamma1|=%f,angle1=%f\n'...
        '    Standing Wave Ratio for Z1 : VSWR1=%f \n'...
        '\n'...
        'Calculated Parameters for Z2 \n'...
        '    Reflection coefficient for Z2:  gamma2 =%f +j(%f)\n'...
        '    Reflection Coefficent for Z2 In Polar form
             :|gamma2|=%f,angle1=%f\n'...
        '    Standing Wave Ratio for Z2 : VSWR2=%f \n'...
        '\n'...
        'Calculated Parameters for Z3 \n'...
        '    Reflection coefficient for Z3:  gamma3 =%f +j(%f)\n'...
        '    Reflection Coefficient for Z3 In Polar form
             :|gamma3|=%f,angle3=%f\n'...
        '    Standing Wave Ratio for Z3: VSWR3=%f \n'...
          '\n']...
,real(gamma1),imag(gamma1),rl1,th1*180/pi,VSWR1,real(gamma2),imag(gamma2),
rl2,th2*180/pi,VSWR2,real(gamma3),imag(gamma3),rl3,th3*180/pi,VSWR3));
```

When the program is executed, a GUI is displayed, as shown in Figure 5.18a, for entering impedances and the result. The results are displayed on the Smith chart in Figure 5.18b.

5.3.1 Input Impedance Determination with Smith Chart

It was shown before that the voltage and current at any point on the transmission line can be expressed as

$$V(z') = \frac{I_L}{2}(Z_L + Z_0)e^{\gamma z'}[1 + \Gamma e^{-2\gamma z'}] \tag{5.83}$$

$$I(z') = \frac{I_L}{2Z_0}(Z_L + Z_0)e^{\gamma z'}[1 - \Gamma e^{-2\gamma z'}] \tag{5.84}$$

where $z' = 1 - z$. Then, the input impedance at a distance d away from the load on the line in terms of reflection coefficient can be obtained as

$$Z(d) = \frac{V(d)}{I(d)} = Z_0\frac{1 + \Gamma(d)}{1 - \Gamma(d)} \tag{5.85}$$

where

$$\Gamma(d) = \Gamma_L e^{-j2\beta d} \tag{5.86}$$

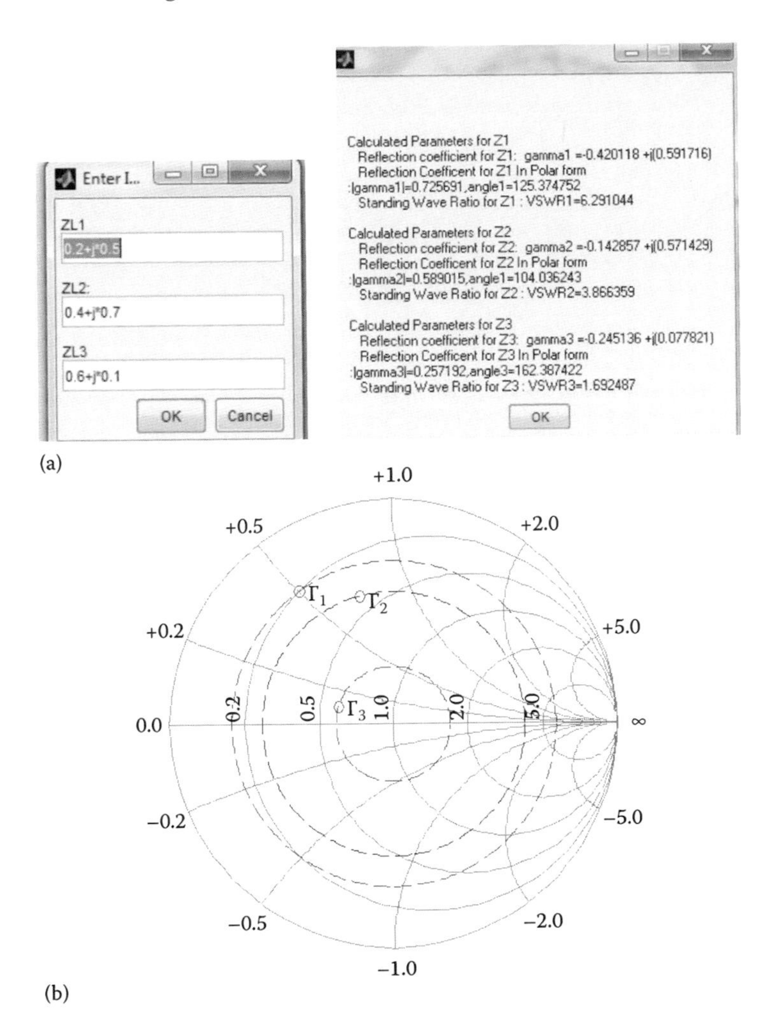

FIGURE 5.18 (a) GUI display for user input and results. (b) Smith chart displaying the calculated values and impedances.

Example

A transmission line of characteristic impedance $Z_0 = 50\ \Omega$ and length $d = 0.2\lambda$ is terminated into a load impedance of $Z_L = (25 - j50)\ \Omega$. Find Γ_L, $Z_{in}(d)$, and SWR using the Smith chart.

Solution

The generic MATLAB code given below is developed to find input impedance by moving toward the generator at any length. The MATLAB code is given below "% on the transmission line at single frequency at any length."

```
%This program find input impedance by moving towards generator
%on the transmission line at single frequency at any length

clear all;
close all;
global Z0;
%Gui Prompt

prompt = {'Enter Load Impedance ZL :', 'Enter the Length (in lambda) d :
','Enter Characteristic Impedance Z0:'};
dlg_title = 'Enter Impedance ';
num_lines = 1;
def = {'25-j*50','.2','50'};
answer = inputdlg(prompt,dlg_title,num_lines,def, 'on');
%convert the strings received from the GUI to numbers
valuearray=str2double(answer);

%Give variable names to the received numbers
ZL=valuearray(1);
d=valuearray(2);
Z0=valuearray(3);


Set_Z0(Z0);
gamma_0=(ZL-Z0)/(ZL+Z0);
[th0,mag_gamma_0]=cart2pol(real(gamma_0),imag(gamma_0));
if th0<0
    th0=th0+2*pi;
end
th_in=th0-2*2*pi*d;
if th_in<0
    th_in=th_in+2*pi;
end
 [x_gamma_in,y_gamma_in]=pol2cart(th_in,mag_gamma_0);
Zin=Z0*(1+x_gamma_in+j*y_gamma_in)/(1-x_gamma_in-j*y_gamma_in);
SWR=(1+abs(gamma_0))/(1-abs(gamma_0));
smith_chart(0);
hold on;
th=th0:(th_in-th0)/29:th_in;
gamma=mag_gamma_0*ones(1,30);
polar(th,gamma,'k');
hold on
s_point(Zin);
text(x_gamma_in+0.04,y_gamma_in-0.03,'\bfZ_{in}');
s_point(ZL);
text(real(gamma_0)+0.04,imag(gamma_0)-0.03,'\bfZ_{L}');

msgbox( sprintf([...
        'Calculated Parameters for Transmission Line \n'...
        '   Load Reflection coefficient :  gamma_0 =%f +j(%f)\n'...
        '   Magnitude of Load Reflection Coefficient
:|gamma_0|=%f,angle=%f\n'...
        '   Input Impedance Zin : Zin=%f +j(%f)\n'...
        '   Standing Wave Ratio : SWR=%f \n'...
        '\n']...
,real(gamma_0),imag(gamma_0),mag_gamma_0,th0*180/pi,real(Zin),
imag(Zin),SWR));
```

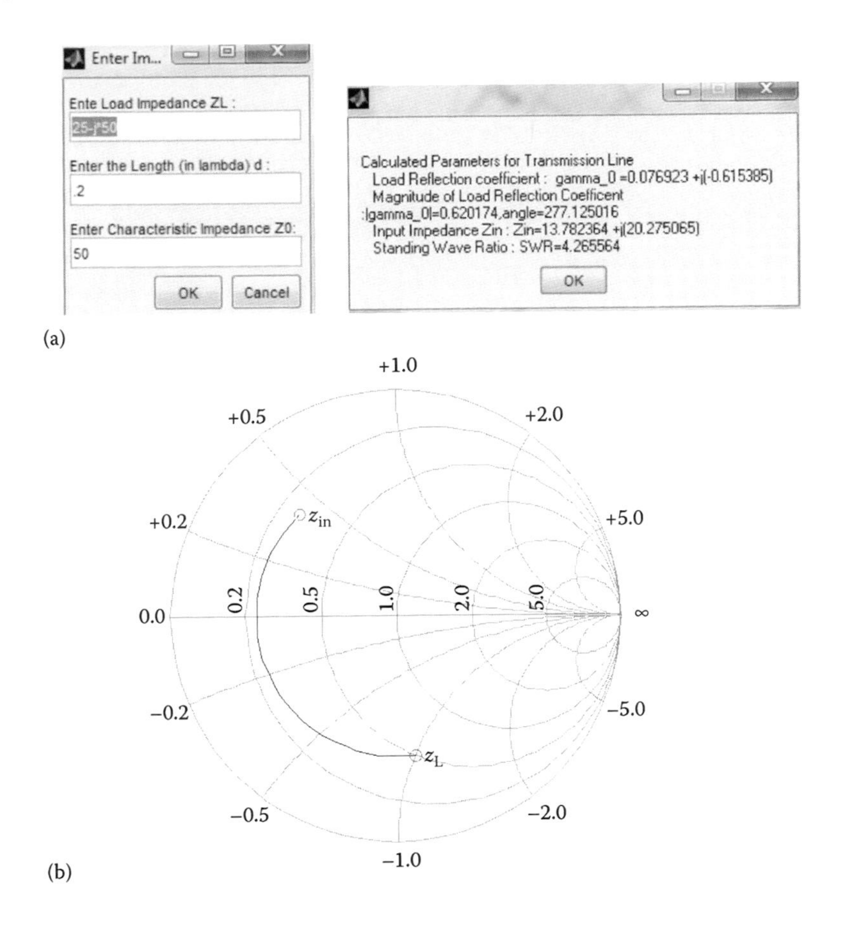

FIGURE 5.19 (a) MATLAB GUI display for user input and results. (b) Input impedance display using Smith chart.

When the program is executed, a GUI and calculated results are displayed, as shown in Figure 5.19a. The program also displays input impedance on the Smith chart, as shown in Figure 5.19b, with the move toward the generator.

5.3.2 Smith Chart as an Admittance Chart

The Smith chart can also be used as an admittance chart by transforming impedances to admittances. Consider the expression for a normalized impedance at any point on the transmission line in terms of reflection coefficient as

$$Z_{\text{in}}(z) = \left(\frac{1+\Gamma(z)}{1-\Gamma(z)}\right) \tag{5.87}$$

The normalized admittance is the reciprocal of impedance and can be written as

$$Y_{\text{in}}(z) = \frac{Y_{\text{in}}(z)}{Y_0} = \frac{1/Z_{\text{in}}(z)}{1/Z_0} = \frac{1}{Z_{\text{in}}(z)/Z_0} = \frac{1}{Z_{\text{in}}(z)} \tag{5.88}$$

Then, the normalized admittance in terms of the reflection coefficient can be expressed as

$$Y_{\text{in}}(z) = \left(\frac{1-\Gamma(z)}{1+\Gamma(z)}\right) \tag{5.89}$$

which can be written as

$$Y_{\text{in}}(z) = \left(\frac{1+\Gamma'(z)}{1-\Gamma'(z)}\right) \tag{5.90}$$

where

$$\Gamma'(z) = -\Gamma(z) = \Gamma(z)e^{-j\pi} \tag{5.91}$$

That means a 180° phase shift for the reflection coefficient gives the value of admittance for the corresponding impedance value. When an impedance point is marked on the Smith chart, moving 180° in the clockwise direction gives the value admittance. Instead of repeating this for each impedance point on the Smith chart, we can keep the location of the impedance fixed and rotate the Smith chart by 180°. This gives the admittance chart as shown in Figure 5.20. When both Z and Y charts are plotted together, we obtain the ZY chart, as shown in Figure 5.21.

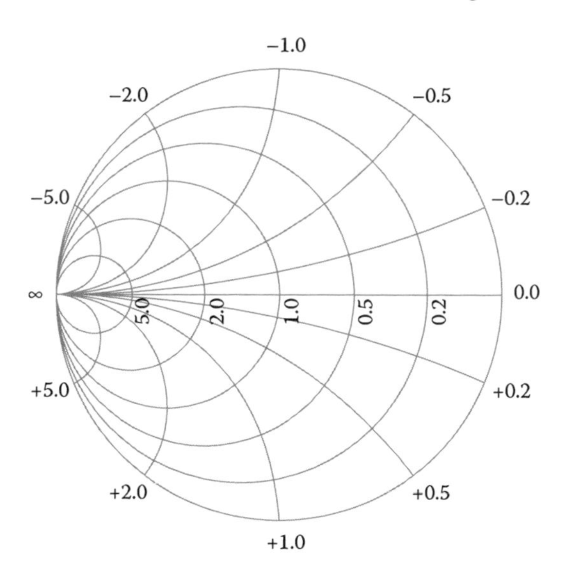

FIGURE 5.20 Admittance, *Y*, Smith chart.

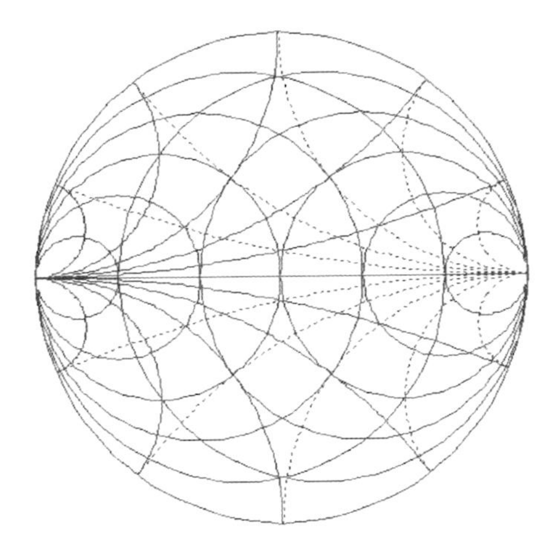

FIGURE 5.21 *ZY* Smith chart.

5.3.3 *ZY* Smith Chart and Its Application

The *ZY* Smith chart gives the ability to implement both impedances and admittances on a single chart. It is a power chart, and it enables designers to make impedance transformation and matching using a unique graphical display when the components are connected in series or in shunt. The effect of adding a single reactive component in series with a complex impedance results in motion along a constant resistance circle in the *ZY* chart. If a single reactive component is added with a complex impedance in shunt, then motion along a constant conductance circle in the *ZY* chart is needed. Whenever an inductor is connected to the network, the direction of movement on the *ZY* chart is toward the upper half, whereas a capacitive involvement results in movement toward the lower part of the chart. All these component motions on the *ZY* chart are illustrated in Figure 5.22.

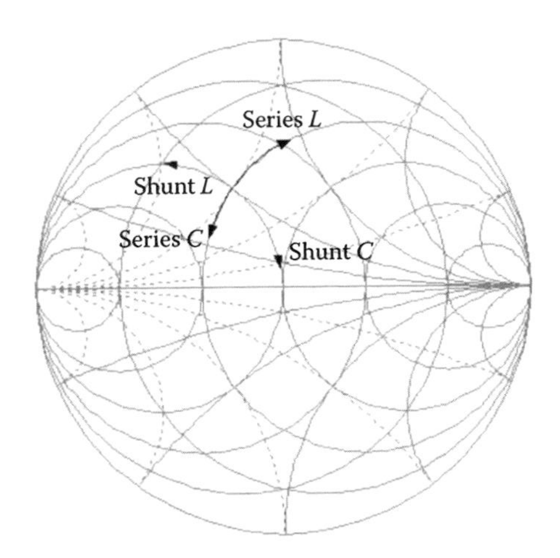

FIGURE 5.22 Adding component using *ZY* Smith chart.

Example

Find the input impedance for the circuit shown in Figure 5.23 at 4 GHz when the load connected is $Z_L = 62.5\ \Omega$ using the Smith chart.

Solution

The process begins with normalizing the load impedance, $Z_L = R = 62.5\ \Omega$.

$$z_L = \frac{Z_L}{Z_0} = \frac{62.5}{50} = 1.25$$

Since the next component is a shunt-connected component, we need to convert this value to a conductance value. That is

$$g_L = \frac{1}{z_L} = \frac{1}{1.25} = 0.8$$

On the *ZY* Smith chart, we mark this point on the conductance circle. The next component is shunt C with a value of 1.59 [pF]. The normalized susceptance value of the capacitor at 4 GHz is found from

$$b_C = B_C Z_0 = \omega C Z_0 = (2\pi 4 \times 10^9)(1.59 \times 10^{-12})50 = 2$$

This corresponds to point B on the Smith chart. This is the amount of rotation that needs to be done on the conductance circle, as shown by point B. The admittance at point B is equal to

$$y_B = 0.8 + j2$$

The next component connected is a series *L* with a value of 8 [nH]. So, we move from conductance circle to resistance circle and read the corresponding impedance value as

$$z_B = 0.17 - j0.43$$

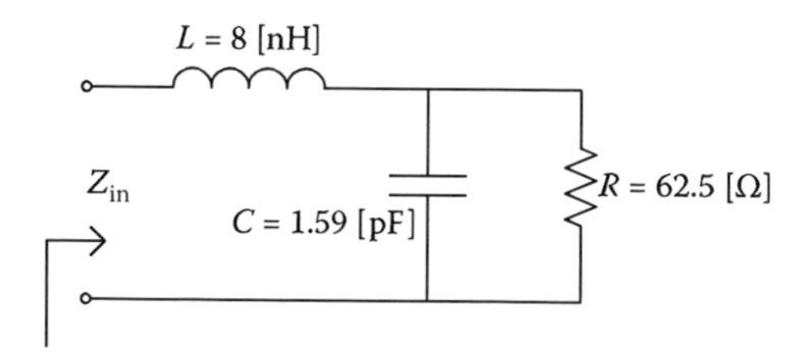

FIGURE 5.23 Impedance transformation.

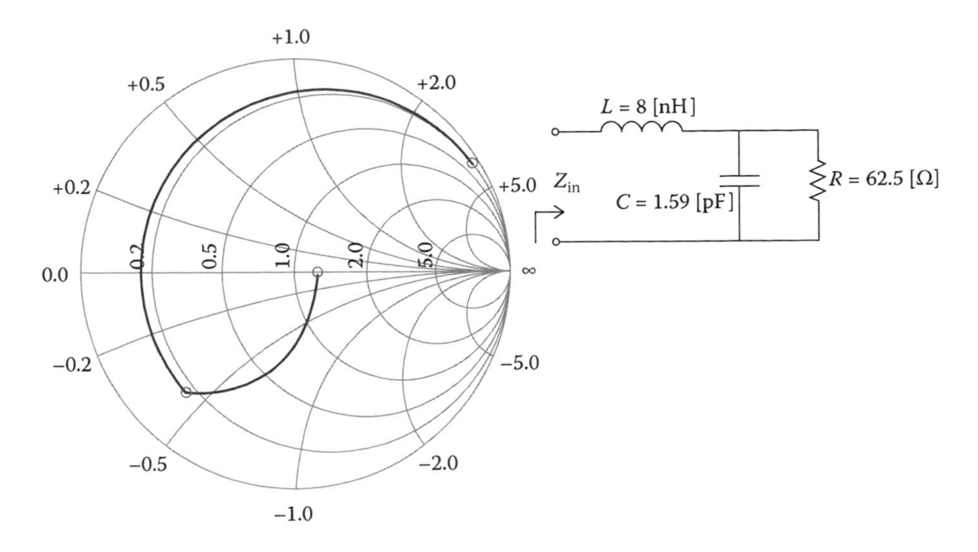

FIGURE 5.24 Impedance transformation using the Smith chart.

The normalized reactance value of the inductor is equal to

$$x_L = \frac{X_L}{Z_0} = \frac{(2\pi 4\times 10^9)(8\times 10^{-9})}{50} = 4$$

This value needs to be added to the impedance at point B to find the impedance value shown as point C on the Smith chart.

$$z_C = z_B + x_L = 0.17 - j0.43 + 4 = 0.17 + 3.57$$

Denormalizing impedance z_C gives the input impedance as

$$Z_{in} = z_C Z_0 = (0.17 + j3.57)\ 50 = (8.5 + j178.5)\ [\Omega]$$

The results are shown on the Smith chart in Figure 5.24.

5.4 IMPEDANCE MATCHING BETWEEN TRANSMISSION LINES AND LOAD IMPEDANCES

Consider the matching network between the load and the transmission line shown in Figure 5.25. The matching network can be implemented using the lumped element *L*-type sections consisting of two reactive elements. There are eight possible *L*-matching networks that are shown in Figure 5.26. These can be illustrated by two generic circuits, as shown in Figure 5.27.

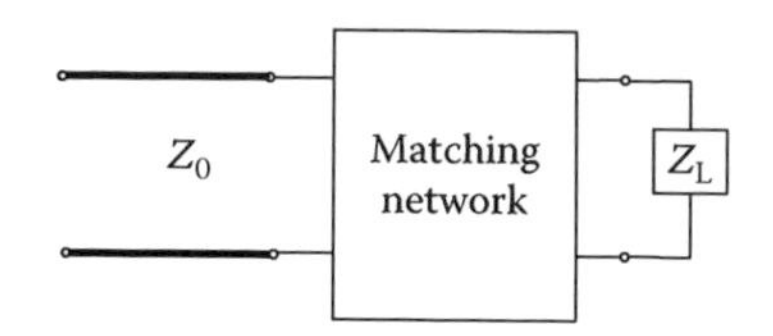

FIGURE 5.25 Matching network between load and transmission line.

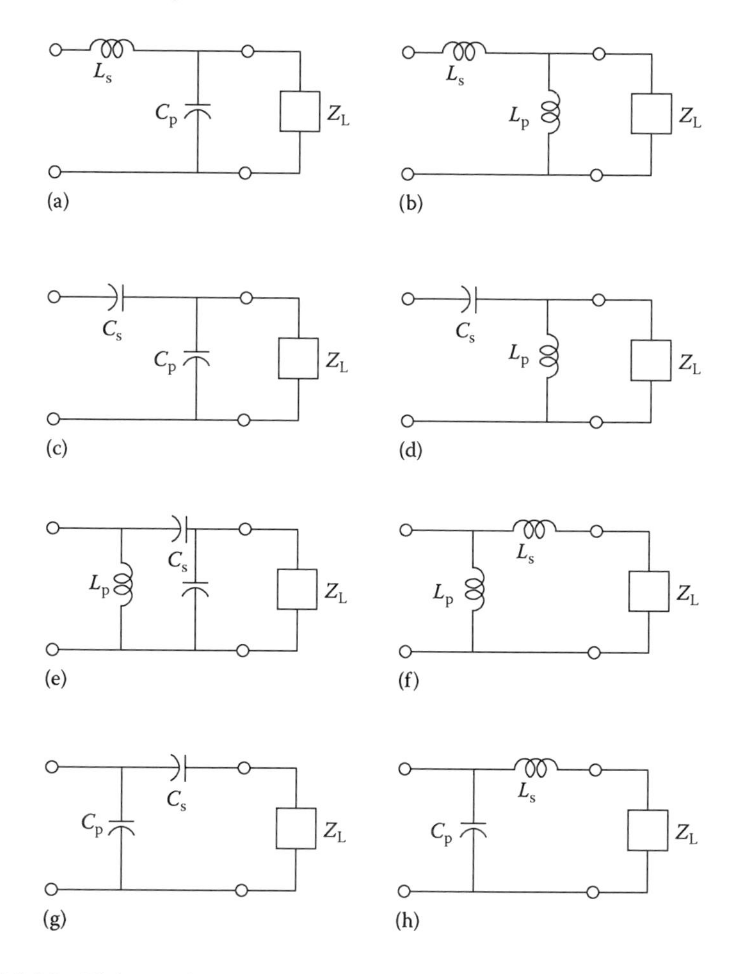

FIGURE 5.26 Eight possible L-matching network sections.

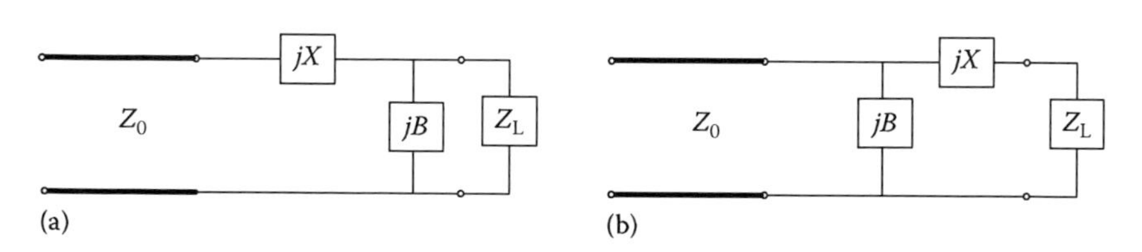

FIGURE 5.27 Generic L-matching network sections to represent eight L sections.

In either of the configurations of Figure 5.27, the reactive elements may be either inductors or capacitors. As a result, there are eight distinct possibilities, as shown in Figure 5.26, for the matching circuit for various load impedances. If the normalized load impedance, $z_L = Z_L/Z_0$, is inside the $1 + jx$ circle on the Smith chart, then the circuit of Figure 5.27a should be used. If the normalized load impedance is outside the $1 + jx$ circle on the Smith chart, the circuit of Figure 5.27b should be used. The $1 + jx$ circle is the resistance circle on the impedance Smith chart for which $r = 1$.

Consider first the circuit given in Figure 5.27a with $Z_L = R_L + jX_L$. It is assumed that $R_L > Z_0$ and $z_L = Z_L/Z_0$ maps inside the $1 + jx$ circle on the Smith chart. For a matched condition, the impedance seen looking into the matching network followed by the load impedance is then equal to Z_0 and can be written as

$$Z_0 = jX + \frac{1}{jB + \left(\dfrac{1}{R_L} + jX_L\right)} \tag{5.92}$$

Separating Equation 5.92 into real and imaginary parts gives two equations with two unknowns, X and B, as

$$B(XR_L - X_L Z_0) = R_L - Z_0 \tag{5.93}$$

$$X(1 - BX_L) = BZ_0 R_L - X_L \tag{5.94}$$

The solution of Equations 5.93 and 5.94 leads to

$$B = \frac{X_L \pm \sqrt{\dfrac{R_L}{Z_0}}\sqrt{R_L^2 + X_L^2 - Z_0 R_L}}{R_L^2 + X_L^2} \tag{5.95}$$

and

$$X = \frac{1}{B} + \frac{X_L Z_0}{R_L} - \frac{Z_0}{BR_L} \tag{5.96}$$

From Equation 5.95, there exist two possible solutions for B and consequently X. Both of these solutions are physically realizable and constitute all the values of B and X. The positive value of X gives an inductor; the negative value of X gives a capacitor. Similarly, the positive value of B gives a capacitor, and the negative value of B gives an inductor.

The same procedure can be repeated for the generic L-matching network shown in Figure 5.27b. This circuit is used when $z_L = Z_L/Z_0$, and it maps outside the $1 + jx$ circle on the Smith chart since it is assumed that $R_L < Z_0$. For a matched condition, the admittance seen looking into the matching network followed by the load impedance $Z_L = R_L + jX_L$ is then equal to $1/Z_0$ and can be written as

$$\frac{1}{Z_0} = jB + \frac{1}{R_L + j(X + X_L)} \tag{5.97}$$

Separating Equation 5.97 into real and imaginary parts gives the following two equations with two unknowns, X and B, as

$$BZ_0(X + X_L) = Z_0 - R_L \tag{5.98}$$

$$(X + X_L) = BZ_0R_L \tag{5.99}$$

The solution of Equations 5.98 and 5.99 leads to

$$X = \sqrt{R_L(Z_0 - R_L)} - X_L \tag{5.100}$$

$$B = \pm\frac{\sqrt{(Z_0 - R_L)/R_L}}{Z_0} \tag{5.101}$$

Equation 5.101 has two possible solutions for B.

In order to match an arbitrary complex load to a line of characteristic impedance Z_0, the real part of the input impedance to the matching network must be Z_0, while the imaginary part must be zero. This implies that a general matching network must have at least two degrees of freedom; in the L-section matching circuit, these two degrees of freedom are provided by the values of the two reactive components.

5.5 SINGLE-STUB TUNING

At high frequencies, it may be desirable to match the given load to the transmission line using transmission lines instead of lumped element components discussed in Section 5.4. Impedance matching can then be done using a single open- or short-circuited length of transmission line called a "stub." It is connected either in parallel or in series with the transmission feed line at a certain distance from the load, as shown in Figure 5.28.

In single-stub tuning, there are two design parameters: the distance, d, from the load to the stub position and the value of susceptance or reactance provided by the shunt or series stub.

5.5.1 Shunt Single-Stub Tuning

When it is a shunt-stub case, as shown in Figure 5.28a, we select d so that the admittance, Y, seen looking into the line at distance d from the load is equal to $Y_0 + jB$. Then, the matching is done by choosing the stub susceptance as $-jB$.

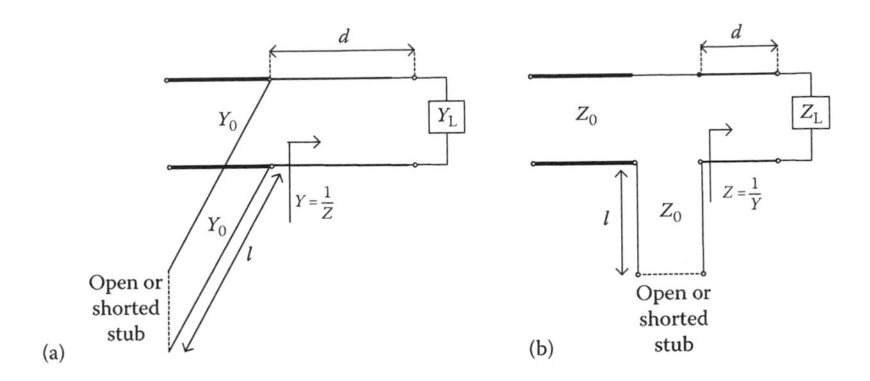

FIGURE 5.28 Single-stub matching: (a) parallel and (b) series.

To obtain the relations for d and l, the input impedance, $Z_L = 1/Y_L = R_L + jX_L$, at a distance d from the load is written as

$$Z = Z_0 \frac{(R_L + jX_L) + jZ_0 \tan \beta d}{Z_0 + j(R_L + jX_L) \tan \beta d} \tag{5.102}$$

The admittance is then wired from Equation 5.102 as

$$Y = G + jB = \frac{1}{Z} \tag{5.103}$$

where

$$G = \frac{R_L(1 + \tan^2 \beta d)}{R_L^2 + (X_L + Z_0 \tan \beta d)^2} \tag{5.104}$$

$$B = \frac{R_L^2 \tan \beta d - (Z_0 - X_L \tan \beta d)(X_L + Z_0 \tan \beta d)}{Z_0 \left[R_L^2 + (X_L + Z_0 \tan \beta d)^2 \right]} \tag{5.105}$$

To have the matching conditions, we need to set G in Equation 5.102 to $G = Y_0 = 1/Z_0$. Hence,

$$Z_0(R_L - Z_0) \tan^2 \beta d - 2X_L Z_0 \tan \beta d + \left(R_L Z_0 - R_L^2 - X_L^2 \right) = 0 \tag{5.106}$$

which leads to two solutions for $\tan \beta d$ as

$$\tan \beta d = \frac{X_L \pm \sqrt{R_L \left[(Z_0 - R_L)^2 + X_L^2 \right] / Z_0}}{R_L - Z_0}, \quad \text{for } R_L \neq Z_0 \tag{5.107}$$

If $R_L = Z_0$, then $\tan\beta d = -X_L/2Z_0$. As a result, we have solutions for d as

$$\frac{d}{\lambda} = \begin{cases} \dfrac{1}{2\pi}\tan^{-1}\left(-\dfrac{X_L}{2Z_0}\right) & \text{for } -\dfrac{X_L}{2Z_0} \geq 0 \\ \dfrac{1}{2\pi}\left(\pi + \tan^{-1}\left(-\dfrac{X_L}{2Z_0}\right)\right) & \text{for } -\dfrac{X_L}{2Z_0} < 0 \end{cases} \tag{5.108}$$

To find the required stub lengths, we first set $B_s = -B$. This leads to the final solutions for open and shorted stubs shown in Figure 5.28a as

$$\frac{l}{\lambda} = \frac{1}{2\pi}\tan^{-1}\left(\frac{B_s}{Y_0}\right) = -\frac{1}{2\pi}\tan^{-1}\left(\frac{B}{Y_0}\right) \quad \text{for open stub} \tag{5.109}$$

$$\frac{l}{\lambda} = -\frac{1}{2\pi}\tan^{-1}\left(\frac{Y_0}{B_s}\right) = \frac{1}{2\pi}\tan^{-1}\left(\frac{Y_0}{B}\right) \quad \text{for shorted stub} \tag{5.110}$$

The Smith chart solution for the matching with open stub is practical and can be described as follows:

- Normalize the load impedance and locate the corresponding admittance on the Z Smith chart.
- Rotate clockwise around the Smith chart from y_L until it intersects the $g = 1$ circle. It intersects the $g = 1$ circle at two points. The "length" of this rotation determines the value d. There are two possible solutions.
- Rotate clockwise from the short/open circuit point around the $g = 0$ circle until the stub b equals $-b$. The "length" of this rotation determines the stub length l.

5.5.2 Series Single-Stub Tuning

For the series stub case shown in Figure 5.28b, d is chosen so that the impedance looking into the line at a distance d from the load is equal to $Z_0 + jX$. Then, the stub reactance is selected to be $-jX$ to match the line.

To obtain the relations for d and l, we write the input admittance, $Y_L = 1/Z_L = G_L + jB_L$, at a distance d from the load as

$$Y = Y_0 \frac{(G_L + jB_L) + jY_0 \tan\beta d}{Y_0 + j(G_L + jB_L)\tan\beta d} \tag{5.111}$$

The impedance is then wired from Equation 5.111 as

$$Z = R + jX = \frac{1}{Y} \tag{5.112}$$

where

$$R = \frac{G_L(1+\tan^2\beta d)}{G_L^2 + (B_L + Y_0\tan\beta d)^2} \tag{5.113}$$

$$X = \frac{G_L^2\tan\beta d - (Y_0 - B_L\tan\beta d)(B_L + Y_0\tan\beta d)}{Y_0\left[G_L^2 + (B_L + Y_0\tan\beta d)^2\right]} \tag{5.114}$$

To have the matching conditions, we need to set G in Equation 5.113 to $R = Z_0 = 1/Y_0$. Hence,

$$Y_0(G_L - Y_0)\tan^2\beta d - 2B_LY_0\tan\beta d + \left(G_LY_0 - G_L^2 - B_L^2\right) = 0 \tag{5.115}$$

which leads to two solutions for tan βd as

$$\tan\beta d = \frac{B_L \pm \sqrt{G_L\left[(Y_0 - G_L)^2 + B_L^2\right]/Y_0}}{G_L - Y_0}, \quad \text{for } G_L \neq Y_0 \tag{5.116}$$

If $G_L = Y_0$, then tan $\beta d = -B_L/2Y_0$. As a result, we have solutions for d as

$$\frac{d}{\lambda} = \begin{cases} \dfrac{1}{2\pi}\tan^{-1}\left(-\dfrac{B_L}{2Y_0}\right) & \text{for } -\dfrac{B_L}{2Y_0} \geq 0 \\[2ex] \dfrac{1}{2\pi}\left(\pi + \tan^{-1}\left(-\dfrac{B_L}{2Y_0}\right)\right) & \text{for } -\dfrac{B_L}{2Y_0} < 0 \end{cases} \tag{5.117}$$

To find the required stub lengths, we first set $X_s = -X$. This leads to the final solutions for open and shorted stubs shown in Figure 5.28b as

$$\frac{l}{\lambda} = \frac{1}{2\pi}\tan^{-1}\left(\frac{X_s}{Z_0}\right) = -\frac{1}{2\pi}\tan^{-1}\left(\frac{X}{Z_0}\right) \tag{5.118}$$

$$\frac{l}{\lambda} = -\frac{1}{2\pi}\tan^{-1}\left(\frac{Z_0}{X_s}\right) = \frac{1}{2\pi}\tan^{-1}\left(\frac{Z_0}{X}\right) \tag{5.119}$$

The Smith chart solution for the matching with series stub is practical and can be described as follows:

- Normalize the load impedance and locate it on the Z Smith chart.
- Rotate clockwise around the Smith chart from z_L until it intersects the $r = 1$ circle. It intersects the $r = 1$ circle at two points. The "length" of this rotation determines the value d. There are two possible solutions.
- Rotate clockwise from the short/open-circuit point around the $r = 0$ circle, until the stub x equals $-x$. The "length" of this rotation determines the stub length l.

5.6 IMPEDANCE TRANSFORMATION AND MATCHING BETWEEN SOURCE AND LOAD IMPEDANCES

Consider the matching network between the source and load, as shown in Figure 5.29. As discussed in Section 5.5, there are eight possible matching networks, as shown in Figure 5.26, which can be represented by generic two types of L-matching networks, as shown in Figure 5.30. We will first derive the analytical equations as we did before in Section 5.5. This time, consider first the generic L-type matching network shown in Figure 5.30b.

Since the source is matched to load impedance, the complex conjugate impedance of the load should be equal to the overall impedance connected to the load impedance. This can be expressed by

$$Z_L^* = \frac{1}{Z_s^{-1} + jB} + jX \tag{5.120}$$

Express

$$Z_s = R_s + jX_s \quad \text{and} \quad Z_L = R_L + jX_L \tag{5.121}$$

Then,

$$Z_{Load}^* = \frac{1}{Z_s^{-1} + jB} + jX = \frac{R_s + jX_s}{1 + jB(R_s + jX_s)} + jX = R_L - jX_L \tag{5.122}$$

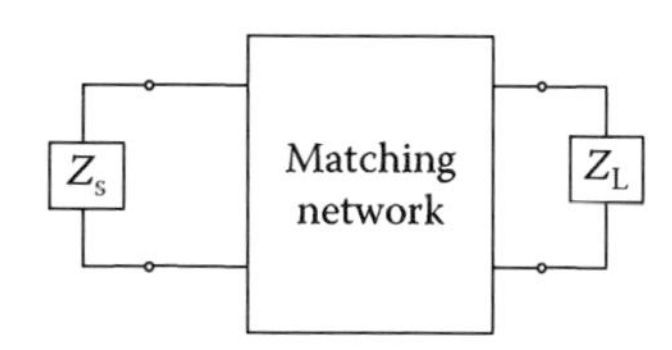

FIGURE 5.29 Matching networks between the load and transmission line.

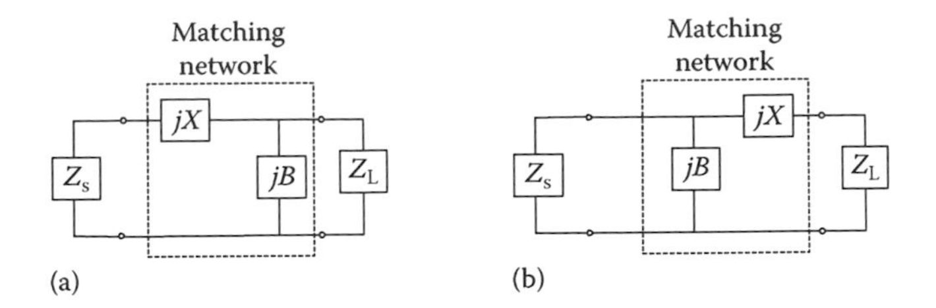

FIGURE 5.30 Generic two *L*-matching networks between source and load impedances.

Separate real and imaginary parts,

$$R_s = R_L(1 - BX_s) + (X_L + X)BR_s \tag{5.123}$$

$$X_s = R_sR_LB - (1 - B_CX_L)(X_L + X) \tag{5.124}$$

Solving for B and X gives

$$X = -X_L \pm \sqrt{R_L(R_s - R_L) + \frac{R_L}{R_s}X_s^2} \tag{5.125}$$

$$B = \frac{R_s - R_L}{R_sX + R_sX_L - R_LX_s} \tag{5.126}$$

The solution given by Equations 5.125 and 5.126 is valid only when $R_s > R_L$. A similar procedure can be applied for the circuit shown in Figure 5.30a. The following equations are obtained for reactance and susceptance by assuming that $R_s < R_L$.

$$B = \frac{R_sX_L \pm \sqrt{R_sR_L\left(R_L^2 + X_L^2 - R_sR_L\right)}}{R_s\left(R_L^2 + X_L^2\right)} \tag{5.127}$$

$$X = \frac{\left(R_L^2 + X_L^2\right)B - X_L + \dfrac{X_s}{R_s}R_L}{\left(R_L^2 + X_L^2\right)B^2 - 2X_LB + 1} \tag{5.128}$$

As can be seen, the analytical calculation of the impedance transformation and matching is tedious. Instead, we can apply the Smith chart to the same task. For this, there is a standard procedure that needs to be followed. The design procedure for matching source impedance to a load impedance using the Smith chart can be outlined as follows:

- Normalize the given source and complex conjugate load impedances, and locate them on the Smith chart.
- Plot constant resistance and conductance circles for the impedances located.
- Identify the intersection points between the constant resistance and conductance circle for the impedances located.
- The number of the intersection point corresponds to the number of possible *L*-matching networks.
- By following the paths that go through intersection points, calculate the normalized reactances and susceptances.
- Calculate the actual values of the inductors and capacitors by denormalizing at the given frequency.

Example

Using the Smith chart, design all possible configurations of two-element matching networks that match source impedance $Z_s = (15 + j50)\ \Omega$ to the load $Z_L = (20 - j30)\ \Omega$. Assume the characteristic impedance of $Z_0 = 50\ \Omega$ and an operating frequency of $f = 4$ GHz.

Solution

The MATLAB program developed previously is modified to plot resistance and conductance circles for the source and complex conjugate of the load impedances. As shown in Figure 5.31, there are four possible *L*-matching networks. These networks are illustrated in Figure 5.32.

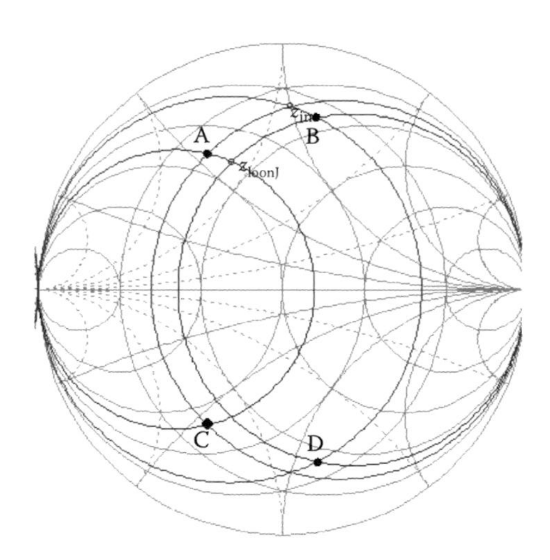

FIGURE 5.31 Number of possible *L*-matching networks to match source and load impedance.

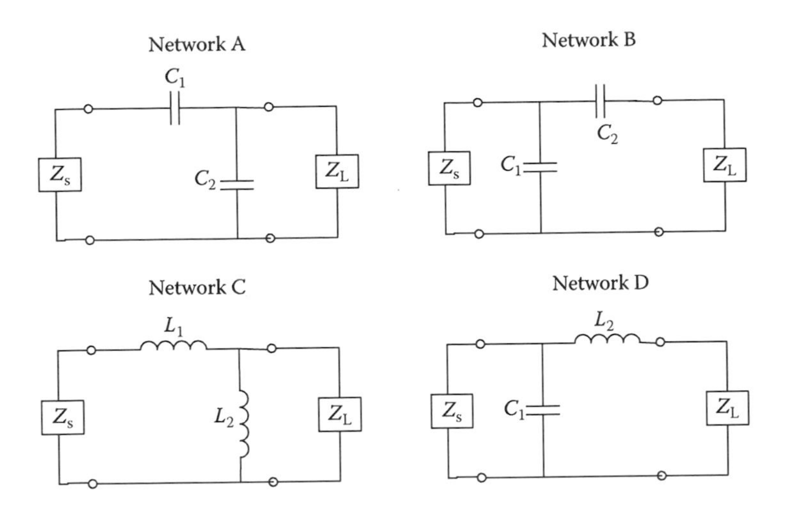

FIGURE 5.32 Possible L-matching networks to match source and load impedance.

5.7 SIGNAL FLOW GRAPHS

Signal flow graphs are used to facilitate analysis of transmission lines in the amplifier design by providing a simplification for the complicated circuits. It is used to determine the critical amplifier design parameters such as reflection coefficients, power, and voltage gains. When a signal flow graph of the circuit is obtained, mathematical relations are developed using Mason's rule. The key elements for the signal flow graph are as follows:

- Each variable is treated as a node.
- The branches represent paths for signal flow.
- The network must be linear.

A node represents the sum of the branches coming into it. The branches are represented by scattering parameters. It is safe to assume that the branches enter dependent variable nodes and leave independent variable nodes. Consider the two-port linear network given in Figure 5.33. This network can be represented using the signal

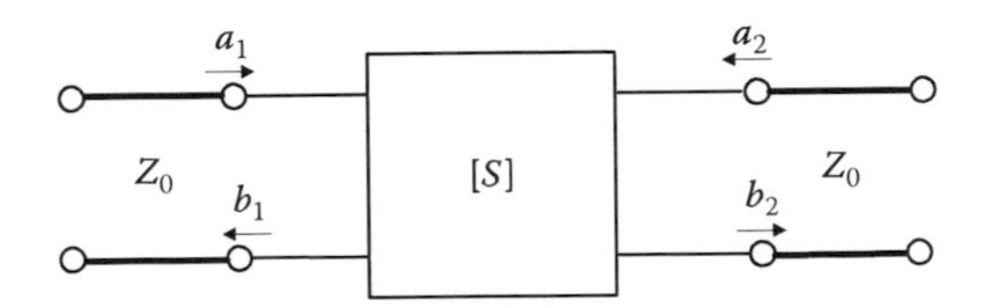

FIGURE 5.33 Two-port linear network illustration.

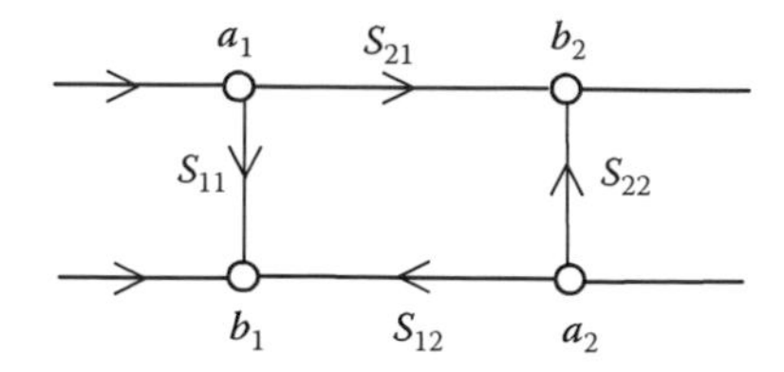

FIGURE 5.34 Signal flow graph implementation of a two-port network.

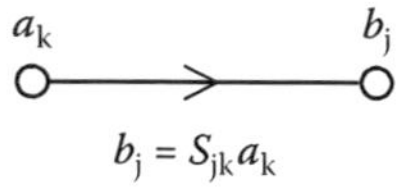

a_k b_j

$b_j = S_{jk} a_k$

FIGURE 5.35 Representation of a branch using scattering parameter.

flow graph, as shown in Figure 5.34. The scattering parameter on each branch is represented by the ratio of the reflected wave to the incident wave:

$$S_{jk} = \frac{b_j}{a_k} \tag{5.129}$$

This can be illustrated with the signal flow graph shown in Figure 5.35 where the node, a_k, from which the wave emanates is assumed to be the incident wave, and the node, b_j, that the wave goes into is assumed to be the reflected wave.

Example

If a signal is given as

$$b = S_{11}a_{11} + S_{12}a_2$$

find its signal flow graph representation.

Solution

Signal b is a dependent node and can be represented as the two incoming branches, as shown in Figure 5.36.

Example

Represent the signal source and source impedance given in Figure 5.37 by a signal flow graph.

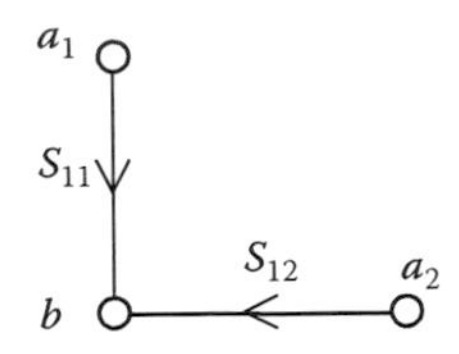

FIGURE 5.36 Representation of dependent node.

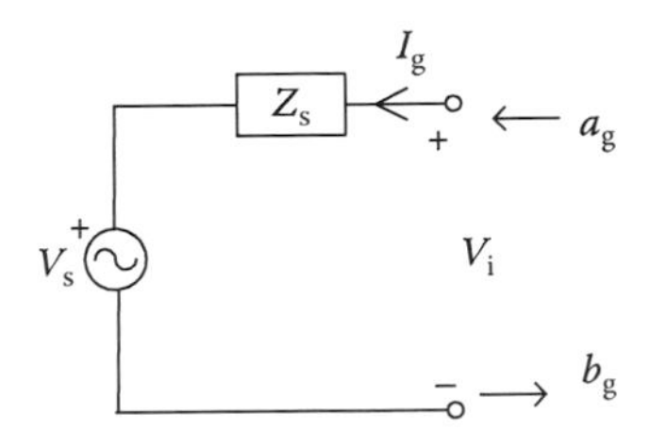

FIGURE 5.37 Source generator and impedance circuit.

Solution

Using the circuit in Figure 5.37, we can write the expression for the voltage at the input as

$$V_i = V_s + I_g Z_s \tag{5.130}$$

which can be written in terms of the incident and reflected waves as

$$V_i^+ + V_i^- = V_s + \left(\frac{V_i^+}{Z_0} - \frac{V_i^-}{Z_0}\right) Z_s \tag{5.131}$$

Solving Equation 5.131 for V_i^- gives

$$b_g = b_s + \Gamma_s a_g \tag{5.132}$$

where

$$b_g = \frac{V_i^-}{\sqrt{Z_0}} \tag{5.133}$$

$$a_g = \frac{V_i^+}{\sqrt{Z_0}} \tag{5.134}$$

$$b_s = \frac{V_s \sqrt{Z_0}}{Z_s + Z_0} \tag{5.135}$$

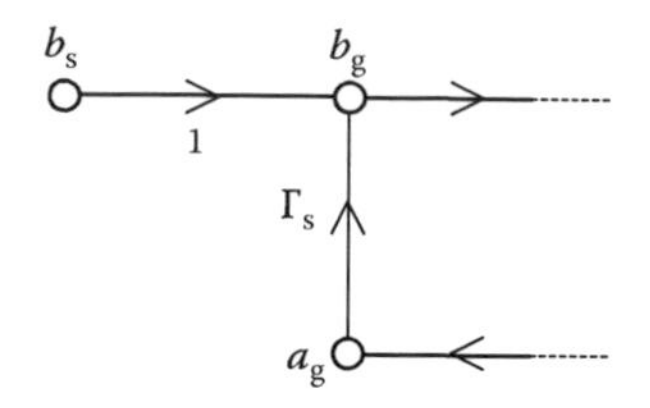

FIGURE 5.38 Signal flow graph representation of source generator and impedance circuit.

Hence, we obtain

$$\Gamma_s = \frac{Z_s - Z_0}{Z_s + Z_0} \tag{5.136}$$

The results can be represented with the signal flow graph shown in Figure 5.38.

Example

Represent the load impedance given in Figure 5.39 by a signal flow graph.

Solution

In Figure 5.39, the load voltage is represented as

$$V_L = Z_L I_L \tag{5.137}$$

Load voltage can be represented in terms of incident and reflected waves as

$$V_L^+ + V_L^- = Z_L + \left(\frac{V_L^+}{Z_0} - \frac{V_L^-}{Z_0}\right) \tag{5.138}$$

Equation 5.138 can be rewritten as

$$b_L = \Gamma_L a_L \tag{5.139}$$

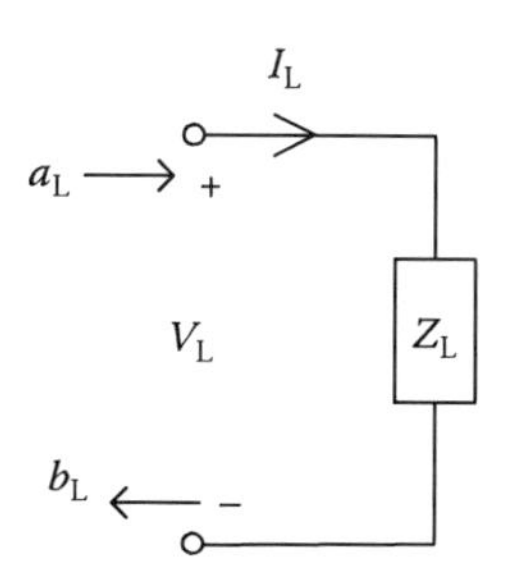

FIGURE 5.39 Load impedance circuit.

where

$$b_L = \frac{V_L^-}{\sqrt{Z_0}} \tag{5.140}$$

$$a_L = \frac{V_L^+}{\sqrt{Z_0}} \tag{5.141}$$

$$\Gamma_L = \frac{Z_L - Z_0}{Z_L + Z_0} \tag{5.142}$$

Using Equation 5.139, we can represent the load impedance in Figure 5.39 with the signal flow graph shown in Figure 5.40.

Example

Represent the two-port transmission circuit shown in Figure 5.38 with the signal flow graph.

Solution

We can now combine the solutions given in Figures 5.38 and 5.40 and obtain the signal flow graph representation for the transmission line circuit shown in Figure 5.41, as illustrated in Figure 5.42.

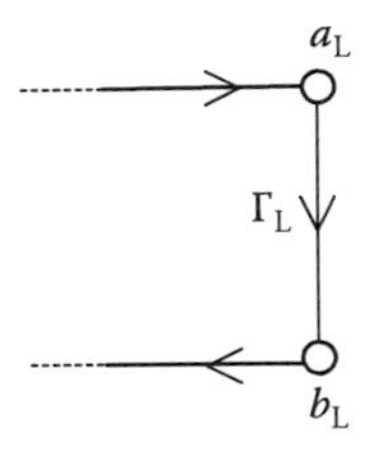

FIGURE 5.40 Signal flow graph representation of load impedance.

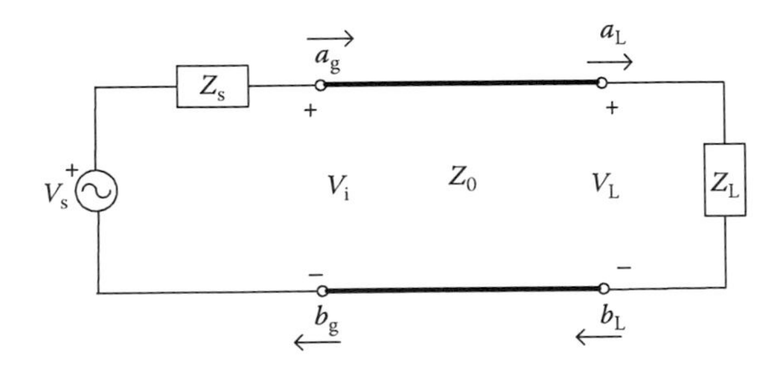

FIGURE 5.41 Transmission line circuit.

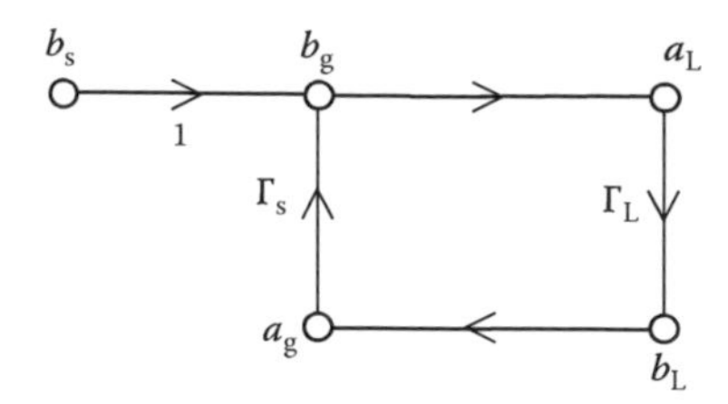

FIGURE 5.42 Signal flow graph representation of transmission line circuit.

PROBLEMS

1. Identify the location of the impedance points given below in the Z Smith chart. Assume that $Z_0 = 50\ \Omega$.

$$Z_1 = 10 + j5, \quad Z_2 = 25 + j15, \quad Y_3 = 0.5 + j1, \quad Y_4 = 2 + j1.6$$

2. A 5-m lossless dielectric-spaced transmission line with $\varepsilon_r = 2.08$ having a characteristic impedance of 50 Ω is terminated with an impedance 50 + j25 (Ω) at an operating frequency of 2 GHz. Find the input impedance.
3. Calculate the following parameters given for the transmission line shown in Figure 5.43 when impedance of 120 + j50 [Ω] is connected.
   a. The VSWR on the line
   b. Load reflection coefficient
   c. Admittance of the load
   d. Impedance at the input of the line of the line
   e. The distance from the load to the first voltage minimum
   f. The distance from the load to the first voltage maximum
4. Find the input impedance for the circuit shown in Figure 5.44 at 3 GHz when the load connected is $Z_L = 75\ \Omega$ using the Smith chart.
5. Using the Smith chart, design all possible configurations of two-element matching networks that match source impedance $Z_s = (25 + j70)\ \Omega$ to the load $Z_L = (10 - j10)\ \Omega$. Assume the characteristic impedance of $Z_0 = 50\ \Omega$ and an operating frequency of $f = 2$ GHz.
6. Using the ZY Smith chart, find the input impedance of the circuit in Figure 5.45 at 3 GHz and 5 GHz.

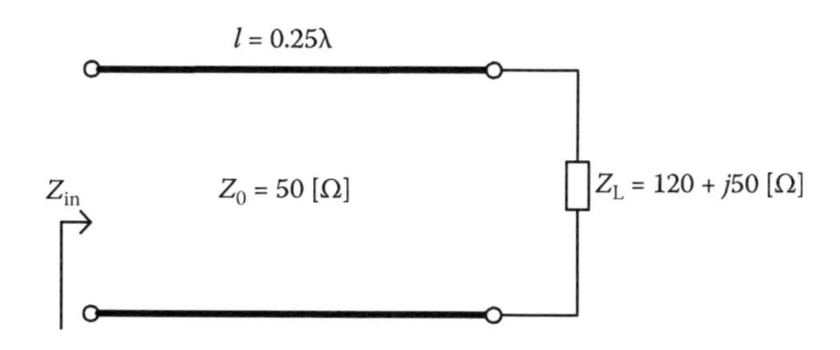

FIGURE 5.43 Transmission line circuit.

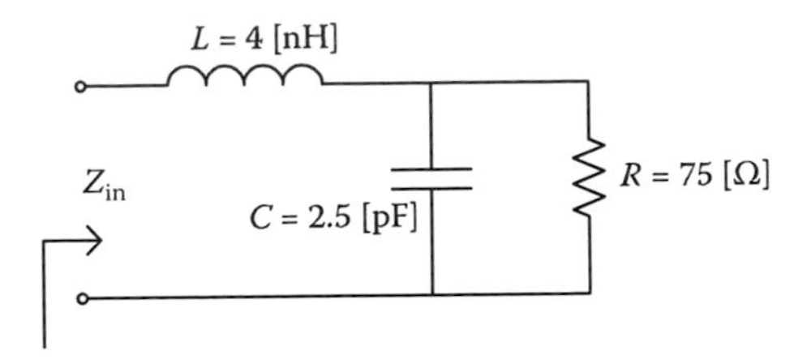

FIGURE 5.44 Impedance transformation.

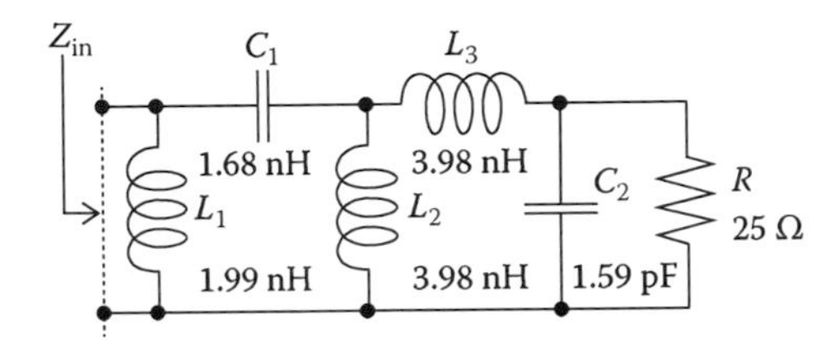

FIGURE 5.45 Input impedance using *ZY* Smith chart.

REFERENCE

1. A. Eroglu. 2013. *RF Circuit Design Techniques for MF-UHF Applications.* CRC Press, Boca Raton, FL.

6 Couplers, Multistate Reflectometers, and RF Power Sensors for Amplifiers

6.1 INTRODUCTION

Radio frequency (RF) power amplifiers have several subsystems and surrounding passive components that include directional couplers, combiners/splitters, impedance, and phase measurement devices such as reflectometers and RF sensors, as shown in Figure 6.1. The complete RF system will only work if all of its subcomponents are designed and interfaced based on the operational requirements. The integration of the subcomponents and assemblies in practice has been done by system engineers in coordination with RF design engineers. In this chapter, the design methods for couplers and reflectometers and RF power sensors will be given.

6.2 DIRECTIONAL COUPLERS

Directional couplers are a critical device in RF amplifiers and used widely as a sampling device for measuring forward and reflected power based on the magnitude of the reflection coefficient. Directional couplers can be implemented as a planar device using transmission lines such as microstrip and stripline or lumped elements with transformers based on the frequency of operation. Conventional directional couplers are four-port devices consisting of main and coupled lines, as shown in Figure 6.2 [1].

Under the matched conditions, when the device is assumed to be lossless, the following relations are valid:

$$S_{11} = S_{22} = S_{33} = S_{44} = 0 \tag{6.1}$$

$$S^{\dagger}S = I \tag{6.2}$$

where † is used for the conjugate transpose of the matrix, and I represents the unit matrix. From Equation 6.2,

$$S_{14}^{*}\left(|S_{13}|^{2} - |S_{24}|^{2}\right) = 0 \tag{6.3}$$

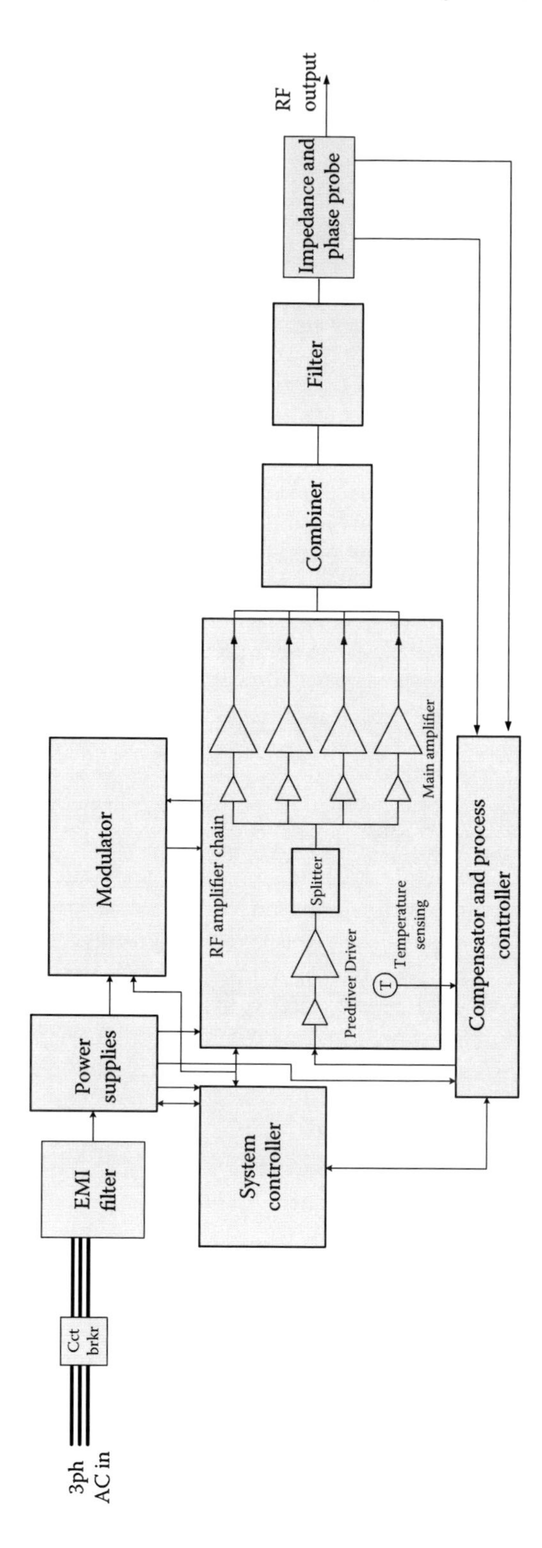

FIGURE 6.1 RF amplifier with its surrounding subsystems.

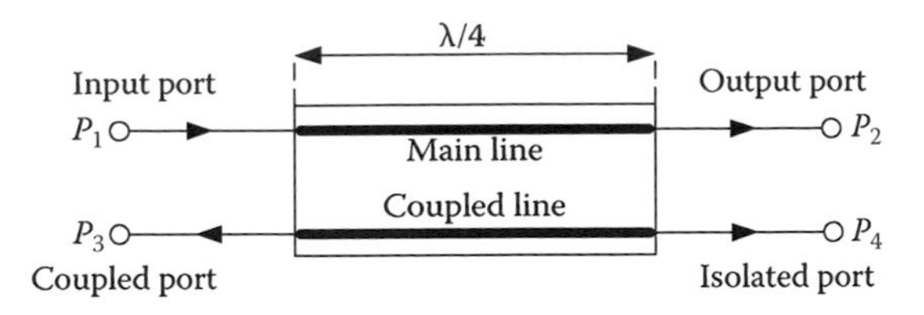

FIGURE 6.2 Directional coupler as a four-port device.

$$S_{23}\left(|S_{12}|^2 - |S_{34}|^2\right) = 0 \tag{6.4}$$

If the network is assumed to be a symmetrical device, then

$$S_{14} = S_{41} = S_{23} = S_{32} = 0 \tag{6.5}$$

Hence,

$$|S_{12}|^2 + |S_{13}|^2 = 1 \tag{6.6}$$

$$|S_{12}|^2 + |S_{24}|^2 = 1 \tag{6.7}$$

$$|S_{13}|^2 + |S_{34}|^2 = 1 \tag{6.8}$$

$$|S_{24}|^2 + |S_{34}|^2 = 1 \tag{6.9}$$

which lead to

$$|S_{13}| = |S_{24}|,\ |S_{12}| = |S_{34}| \tag{6.10}$$

As a result, the scattering matrix for symmetrical, lossless directional coupler can be obtained as

$$S = \begin{bmatrix} 0 & \alpha & j\beta & 0 \\ \alpha & 0 & 0 & j\beta \\ j\beta & 0 & 0 & \alpha \\ 0 & j\beta & \alpha & 0 \end{bmatrix} \tag{6.11}$$

where

$$S_{12} = S_{34} = \alpha, \tag{6.12}$$

$$S_{13} = S_{24} = j\beta \tag{6.13}$$

Important directional coupler performance parameters, i.e., the coupling level, isolation level, and directivity level, can be found from

$$\text{Coupling level (dB)} = 10\log\left(\frac{P_1}{P_3}\right) = -20\log(\beta) \tag{6.14}$$

$$\text{Isolation level (dB)} = 10\log\left(\frac{P_1}{P_4}\right) = -20\log\left(|S_{14}|\right) \tag{6.15}$$

$$\text{Directivity level (dB)} = 10\log\left(\frac{P_3}{P_4}\right) = 20\log\left(\frac{\beta}{|S_{14}|}\right) \tag{6.16}$$

6.2.1 Microstrip Directional Couplers

The design of microstrip directional couplers has been discussed in Refs. [1–3]. In this section, two-line, three-line, and multilayer planar directional coupler designs will be discussed.

6.2.1.1 Two-Line Microstrip Directional Couplers

Consider the geometry of a symmetrical microstrip directional coupler as shown in Figure 6.3.

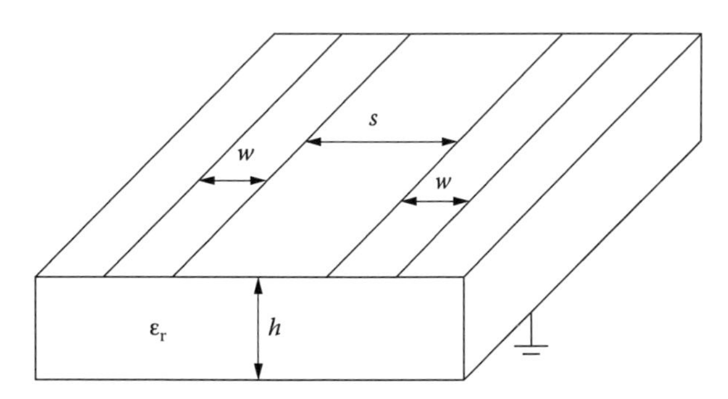

FIGURE 6.3 Symmetrical two-line microstrip directional coupler.

In practice, port termination impedances, coupling level, and operational frequency are input design parameters that are being used to realize couplers. The matched system is accomplished when the characteristic impedance,

$$Z_o = \sqrt{Z_{oe} Z_{oo}} \tag{6.17}$$

is equal to the port impedance. In Equation 6.17, Z_{oe} and Z_{oo} are the even- and odd-mode impedances, respectively. The even and odd impedances, Z_{oe} and Z_{oo}, of the microstrip coupler given in Figure 6.3 can be found from

$$Z_{oe} = Z_0 \sqrt{\frac{1+10^{C/20}}{1-10^{C/20}}} \tag{6.18}$$

$$Z_{oo} = Z_0 \sqrt{\frac{1-10^{C/20}}{1+10^{C/20}}} \tag{6.19}$$

where C is the forward coupling requirement and given in decibels. The physical dimensions of the directional coupler are found using the synthesis method. Application of the synthesis method gives the spacing ratio s/h of the coupler in Figure 6.3 as

$$s/h = \frac{2}{\pi} \cosh^{-1} \left[\frac{\cosh\left[\frac{\pi}{2}\left(\frac{w}{h}\right)_{se}\right] + \cosh\left[\frac{\pi}{2}\left(\frac{w}{h}\right)'_{so}\right] - 2}{\cosh\left[\frac{\pi}{2}\left(\frac{w}{h}\right)'_{so}\right] - \cosh\left[\frac{\pi}{2}\left(\frac{w}{h}\right)_{se}\right]} \right] \tag{6.20}$$

$(w/h)_{se}$ and $(w/h)_{so}$ are the shape ratios for the equivalent single case corresponding to even-mode and odd-mode geometry, respectively. $(w/h)'_{so}$ is the second term for the shape ratio. (w/h) is the shape ratio for the single microstrip line, and it is expressed as

$$\frac{w}{h} = \frac{8\sqrt{\left[\exp\left(\frac{R}{42.4}\sqrt{(\varepsilon_r+1)}\right)-1\right]\frac{7+(4/\varepsilon_r)}{11} + \frac{1+(1/\varepsilon_r)}{0.81}}}{\left[\exp\left(\frac{R}{42.4}\sqrt{\varepsilon_r+1}\right)-1\right]} \tag{6.21}$$

where

$$R = \frac{Z_{oe}}{2} \quad \text{or} \quad R = \frac{Z_{oo}}{2} \tag{6.22}$$

Z_{ose} and Z_{oso} are the characteristic impedances corresponding to single microstrip shape ratios $(w/h)_{se}$ and $(w/h)_{so}$, respectively. They are given as

$$Z_{ose} = \frac{Z_{oe}}{2} \tag{6.23}$$

$$Z_{oso} = \frac{Z_{oo}}{2} \tag{6.24}$$

and

$$(w/h)_{se} = (w/h)\Big|_{R=Z_{ose}} \tag{6.25}$$

$$(w/h)_{so} = (w/h)\Big|_{R=Z_{oso}} \tag{6.26}$$

The term $(w/h)'_{so}$ in Equation 6.20 is given as

$$\left(\frac{w}{h}\right)'_{so} = 0.78\left(\frac{w}{h}\right)_{so} + 0.1\left(\frac{w}{h}\right)_{se} \tag{6.27}$$

After the spacing ratio s/h for the coupled lines is found, we can proceed to find w/h for the coupled lines. The shape ratio for the coupled lines is

$$\left(\frac{w}{h}\right) = \frac{1}{\pi}\cosh^{-1}(d) - \frac{1}{2}\left(\frac{s}{h}\right) \tag{6.28}$$

where

$$d = \frac{\cosh\left[\frac{\pi}{2}\left(\frac{w}{h}\right)_{se}\right](g+1) + g - 1}{2} \tag{6.29}$$

$$g = \cosh\left[\frac{\pi}{2}\left(\frac{s}{h}\right)\right] \tag{6.30}$$

The physical length of the directional coupler is obtained using

$$l = \frac{\lambda}{4} = \frac{c}{4f\sqrt{\varepsilon_{eff}}} \tag{6.31}$$

where $c = 3 \times 10^8$ m/s, and f is the operational frequency in hertz. Hence, the length of the directional coupler can be found if the effective permittivity constant ε_{eff} of the coupled structure shown in Figure 6.1 is known. ε_{eff} can be found from

$$\varepsilon_{\text{eff}} = \left[\frac{\sqrt{\varepsilon_{\text{effe}}} + \sqrt{\varepsilon_{\text{effo}}}}{2}\right]^2 \tag{6.32}$$

$\varepsilon_{\text{effe}}$ and $\varepsilon_{\text{effo}}$ are the effective permittivity constants of the coupled structure for odd and even modes, respectively. $\varepsilon_{\text{effe}}$ and $\varepsilon_{\text{effo}}$ depend on even- and odd-mode capacitances C_{e} and C_{o} as

$$\varepsilon_{\text{effe}} = \frac{C_{\text{e}}}{C_{\text{el}}} \tag{6.33}$$

$$\varepsilon_{\text{effo}} = \frac{C_{\text{o}}}{C_{\text{ol}}} \tag{6.34}$$

$C_{\text{el,ol}}$ is the capacitance with air as dielectric. All the capacitances are given as capacitance per unit length. The even-mode capacitance C_{e} is

$$C_{\text{e}} = C_{\text{p}} + C_{\text{f}} + C_{\text{f}}' \tag{6.35}$$

The capacitances in the even mode for the coupled lines can be visualized as shown in Figure 6.4.

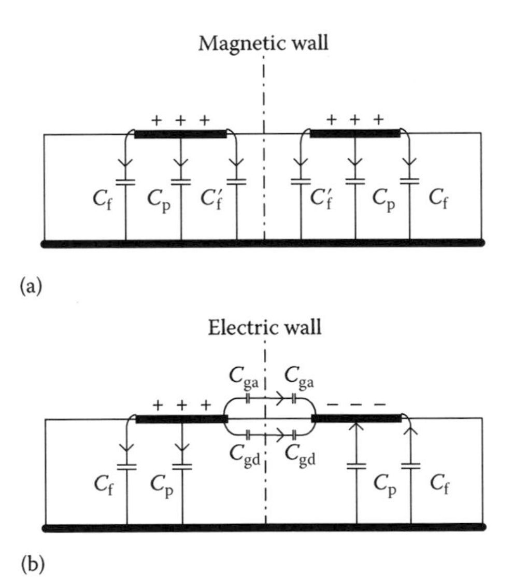

FIGURE 6.4 Coupled line mode representation: (a) even mode; (b) odd mode.

C_p is the parallel plate capacitance and is defined as

$$C_p = \varepsilon_0 \varepsilon_r \frac{w}{h} \tag{6.36}$$

where w/h is found in Section 6.1. C_f is the fringing capacitance due to the microstrip being taken alone as if it were a single strip. That is equal to

$$C_f = \frac{\sqrt{\varepsilon_{seff}}}{2cZ_0} - \frac{C_p}{2} \tag{6.37}$$

Here, ε_{seff} is the effective permittivity constant of a single-strip microstrip. It can be expressed as

$$\varepsilon_{seff} = \frac{\varepsilon_r + 1}{2} - \frac{\varepsilon_r - 1}{2} F(w/h) \tag{6.38}$$

where

$$F(w/h) = \begin{bmatrix} (1+12h/w)^{-1/2} + 0.041(1-w/h)^2 & \text{for } \left(\dfrac{w}{h} \le 1\right) \\ (1+12h/w)^{-1/2} & \text{for } \left(\dfrac{w}{h} \ge 1\right) \end{bmatrix} \tag{6.39}$$

C'_f is given by the following equation:

$$C'_f = \frac{C_f}{1 + A\left(\dfrac{h}{s}\right)\tanh\left(\dfrac{10s}{h}\right)} \left(\frac{\varepsilon_r}{\varepsilon_{seff}}\right)^{1/4} \tag{6.40}$$

and

$$A = \exp\left[-0.1\exp\left(2.33 - 1.5\frac{w}{h}\right)\right] \tag{6.41}$$

The odd-mode capacitance C_o is

$$C_o = C_p + C_f + C_{ga} + C_{gd} \tag{6.42}$$

The capacitances in the odd mode for the coupled lines can be visualized as shown in Figure 6.4. C_{ga} is the capacitance term in the odd mode for the fringing field across the gap in the air region. It can be written as

$$C_{ga} = \varepsilon_0 \frac{K(k')}{K(k)} \tag{6.43}$$

where

$$\frac{K(k')}{K(k)} = \begin{cases} \dfrac{1}{\pi}\ln\left[2\dfrac{1+\sqrt{k'}}{1-\sqrt{k'}}\right], & 0 \le k^2 \le 0.5 \\ \dfrac{\pi}{\ln\left[2\dfrac{1+\sqrt{k'}}{1-\sqrt{k'}}\right]}, & 0.5 \le k^2 \le 1 \end{cases} \tag{6.44}$$

and

$$k = \frac{\left(\dfrac{s}{h}\right)}{\left(\dfrac{s}{h}\right)+\left(\dfrac{2w}{h}\right)} \tag{6.45}$$

$$k' = \sqrt{1-k^2} \tag{6.46}$$

C_{gd} represents the capacitance in the odd mode for the fringing field across the gap in the dielectric region. It can be found using

$$C_{gd} = \frac{\varepsilon_0\varepsilon_r}{\pi}\ln\left\{\coth\left(\frac{\pi}{4}\frac{s}{h}\right)\right\} + 0.65C_f\left[\frac{0.02}{\left(\dfrac{s}{h}\right)}\sqrt{\varepsilon_r} + \left(1-\frac{1}{\varepsilon_r^2}\right)\right] \tag{6.47}$$

Since

$$Z_{oe} = \frac{1}{c\sqrt{C_e C_{e1}}} \tag{6.48}$$

$$Z_{oo} = \frac{1}{c\sqrt{C_o C_{o1}}} \tag{6.49}$$

then we can write

$$C_{e1} = \frac{1}{c^2 C_e Z_{oe}^2} \tag{6.50}$$

$$C_{o1} = \frac{1}{c^2 C_o Z_{oo}^2} \tag{6.51}$$

Substituting Equations 6.32, 6.35, 6.40, and 6.51 into Equations 6.34 and 6.35 gives the even- and odd-mode effective permittivities ε_{effe} and ε_{effo}. When Equations 6.34 and 6.35 are substituted into Equation 6.33, we can find the effective permittivity constant ε_{eff} of the coupled structure. Now, Equation 6.31 can be used to calculate the physical length of the directional coupler at the operational frequency. The design tables to design two-line microstrip couplers with various commonly used RF materials and several application examples are given in Ref. [1].

6.2.1.2 Three-Line Microstrip Directional Couplers

Three-line, six-port microstrip directional couplers can be used for several purposes in RF applications including voltage, current, impedance, and voltage standing wave ratio (VSWR) measurements. As a result, six-port microstrip directional couplers are cost-effective alternatives to existing reflectometers. The design and performance of six-port reflectometers based on microstrip-type couplers have been analyzed and given in Ref. [4]. The method described in Refs. [1,2] gives the complete design method of two-line symmetrical directional couplers with closed-form relations using the synthesis technique, as described in Section 6.2.1.1. The method used in Refs. [1,2] reflects the design practice since the physical dimensions of the coupler are not known prior to the design of the coupler. Their design procedure requires only the knowledge of port impedances, the desired coupling level, and the operational frequency. The physical dimensions of the coupler including the width of the trace, the spacing between them, and the thickness of the dielectric substrate are then determined using the closed relations based on the given three design requirements.

Three-line, six-port directional couplers shown in Figure 6.5 give cost-effective alternatives to existing reflectometers and can be used for diagnostic purposes. There are some analytical formulations to design the six-port couplers, which are given in Refs. [4–7], but none of them in the literature uses the practical approach for two-line coupler design, which reflects the engineering practice outlined in Section 6.1.

In this section, we present closed-form relations to design three-line microstrip directional couplers using the method implemented in practice. The design method given here again requires knowledge of only coupling level, port impedances, and operational frequency. A three-step design procedure with accurate closed formulas is given to have a complete design of symmetrical three-line microstrip directional couplers at the desired operational frequency. The physical dimensions of the coupler including the physical length are obtained with the method presented, and the

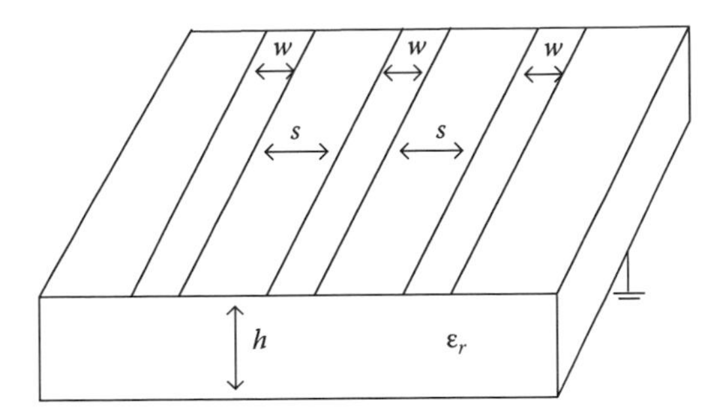

FIGURE 6.5 Three-line microstrip directional coupler.

coupler performance is compared with the planar electromagnetic simulators such as Sonnet and Ansoft Designer. It is shown that the results are in close agreement, and the method can be used for applications that require accuracy. The step-by-step design procedure to implement three-line couplers is as follows:

Step 1: Generate the design specifications for a two-line coupler using equations in Refs. [1,2].

The simulations proved that a three-line coupler can be designed by placing the third coupled line symmetrically on the other side of a two-line coupler for the given coupling level of coupling K_2, port impedance, dielectric constant, and operational frequency. Hence, the three-line coupler can be designed using the same physical dimensions s/h, w/h ratios, and length obtained using the equations in Refs. [1,2].

Step 2: Estimation of the coupling K_{13} between the two coupled lines through the main line using the reverse analysis of Refs. [1,2].

Step 3: Calculation of the mode impedances Z_{oe}, Z_{oo}, and Z_{ee} for the three-line coupler designed using the coupling levels K_2 and K_{13}.

The coupling levels of a three-line microstrip coupler are given by the following equations [1,2]:

$$K_2 = \frac{Z_{ee} - Z_{oo}}{Z_{ee} + Z_{oo}} \tag{6.52}$$

$$K_{13} = \frac{\sqrt{Z_{ee}Z_{oo}} - Z_{oe}}{\sqrt{Z_{ee}Z_{oo}} + Z_{oe}} \tag{6.53}$$

where K_2 (not in decibels) represents the coupling from the side lines into the center line, which is known, and K_{13} (not in decibels) represents the coupling between the side lines through the center line calculated from the reverse analysis using MATLAB® and verified by Ansoft Designer. It can be assumed that $Z_{oe} = Z_o$, and thereby equations for Z_{ee} and Z_{oo} are found by solving Equations 6.52 and 6.53 as

$$Z_{oo} = Z_o\left(\frac{1+K_{13}}{1-K_{13}}\right)\sqrt{\frac{1-K_2}{1+K_2}} \tag{6.54}$$

$$Z_{ee} = Z_o\left(\frac{1+K_{13}}{1-K_{13}}\right)\sqrt{\frac{1+K_2}{1-K_2}} \tag{6.55}$$

Design Example

Design a 15-dB three-line coupler using Teflon at 300 MHz with the method introduced.

Solution

The design procedure for two-line conventional directional couplers and three-line directional couplers for Teflon with relative permittivity constant 2.08 has been applied at 300 MHz to realize a 15-dB coupler. Two-line microstrip is first designed using a MATLAB graphical user interface (GUI) developed with the formulation given in this chapter, and the results showing the physical dimensions of the microstrip coupler are illustrated as shown in Figure 6.6.

Based on the results obtained, Ansoft Designer is used to simulate the same coupler, as shown in Figure 6.7.

The simulation results showing the coupling level of Teflon at 300 MHz for a two-line microstrip directional coupler are illustrated in Figure 6.8 and are equal to 14.66 dB. Following the design procedure to realize a three-line microstrip coupler and obtain its physical dimensions as has been done with a MATLAB GUI and shown in Figure 6.9, a three-line microstrip directional coupler is then simulated with Ansoft Designer, as shown in Figure 6.10.

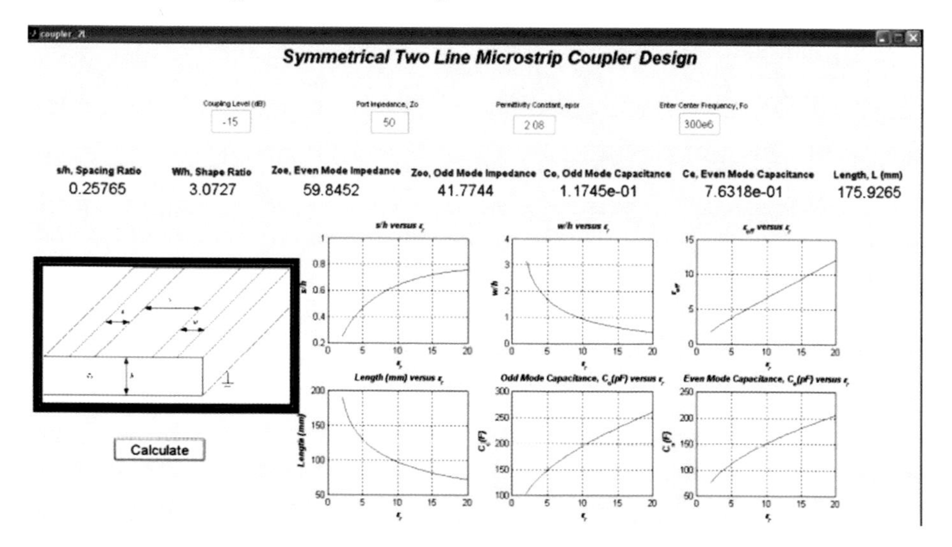

FIGURE 6.6 MATLAB GUI for a two-line microstrip directional coupler.

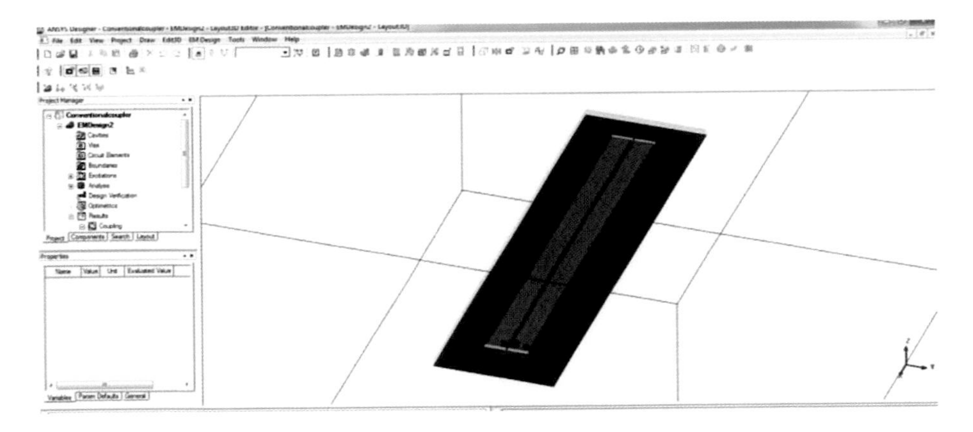

FIGURE 6.7 Simulated two-line microstrip directional coupler.

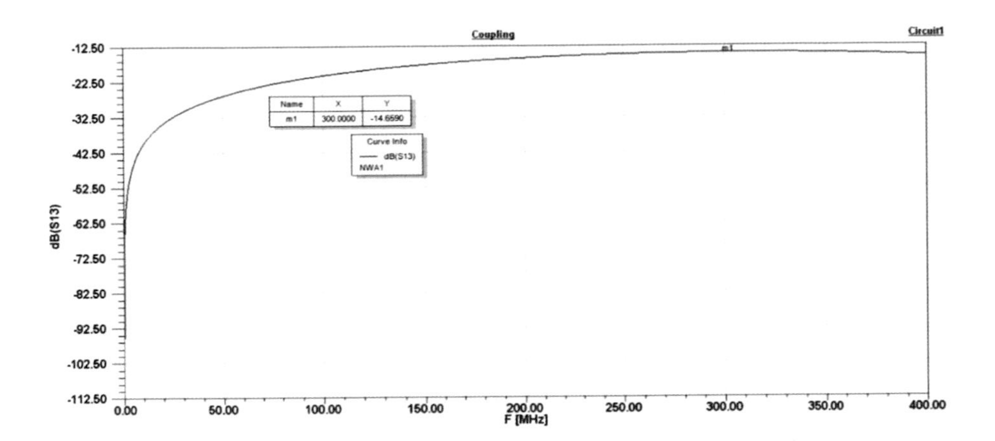

FIGURE 6.8 Simulated results for the coupling level for a two-line symmetrical coupler.

FIGURE 6.9 MATLAB GUI for a three-line microstrip directional coupler.

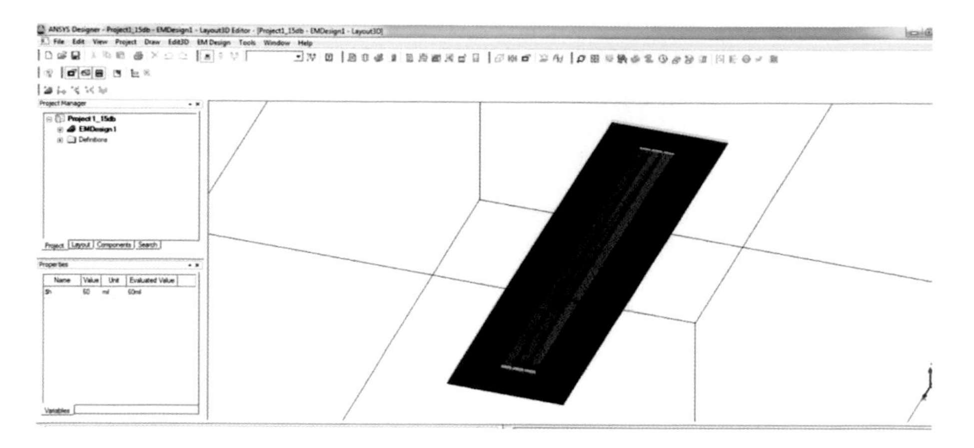

FIGURE 6.10 Simulated three-line microstrip directional coupler.

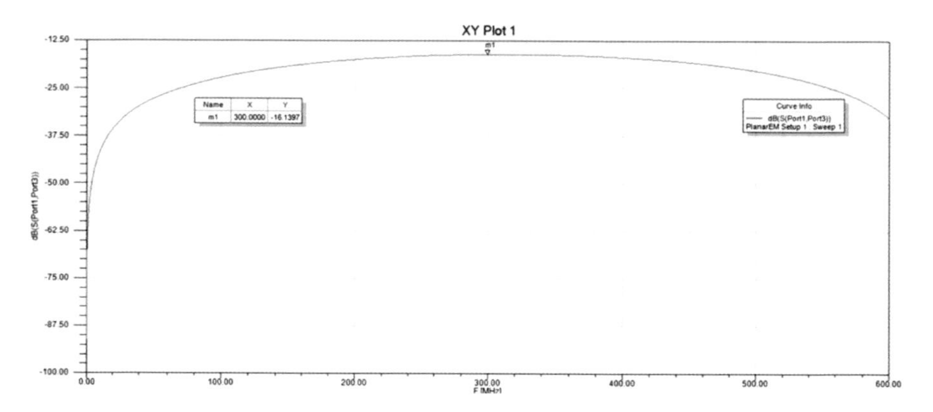

FIGURE 6.11 Simulated results for the coupling level for a two-line symmetrical coupler.

The simulation results showing the coupling level between the main line and coupled line for Teflon at 300 MHz for a three-line microstrip directional coupler are illustrated in Figure 6.11 and are equal to 16.1397 dB. Hence, the given closed-form relations and design procedure can be successfully designed to implement three-line microstrip directional couplers.

6.2.2 Multilayer Planar Directional Couplers

In this section, the design of multilayer microstrip two-line and three-line directional couplers is given. A step-by-step design procedure reflecting the design practice of directional couplers, which requires only information on coupling level, port impedances, and operational frequency, is given. The method based on the synthesis technique applied in the design of two-line microstrip symmetrical directional couplers by Eroglu [1,2] is adapted to design multilayer directional couplers with the aid of electromagnetic simulators and curve fitting. The proposed design method is compared with the existing measurement results, and the accuracy is verified. It also has been shown that the directivity of the couplers designed using the multilayer structure is improved significantly. A method such as the one presented in this chapter can be used to design multilayer two-line and three-line directional couplers with ease and high accuracy where better performance is needed.

The geometry of the four-port coupler for which the design method is proposed here is shown in Figure 6.12. The concept of improved directivity in comparison to the two-layer structure has been shown in Ref. [8]. The information required to design the proposed model consists of the required coupling level, port impedances, permittivity of the material, thickness of the material, and the operational frequency. The methods used in Refs. [9–13] involve compensation using the shunt inductors for the improvement in the directivity, but in the proposed model, the coupled line has been embedded into the material for the improvement of the directivity, which does not need any other components. The directivity improvement has been achieved in Ref. [14] by increasing the even-mode phase velocity by meandering the coupler, which is a very complex design technique. A method to improve the directivity by designing asymmetric microstrip couplers based on the concept of equalization of the even- and odd-mode phase velocities is given in Ref. [15].

The phase velocity compensation, which involves complicated structures and substrates with specified physical constants as shown in Ref. [16], also helps in directivity improvement. Experimental results with no closed-form relations showing improvement in directivity are also given in Ref. [17].

The step-by-step design procedure to realize two-line and three-line multilayer planar directional couplers is as follows:

Step 1: Generate the design parameters for a two-line coupler using equations in Refs. [1,2] for −10-dB coupling and 120-mils thickness and obtained spacing and shape ratios.

Step 2: Use simulation with parametric analysis and move the coupled line with fine predetermined distance inside the dielectric.

Step 3: Obtain the new coupling for the predetermined distance and repeat this until the full thickness of the dielectric is reached.

Step 4: Use curve fitting and obtain the equation for the material used and relate the coupling level and the height of the coupled line.

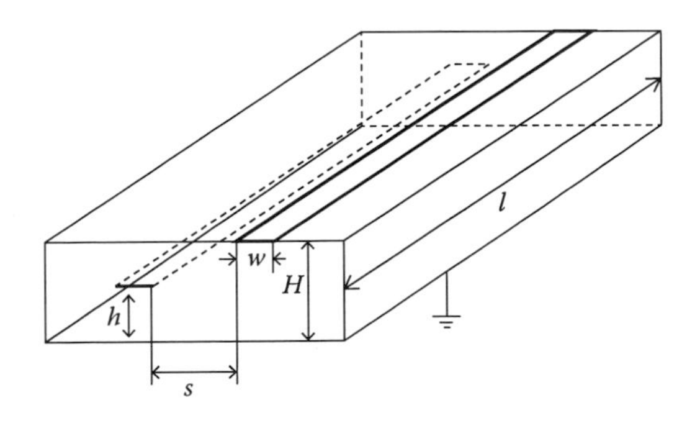

FIGURE 6.12 Multilayer two-line, four-port microstrip directional coupler.

Design Example

Design a 15-dB two-line, multilayer directional coupler using FR4 at 300 MHz with the method introduced. Compare the directivity of the results for the multilayer directional coupler with the directivity of the conventional two-line directional coupler.

Solution

The design procedure begins with finding the physical dimension of the conventional two-line microstrip directional coupler with the method introduced in Refs. [1,2]. A MATLAB GUI has been developed to design any multilayer two-line directional coupler, as shown in Figure 6.13.

The multilayer directional coupler is then implemented using Ansoft Designer, as shown in Figure 6.14.

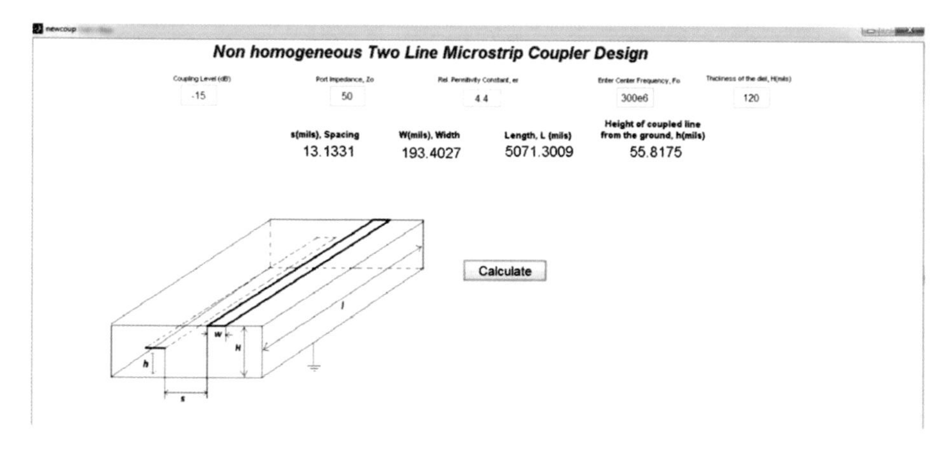

FIGURE 6.13 MATLAB GUI for multilayer two-line, four-port microstrip directional coupler.

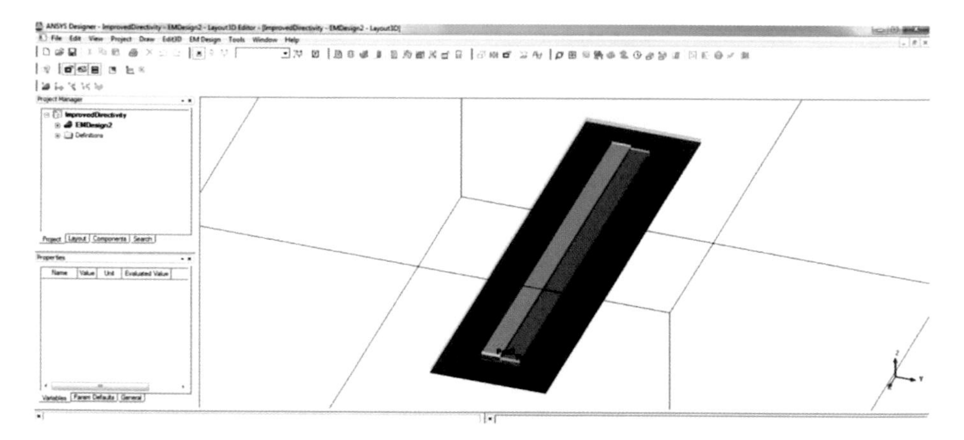

FIGURE 6.14 Simulated three-line, multilayer planar directional coupler.

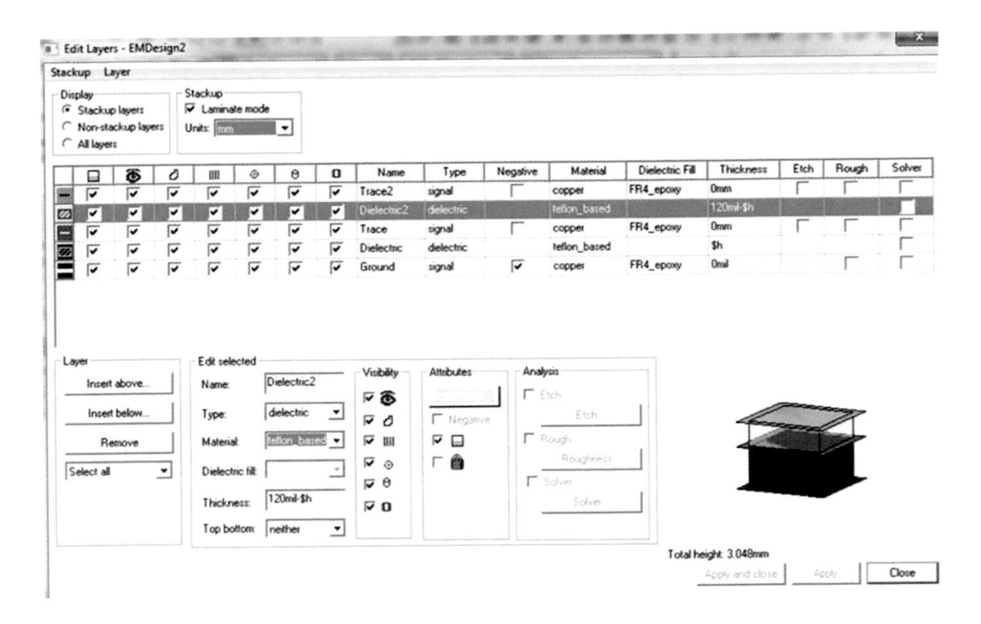

FIGURE 6.15 Multilayer configuration setup for simulation.

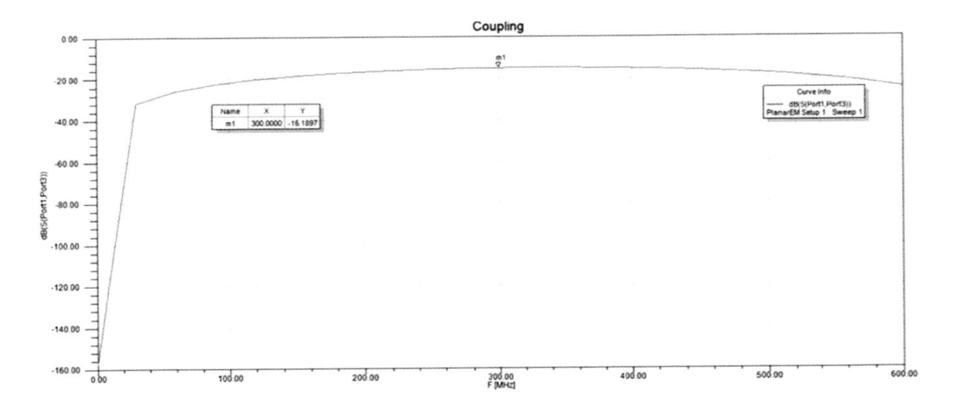

FIGURE 6.16 Simulated results for the coupling level for multilayer coupler.

The multilayer configuration setup detail used in the simulation of the coupler is given in Figure 6.15. The simulation results for coupling and directivity are given in Figures 6.16 and 6.17, respectively. The simulation results show that more than 6-dB improvement is obtained vs. conventional two-line microstrip directional couplers. The coupling and directivity levels of the coupler are found to be –15.1897 and –18.5386 dB, respectively.

6.2.3 Transformer-Coupled Directional Couplers

Couplers can be implemented using distributed elements or lumped elements as discussed before. The type of application, the operational frequency, and the

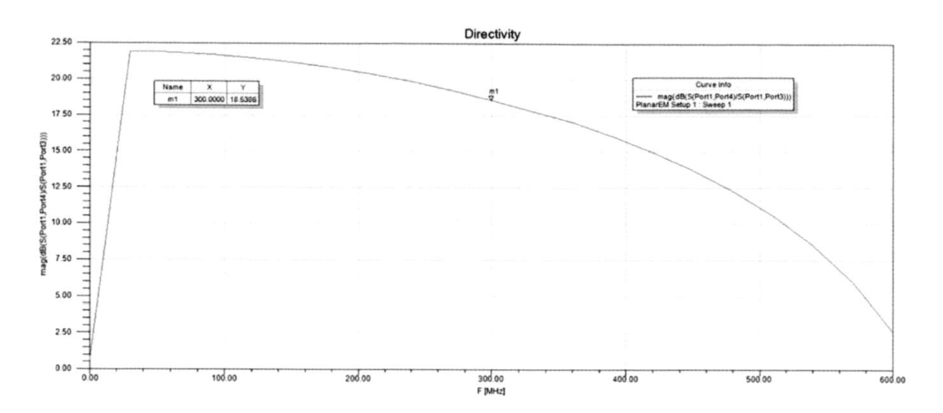

FIGURE 6.17 Simulated results for the directivity level for multilayer coupler.

power-handling capability are among the important factors that dictate the type of the directional coupler that will be used in the RF system. Conventional directional couplers are designed as four-port couplers and have been studied extensively. However, better performance and more functionality from couplers can be obtained when they are implemented as six-port couplers.

It is possible to use a six-port coupler for VSWR measurement [18]. It also plays a very important role in measuring the voltage, current, power, impedance, and phase, as discussed in Ref. [19]. A detailed study of a wideband impedance measurement using a six-port coupler is given in Ref. [20]. The design and analysis of a six-port stripline coupler with a high phase and amplitude balance has been studied in Ref. [21]. One of the important applications of six-port couplers is their implementation as reflectometers. The theory of six-port reflectometer is detailed in Refs. [22,23]. In Refs. [24,25], the six-port reflectometer based on four-port coplanar-waveguide couplers has been modified to meet optimum design specifications. Similar studies to realize reflectometers using couplers are reported in Refs. [26–28]. Theoretical analysis of the impedance measurement using a six-port coupler as the reflectometer has been introduced in Refs. [5,7]. The design and performance of six-port reflectometers based on microstrip-type couplers have been analyzed in Refs. [4,29]. Hansson and Riblet [30] managed to realize a six-port network of an ideal q-point distribution by using a matched reciprocal lossless five-port and a directional coupler. An improved complex reflection coefficient measurement device consisting of two six-port couplers is presented in Ref. [31]. Similarly, a six-port device has been designed for power measurement with two six-port directional couplers and discussed in Ref. [22]. Six-port couplers can also be used in designing power splitting and combining networks [32]. Six-port devices have also been commonly used for source pull and load pull characterization of active devices and systems [33,34]. As a result, the design of the six-port coupler reported in the literature is based on the planar structures involving microstrips, striplines, or different waveguide structures. For high-power and low-cost applications, directional couplers can be implemented by means of RF transformers. Four-port directional coupler design using transformer coupling is given in Refs. [35,36].

In this section, a detailed analysis of four-port and six-port directional couplers using ideal RF transformers is presented. Closed-form expressions at each port are obtained, and coupling, isolation, and directivity levels of the six-port coupler using transformer coupling are given. The S parameters for four-port and six-port couplers are derived, and coupler performance parameters are expressed in terms of S parameters. Based on the analytical model, a MATLAB GUI has been developed and used for the design, simulation, and analysis of four-port and six-port couplers using transformer coupling. The directional coupler is then simulated using frequency-domain and time-domain simulators such as Ansoft Designer and PSpice, and the simulation results are compared with the analytical results. The six-port coupler is then implemented and measured with Network Analyzer HP 8753ES. The proposed model can be used as a building block in various applications such as reflectometers, high-power impedance and power measurements, VSWR measurement, and load pull or source pull of active devices.

6.2.3.1 Four-Port Directional Coupler Design and Implementation

The design, simulation, and implementation of a four-port coupler are given in Ref. [1] and will be only briefly described here. The S parameters of the four-port directional coupler shown in Figure 6.18 can be represented in matrix form as

$$S = \begin{bmatrix} S_{11} & S_{12} & S_{13} & S_{14} \\ S_{21} & S_{22} & S_{23} & S_{24} \\ S_{31} & S_{31} & S_{33} & S_{34} \\ S_{41} & S_{42} & S_{43} & S_{44} \end{bmatrix} \tag{6.56}$$

The performance of a four-port coupler can be calculated using S_{13} and S_{14} for coupling, isolation, and directivity levels when the excitation is from port 1 on the main line. In Figure 6.18, T_1 is the transformer with turns ratio N_1:1, and T_2 is the transformer with turns ratio N_2:1. The transformers are assumed to be ideal and

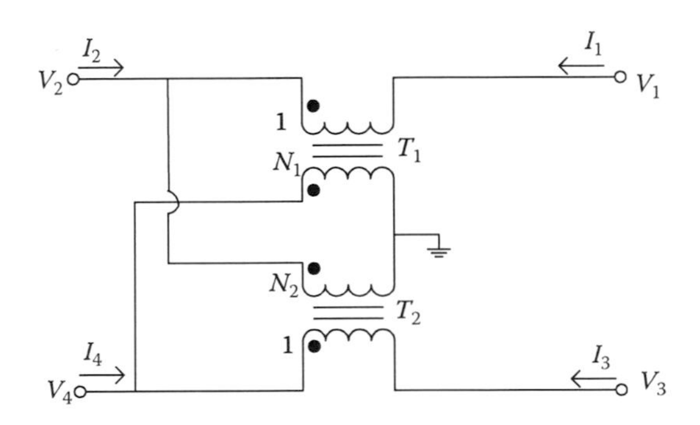

FIGURE 6.18 Four-port transformer directional coupler.

lossless. The relations between voltages and currents through turn ratios of the directional coupler at the ports can be obtained as

$$V_2 = N_2(V_4 - V_3) \tag{6.57}$$

$$V_4 = N_1(V_2 - V_1) \tag{6.58}$$

and

$$I_1 = N_1(I_3 + I_4) \tag{6.59}$$

$$I_3 = N_1(I_1 + I_2) \tag{6.60}$$

The scattering parameters of the coupler can be obtained by using the incident and reflected waves, which are designated by a_i and b_i. Then, the voltages and currents can be expressed in terms of waves as

$$V_i = \sqrt{Z}(a_i + b_i) \tag{6.61}$$

$$I_i = \frac{1}{\sqrt{Z}}(a_i - b_i) \tag{6.62}$$

Z is the characteristic impedance at the ports of the directional coupler. The scattering parameters of the coupler are obtained by relating the incident and reflected waves using

$$S_{ij} = \left.\frac{b_i}{a_j}\right|_{a_k=0 \text{ for } k\neq j} \tag{6.63}$$

The scattering parameters that are required to calculate the coupler performance parameters are

$$S_{13} = \frac{(-2N_1N_2)(N_1 + N_2)}{\left(4N_1^2N_2^2 + 1 + \left(N_1^2 - N_2^2\right)\right)} \tag{6.64}$$

$$S_{14} = \frac{(-2N_1)\left(-N_1N_2 + N_2^2 + 1\right)}{\left(4N_1^2N_2^2 + 1 + \left(N_1^2 - N_2^2\right)\right)} \tag{6.65}$$

The coupling, isolation, and directivity levels of a four-port coupler are then expressed using S parameters in Section 6.2. Equations 6.64 and 6.65 lead to directional coupler performance parameter calculations through knowledge of only turns

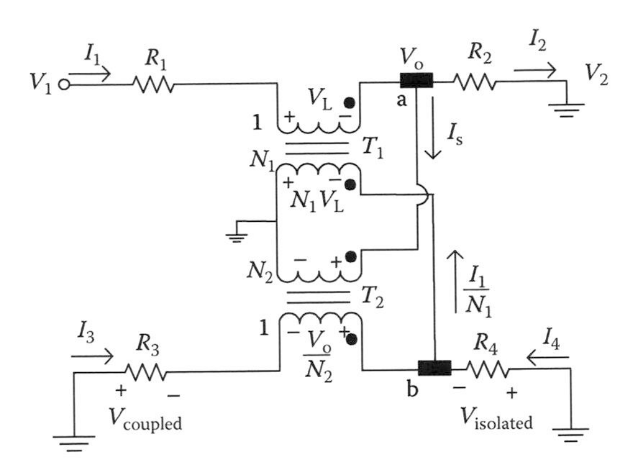

FIGURE 6.19 Four-port transformer directional coupler for circuit analysis.

ratios under the assumption that all ports are matched. However, one other important aspect of the directional coupler in practical applications is the real operating conditions including voltage, current, and power ratings and operational frequency. These parameters dictate the type of core, the winding and the wire, the coax line, and the insulation that will be used in the design. As a result, circuit analysis is needed to determine the operating conditions on the coupler at each node. This analysis is detailed in Ref. [1] using the circuit analysis of the four-port coupler shown in Figure 6.19.

The application of nodal analysis for the coupler circuit in Figure 6.19 gives the performance parameters for the coupler as

$$\text{Coupling level (dB)} = 20\log\left(\frac{V_{\text{coupled}}}{\left(\frac{R_2}{R_1+R_2}\right)V_1}\right) \tag{6.66}$$

$$\text{Isolation level (dB)} = 20\log\left(\frac{V_{\text{isolated}}}{\left(\frac{R_2}{R_1+R_2}\right)V_1}\right) \tag{6.67}$$

Similarly, directivity level is found from

$$\text{Directivity level (dB)} = \text{coupling level (dB)} - \text{isolation level (dB)} \tag{6.68}$$

6.2.3.2 Six-Port Directional Coupler Design and Implementation

The four-port coupler that is introduced in Section 6.2.3.1 is used as a basic element to realize a six-port coupler using transformer coupling for high-power RF applications. Six-port coupler design and analysis using transformer coupling have not been

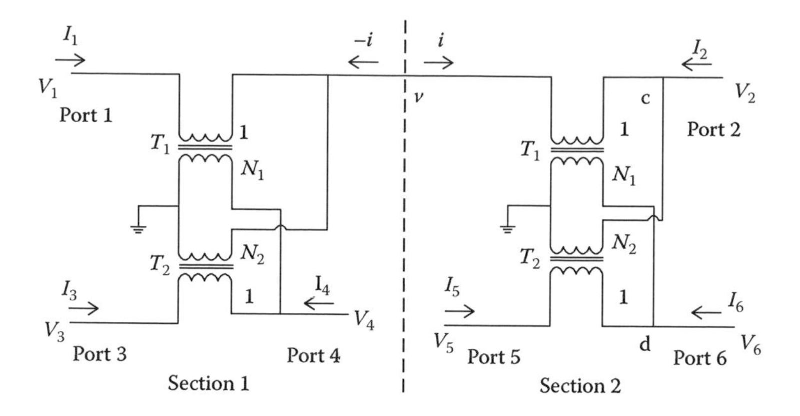

FIGURE 6.20 Six-port transformer directional coupler.

reported before in the literature according to authors' knowledge. In Figure 6.20, T_1 is the transformer with turns ratio N_1:1, and T_2 is the transformer with turns ratio N_2:1. The transformers are assumed to be ideal and lossless. The S parameters of the six-port directional coupler shown in Figure 6.20 can be obtained from

$$\begin{pmatrix} b_1 \\ b_2 \\ \vdots \\ \\ b_6 \end{pmatrix} = \begin{pmatrix} S_{11} & S_{12} & & S_{16} \\ S_{21} & \ddots & & \vdots \\ S_{31} & & \ddots & \vdots \\ \\ S_{61} & \cdots & \cdots & S_{66} \end{pmatrix} \begin{pmatrix} a_1 \\ a_2 \\ \vdots \\ \\ a_6 \end{pmatrix} \tag{6.69}$$

The relations between voltages and currents through turn ratios of the directional coupler at each port can be obtained as

$$v = N_2(V_4 - V_3) \tag{6.70}$$

$$V_4 = N_1(v - V_1) \tag{6.71}$$

$$V_2 = N_2(V_6 - V_5) \tag{6.72}$$

$$V_6 = N_1(V_2 - v) \tag{6.73}$$

and

$$I_1 = N_1(I_4 + I_3) \tag{6.74}$$

$$I_3 = N_2(-i + I_1) \tag{6.75}$$

$$i = N_1(I_5 + I_6) \tag{6.76}$$

$$I_5 = N_2(I_2 + i) \tag{6.77}$$

Then, the voltages and currents are expressed in terms of waves using relations 6.61 and 6.62, and the scattering parameters for coupler performance are then obtained with application of Equation 6.63 as

$$S_{62} = S_{31} = \frac{\left(-4N_1^4N_2^3 - 4N_1^3N_2^4 + 2N_1^3N_2^2 - 2N_1^2N_2^3 - 2N_1^2N_2\right)}{8N_1^4N_2^4 + 4N_1^4N_2^2 - 8N_1^3N_2^3 - 2N_1^3N_2 + 4N_1^2N_2^4 + 10N_1^2N_2^2 + N_1^2 + N_2^2 - 2N_1N_2^3 - 4N_1N_2 + 1} \tag{6.78}$$

$$S_{42} = S_{51} = \frac{\left(-4N_1^4N_2^3 - 4N_1^3N_2^4 + 2N_1^3N_2^2 + 2N_1^2N_2^3\right)}{8N_1^4N_2^4 + 4N_1^4N_2^2 - 8N_1^3N_2^3 - 2N_1^3N_2 + 4N_1^2N_2^4 + 10N_1^2N_2^2 + N_1^2 + N_2^2 - 2N_1N_2^3 - 4N_1N_2 + 1} \tag{6.79}$$

$$S_{52} = S_{41} = \frac{\left(4N_1^4N_2^3 - 4N_1^3N_2^4 - 10N_1^3N_2^2 + 2N_1^2N_2^3 + 6N_1^2N_2 - 2N_1N_2^2 - 2N_1\right)}{8N_1^4N_2^4 + 4N_1^4N_2^2 - 8N_1^3N_2^3 - 2N_1^3N_2 + 4N_1^2N_2^4 + 10N_1^2N_2^2 + N_1^2 + N_2^2 - 2N_1N_2^3 - 4N_1N_2 + 1} \tag{6.80}$$

$$S_{32} = S_{61} = \frac{\left(4N_1^4N_2^3 - 4N_1^3N_2^4 - 6N_1^3N_2^2 + 2N_1^2N_2^3 + 2N_1^2N_2\right)}{8N_1^4N_2^4 + 4N_1^4N_2^2 - 8N_1^3N_2^3 - 2N_1^3N_2 + 4N_1^2N_2^4 + 10N_1^2N_2^2 + N_1^2 + N_2^2 - 2N_1N_2^3 - 4N_1N_2 + 1} \tag{6.81}$$

The coupling and isolation levels of the six-port directional coupler when operating in forward and reverse modes are then expressed using S parameters as

$$\text{First coupling level (port 3)} = 20 \log(-S_{13}) \text{ dB} \tag{6.82}$$

$$\text{Second coupling level (port 5)} = 20 \log(-S_{15}) \text{ dB} \tag{6.83}$$

$$\text{First isolation level (port 4)} = 20 \log(-S_{14}) \text{ dB} \tag{6.84}$$

$$\text{Second isolation level (port 6)} = 20 \log(-S_{16}) \text{ dB} \tag{6.85}$$

The directivity level can be obtained again from Equation 6.78 accordingly. Equations 6.78 through 6.85 with Equation 6.78 lead to directional coupler performance parameter calculations through knowledge of only turns ratios under the assumption that all ports are matched. The real operating conditions require a six-port directional coupler to be analyzed with circuit analysis techniques. The complete analysis of the six-port coupler has been performed using forward and reverse modes for the circuit shown in Figures 6.21 and 6.22, respectively.

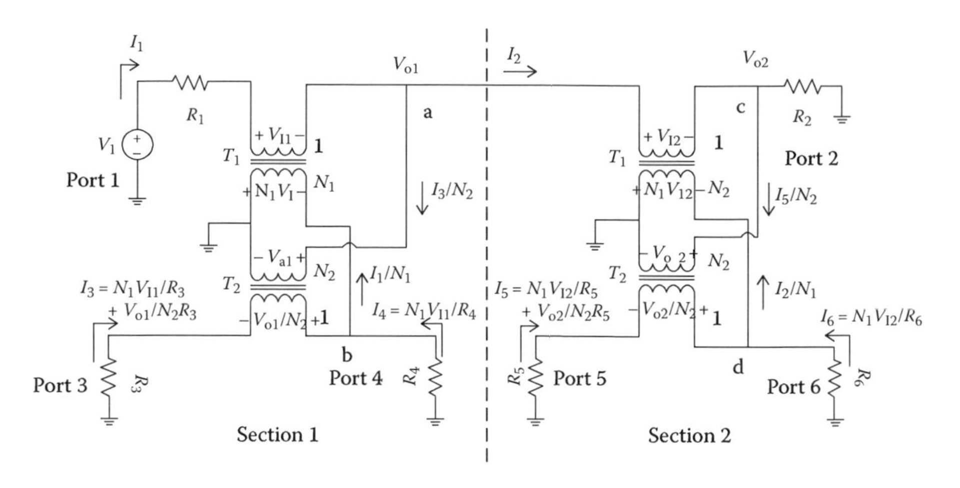

FIGURE 6.21 Forward-mode analysis of six-port coupler when $V_2 = V_3 = V_4 = V_5 = V_6 = 0$.

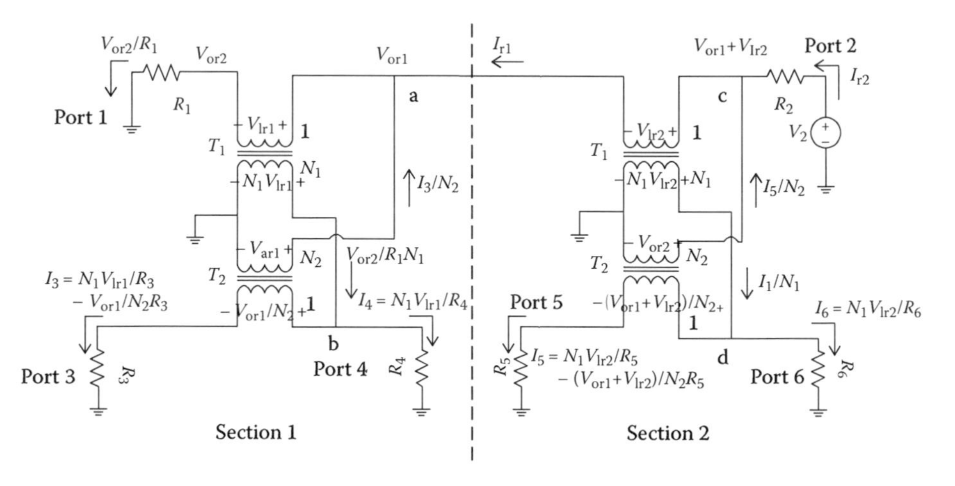

FIGURE 6.22 Reverse-mode analysis of six-port coupler when $V_1 = V_3 = V_4 = V_5 = V_6 = 0$.

6.2.3.2.1 Forward-Mode Analysis

In the forward-mode operation, V_1 is the excitation voltage with the other port voltages replaced by shorts, i.e., $V_2 = V_3 = V_4 = V_5 = V_6 = 0$. V_{o2} is the output voltage. Ports 3 and 5 are the first and second coupled ports, respectively, and ports 4 and 6 are the first and second isolated ports, respectively. The circuit analysis of the six-port coupler in forward mode then gives the following relations:

$$I_1 - \left(\frac{N_1 V_{11}}{R_3 N_2}\right) + \left(\frac{V_{o1}}{R_3 N_2^2}\right) + I_2 \text{ at the "node a"} \tag{6.86}$$

$$\frac{I_1}{N_1} = \left(\frac{N_1 V_{11}}{R_3}\right) + \left(\frac{V_{o1}}{R_3 N_2}\right) + \left(\frac{N_1 V_{11}}{R_4}\right) \text{ at the "node b"} \tag{6.87}$$

$$I_2 = \left(\frac{N_1 V_{12}}{R_5 N_2}\right) + \left(\frac{V_{o2}}{R_5 N_2^2}\right) + \frac{V_{o2}}{R_2} \text{ at the "node c"} \tag{6.88}$$

$$\frac{I_2}{N_1} = \left(\frac{N_1 V_{12}}{R_5}\right) + \left(\frac{V_{o2}}{R_5 N_2}\right) + \left(\frac{N_1 V_{12}}{R_6}\right) \text{ at the "node d"} \tag{6.89}$$

Furthermore, the voltages can be related as

$$V_1 - I_1 R_1 - V_{11} = V_{o1} \tag{6.90}$$

$$V_{o1} - V_{12} = V_{o2} \tag{6.91}$$

In practical applications, the terminal resistances are assumed to be equal, i.e., $R_1 = R_2 = R_3 = R_4 = R_5 = R_6 = r$. This leads to two important equations as

$$V_{o2} = aV_{12} \tag{6.92}$$

$$bV_{12} = cV_{o2} + V_1 \tag{6.93}$$

which leads to

$$V_{o2} = \left(\frac{a}{b - ca}\right) V_1 \tag{6.94}$$

where

$$a = \frac{N_1 N_2 (1 - 2N_1 N_2)}{\left(N_1 N_2 - N_1^2 - 1\right)} \tag{6.95}$$

$$b = \left(\frac{N_1 + N_2}{N_2}\right) + 2N_1^2 + 1 - \left(\frac{2N_1^2 + 1}{N_1 N_2 - 2(N_1 N_2)^2}\right) \tag{6.96}$$

and

$$c = \left(\frac{2N_1^2 + 1}{N_1 N_2 - 2(N_1 N_2)^2}\right) - \left(\frac{N_1 + N_2}{N_2}\right) \tag{6.97}$$

The input resistance and the coupled port resistance can now be found from

$$R_{in} = \frac{V_1}{I_1} - r = R_{coupled} \tag{6.98}$$

6.2.3.2.2 *Reverse-Mode Analysis*

In the reverse-mode operation, V_2 is the applied input voltage with the other port voltages replaced by shorts, i.e., $V_1 = V_3 = V_4 = V_5 = V_6 = 0$. V_{or2} is the output voltage. Ports 4 and 6 are the first and second reverse coupled ports, and ports 3 and 5 are the first and second reverse isolated ports, respectively. The circuit analysis in reverse mode gives the following relations:

$$I_{r1} + \left(\frac{N_1 V_{lr1}}{R_3 N_2}\right) - \left(\frac{V_{or1}}{R_3 N_2^2}\right) = \frac{V_{or2}}{R_1} \text{ at the "node a"} \tag{6.99}$$

$$\frac{V_{or2}}{R_1 N_1} = \left(\frac{N_1 V_{lr1}}{R_3}\right) - \left(\frac{V_{or1}}{R_3 N_2}\right) + \left(\frac{N_1 V_{lr1}}{R_4}\right) \text{ at the "node b"} \tag{6.100}$$

$$I_{r2} + \left(\frac{N_1 V_{lr2}}{R_5 N_2}\right) - \left(\frac{V_{or1} + V_{lr2}}{R_5 N_2^2}\right) = I_{r1} \text{ at the "node c"} \tag{6.101}$$

$$\frac{I_{r1}}{N_1} = \left(\frac{N_1 V_{lr2}}{R_6}\right) - \left(\frac{V_{or1} + V_{lr2}}{R_5 N_2}\right) + \left(\frac{N_1 V_{lr2}}{R_5}\right) \text{ at the "node d"} \tag{6.102}$$

Similar to forward-mode analysis, the voltage relations for reverse mode can be written as

$$V_{or2} + V_{lr1} = V_{or1} \tag{6.103}$$

$$V_2 - I_{r2} R_2 = V_{or1} + V_{lr2} \tag{6.104}$$

When $R_1 = R_2 = R_3 = R_4 = R_5 = R_6 = r$, we obtain

$$a_r V_{lr2} + b_r V_{or1} = V_2 \tag{6.105}$$

$$c_r V_{lr1} = V_{or1} \tag{6.106}$$

$$d_r V_{lr1} - e_r V_{lr1} + b_r V_{lr2} = V_2 \tag{6.107}$$

which leads to

$$V_{or1} = \left[\frac{c_r (a_r - b_r)}{a_r (c_r d_r - e_r) - c_r b_r^2}\right] V_2 \tag{6.108}$$

where

$$a_r = \frac{N_2^2 + 1 - 2N_1N_2 + 2(N_1N_2)^2}{N_2^2} \tag{6.109}$$

$$b_r = \frac{N_2^2 + 1 - N_1N_2}{N_2^2} \tag{6.110}$$

$$c_r = \frac{N_2\left(1 + 2N_1^2\right)}{N_1 + N_2} \tag{6.111}$$

$$d_r = \frac{2\left(1 + N_2^2\right)}{N_2^2} \tag{6.112}$$

and

$$e_r = \frac{(N_1 + N_2)}{N_2} \tag{6.113}$$

The output and the isolated port resistances can be calculated as

$$R_{out} = \frac{V_2}{I_{r2}} - r = R_{isolated} \tag{6.114}$$

The summary of the analytical results giving design parameters is illustrated in Table 6.1.

Design Example

Design a six-port transformer-coupled directional coupler with 20-dB coupling and better than 30-dB directivity at 27.12 MHz when the input voltage is $V_{in,peak}$ = 100[V]. The port impedances are matched and given to be equal to R = 50[Ω].

Solution

The MATLAB GUI that has been developed using the analytical formulation and its results have been compared with the simulated results. The couplers have been simulated with time-domain and frequency-domain simulators using Ansoft Designer and PSpice to verify the results obtained using the MATLAB GUI. The MATLAB GUI results for the six-port coupler are illustrated in Figure 6.23. This GUI for the six-port coupler gives performance parameters including coupling, isolation, and directivity in reverse and forward modes. In addition, voltages, currents, and equivalent port impedances in forward and reverse modes are also calculated and displayed using the GUI display. The GUI window is divided into two sections to display forward- and display-mode performances of six-port coupler. It is important to note that the directivity is improved by 6 dB between the section of

TABLE 6.1 Design Equations for Six-Port Transformer-Coupled Directional Coupler

Forward Mode		Reverse Mode	
Output voltage	V_{o2}	Reverse output voltage	V_{or2}
First coupled voltage	$N_1V_{l1}+\frac{V_{o1}}{N_2}$	First reverse coupled voltage	N_1V_{lr2}
Second coupled voltage	$N_1V_{l2}+\frac{V_{o2}}{N_2}$	Second reverse coupled voltage	N_1V_{lr1}
First isolated voltage	N_1V_{l1}	First reverse isolated voltage	$-N_1V_{lr2}+\left(\frac{V_{or1}+V_{lr2}}{N_2}\right)$
Second isolated voltage	N_1V_{l2}	Second reverse isolated voltage	$-N_1V_{lr1}+\frac{V_{or1}}{N_2}$
Input return loss	$-20\log\left(\frac{r-R_{in}}{r+R_{in}}\right)$	Output return loss	$-20\log\left(\frac{r-R_{out}}{r+R_{out}}\right)$
Coupled port return loss	$-20\log\left(\frac{r-R_{coupled}}{r+R_{coupled}}\right)$	Isolated port return loss	$-20\log\left(\frac{r-R_{isolated}}{r+R_{isolated}}\right)$
Insertion loss	$-20\log\left(\frac{V_{o2}}{0.5*V_1}\right)$	Reverse insertion loss	$-20\log\left(\frac{V_{o2}}{0.5*V_1}\right)$
First coupled port loss	$-20\log\left(\frac{N_1V_{l1}+\frac{V_{o1}}{N_2}}{0.5*V_1}\right)$	First reverse coupled port loss	$-20\log\left(\frac{N_1V_{lr2}}{0.5*V_2}\right)$
Second coupled port loss	$-20\log\left(\frac{N_1V_{l2}+\frac{V_{o2}}{N_2}}{0.5*V_1}\right)$	Second reverse coupled port loss	$-20\log\left(\frac{N_1V_{lr1}}{0.5*V_2}\right)$
First isolated port loss	$-20\log\left(\frac{N_1V_{l1}}{0.5*V_1}\right)$	First reverse isolated port loss	$-20\log\left(\frac{N_1V_{lr2}+\left(\frac{V_{o1}+V_{lr2}}{N_2}\right)}{0.5*V_2}\right)$
Second isolated port loss	$-20\log\left(\frac{N_1V_{l2}}{0.5*V_1}\right)$	Second reverse isolated port loss	$-20\log\left(\frac{N_1V_{lr2}+\frac{V_{or1}}{N_2}}{0.5*V_2}\right)$

the coupler and the excitation port. This improvement is illustrated by calculated levels in coupling and isolation, which are shown as coupling 2 and isolation 2 for each mode of the operation in Figure 6.23. The difference between coupling and isolation levels gives the amount of directivity as discussed earlier.

The six-port coupler is simulated by the frequency-domain simulator Ansoft Designer, as shown in Figure 6.24. The simulation results are illustrated in Figures 6.25 and 6.26 for forward and reverse modes, respectively.

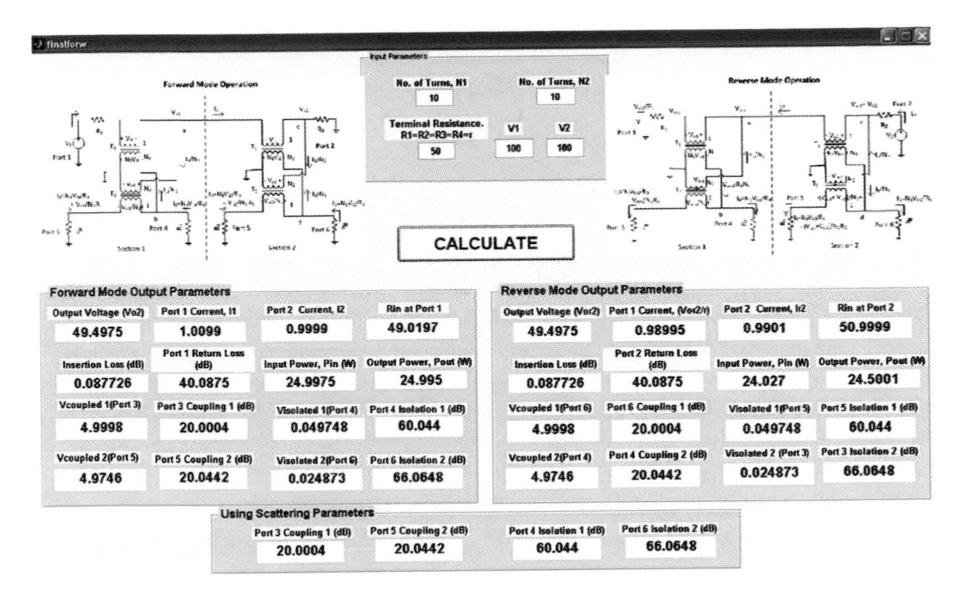

FIGURE 6.23 MATLAB GUI results for six-port coupler.

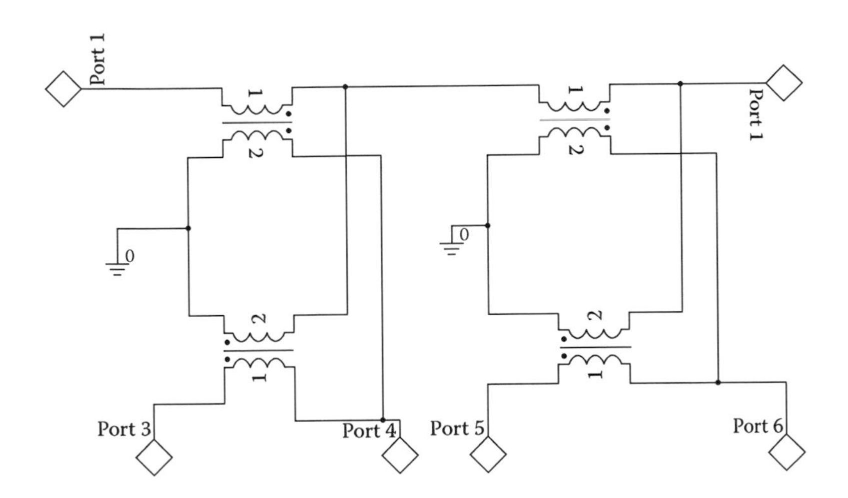

FIGURE 6.24 Simulated circuit of six-port coupler using Ansoft Designer for frequency-domain analysis.

Figure 6.25 shows the forward coupling and isolation levels in forward mode, whereas Figure 6.26 gives similar performance parameters of the coupler in reverse mode. The simulation results of the frequency-domain circuit simulator are in agreement with the results obtained by the MATLAB GUI as illustrated.

The time-domain analysis is performed for the six-port coupler using the same interfacing impedance and power requirements. The inductor design for the transformer is detailed in the following. The time-domain six-port transformer coupling circuit operating in forward mode is shown in Figure 6.27.

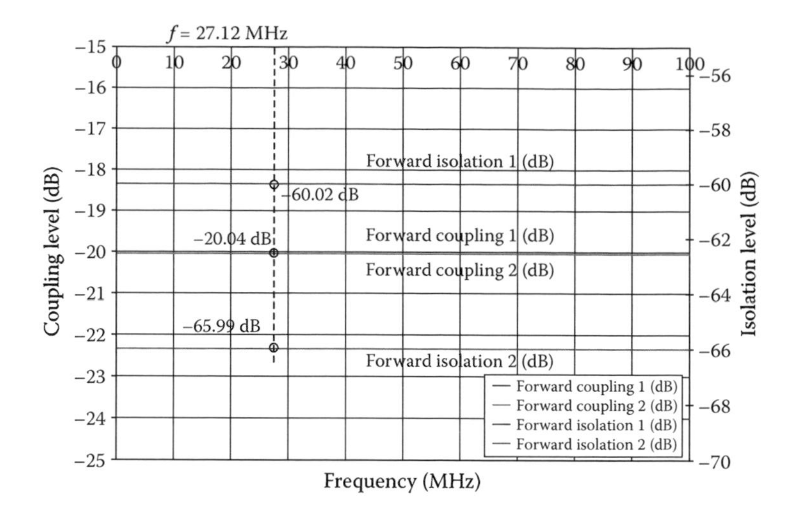

FIGURE 6.25 Simulated circuit of six-port coupler using Ansoft Designer for frequency-domain analysis in forward mode.

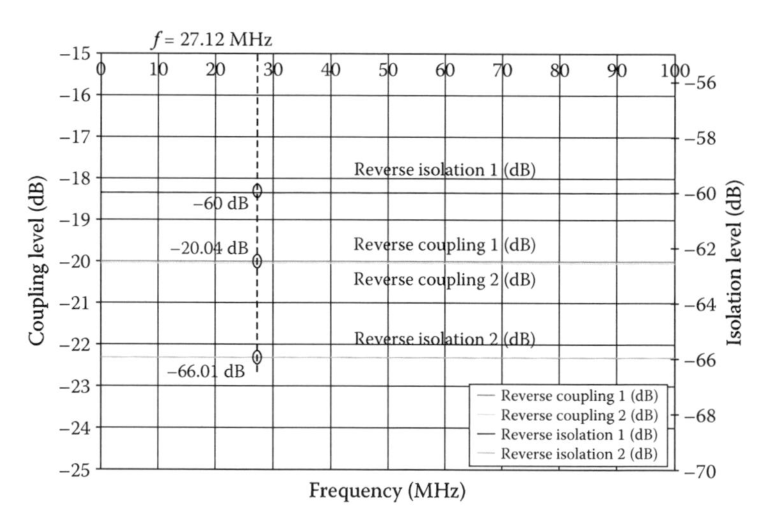

FIGURE 6.26 Simulated circuit of six-port coupler using Ansoft Designer for frequency-domain analysis in reverse mode.

The simulation results showing the coupling and isolation in forward mode are given in Figure 6.28. The coupling level and isolation are found to be –20.003 and –57.5 dB, respectively. The directivity is found to be 37.497 dB based on the time-domain simulation. The simulated values are in agreement with the frequency-domain simulator and the MATLAB GUI program. Furthermore, the individual port voltages are found using PSpice, and the results are illustrated in Figure 6.29. They are also in agreement with the illustrated results in the MATLAB GUI.

The details about the transformer that will be used in the implementation of the coupler in Figure 6.27 have to be determined. This includes the type of material that will be used as magnetic core, winding information, inductance information, etc.

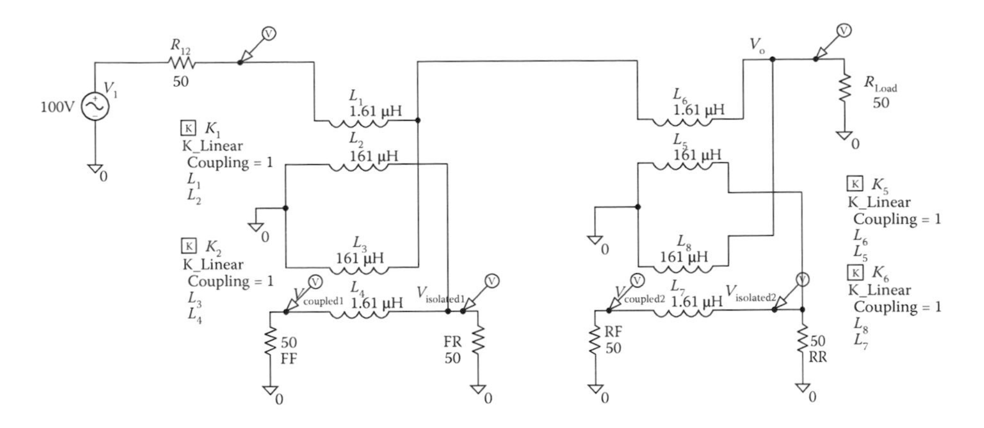

FIGURE 6.27 Time-domain simulation of six-port transformer coupler using PSpice in forward mode.

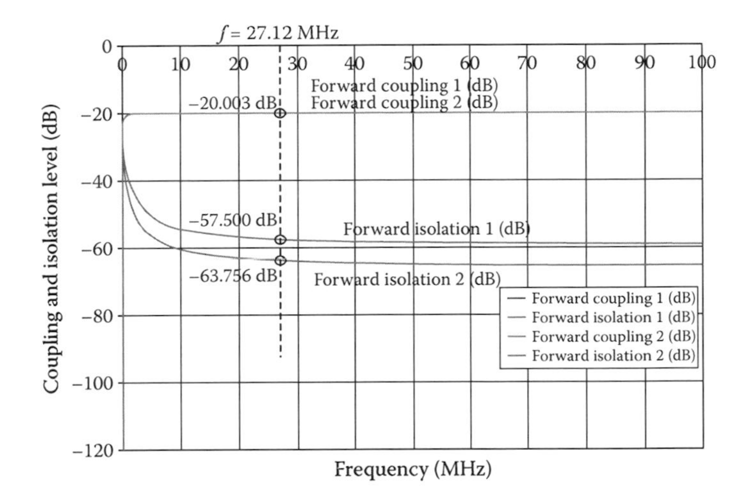

FIGURE 6.28 Time-domain simulation results for coupling and isolation levels of six-port coupler using PSpice in forward mode.

The core material is chosen to be a –7 material, which is carbonyl TH with permeability of mi = 9 and has good performance for applications when the frequency of operation is between 3 and 35 MHz based on the manufacturer-measured performance data, including saturation magnetic flux density, loss, and thermal profile. The core dimensions are given to be OD = 1.75 cm, ID = 0.94 cm, and h = 0.48 cm. This core is designated as T-68-7 with white color code by the manufacturer [37]. The geometry of the core is shown in Figure 6.30a. Three cores are stacked, and 20 American wire gauge (AWG) is used for winding, as illustrated in Figure 6.30b. The final constructed inductor configuration is shown in Figure 6.30c. The inductance value and configuration are obtained using the method and GUI developed in Ref. [38]. Based on the method in Ref. [38], 10 turns give an inductance value of 1.61 mH and an impedance value that is more than five times higher than the impedance termination, which is 50 W.

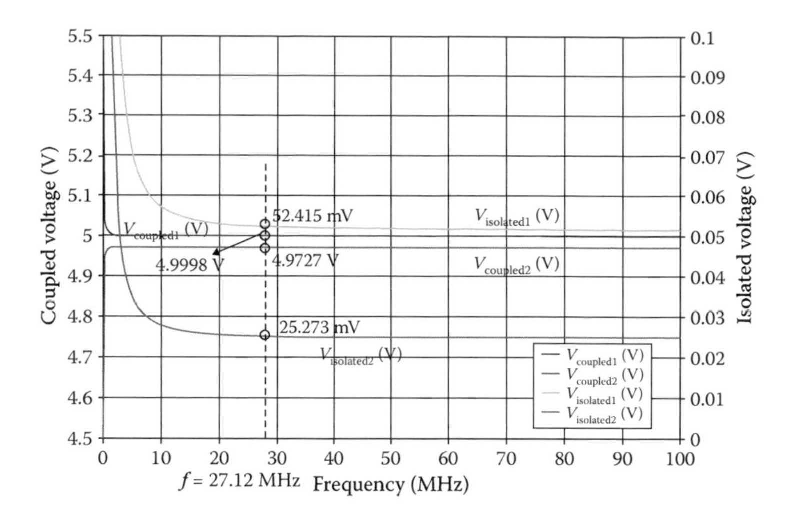

FIGURE 6.29 Time-domain simulation results showing port voltages for six-port coupler using PSpice in forward mode.

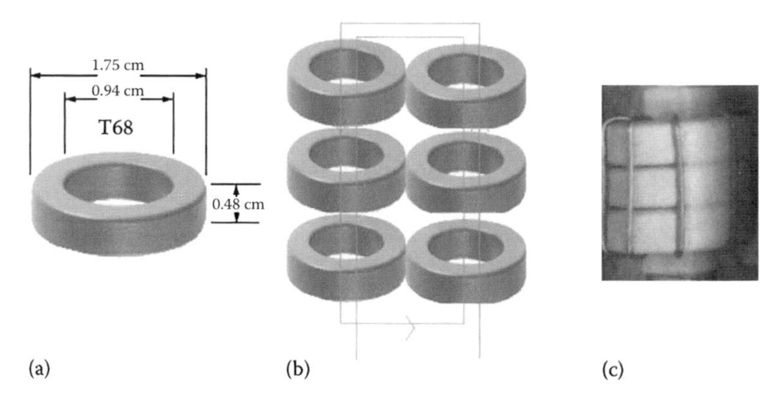

FIGURE 6.30 Geometry of the core used in construction of transformer coupler. (a) Single core, (b) layout of inductor with multiple core, and (c) constructed multiple core inductor.

The GUI window showing the parameters, inductance value, and all other design parameters including the physical length of the winding wire required to construct stacked inductor configuration is shown in Figure 6.31. Hence, the transformer that is designed should now be able to produce the required coupling level for the operating conditions in the forward and reverse modes at the center

toroid

TOROIDAL INDUCTOR CHARACTERIZATION PROGRAM

Enter N, Number of Turns	Enter OD in cm	Enter ID in cm	Enter h in cm	n, Number of Stacked Cores	mui, Initial Permeability	d, Wire Diam with No Coating in cm	s, Coating Thickness in mm	epsr, Coating Material Permittivity	Fo, Frequency of Interest in Hz
10	1.75	0.94	0.48	3	9	0.0812	0.07	3.5	27.12e6

Ls(H) at Fo	Rs(Ohms) at Fo	Q at Fo	Rw (Ohms) at Fo	Re(Ohms) at Fo	L(H)	C(F)	Fr, Resonant Frequency (MHz)
3.6433e-006	9.6759	64.1616	1.8886	0.0026235	1.6109e-006	1.1928e-011	36.308

le, Magnetic Path Length in cm	lw, Wire Length in cm	Ae, Cross Sectional Area in cm^2
4.0945	73.2173	0.5832

FIGURE 6.31 Toroidal design and characterization program for six-port coupler inductor design.

frequency, which is given to be f = 27.12 MHz. The assembly details for the six-port coupler are illustrated using the configuration shown in Figure 6.32a, and the constructed coupler is illustrated in Figure 6.32b. The six-port transformer coupler is measured by the network analyzer, HP8753ES.

The measurement results using network analyzer and input impedance vs. frequency on the Smith chart when excitation is done from port 1, which corresponds to a forward-mode operation, are illustrated in Figure 6.33. All other

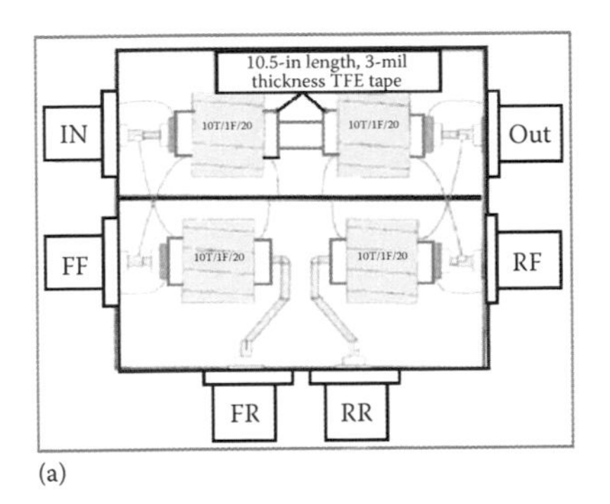

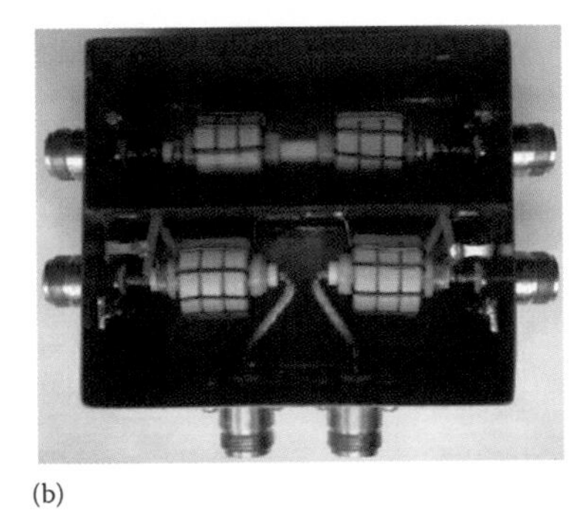

FIGURE 6.32 Toroidal design and characterization program for the six-port coupler inductor design. (a) Layout of six-port coupler and (b) constructed six-port coupler.

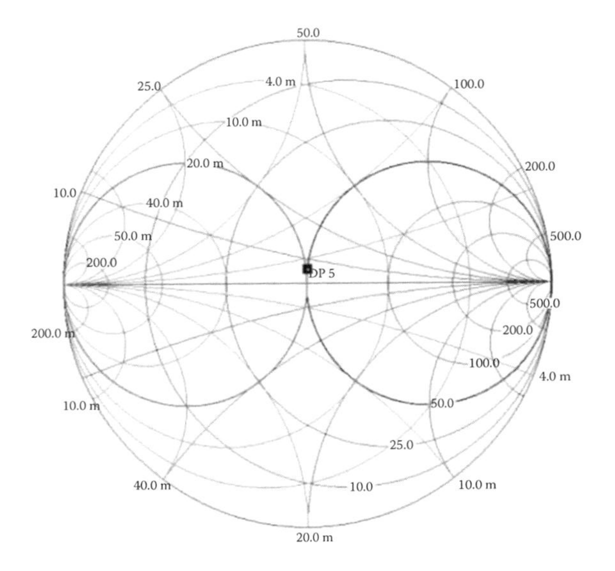

Point	Z	Q	Frequency
DP 1	(50.000 + j5.914) Ω	0.118	24.4 MHz
DP 2	(50.025 + j5.991) Ω	0.120	25.8 MHz
DP 3	(50.055 + j6.091) Ω	0.122	27.1 MHz
DP 4	(50.080 + j6.201) Ω	0.124	28.5 MHz
DP 5	(50.104 + j6.305) Ω	0.126	29.8 MHz

FIGURE 6.33 Network analyzer outputs for measuring the forward impedance for frequencies within ±10% of the center frequency, 27.12 MHz.

six-port transformer coupler performance parameters are also measured in forward and reverse modes by the network analyzer, and they are given vs. frequency within 10-MHz bandwidth of the center frequency in Figure 6.34a and b, respectively. The broadband frequency responses of the coupler in forward and reverse modes from 1 to 100 MHz are given in Figure 6.35a and b, respectively.

The analytical, simulation, and measurement results are tabulated and shown in Table 6.2. As seen from the results, the analytical and frequency-domain model for the transformer is used in the frequency domain. In the time-domain simulation, the accuracy of the system is increased by implementing the inductor model that is close to the one that is used in the construction of the coupler. The measured coupling level is in agreement with analytical, frequency-domain, and time-domain simulators. The measured isolation and, as a result, the directivity level are closer with the time-domain simulator since a more accurate inductor model is used for the transformer.

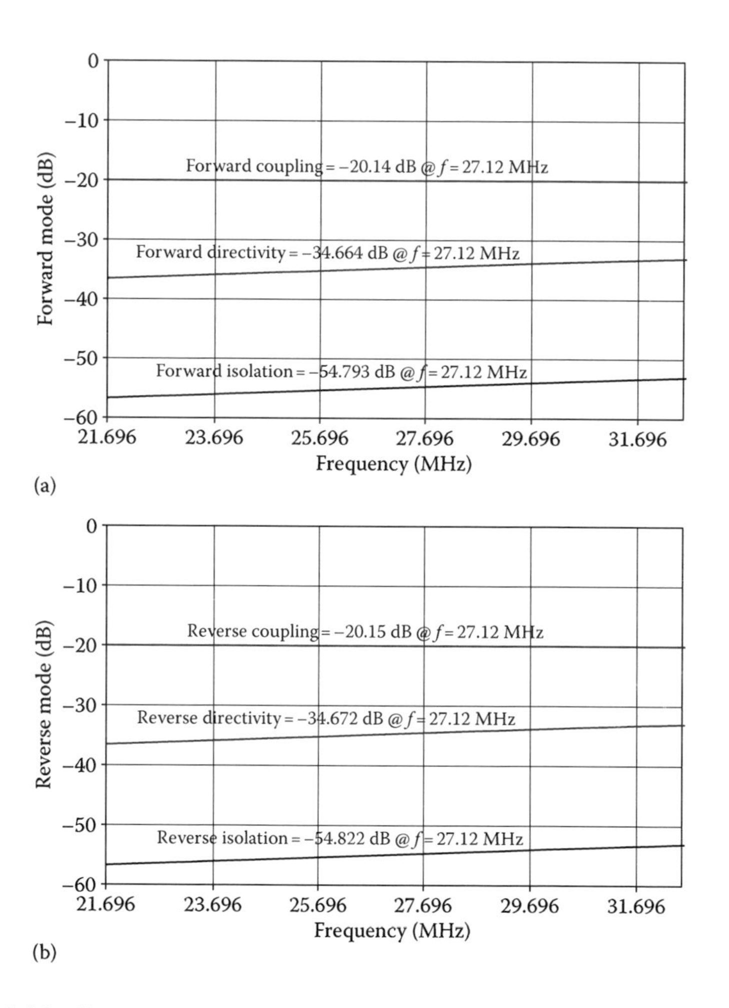

FIGURE 6.34 Six-port coupler coupling, isolation, and directivity measurements within 10-MHz bandwidth of the center frequency for (a) forward and (b) reverse modes.

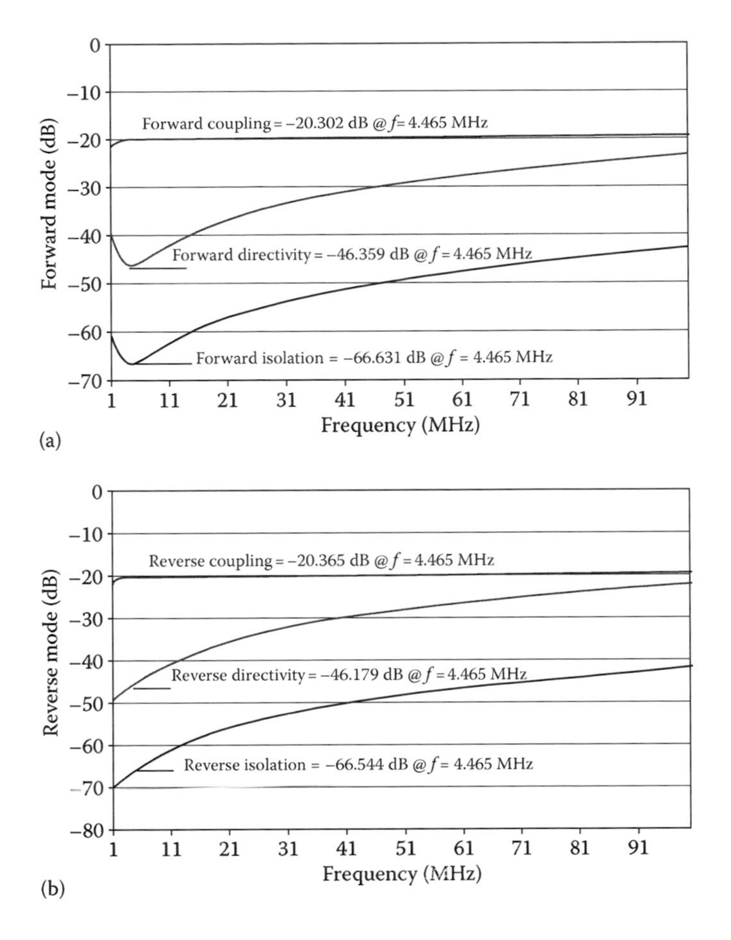

FIGURE 6.35 The broadband frequency response of six-port coupler coupling, isolation, and directivity measurements in (a) forward and (b) reverse modes.

TABLE 6.2
Six-Port Transformer Coupler Performance Parameter Comparison

	Analytical	Frequency-Domain Simulation	Time-Domain Simulation	Measurement
		Forward Mode (dB)		
Coupling	20.004	20.00	20.03	20.14
Isolation	60.044	60.02	57.50	54.793
Directivity	40.04	40.02	37.47	34.664
		Reverse Mode (dB)		
Coupling	20.004	20.00	20.06	20.15
Isolation	60.044	60.00	57.45	54.822
Directivity	40.04	40.00	37.39	34.672

6.3 MULTISTATE REFLECTOMETERS

The concept of the six-port reflectometer theory was first introduced by Hoer and Engen [22,23] and then became an attractive method for the measurement of voltage, current, impedance, phase, and power. Five-port reflectometers have also been investigated for complex reflection coefficient measurement [39]. Although six-port and five-port reflectometers are attractive and low-cost alternatives to network analyzers, it is not convenient to use them in practice since commercially available devices are mainly four-port directional couplers.

Four-port networks such as couplers can also be used to detect power and measure the magnitude of the reflection coefficient. However, it is not possible to measure the complex reflection coefficient with four-port couplers using conventional techniques. There has been research on the use of four-port couplers as multistate reflectometers to be implemented in an automated environment and be used for the measurement of several important parameters such as complex reflection coefficient [40,41]. The analysis of multistate reflectometers that is applicable in practice is given in Ref. [42] using a variable attenuator concept with a four-port network. However, no equations or solutions have been obtained or presented in Ref. [42] for the reflection coefficient calculation where power circles are constructed, intersection point is obtained, and impedance is determined accurately.

In this section, the analysis of multistate reflectometers based on a four-port network and a variable attenuator proposed in Ref. [42] and shown in Figure 6.36 is extended to examine the more general theory of using scalar power measurements to determine the complex reflection coefficient. The explicit closed-form relations and solutions for the system of equations are derived and used to calculate

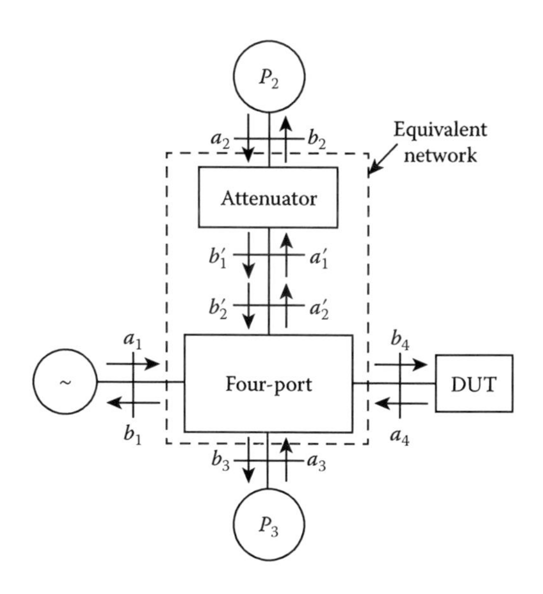

FIGURE 6.36 Multistate reflectometer based on four-port coupler and attenuator.

the complex reflection coefficient with the concept of the radical center for three power circles.

Analytical results based on the derivations have been obtained; the general theory is verified, and calibration of multistate reflectometers has been discussed.

6.3.1 Multistate Reflectometer Based on Four-Port Network and Variable Attenuator

The multistate reflectometer based on the four-port network and variable attenuator proposed in Ref. [42] is shown in Figure 6.36. It consists of one arbitrary four-port network, a power source, the device under test (DUT), a variable attenuator, and two scalar power detectors. The terms a_i represent the complex amplitude of the voltage wave incident to port i, and the b_i terms represent the emergent voltage wave amplitude from port i. The general operating principles of the device shown in Figure 6.36 are based on those of the more well-known six-port reflectometer designs developed by Engen [23]. The key difference is that in the four-port design shown in Figure 6.36, the system should be measured under two different attenuator settings to obtain the necessary number of equations to solve for the value of the complex reflection coefficient for the DUT.

The initial analysis of the multistate reflectometer system with the attenuator shown in Figure 6.36 is conducted in Ref. [42], and the measured powers at ports 2 and 3 in terms of reflection coefficient Γ were given as

$$P_i = |b_i|^2 = q_i \left|\frac{1+A_i\Gamma}{1+A_0'\Gamma}\right|^2, \quad (i=2,3) \tag{6.115}$$

Since, in the system illustrated in Figure 6.36, there are two different network statuses, then Equation 6.115 leads to two sets of equations as

$$P_i' = |b_i'|^2 = q_i' \left|\frac{1+A_i'\Gamma}{1+A_0'\Gamma}\right|^2, \quad (i=2,3) \tag{6.116}$$

$$P_i'' = |b_i''|^2 = q_i'' \left|\frac{1+A_i''\Gamma}{1+A_0''\Gamma}\right|^2, \quad (i=2,3) \tag{6.117}$$

where q_i' and q_i'' for $i = 2, 3$ are real constants, A_i' and A_i'' for $i = 0, 2, 3$ are complex constants, and P_i' and P_i'' for $i = 2, 3$ are the power meter readings for each of the two network statuses. In Ref. [42], no further analysis has been performed nor were explicit relations and solutions for the reflection coefficient calculation given. Hence, it is impossible to construct the power circles to identify the intersection point for the complex reflection coefficient calculation. This analysis has been performed; explicit relations and solutions are obtained, and analytical results are presented in this chapter.

Equations 6.116 and 6.117 present a set of bilinear equations that need to be solved for complex reflection coefficient, Γ. The procedure that is used in Ref. [39] to find power ratio equations for five-port networks can be implemented for the system shown in Figure 6.36. The analysis begins with separating Equation 6.115 into its corresponding real and imaginary parts as

$$P_i = q_i \left| \frac{1+(c_i + jd_i)(x+jy)}{1+(c_0 + jd_0)(x+jy)} \right|^2 \tag{6.118}$$

or

$$P_i = q_i \frac{\left(c_i^2 + d_i^2\right)x^2 + \left(c_i^2 + d_i^2\right)y^2 + 2c_i x - 2d_i y + 1}{\left(c_0^2 + d_0^2\right)x^2 + \left(c_0^2 + d_0^2\right)y^2 + 2c_0 x - 2d_0 y + 1} \tag{6.119}$$

where

$$\Gamma = |\Gamma| \angle \psi^\circ = x + jy \tag{6.120}$$

and

$$A_i = \alpha_i \angle \phi_i^\circ = c_i + jd_i \tag{6.121}$$

Equation 6.119 can be expressed as

$$\left(P_i \alpha_0^2 - q_i \alpha_i^2\right)(x^2 + y^2) + 2(P_i c_0 - q_i c_i)x + 2(q_i d_i - P_i d_0)y = q_i - P_i \tag{6.122}$$

Furthermore, Equation 6.111 can be put in the following form:

$$x^2 + 2u_i x + y^2 + 2v_i y = 2r_i, \ (i = 2,3) \tag{6.123}$$

where

$$\omega_i = P_i \alpha_0^2 - q_i \alpha_i^2 \tag{6.124}$$

$$u_i = \frac{P_i c_0 - q_i c_i}{\omega_i} \tag{6.125}$$

$$v_i = \frac{q_i d_i - P_i c_0}{\omega_i} \tag{6.126}$$

$$r_i = \frac{q_i - P_i}{2\omega_i} \tag{6.127}$$

It is now clear from Equation 6.123 that it represents the general form of the equation for a circle with center $(-u_i, -v_i)$. Hence, Equation 6.123 can be written in center-radius form as

$$\left(x-\left(-u_{\mathrm{i}}\right)\right)^{2}+\left(y-\left(-v_{\mathrm{i}}\right)\right)^{2}=R_{\mathrm{i}}^{2}, \quad (i=2,3) \tag{6.128}$$

where $R_{\mathrm{i}}^2 = 2r_{\mathrm{i}} + u_{\mathrm{i}}^2 + v_{\mathrm{i}}^2$. Equation 6.118 defines a set of circles in the complex plane indicating possible values of complex reflection coefficient, Γ. In order to solve this system for Γ, at least two independent circle equations must be solved for their intersection points. Most solution methods for six-port reflectometer designs utilize a ratio of power readings as opposed to each independent power reading [23]. The resulting equation is of identical form to Equation 6.115, and this approach yields many benefits. Additionally, if the power reading being normalized to is highly independent of a_2, it acts to stabilize the system against power fluctuations.

In multistate reflectometers, the approach in Refs. [40,41] is to use one variable-state port and one reference port. The reference port is coupled to forward power, whereas the variable-state port is connected to a phase-shifting network. The method is more complicated than that proposed in Ref. [42] but approaches the ideal behavior proposed by Engen [23]. Three power ratios are measured by dividing the variable reading by the forward-coupled reading.

Regardless of the specific calibration/measurement scheme being used, the general solution for Γ is described by the intersection of three circles. In reality, however, the circles will not intersect due to noise and inaccuracies, but this can be overcome quite simply by using the concept of the radical center.

The radical center of three circles is the unique point, which possesses equal power with respect to all three circles. In other words, it is the point where the tangent lines to all circles are of equal length. For three overlapping circles, the radical center is given by the intersection point of the three common chords between all three circles [23]. Additionally, the radical center is still defined when no circle intersections occur as the intersection point of the three radical axes. If the measured location of Γ is interpreted as the radical center of three power or power-ratio circles, then the bilinear equations of the form of Equation 6.123 may be reduced to a simple system of linear equations by subtraction:

$$2(u_{\mathrm{i}} - u_{\mathrm{j}})x + 2(v_{\mathrm{i}} - v_{\mathrm{j}})\,y = 2(r_{\mathrm{i}} - r_{\mathrm{j}}) \tag{6.129}$$

where the equation of circle j is subtracted from circle i. Equation 6.129 is the equation of the radical axis between circles i and j. In a three-circle system, two more such equations exist between circles i and k, and between circles j and k, giving a system of linear equations, which may be solved for x and y, which are the real and imaginary components of the complex reflection coefficient.

The analytical results have been obtained with MATLAB using the formulation and solutions discussed and are illustrated in Figures 6.37 and 6.38. Figure 6.37 demonstrates an imperfect power circle intersection, which could be the result of measurement noise and/or calibration inaccuracies. The three large circles represent the power measurement circles, which are separated by approximately 120° in phase and at an equal distance from the origin. The radical center is located at (0.27–j0.21) inside the unit circle. The error bound of the measurement can be considered as the unique triangle, which has the three power circles as its excircles. Once again, the radical center is the unique point with equal power to all three measurement circles.

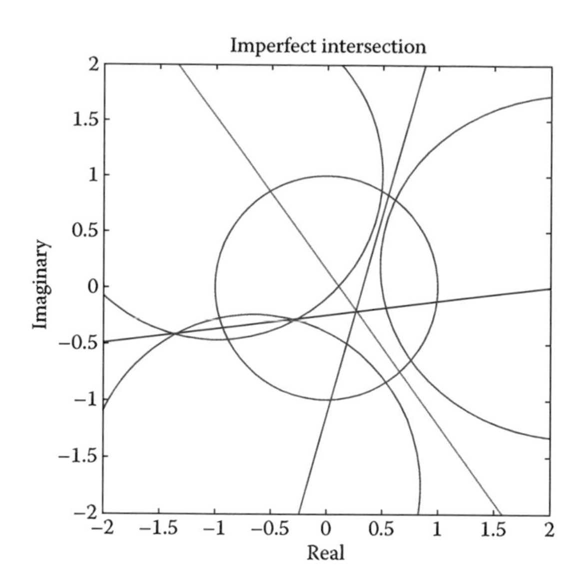

FIGURE 6.37 Illustration of imperfect power circle for complex reflection coefficient determination.

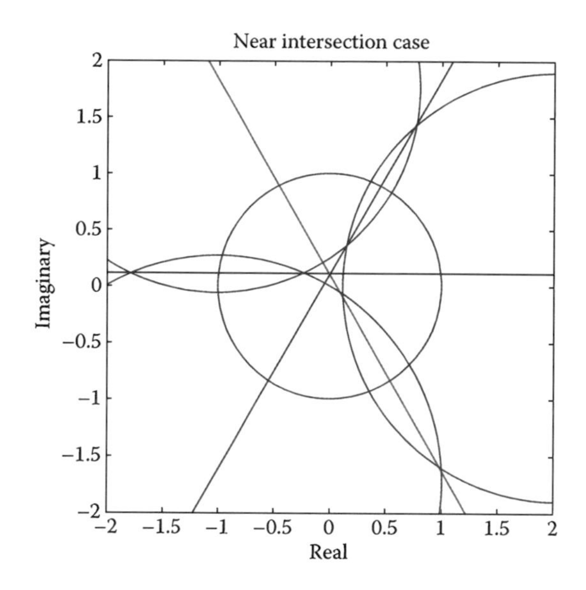

FIGURE 6.38 Illustration of near intersection power circle for complex reflection coefficient determination.

Figure 6.38 illustrates the case of a near intersection of the three power circles for complex reflection coefficient determination. The radical axes are clearly shown as the lines passing through the common chords of each circle intersection. The error triangle in this case is much smaller and represents the most likely measurement scenario when using accurately calibrated detectors.

One additional area of concern is in the calibration of multiport and multistate reflectometers. In calibration, there is no simple linearization, and it is often the case that numerical methods are required to calculate the calibration constants. Since the inception of the first six-port reflectometers, there have been several breakthroughs in reducing the complexity of the calibration process. Most of these improvements come in the form of realizing hidden relationships between the complex constant parameters of the six-port or multiport network. One notable result shown in Ref. [43] is that the number of required calibration standards can be reduced to four by utilizing reflective standards with a certain phase relationship. These four standards lead to a system of 12 circles divided between four complex planes. As demonstrated in Ref. [43], a numerical error function minimization approach is usually taken to find the calibration constants, which give circle intersections in all four planes simultaneously. With such methods, convergence issues may arise depending on the properties of the multiport network being calibrated and on the standards of calibration themselves.

6.4 RF POWER SENSORS

RF power sensors measure the forward and reflected power of a signal connected to a load. The high-level overview can be seen in Figure 6.39. The signal is received at the input and travels along a 50-Ω microstrip transmission line. It is then split using a coupled-line directional coupler, sending the signal through matching networks and into the logarithmic diode RF power detector, as well as the module and phase detector. The output at both steps is filtered for anti-aliasing before entering the controller unit. The analog signal is converted to a digital signal, so the digital signal processor can perform the calibration techniques required to compensate for inaccuracies. The microcontroller should be able to communicate the calibrated information to a GUI and an Internet server, where it can be displayed both numerically and graphically. The block diagram of the implementation of the power sensor is given in Figure 6.40. The system is powered by a supply, which can come from an alternating current or battery source, and is regulated for use by the components in the design.

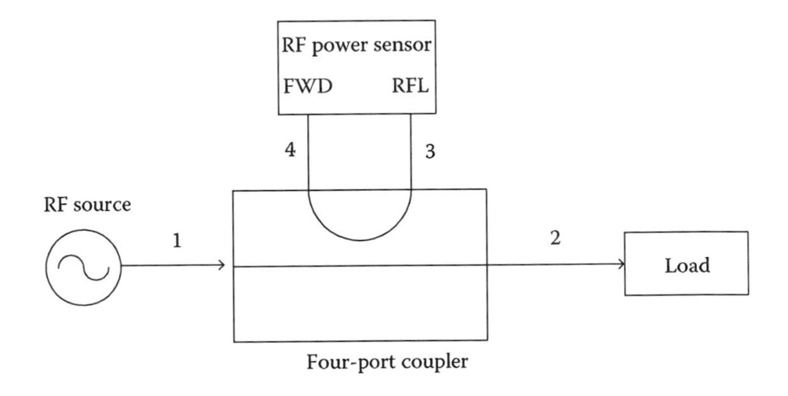

FIGURE 6.39 Illustration of RF power sensor.

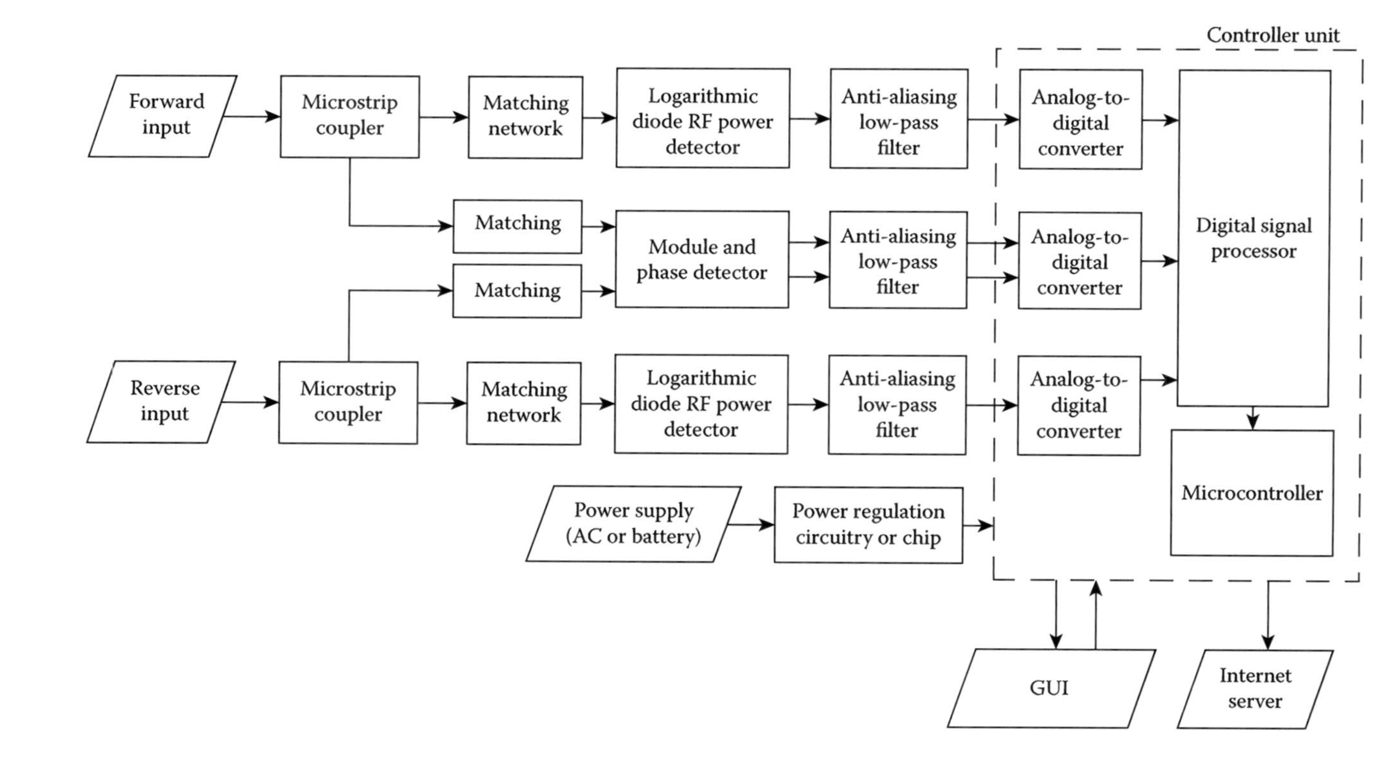

FIGURE 6.40 Block diagram of the implementation of RF power sensor.

The magnitude and phase detection in the RF power sensor can be done with an integrated circuit (IC), which has two six-stage logarithmic amplifiers, a magnitude comparator, and a phase discriminator to measure the relative magnitude and phase between two input signals. The input to the IC for magnitude and phase detection is obtained by coupling a portion of the input power using a microstrip coupler. This method achieves low insertion loss and ensures that the scalar power measurements will not be adversely affected. However, this coupling has very low flatness and requires a more robust calibration procedure to account for the errors introduced by the system.

The magnitude and phase measurement should be calibrated with the algorithm developed based on the frequency of operation and the components used. This calibration is based on the use of a four-port directional coupler with the source connected to port 1, the load connected to port 2, and the probes connected to ports 3 and 4. It has been shown that the incident wave at port 1 can be represented by Equation 6.130, where a_L is the reflected voltage wave from the load, Γ_L is the load reflection coefficient, and A_i, B_i, and C are complex constant parameters describing the network:

$$b_i = \frac{A_i + B_i\Gamma_L}{1 + C\Gamma_L} a_L \tag{6.130}$$

The ratio of two input signals, so an equation defining the ratio of reflected (port 3) to incident (port 4), is given by

$$N = \frac{b_3}{b_4} = \frac{A_3 + B_3\Gamma_L}{A_4 + B_4\Gamma_L} = \frac{\dfrac{A_3}{A_4} + \dfrac{B_3}{A_4}\Gamma_L}{1 + \dfrac{B_4}{A_4}\Gamma_L} = \frac{A + B\Gamma_L}{A + C\Gamma_L} \tag{6.131}$$

where A, B, and C are the three new complex constants. Equation 6.131 can be rearranged to give

$$A + \Gamma_i B + (-N_i\Gamma_i)\, C = N_i \tag{6.132}$$

Since there are three unknown constants defining the system (A, B, C), a minimum of three independent equations are required to solve for the constants. Also, note that each complex constant consists of two real constants, but the simplification may be made since each measurement consists of two real values. The system of equations used to solve for the constants A, B, and C can be obtained using

$$\begin{bmatrix} 1 & \Gamma_1 & -N_1\Gamma_1 \\ 1 & \Gamma_2 & -N_2\Gamma_2 \\ 1 & \Gamma_3 & -N_3\Gamma_3 \end{bmatrix} \begin{bmatrix} A \\ B \\ C \end{bmatrix} = \begin{bmatrix} N_1 \\ N_2 \\ N_3 \end{bmatrix} \tag{6.133}$$

Equation 6.132 can be solved in a number of ways, but one common approach is to use Cramer's rule, where the constant values are given by Equation 6.134. The Δ represents the determinant of the system, and Δ_i represents the determinant where the ith column has been replaced by the vector of measured values N:

$$A = \frac{\Delta_1}{\Delta},\ B = \frac{\Delta_2}{\Delta},\ \text{and } A = \frac{\Delta_3}{\Delta}, \tag{6.134}$$

The details of the calculation of Δ, Δ_1, Δ_2, and Δ_3 are given by Equations 6.135 through 6.137. Note that several common terms exist between these equations, which can be used to reduce the amount of actual calculation needed overall.

$$\Delta = \begin{bmatrix} 1 & \Gamma_1 & -N_1\Gamma_1 \\ 1 & \Gamma_2 & -N_2\Gamma_2 \\ 1 & \Gamma_3 & -N_3\Gamma_3 \end{bmatrix} = \Gamma_2\Gamma_3(N_2 - N_3) + \Gamma_1\Gamma_3(N_3 - N_1) + \Gamma_1\Gamma_2(N_1 - N_2) \tag{6.135}$$

$$\Delta_1 = \begin{bmatrix} N_1 & \Gamma_1 & -N_1\Gamma_1 \\ N_2 & \Gamma_2 & -N_2\Gamma_2 \\ N_3 & \Gamma_3 & -N_3\Gamma_3 \end{bmatrix} = N_1\Gamma_2\Gamma_3(N_2 - N_3) + N_2\Gamma_1\Gamma_3(N_3 - N_1) + N_3\Gamma_1\Gamma_2(N_1 - N_2) \tag{6.136}$$

$$\Delta_2 = \begin{bmatrix} 1 & N_1 & -N_1\Gamma_1 \\ 1 & N_2 & -N_2\Gamma_2 \\ 1 & N_3 & -N_3\Gamma_3 \end{bmatrix} = N_2N_3(\Gamma_2 - \Gamma_3) + N_1N_3(\Gamma_3 - \Gamma_1) + N_1N_2(\Gamma_1 - \Gamma_2) \tag{6.137}$$

$$\Delta_3 = \begin{bmatrix} 1 & \Gamma_1 & N_1 \\ 1 & \Gamma_2 & N_2 \\ 1 & \Gamma_3 & N_3 \end{bmatrix} = \Gamma_1(N_2 - N_3) + \Gamma_2(N_3 - N_2) + \Gamma_3(N_1 - N_2) \tag{6.138}$$

The solution of Equations 6.135 through 6.138 yields three complex constants A, B, and C, which define the system. However, these calculations are performed in an embedded application using a microcontroller in power sensor, so it is desirable to minimize the total number of operations needed to compute Γ_L. This is done by plugging the results of the determinant calculations into the equation for Γ_L, as shown in Equation 6.139:

$$\Gamma_L = \frac{N - A}{B - NC} = \frac{N - \frac{\Delta_2}{\Delta}}{\frac{\Delta_2}{\Delta} - N\frac{\Delta_3}{\Delta}} = \frac{N\Delta - \Delta_2}{\Delta_2 - N\Delta_3} \tag{6.139}$$

Equation 6.139 shows the relationship between the four calibration parameters Δ, Δ_1, Δ_2, and Δ_3 and the measured value N. This simplification eliminates the need to perform three divisions in calculating A, B, and C, which are costly operations in a microcontroller. The load reflection coefficient may now be obtained from the load reflection coefficient by Equation 6.140, where Z_0 is the system reference impedance, which is generally 50 Ω.

$$Z_L = Z_o \frac{1+\Gamma_L}{1-\Gamma_L} \tag{6.140}$$

The typical sensor calibration that takes place before the measurement is shown in Figure 6.41. It consists of a sequence of connection and calibration based on the known load standards such as short, open, and 50-Ω load. Calibration points for these impedances are shown in Figure 6.42.

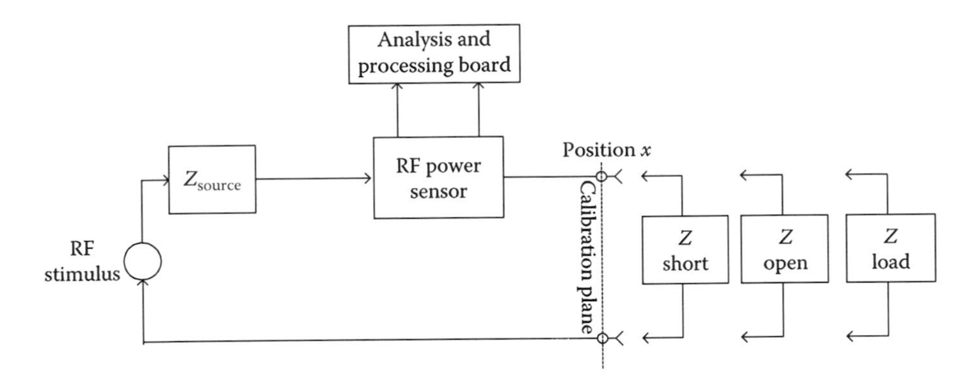

FIGURE 6.41 Illustration of RF power sensor calibration.

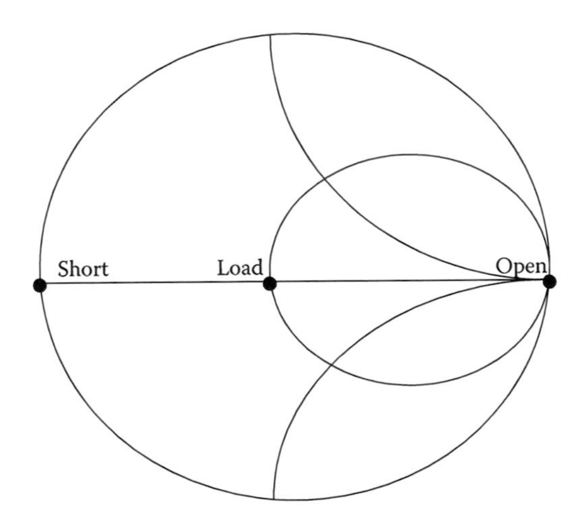

FIGURE 6.42 Calibration points on the Smith chart.

PROBLEMS

1. Design a 20-dB two-line directional coupler using ROGERS 4003 at 1 GHz using the formulation given. Compare your results using a planar electromagnetic simulator.
2. Design a 10-dB three-line coupler using FR4 at 1 GHz with the formulation given. Compare your results using a planar electromagnetic simulator.
3. Calculate and compare the directivity of the two-line and three-line couplers in Problem 1 for 20-dB directivity.
4. Design the multilayer configuration of the coupler given in Problem 1, and calculate and compare the directivity of the coupler for two-line, three-line, and multilayer configurations.

REFERENCES

1. A. Eroglu. 2013. *RF Circuit Design Techniques for MF-UHF Applications*. CRC Press, Boca Raton, FL.
2. A. Eroglu, and J.K. Lee. 2008. The complete design of microstrip directional couplers using the synthesis technique. *IEEE Transactions on Instrumentation and Measurement*, Vol. 57, No. 12, pp. 2756–2761, December.
3. G.D. Monteath. 1955. Coupled transmission lines as symmetrical directional couplers. *Proceedings of the IEE—Part B: Radio and Electronic Engineering*, Vol. 102, Pt. B, pp. 383–392, May.
4. N.A. El-Deeb. 1983. The calibration and performance of a microstrip six-port reflectometer. *IEEE Transactions on Microwave Theory and Techniques*, Vol. MTT-31, No. 7, pp. 509–514.
5. R.J. Collier, and N.A. El-Deeb. 1980. Microstrip coupler suitable for use as a six-port reflectometer. *IEE Proceedings, Microwaves, Optics and Antennas*, Vol. 127, No. 2, Pt. H, pp. 87–91, April.
6. M. Pavlidis, and H.L. Hartnagel. 1976. The design and performance of three-line microstrip couplers. *IEEE Transactions on Microwave Theory and Techniques*, Vol. MTT-24, pp. 631–640.
7. R.J. Collier, and N.A. El-Deeb. 1979. On the use of a microstrip three-line system as a six-port reflectometer. *IEEE Transactions on Microwave Theory and Techniques*, Vol. MTT-27, pp. 847–853.
8. A. Eroglu, R. Goulding, P. Ryan, J. Caughman, and D. Rasmussen. 2010. Novel broadband multilayer microstrip directional couplers. *2010 IEEE AP-S*, pp. 1–4, July 11–17.
9. H. Park, and Y. Lee. 2010. Asymmetric coupled line directional coupler loaded with shunt inductors for directivity enhancement. *Electronic Letters*, Vol. 46, No. 6, pp. 425–426, March 18.
10. S. Lee, and Y. Lee. 2009. An inductor-loaded microstrip directional coupler for directivity enhancement. *IEEE Microwave and Wireless Components Letters*, Vol. 19, No. 6, pp. 362–364, June.
11. S. Lee, and Y. Lee. 2010. A design method for microstrip directional couplers loaded with shunt inductors for directivity enhancement. *IEEE Transactions on Microwave Theory and Techniques*, Vol. 58, No. 4, pp. 994–1002, April.
12. R. Phromloungsri, V. Chamnanphrai, and M. Chongcheawchamnan. 2006. Design high-directivity parallel-coupled lines using quadrupled inductive-compensated technique. *Proceedings of Asia-Pacific Microwave Conference*, pp. 1380–1383.

13. D. Kajfez. 1978. Raise coupler directivity with lumped compensation. *Microwaves*, Vol. 27, pp. 64–70, March.
14. S.-M. Wang, C.-H. Chen, and C.-Y. Chang. 2003. A study of meandered microstrip coupler with high directivity. *Microwave Symposium Digest, 2003 IEEE MTT-S International*, Vol. 1, pp. 63–66.
15. K. Sachse. 1988. Analysis of asymmetrical inhomogeneous coupled line directional coupler. *Microwave Conference*, pp. 985–990.
16. S.L. March. 1982. Phase velocity compensation in parallel-coupled microstrip. *1982 MTT-S Int'l Microwave Symposium Digest*, pp. 410–412, May.
17. A. Eroglu. 2007. Design of three-line multi-layer microstrip directional couplers at HF for high power applications. *Microwave Journal*, Vol. 50, pp. 150–154, April.
18. J. Huang, Q. Hao, J. She, and Z. Feng. 2005. A six-port coupler with high directivity for VSWR measurement. *APMC 2005 Proceedings*.
19. C.A. Hoer. 1972. The six-port coupler: A new approach to measuring voltage, current, power, impedance, and phase. *IEEE Transactions on Instrumentation and Measurement*, Vol. IM-21, No. 4, pp. 466–470.
20. A.L. Cullen, S.K. Judah, and F. Nikravesh. 1980. Impedance measurement using a six-port directional coupler. *IEE Proceedings, Microwaves, Optics and Antennas*, Vol. 127, No. 2, Pt. H, pp. 92–98.
21. T.S. Cooper, G. Baldwin, and R. Farrell. 2006. Six-port precision directional coupler. *Electronics Letters*, Vol. 42, No. 21, pp. 1232–1233.
22. G.F. Engen, and C.A. Hoer. 1972. Application of an arbitrary six-port junction to power measurement problems. *IEEE Transactions on Instrumentation and Measurement*, Vol. 21, No. 4, pp. 470–474.
23. G.F. Engen. 1977. The six-port reflectometer; an alternative network analyser. *IEEE Transactions on Microwave Theory and Techniques*, Vol. MTT-25, pp. 1075–1080.
24. J.J. Yao, and S.P. Yeo. 2008. Six-port reflectometer based on modified hybrid couplers. *IEEE Transactions on Microwave Theory and Techniques*, Vol. 56, No. 2, pp. 493–498.
25. J.J. Yao, S.P. Yeo, and M.E. Bialkowski. 2009. Modifying branch-line coupler design to enhance six-port reflectometer performance. *IEEE, International Microwave Symposium*, pp. 1669–1672.
26. E.J. Griffin. 1982. Six-port reflectometer circuit comprising three directional couplers. *Electronics Letters*, Vol. 18, No. 12, pp. 491–493.
27. U. Stumper. 1990. Simple millimeter-wave six-port reflectometers. *IEEE Conference on Precision Electromagnetic Measurements*, pp. 51–52.
28. C. Akyel, and F.M. Ghannouchi. 1994. A new design for high-power six-port reflectometers using hybrid stripline/waveguide technology. *IEEE Transactions on Instrumentation and Measurement*, Vol. 43, No. 2, pp. 316–321.
29. A.S.S. Mohra. 2001. Six-port reflectometer realization using two microstrip three-section couplers. *18th National Radio Science Conference*, March, Egypt.
30. E.R.B. Hansson, and G.P. Riblet. 1983. An ideal six-port network consisting of a matched reciprocal lossless five-port and a perfect directional coupler. *IEEE Transactions on Microwave Theory and Techniques*, Vol. 31, pp. 284–288.
31. J.A. Dobrowolski. 1982. Improved six-port circuit for complex reflection coefficient measurements. *Electronics Letters*, Vol. 18, No. 17, pp. 748–750.
32. F. Alessandri. 2003. A new multiple-tuned six-port Riblet-type directional coupler in rectangular waveguide. *IEEE Transactions on Microwave Theory and Techniques*, Vol. MTT-51, No. 5, pp. 1441–1448.
33. F.M. Ghannouchi, R.G. Bosisio, and Y. Demers. 1989. Load-pull characterization method using six-port techniques. *6th IEEE Instrumentation and Measurement Technology Conference*, IMTC-89, pp. 536–539.

34. D. Lê, P. Poiré, and F.M. Ghannouchi. 1998. Six-port-based active source-pull measurement technique. *Measurement Science and Technology*, Vol. 9, No. 8, pp. 1336–1342.
35. J. Carr. 1999. Directional couplers. *RF Design*, pp. 676–677, August.
36. D. Kajfez. 1999. Scattering matrix of a directional coupler with ideal transformers. *IEEE Microwaves, Antennas and Propagation*, Vol. 146, No. 4, pp. 295–297, August.
37. Micrometals. Available at http://www.micrometals.com/rfparts/rftoroid4.html.
38. A. Eroglu. 2012. Complete modeling of toroidal inductors for high power RF applications. *IEEE Transactions on Magnetics*, Vol. 48, No. 11, pp. 4526–4529, November.
39. L. Shihe, and R.G. Bosisio. 1983. The measurement of complex reflection coefficient by means of a five-port reflectometer. *IEEE Transactions on Microwave Theory and Techniques*, Vol. 31, No. 4, pp. 321–326, April.
40. L.C. Oldfield, J.P. Ide, and E.J. Griffin. 1985. A multistate reflectometer. *IEEE Transactions on Instrumentation and Measurement*, Vol. 34, No. 2, pp. 198–201, June.
41. S.P. Yeo, and S.T. Tay. 2000. Improved design for multistate reflectometer (with two power detectors) for measuring reflection coefficients of microwave devices. *IEEE Transactions on Instrumentation and Measurement*, Vol. 49, No. 1, pp. 61–65, February.
42. Q. Sui, K. Wang, and L. Li. 2010. The measurement of complex reflection coefficient by means of an arbitrary four-port network and a variable attenuator. *2010 International Symposium on Signals Systems and Electronics (ISSSE)*, Vol. 2, pp. 1–3, September 17–20.
43. L. Qiao, and S.P. Yeo. 1995. Improved implementation of four-standard procedure for calibrating six-port reflectometers. *IEEE Transactions on Instrumentation and Measurement*, Vol. 44, No. 3, pp. 632–636, June.

7 Filter Design for RF Power Amplifiers

7.1 INTRODUCTION

In radio frequency (RF) power amplifiers, the main purpose of filters is to eliminate spurious and harmonic contents from the frequency of interest. Depending on the frequency of operation and application, a filter can be implemented using lumped elements or distributed elements, or it can comprise a mix of lumped and distributed elements. Filters that are implemented with distributed elements are based on lumped-element filter prototype models. The lumped-element low-pass filter (LPF) prototypes are the basis of all the filter prototypes. Most of the filters are implemented using four conventional filter types: low pass, high pass, bandpass, and bandstop filters (BSFs). LPFs provide maximum power transfer at frequencies below cut-off or corner frequency, f_c. The frequency is stopped above f_c for LPFs. High-pass filters (HPFs) behave opposite to LPFs and pass signals at frequencies above f_c. Bandpass filters (BPFs) pass signals at frequencies between lower and upper cut-off frequencies and stop everything else out of this frequency band. BSFs function opposite to BPFs and stop signals between lower and upper cut-off frequencies and pass everything else. The commonly used basic filter types are shown with their ideal frequency characteristics in Figure 7.1. There are also other filter types that can be designed and implemented for specific applications. Filters for amplifiers are interfaced as off-line filters or in-line filters. When they are implemented as off-line filters, they present a matched impedance to the amplifier signal at the operational frequency and present very high impedance at all other frequencies.

If the filters are implemented in-line filters, then they need to present a match impedance at the operational frequency and block signals at any other frequency depending on the application.

The common method to analyze filters is to treat them as lossless, linear two-port networks. The conventional filter design procedure for LPFs, HPFs, BPFs, or BSFs begins from LPF prototype and then involves impedance and frequency scaling, and filters transformation to HPFs, BPFs or BSFs to obtain final component values at the frequency of operation [1]. The design is then simulated and compared with specifications. The final step in the design of filters involves implementation and measurement of the filter response.

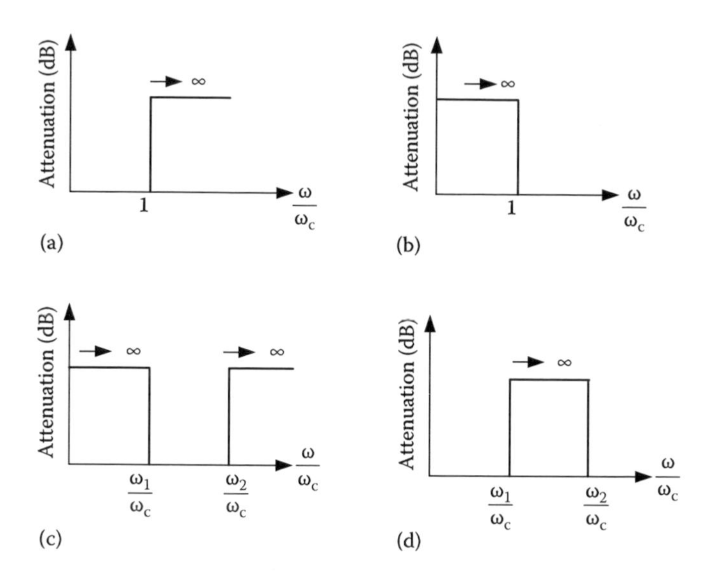

FIGURE 7.1 Ideal filter characteristics: (a) low-pass filter, (b) high-pass filter, (c) bandpass filter, and (d) bandstop filter.

The attenuation profiles of the LPF can be binomial (Butterworth), Chebyshev, or elliptic (Cauer), as shown in Figure 7.2. Binomial filters provide monotonic attenuation profile and need more components to achieve steep attenuation transition from passband to stopband, whereas Chebyshev filters have steeper slope and equal amplitude ripples in the passband.

Elliptic filters have steeper transition from passband to stopband similar to Chebyshev filters and exhibit equal amplitude ripples in the passband and stopband. RF/microwave filters and filter components can be represented using a two-port network shown in Figure 7.3. The network analysis can be conducted using *ABCD* parameters for each filter. The filter elements can be considered as cascaded components, and hence, the overall *ABCD* parameter of the network is just a simple matrix multiplication of the *ABCD* parameter for each element. The characteristics of the filter in practice are determined via insertion loss, S_{21}, and return loss, S_{11}. *ABCD* parameters can be converted to scattering parameters, and insertion loss and return loss for the filter can be determined. The detailed analysis procedure for filters has been given in Ref. [1].

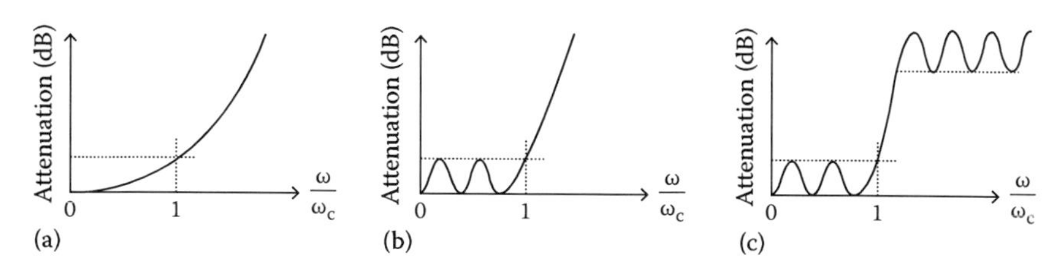

FIGURE 7.2 Attenuation profiles of LPF: (a) binominal filter, (b) Chebyshev filter, and (c) elliptic filter.

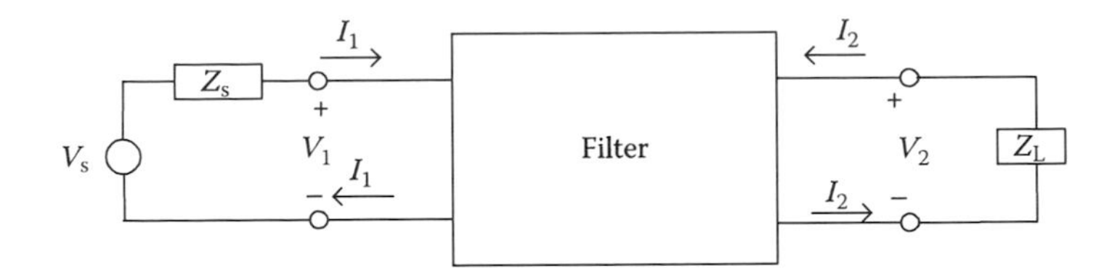

FIGURE 7.3 Two-port network representation.

7.2 FILTER DESIGN BY INSERTION LOSS METHOD

There are two main filter synthesis methods in the design of RF filters: image parameter method and insertion loss method. Although the design procedure with the image parameter method is straightforward, it is not possible to realize an arbitrary frequency response with the use of that method. The insertion loss method will be applied to design and implement the filters in this section. Filter design with the insertion loss method begins with complete filter specifications. Filter specifications are used to identify the prototype filter values, and prototype filter circuit is synthesized. Scaling and transformation of the prototype values are performed to have the final filter component values. The prototype element values of the LPF circuit are obtained using power loss ratio.

7.2.1 Low Pass Filters

Consider the two-element LPF prototype shown in Figure 7.4.

In the insertion loss method, the filter response is defined by the power loss ratio, P_{LR}, which is given by

$$P_{LR} = \frac{P_{incident}}{P_{load}} = \frac{1}{1-|S_{11}|^2} \tag{7.1}$$

where

$$S_{11} = \frac{Z_{in}-1}{Z_{in}+1} \tag{7.2}$$

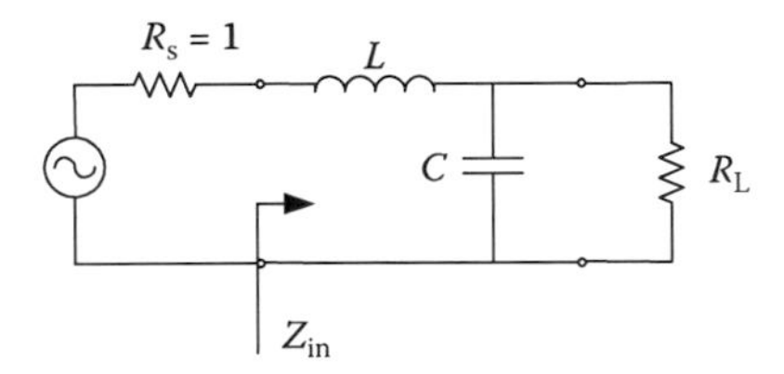

FIGURE 7.4 Two-element, low-pass prototype circuit.

$P_{incident}$ refers to the available power from the source, and P_{load} represents the power delivered to the load. As explained before, the attenuation characteristics of the filter can fall into one of these three categories: binomial (Butterworth), Chebyshev, or elliptic. The binomial or Butterworth response provides the flattest passband response for a given filter and is defined by

$$P_{LR} = 1 + k^2 \left(\frac{\omega}{\omega_c}\right)^{2N} \tag{7.3}$$

where $k = 1$, and N is the order of the filter. Chebyshev filters provide steeper transition from passband to stopband while they have equal ripples in the passband, and their attenuation characteristics are defined by

$$P_{LR} = 1 + k^2 T_N^2 \left(\frac{\omega}{\omega_c}\right) \tag{7.4}$$

where T_N is the Chebyshev polynomial. The prototype values of the filter circuit, L and C, shown in Figure 7.4 are found by solving Equations 7.1 through 7.4.

7.2.1.1 Binomial Filter Response

Binomial filter response can be explained using Figure 7.4 for a two-element LPF network. Source impedance and cut-off frequency for the circuit in Figure 7.4 are assumed to be 1 Ω and 1 rad/s, respectively. In the LPF network, $N = 2$, and power loss becomes

$$P_{LR} = 1 + \omega^4 \tag{7.5}$$

The input impedance is found as

$$Z_{in} = \frac{j\omega L(1+\omega^2 R^2 C^2) + R(1 - j\omega RC)}{(1+\omega^2 R^2 C^2)} \tag{7.6}$$

which leads to

$$S_{11} = \frac{\left[\dfrac{j\omega L(1+\omega^2 R^2 C^2) + R(1 - j\omega RC)}{(1+\omega^2 R^2 C^2)}\right] - 1}{\left[\dfrac{j\omega L(1+\omega^2 R^2 C^2) + R(1 - j\omega RC)}{(1+\omega^2 R^2 C^2)}\right] + 1} \tag{7.7}$$

The L and C values satisfying the equation are found to be

$$L = C = 1.4142 \tag{7.8}$$

TABLE 7.1
Component Values for Binomial LPF Response with $g_0 = 1$, $\omega_c = 1$

N	g_1	g_2	g_3	g_4	g_5	g_6	g_7	g_8	g_9
1	2.0000	1.0000							
2	1.4142	1.4142	1.0000						
3	1.0000	2.0000	1.0000	1.0000					
4	0.7654	1.8478	1.8478	0.7654	1.0000				
5	0.6180	1.6180	2.0000	1.6180	0.6180	1.0000			
6	0.5176	1.4142	1.9318	1.9318	1.4142	0.5176	1.0000		
7	0.4450	1.2470	1.8019	2.0000	1.8019	1.2470	0.4450	1.0000	
8	0.3902	1.1111	1.6629	1.9615	1.9615	1.6629	1.1111	0.3902	1.0000

The same procedure is applied for LPF circuit with any number. The values obtained using this method are tabulated and given in Table 7.1 [1]. In essence, the two-element low-pass proto circuit analyzed is called a ladder network. Although the LPF circuit in Figure 7.4 begins with a series inductor, the analysis applies when it is switched with a shunt capacitor. In addition, the number of elements can be increased to N, and the ladder network can be generalized, as shown in Figure 7.5, where the values shown in Table 7.1 can be used for binomial response. In the low-pass prototype circuits in Figure 7.5, g_0 represents the source resistance or conductance, whereas g_{N+1} represents the load resistance or conductance. g_N is an inductor for a series-connected component and a capacitor for a parallel-connected component. Attenuation curves for the low-pass prototype filters can be found from

$$\text{Attenuation (dB)} = 10 \log(P_{LR}) \tag{7.9}$$

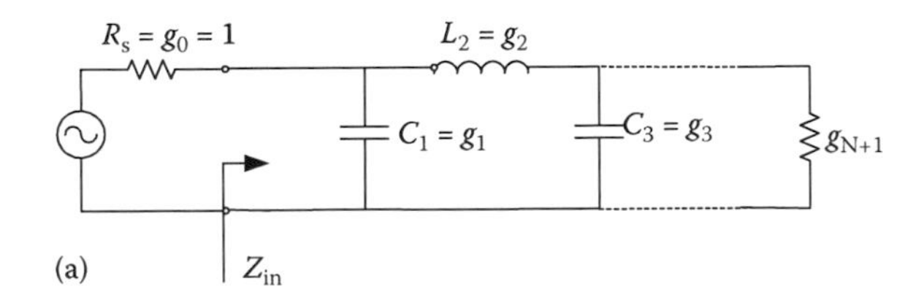

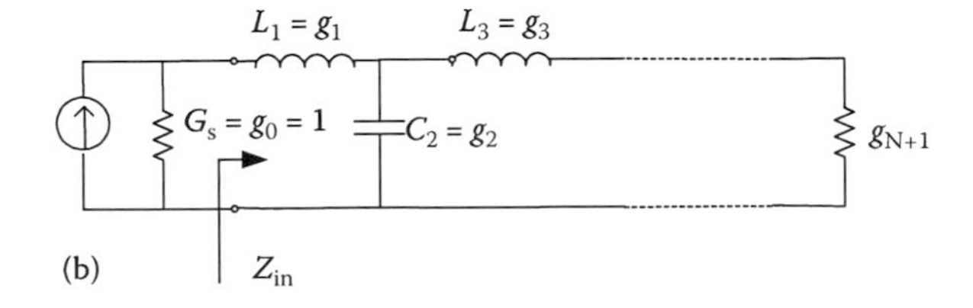

FIGURE 7.5 Low-pass prototype ladder networks: (a) first element shunt C; (b) first element series L.

Attenuation curves for binomial response using Equation 7.9 are obtained by MATLAB® and are given in Figure 7.6 [1]. Once the design filter specifications are given, the number of required elements to have the required attenuation is determined from attenuation curves. In the second step, the table is used to determine the normalized component values for the required number of elements found in the previous stage. Then, scaling and transformation step are performed, and the final filter component values are obtained.

Since the original normalized component values of the LPF are designated as L, C, and R_L, the final scaled component values of the filter with source impedance R_0 are found from

$$R_s' = R_0 \tag{7.10}$$

$$R_L' = R_0 R_L \tag{7.11}$$

$$L' = R_0 \frac{L_n}{\omega_c} \tag{7.12}$$

$$C' = \frac{C_n}{R_0 \omega_c} \tag{7.13}$$

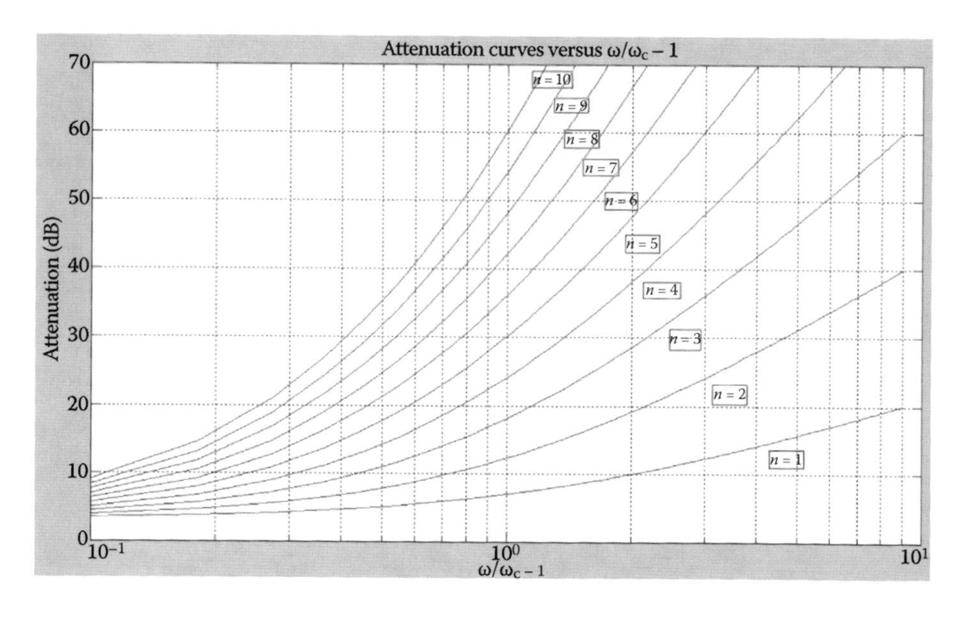

FIGURE 7.6 Attenuation curves for binomial filter response for low-pass prototype circuits.

7.2.1.2 Chebyshev Filter Response

The Chebyshev filter response can be obtained similarly using the two-element, low-pass prototype circuit given in Figure 7.4. The power loss takes the following form when $N = 2$:

$$P_{LR} = 1 + k^2 T_2^2\left(\frac{\omega}{\omega_c}\right) \quad (7.14)$$

where

$$T_2\left(\frac{\omega}{\omega_c}\right) = 2\left(\frac{\omega}{\omega_c}\right)^2 - 1 \quad (7.15)$$

Substituting Equation 7.15 into Equation 7.14 gives

$$P_{LR} = 1 + k^2(4\omega^4 - 4\omega^2 + 1) \quad (7.16)$$

The input impedance and S_{11} are given by Equations 7.6 and 7.7. When Equation 7.7 is substituted into Equation 7.1 with Equation 7.16, the L and C values satisfying the equation are found. Chebyshev polynomials up to the seventh order are given in Table 7.2 [1]. The polynomials given in Table 7.2 can be used to obtain design tables giving component values for various ripple values, as shown in Figure 7.7. Chebyshev polynomials are defined by a three-term recursion where

$$T_0(x) = 1, \quad T_1(x) = x, \quad T_{n+1}(x) = 2xT_n(x) - T_{n-1}(x), \quad n = 1, 2\ldots \quad (7.17)$$

where $x = \omega/\omega_c$. The attenuation curves for Chebyshev LPF response are obtained from

$$\text{Attenuation (dB)} = 10\log\left(1 + \varepsilon^2 T_N^2\left(\frac{\omega}{\omega_c}\right)'\right) \quad (7.18)$$

where

$$\left(\frac{\omega}{\omega_c}\right)' = \left(\frac{\omega}{\omega_c}\right)\cosh(B) \quad (7.19)$$

$$B = \frac{1}{N}\cosh^{-1}\left(\frac{1}{\varepsilon}\right) \quad (7.20)$$

$$\varepsilon = \sqrt{10^{\frac{\text{ripple(dB)}}{10}} - 1} \quad (7.21)$$

TABLE 7.2
Chebyshev Polynomials up to Seventh Order

Order of Polynomial, N	$T_N\left(\frac{\omega}{\omega_c}\right)$
1	$\frac{\omega}{\omega_c}$
2	$2\left(\frac{\omega}{\omega_c}\right)^2 - 1$
3	$4\left(\frac{\omega}{\omega_c}\right)^3 - 3\left(\frac{\omega}{\omega_c}\right)$
4	$8\left(\frac{\omega}{\omega_c}\right)^4 - 8\left(\frac{\omega}{\omega_c}\right)^2 + 1$
5	$16\left(\frac{\omega}{\omega_c}\right)^5 - 20\left(\frac{\omega}{\omega_c}\right)^3 + 5\left(\frac{\omega}{\omega_c}\right)$
6	$32\left(\frac{\omega}{\omega_c}\right)^6 - 48\left(\frac{\omega}{\omega_c}\right)^4 + 18\left(\frac{\omega}{\omega_c}\right)^2 - 1$
7	$64\left(\frac{\omega}{\omega_c}\right)^7 - 112\left(\frac{\omega}{\omega_c}\right)^5 + 58\left(\frac{\omega}{\omega_c}\right)^3 - 7\left(\frac{\omega}{\omega_c}\right)$

The attenuation curves are obtained and given for several ripple values using MATLAB in Ref. [1] with examples.

Example—F/2 LPF Design for RF Power Amplifiers

Design an F/2 filter for an RF power amplifier that is operating at 13.56 MHz. The filter should have no impact during the normal operation of the amplifier. It should have at least 20-dB attenuation at F/2 frequency. The passband ripple should not exceed 0.1-dB ripple. It is given that the amplifier is presenting 30-Ω impedance to load line.

Solution

In RF power amplifier applications, signals having frequency of F/2 may become an important problem that affects the signal purity and the amount of power delivered to the load. This problem can be resolved by eliminating signals using LPFs commonly called the F/2 filter. The F/2 filter is connected off-line to the load line

Ripple = 0.01 dB								
N	g_1	g_2	g_3	g_4	g_5	g_6	g_7	g_8
1	0.096	1						
2	0.4488	0.4077	1.1007					
3	0.6291	0.9702	0.6291	1				
4	0.7128	1.2003	1.3212	0.6476	1.1007			
5	0.7563	1.3049	1.5773	1.3049	0.7563	1		
6	0.7813	1.36	1.6896	1.535	1.497	0.7098	1.1007	
7	0.7969	1.3924	1.7481	1.6331	1.7481	1.3924	0.7969	1
Ripple = 0.1 dB								
N	g_1	g_2	g_3	g_4	g_5	g_6	g_7	g_8
1	0.3052	1						
2	0.843	0.622	1.3554					
3	1.0315	1.1474	1.0315	1				
4	1.1088	1.3061	1.7703	0.818	1.3554			
5	1.1468	1.3712	1.975	1.3712	1.1468	1		
6	1.1681	1.4039	2.0562	1.517	1.9029	0.8618	1.3554	
7	1.1811	1.4228	2.0966	1.5733	2.0966	1.4228	1.1811	1
Ripple = 0.5 dB								
N	g_1	g_2	g_3	g_4	g_5	g_6	g_7	g_8
1	0.6986	1						
2	1.4029	0.7071	1.9841					
3	1.5963	1.0967	1.5963	1				
4	1.6703	1.1926	2.3661	0.8419	1.9841			
5	1.7058	1.2296	2.5408	1.2296	1.7058	1		
6	1.7254	1.2479	2.6064	1.3137	2.4758	0.8696	1.9841	
7	1.7372	1.2583	2.6381	1.3444	2.6381	1.2583	1.77372	1
Ripple = 1 dB								
N	g_1	g_2	g_3	g_4	g_5	g_6	g_7	g_8
1	1.0177	1						
2	1.8219	0.685	2.6599					
3	2.0236	0.9941	2.0236	1				
4	2.0991	1.0644	2.8311	0.7892	2.6599			
5	2.1349	1.0911	3.0009	1.0911	2.1349	1		
6	2.1546	1.1041	3.0634	1.1518	2.9367	0.8101	2.6599	
7	2.1664	1.1116	3.0934	1.1736	3.0934	1.1116	2.1664	1
Ripple = 3 dB								
N	g_1	g_2	g_3	g_4	g_5	g_6	g_7	g_8
1	1.9953	1						
2	3.1013	0.5339	5.8095					
3	3.3487	0.7117	3.3487	1				
4	3.4389	0.7483	4.3471	0.592	5.8095			
5	3.4817	0.7618	4.5381	0.7618	3.4817	1		
6	3.5045	0.7685	4.6061	0.7929	4.4641	0.6033	5.8095	
7	3.5182	0.7723	4.6386	0.8039	4.6386	0.7723	3.5182	1

FIGURE 7.7 Component values for Chebyshev LPF response with $g_0 = 1$, $\omega_c = 1$, $N = 1$ to 7.

of an amplifier and presents high impedance at the center frequency but matched impedance at F/2. The analysis begins with identifying F/2 frequency as

$$\frac{F}{2} = 6.78\,[\text{MHz}]$$

The cut-off frequency of the filter is selected to be 25%–35% higher than F/2 as a rule of thumb. The attenuation at the cut-off frequency is expected to be 3 dB, as shown below:

$$\text{Attenuation} = 3\text{ dB @ } f_c = 9\ [\text{MHz}]$$

Now, the steps that were used before can be applied to design the filter. Since the ripple requirement in the passband is mentioned, the Chebyshev filter will be used for design and implementation.

Step 1. Use Figure 6.25 to determine the required number of elements to get a minimum 20-dB attenuation at $f = 13.56$ MHz.

$$\left|\frac{\omega}{\omega_c}\right| - 1 = \frac{13.56}{9} - 1 = 0.5 \rightarrow N = 5$$

Step 2. Use Figure 7.7 to determine the normalized LPF component values as

Ripple = 0.01 dB						
N	g_1	g_2	g_3	g_4	g_5	g_6
5	1.1468	1.3712	1.975	1.3712	1.1468	1

The component values of the filter shown in Figure 7.8 are obtained from the table as

$$L_1 = L_5 = 1.1468, \quad L_3 = 1.975, \quad C_2 = C_4 = 1.3712$$

Step 3. Apply impedance and frequency scaling:

$$R_S' = R_0 = 30\,[\Omega]$$

$$R_L' = R_0 R_L = 30(1) = 30\,[\Omega]$$

$$L_1' = L_5' = R_0 \frac{L_n}{\omega_c} = 30 \frac{1.1468}{(2\pi \times 8 \times 10^6)} = 684.44\,[\text{nH}]$$

$$C_2' = C_4' = \frac{C_n}{R_0 \omega_c} = \frac{1.3712}{30(2\pi \times 8 \times 10^6)} = 909.3\ [\text{pF}]$$

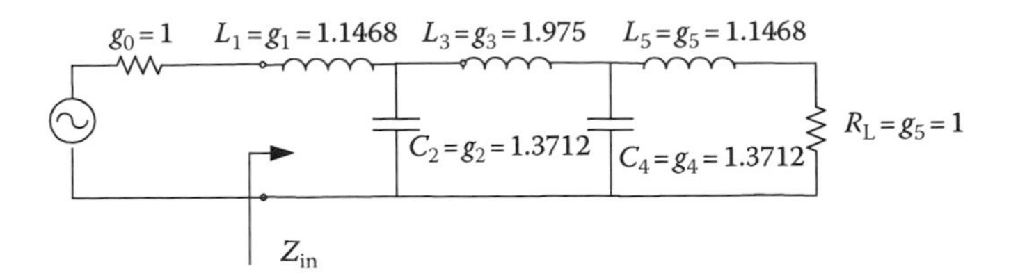

FIGURE 7.8 Fifth-order normalized LPF for Chebyshev response.

$$L_3' = R_0 \frac{L_n}{\omega_c} = 30 \frac{1.975}{(2\pi \times 8 \times 10^6)} = 1178.7 \text{ [nH]}$$

The final LPF circuit having Chebyshev filter response is shown in Figure 7.9. The final circuit shown in Figure 7.9 is analyzed using network parameters, and insertion loss is obtained using *ABCD* parameters for the cascaded components as previously discussed:

$$\begin{bmatrix} A & B \\ C & D \end{bmatrix} = \begin{bmatrix} 1 & Z_S \\ 0 & 1 \end{bmatrix} \begin{bmatrix} 1 & j\omega L_1' \\ 0 & 1 \end{bmatrix} \begin{bmatrix} 1 & 0 \\ j\omega C_2' & 1 \end{bmatrix} \begin{bmatrix} 1 & j\omega L_3' \\ 0 & 1 \end{bmatrix} \begin{bmatrix} 1 & 0 \\ j\omega C_4' & 1 \end{bmatrix} \begin{bmatrix} 1 & j\omega L_5' \\ 0 & 1 \end{bmatrix} \begin{bmatrix} 1 & 0 \\ \frac{1}{Z_L} & 1 \end{bmatrix} \tag{7.22}$$

The insertion loss in the passband and stopband is obtained using MATLAB and is shown in Figures 7.10 and 7.11.

The passband ripple is less than 0.1 dB, and the cut-off frequency is around 9 MHz, as shown in Figure 7.10. In addition, we have more than 25-dB attenuation at 13.56 MHz, as illustrated in Figure 7.11. The circuit is simulated with Ansoft Designer for accuracy using the circuit shown in Figure 7.12. The passband ripple, attenuation at cut-off frequency, and operational frequency are given in Figure 7.13 and are in agreement with the MATLAB results obtained.

The input impedance for the filter designed is given in Figure 7.14. Based on the results on the Smith chart, the filter input impedance is (29.58 – *j*8.48) Ω at F/2 and (0.06 + *j*43.02) Ω. Hence, the filter presents a very closely matched load to the amplifier at F/2 and terminates the F/2 frequency content; moreover, it presents inductance, acts like an open load at the operational frequency, and does not have any impact on amplifier performance.

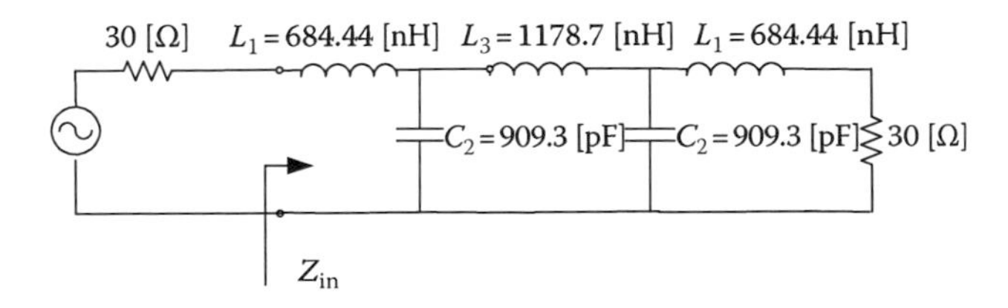

FIGURE 7.9 Final LPF with Chebyshev response.

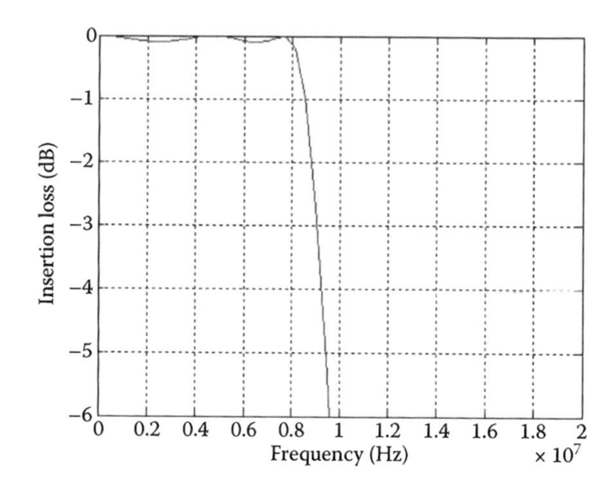

FIGURE 7.10 Passband ripple response for fifth-order LPF with Chebyshev filter response.

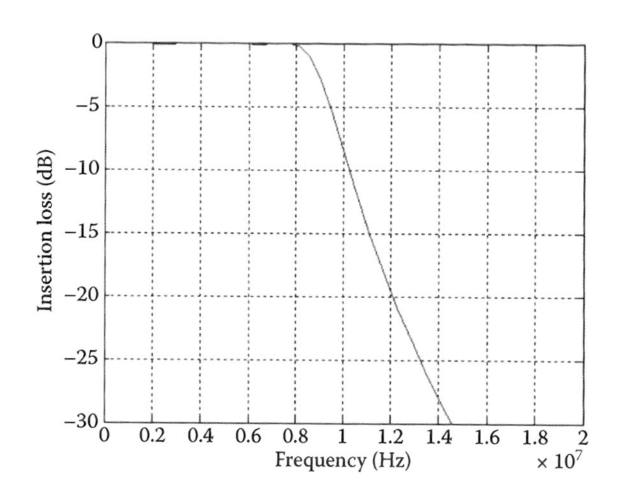

FIGURE 7.11 Attenuation response for fifth-order LPF with Chebyshev filter response.

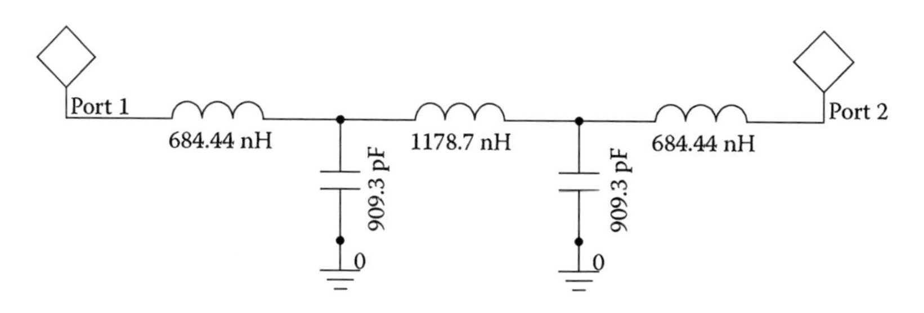

FIGURE 7.12 Simulated fifth-order LPF.

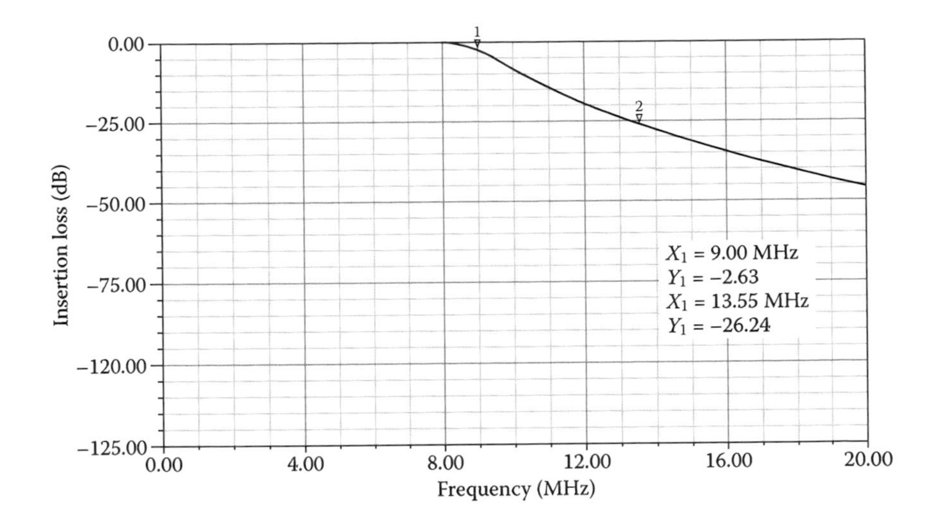

FIGURE 7.13 Simulation results for fifth-order LPF.

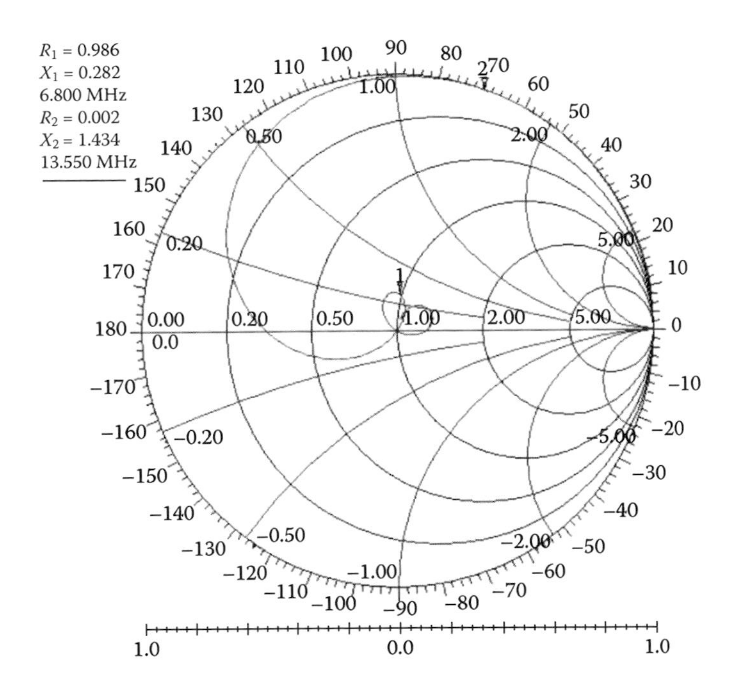

FIGURE 7.14 Input impedance of fifth-order LPF.

7.2.2 High-Pass Filters

HPFs are designed from LPF prototypes using the frequency transformation given by

$$-\frac{\omega_c}{\omega} \rightarrow \omega \tag{7.23}$$

This transformation converts an LPF to an HPF with the following frequency and impedance scaling relations for L and C:

$$L'_n = \frac{R_0}{\omega_c C_n} \tag{7.24}$$

$$C'_n = \frac{1}{\omega_c R_0 L_n} \tag{7.25}$$

The design begins with the low-pass prototype by finding L_n and C_n and then applying Equations 7.24 and 7.25. The transformation of the components from an LPF to an HPF is illustrated in Figure 7.15.

7.2.3 Bandpass Filters

BPFs are designed from LPF prototypes using the frequency transformation given by

$$\frac{\omega_0}{\omega_{c2} - \omega_{c1}}\left(\frac{\omega}{\omega_0} - \frac{\omega_0}{\omega}\right) \rightarrow \omega \tag{7.26}$$

The term $(\omega_{c2} - \omega_{c1})/\omega_0$ is called the fractional bandwidth, and ω_0 is called the resonant or center frequency, and ω_{c2} and ω_{c1} are the upper and lower cut-off frequencies, respectively. The transformation given by Equation 7.26 maps the series component of an LPF prototype circuit to a series LC circuit and the shunt component of an LPF prototype circuit to a shunt LC circuit in the BPF. The component values of the series LC circuit are calculated as

$$L'_n = \frac{R_0 L_n}{\left(\dfrac{\omega_0}{\omega_{c2} - \omega_{c1}}\right)^{-1} \omega_0} \tag{7.27}$$

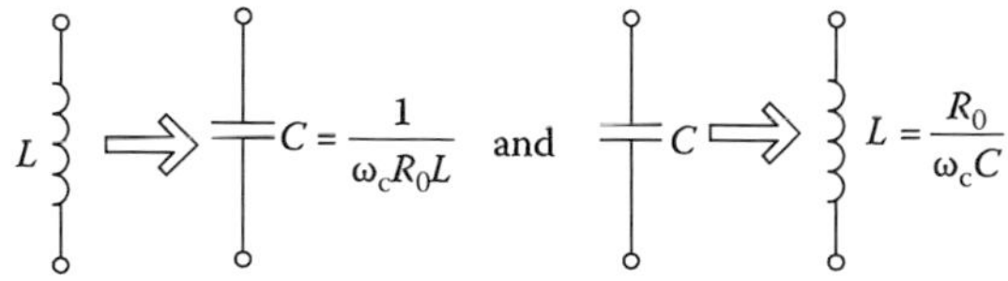

$L \Rightarrow C = \frac{1}{\omega_c R_0 L}$ and $C \Rightarrow L = \frac{R_0}{\omega_c C}$

FIGURE 7.15 LPF to HPF component transformation.

$$C'_n = \frac{\left(\dfrac{\omega_0}{\omega_{c2} - \omega_{c1}}\right)^{-1}}{\omega_0 R_0 L_n} \tag{7.28}$$

The component values of the shunt LC circuit are calculated as

$$L'_n = \frac{\left(\dfrac{\omega_0}{\omega_{c2} - \omega_{c1}}\right)^{-1} R_0}{\omega_0 C_n} \tag{7.29}$$

$$C'_n = \frac{C_n}{\left(\dfrac{\omega_0}{\omega_{c2} - \omega_{c1}}\right)^{-1} R_0 \omega_0} \tag{7.30}$$

The transformation of the components from an LPF to a BPF is illustrated in Figure 7.16.

7.2.4 Bandstop Filters

BSFs are designed from LPF prototypes using the frequency transformation given by

$$\frac{\omega_{c2} - \omega_{c1}}{\omega_0}\left(\frac{\omega}{\omega_0} - \frac{\omega_0}{\omega}\right)^{-1} \rightarrow \omega \tag{7.31}$$

This transformation maps the series component of an LPF prototype circuit to a shunt LC circuit and a shunt component of an LPF prototype circuit to a series LC circuit in the BSF. The component values of the shunt LC circuit are calculated as

$$L'_n = \frac{\left(\dfrac{\omega_0}{\omega_{c2} - \omega_{c1}}\right)^{-1} L_n R_0}{\omega_0} \tag{7.32}$$

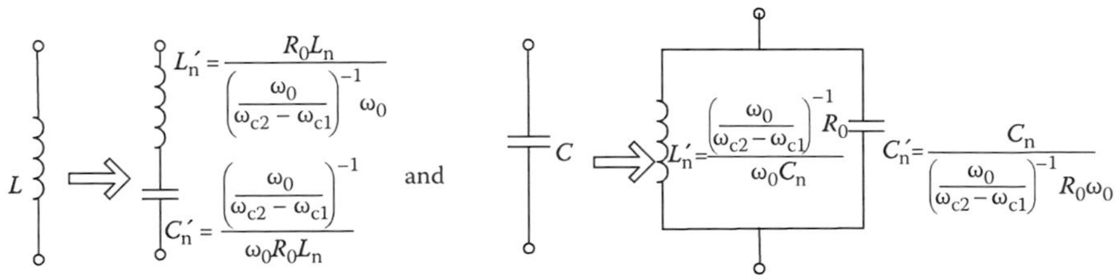

FIGURE 7.16 LPF to BPF component transformation.

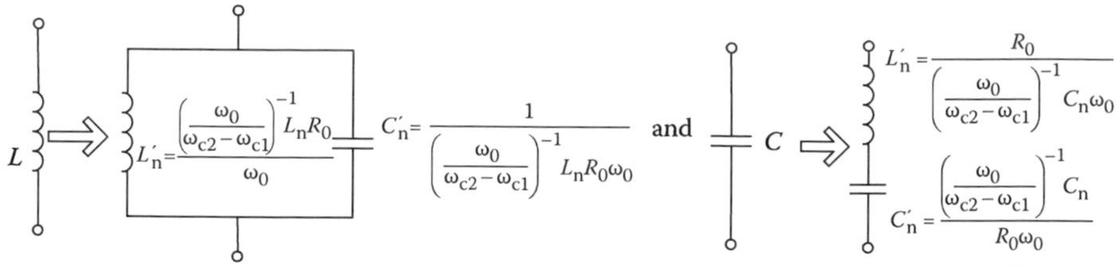

FIGURE 7.17 LPF to BSF component transformation.

$$C'_n = \frac{1}{\left(\dfrac{\omega_0}{\omega_{c2} - \omega_{c1}}\right)^{-1} L_n R_0 \omega_0} \tag{7.33}$$

The component values of the series LC circuit are calculated as

$$L'_n = \frac{R_0}{\left(\dfrac{\omega_0}{\omega_{c2} - \omega_{c1}}\right)^{-1} C_n \omega_0} \tag{7.34}$$

$$C'_n = \frac{\left(\dfrac{\omega_0}{\omega_{c2} - \omega_{c1}}\right)^{-1} C_n}{R_0 \omega_0} \tag{7.35}$$

The transformation of the components from an LPF to a BSF is illustrated in Figure 7.17.

7.3 STEPPED-IMPEDANCE LPFs

A stepped-impedance filter is made up of high- and low-impedance sections of transmission line shown in Figure 7.18. Using transmission line theory, the high- and low-impedance sections are implemented to realize LPFs.

The two-port Z-parameter matrix for a transmission line in Figure 7.18 is

$$Z = \begin{bmatrix} -jZ_0 \cot(\beta\ell) & -jZ_0 \csc(\beta\ell) \\ -jZ_0 \csc(\beta\ell) & -jZ_0 \cot(\beta\ell) \end{bmatrix} \tag{7.36}$$

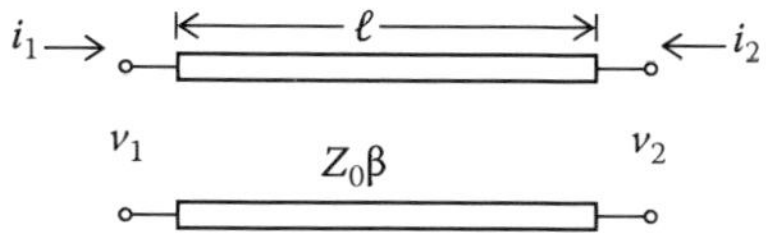

FIGURE 7.18 Transmission line model.

where

$$Z_{11} = Z_{22} = -jZ_0 \cot(\beta\ell) \quad \text{and} \quad Z_{12} = Z_{21} = -jZ_0 \csc(\beta\ell) \tag{7.37}$$

An equivalent *T*-connected network can be used to represent the two-port transmission line network in Figure 7.18. The equivalent *T*-connected network representing the transmission line is shown in Figure 7.19, where the components of the *T*-network are defined as

$$Z_A = Z_B = Z_{11} - Z_{12} = jZ_0 \tan\left(\frac{\beta l}{2}\right) \tag{7.38}$$

$$Z_c = Z_{12} = -jZ_0 \csc(\beta\ell) \tag{7.39}$$

When the electrical length, $\beta\ell$, is small, then the following approximation can be made:

$$\sin(\beta l) \approx \beta l,\ \cos(\beta l) \approx 1,\ \text{and}\ \tan(\beta l) \approx (\beta l) \tag{7.40}$$

Approximations given by Equation 7.40 lead to the following element values for the *T*-network shown in Figure 7.20:

$$Z_A = Z_B = Z_{11} - Z_{12} \approx jZ_0\left(\frac{\beta l}{2}\right) \tag{7.41}$$

$$Z_C = Z_{12} \approx \frac{Z_0}{j\beta\ell} \tag{7.42}$$

Consider the case when characteristic impedance Z_0 is very high. This impedance is denoted as Z_{High}. For the shunt component, since $\beta\ell$ is very small, the impedance will be very large. In fact, it can be considered an open circuit. This results in an approximate circuit impedance of series component, $jZ_{High}\beta\ell$, as

$$\frac{Z_{High}}{j\beta\ell} \rightarrow \infty \quad \text{when } \beta\ell \ll 1 \text{ and } Z_{High} \gg Z_0 \tag{7.43}$$

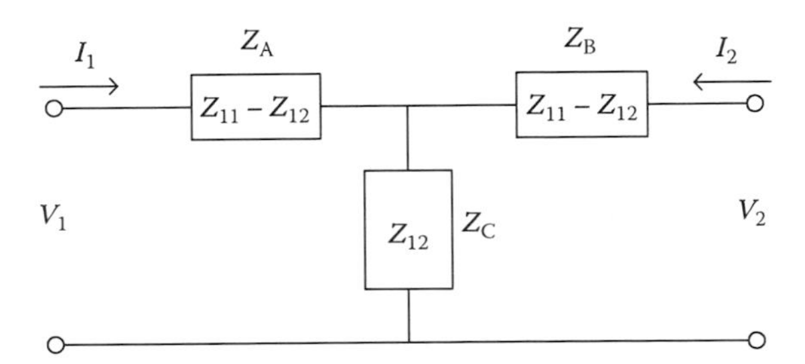

FIGURE 7.19 *T*-network equivalent circuit.

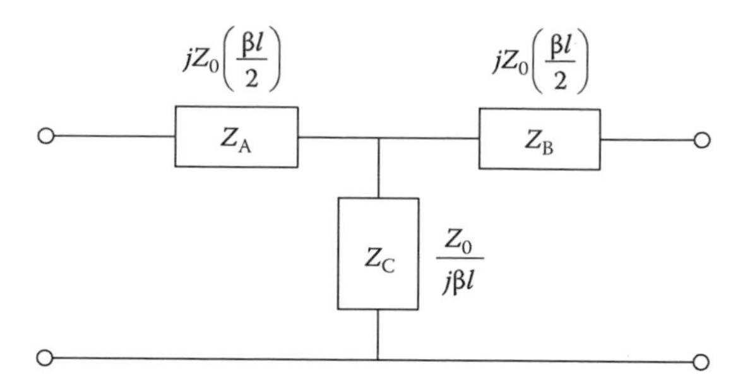

FIGURE 7.20 *T*-network representation with transmission lines.

High impedance condition transforms the *T*-network to equivalent series-connected *L*-network, as shown in Figure 7.21. Now, consider the case when the characteristic impedance is low (Z_{Low}). This time, the series components have a very low impedance and can be considered shorted. The resulting approximate circuit impedance is that of the shunt component alone or $Z_{Low}/j\beta\ell$ as

$$jZ_{Low}\left(\frac{\beta\ell}{2}\right)\to 0 \quad \text{when } \beta\ell \ll 1 \text{ and } Z_{Low} \ll Z_0 \tag{7.44}$$

As a result, low impedance condition transforms the *T*-network to an equivalent shunt-connected *C*-network, as shown in Figure 7.22. The physical length of component values for series- and shunt-connected elements is found from Equations 7.43 and 7.44. The length for an inductive element can be obtained from

$$X_L = j\omega L = jZ_{High}\beta\ell \tag{7.45}$$

So,

$$\ell_{High} = \frac{\omega L}{Z_{High}\beta} \tag{7.46}$$

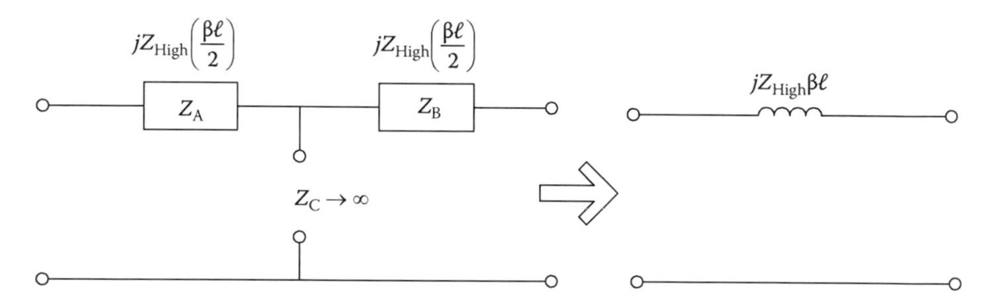

FIGURE 7.21 High-impedance transformation of *T*-network.

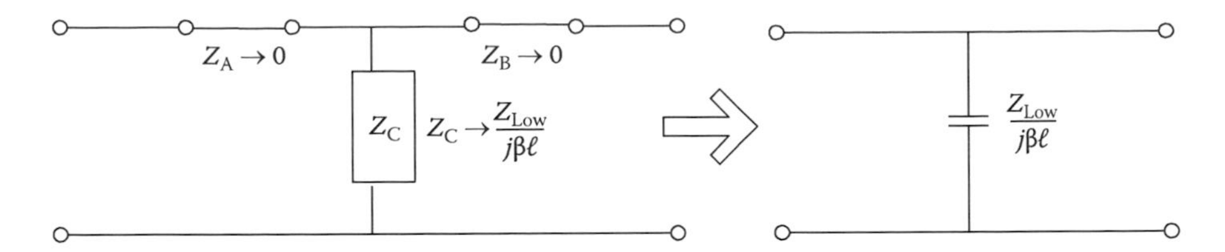

FIGURE 7.22 Low-impedance transformation of *T*-network.

The length for capacitive elements is found from

$$X_C = \frac{1}{j\omega C} = \frac{Z_{Low}}{j\beta\ell} \tag{7.47}$$

and the length is

$$\ell_{Low} = \frac{Z_{Low}\omega C}{\beta}. \tag{7.48}$$

L and *C* values are the values obtained using an LPF prototype circuit based on the filter specifications. In Equations 7.45 and 7.48, phase constant is defined as

$$\beta = \frac{\omega}{v_p} \tag{7.49}$$

where v_p is a phase velocity as defined by

$$v_p = \frac{c}{\sqrt{\varepsilon_e}} \tag{7.50}$$

ε_e is the effective permittivity constant of the microstrip line. The high and low impedance values are desired to be

$$Z_{Low} < Z_0 < Z_{High} \tag{7.51}$$

The selection of Z_{Low} and Z_{High} values carries importance for the response of the filter. The ratio of Z_{High} to Z_{Low} should be kept as large as possible to get more accurate results. We can define approximate limits for Z_{Low} and Z_{High} based on the assumption that electrical length is small if

$$\beta\ell < \frac{\pi}{4} \tag{7.52}$$

Then, the impedance limit for Z_{Low} is

$$Z_{\text{Low}} < \frac{\pi}{4\omega_c C} \tag{7.53}$$

and the impedance limit for Z_{High} is

$$Z_{\text{High}} > \frac{4\omega_c L}{\pi} \tag{7.54}$$

Once Z_{Low} and Z_{High} are defined, the width of each line can be obtained using the microstrip line equation defined by

$$\frac{W}{d} = \begin{cases} \dfrac{8e^{\text{A}}}{e^{2\text{A}} - 2} & \text{for} \quad W/d < 2 \\ \dfrac{2}{\pi}\left[B - 1 - \ln(2B-1) + \dfrac{\varepsilon_r - 1}{2\varepsilon_r}\left\{\ln(B-1) + 0.39 - \dfrac{0.61}{\varepsilon_r}\right\}\right] & \text{for} \quad W/d > 2 \end{cases} \tag{7.55}$$

where

$$A = \frac{Z_0}{60}\sqrt{\frac{\varepsilon_r + 1}{2}} + \frac{\varepsilon_r - 1}{\varepsilon_r + 1}\left(0.23 + \frac{0.11}{\varepsilon_r}\right) \tag{7.56}$$

$$B = \frac{377\pi}{2Z_0\sqrt{\varepsilon_r}} \tag{7.57}$$

7.4 STEPPED-IMPEDANCE RESONATOR BPFs

Conventional parallel-coupled BPFs suffer drastically from spurious harmonics. The stepped-impedance resonator (SIR) filters can be used to realize high-performance BPFs by suppressing the spurious harmonics to overcome this problem. One of the key features of an SIR is that its resonant frequencies can be tuned by adjusting the impedance ratios of the high-*Z* and low-*Z* sections. The symmetrical tri-section SIR used in the BPF design is shown in Figure 7.23.

In the symmetrical SIR structure, each section is desired to have the same electrical length. Then, it can be shown that the resonance occurs when θ is equal to

$$\theta = \tan^{-1}\left(\sqrt{\frac{K_1 K_2}{K_1 + K_2 + 1}}\right) \tag{7.58}$$

where

$$K_1 = \frac{-(\cos\alpha)(\cos\beta) + \sqrt{(\cos\alpha)^2(\cos\beta)^2 + 4(\sin b)^2(\cos(ab))^2}}{2(\cos(ab))^2} \quad (7.59)$$

$$K_2 = \frac{1 + K_1}{\tan^2(ab) - K_1} \quad (7.60)$$

The design parameters in Equations 7.59 and 7.60 are found from

$$a = \frac{f_{s1}}{f_o} \quad (7.61)$$

$$b = \frac{\pi}{2}\frac{f_o}{f_{s2}} \quad (7.62)$$

$$\alpha = \frac{\pi}{2}\frac{f_{s1} + f_o}{f_{s2}} \quad (7.63)$$

$$\beta = \frac{\pi}{2}\frac{f_{s1} - f_o}{f_{s2}} \quad (7.64)$$

The terminating impedance of the SIR at the input and output is desired to be Z_3 = 50 [Ω]. Once the operating frequencies, f_o, f_{s1}, f_{s2}, of the BPF and the terminating impedance, Z_3, of the SIR are identified, the line impedances, Z_1, and Z_2, are found from

$$Z_2 = \frac{Z_3}{K_1} \quad (7.65)$$

$$Z_1 = \frac{Z_2}{K_2} \quad (7.66)$$

The physical length and the width of the transmission lines in the tri-section SIR are found using microstrip line equations. The symmetrical SIR illustrated in Figure 7.23 has

$$\theta_1 = 2\theta, \quad \theta_2 = \theta, \quad \theta_3 = \theta \quad (7.67)$$

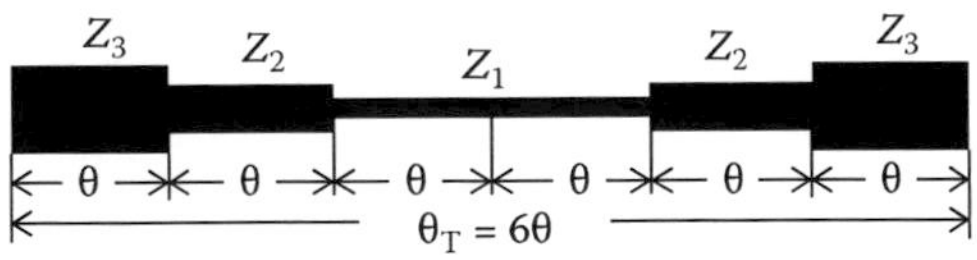

FIGURE 7.23 Tri-section SIR.

The physical length for each section in the SIR can be found from

$$l_n = \frac{\lambda_n \theta_n}{2\pi}, \quad n = 1,2,3 \tag{7.68}$$

The width of the sections in the SIR is obtained from Equations 7.55 through 7.57. The performance of BPFs with SIRs can be improved by using the configuration given in Figure 7.24. The BPF in Figure 7.24 provides triple-band filter characteristics with the coupling scheme shown in Figure 7.25. In Figure 7.24, the coupled lines' equivalent circuit is represented by two single transmission lines of electrical length θ, characteristic impedance Z_0, and admittance inverter parameter J, as shown in Figure 7.26. Inverter parameter J is an important design parameter because it is directly proportional to the coupling strength of the coupled lines.

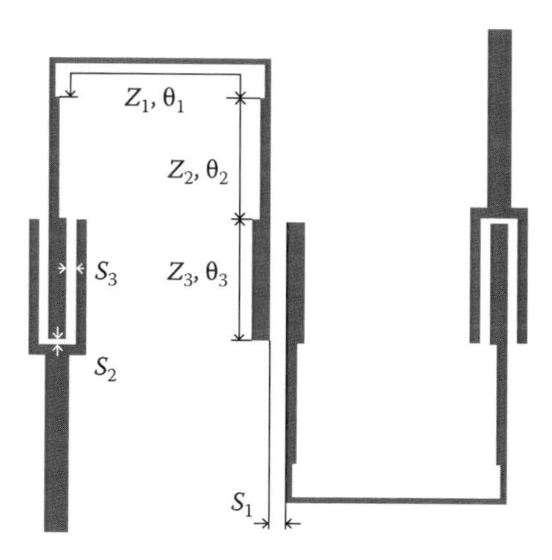

FIGURE 7.24 Triple-band BPF using SIRs.

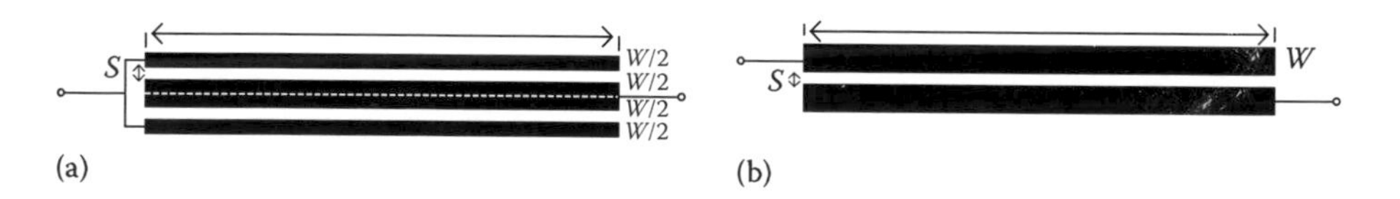

FIGURE 7.25 Coupling schemes: (a) improved coupling scheme; (b) conventional coupling scheme.

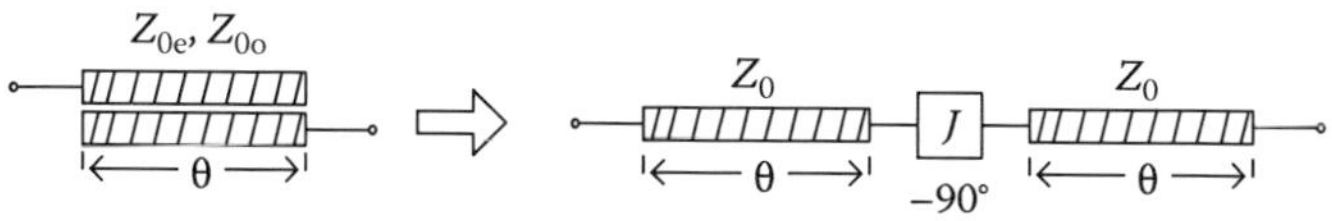

FIGURE 7.26 Equivalent circuit of parallel-coupled lines.

This parameter is found using network synthesis from the equivalent circuit and is given by

$$J_{01} = Y_0 \sqrt{\frac{2k\theta_0}{g_0 g_1}} \tag{7.69}$$

$$J_{j,j+1} = Y_0 \frac{2k\theta_0}{\sqrt{g_j g_{j+1}}} \tag{7.70}$$

$$J_{n,n+1} = Y_0 \sqrt{\frac{2k\theta_0}{g_n g_{n+1}}} \tag{7.71}$$

As the ratio J/Y between the coupled lines increases, the coupling strength also increases.

7.5 EDGE/PARALLEL-COUPLED, HALF-WAVELENGTH RESONATOR BPFs

A parallel-coupled, half-wavelength BPF is shown in Figure 7.27. For a filter of order n, there are $n + 1$ couplings between half-wavelength resonators. The first step of the design begins with finding the low-pass prototype circuit component values.

Then, the low-pass prototype filter undergoes transformations to have the desired BPF characteristics, as described in Section 7.2.3. To configure the circuit in a parallel microstrip implementation, the filter must first be represented as a series of cascaded J inverters. The equations to perform this are given by

$$\frac{J_{01}}{Y_0} = \sqrt{\frac{\pi}{2} \frac{FBW}{g_0 g_1}} \tag{7.72}$$

$$\frac{J_{j,j+1}}{Y_0} = \frac{\pi FBW}{2} \frac{1}{\sqrt{g_j g_{j+1}}} \quad \text{for } j = 1 \text{ to } n-1 \tag{7.73}$$

$$\frac{J_{n,n+1}}{Y_0} = \sqrt{\frac{\pi}{2} \frac{FBW}{g_n g_{n+1}}} \tag{7.74}$$

where $g_0 - g_n$ are the normalized impedance elements of the low-pass prototype filter, the FBW is the fractional bandwidth of the filter of 15%, and Y_0 is the characteristic admittance. These coefficients can be calculated in MATLAB. The next important step in the filter design is to calculate the even- and odd-mode impedance values for the

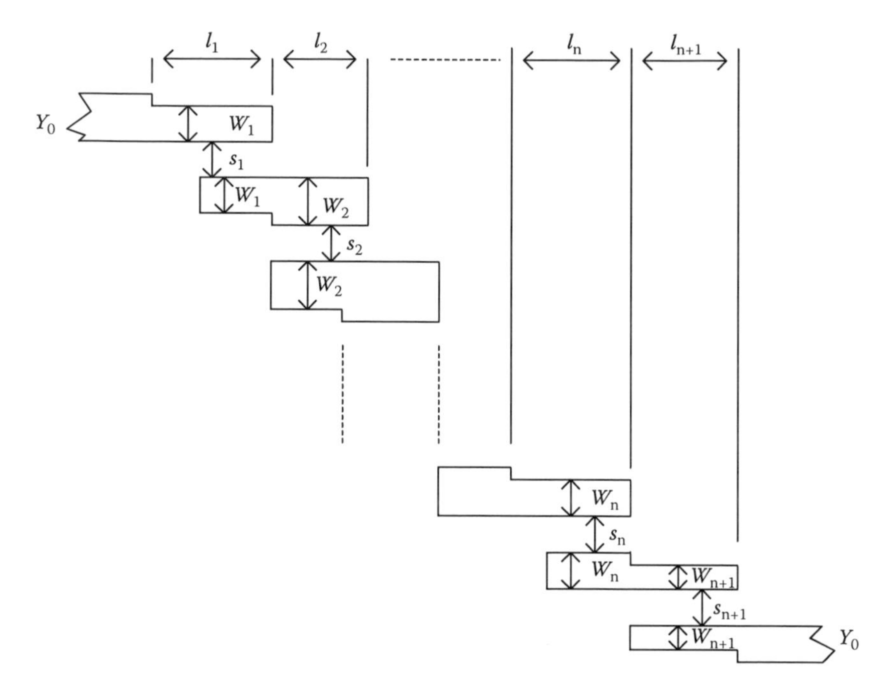

FIGURE 7.27 General setup for implementation of a microstrip edge-coupled BPF.

coupled microstrip lines. These values are also calculated in the MATLAB script from the following equations:

$$(Z_{0e})_{j,j+1} = \frac{1}{Y_0}\left[1 + \frac{J_{j,j+1}}{Y_0} + \left(\frac{J_{j,j+1}}{Y_0}\right)^2\right] \quad \text{for } j = 0 - n \tag{7.75}$$

$$(Z_{0o})_{j,j+1} = \frac{1}{Y_0}\left[1 - \frac{J_{j,j+1}}{Y_0} + \left(\frac{J_{j,j+1}}{Y_0}\right)^2\right] \quad \text{for } j = 0 - n \tag{7.76}$$

These even- and odd-mode impedances are directly used to find the dimensions of the microstrips in the BPF. After the even- and odd-mode impedances are determined, the synthesis technique introduced in Refs. [1,2] and in Chapter 6 is used to determine accurate physical dimensions. Based on the synthesis method, the spacing ratio between coupled lines is found using

$$s/h = \frac{2}{\pi}\cosh^{-1}\left[\frac{\cosh\left[\frac{\pi}{2}\left(\frac{w}{h}\right)_{se}\right] + \cosh\left[\frac{\pi}{2}\left(\frac{w}{h}\right)'_{so}\right] - 2}{\cosh\left[\frac{\pi}{2}\left(\frac{w}{h}\right)'_{se}\right] - \cosh\left[\frac{\pi}{2}\left(\frac{w}{h}\right)_{se}\right]}\right] \tag{7.77}$$

$(w/h)_{se}$ and $(w/h)_{so}$ are the shape ratios for the equivalent single case corresponding to even-mode and odd-mode geometry, respectively. $(w/h)'_{so}$ is the second term for the shape ratio. (w/h) is the shape ratio for the single microstrip line, and it is expressed as

$$\frac{w}{h}=\frac{8\sqrt{\left[\exp\left(\frac{R}{42.4}\sqrt{(\varepsilon_r+1)}\right)-1\right]\frac{7+(4/\varepsilon_r)}{11}+\frac{1+(1/\varepsilon_r)}{0.81}}}{\left[\exp\left(\frac{R}{42.4}\sqrt{\varepsilon_r+1}\right)-1\right]} \tag{7.78}$$

where

$$R=\frac{Z_{0e}}{2} \tag{7.79}$$

or

$$R=\frac{Z_{0o}}{2} \tag{7.80}$$

Z_{0se} and Z_{0so} are the characteristic impedances corresponding to single microstrip shape ratios $(w/h)_{se}$ and $(w/h)_{so}$, respectively. They are given as

$$Z_{0se}=\frac{Z_{0e}}{2} \tag{7.81}$$

$$Z_{0so}=\frac{Z_{0o}}{2} \tag{7.82}$$

and

$$(w/h)_{se}=(w/h)\Big|_{R=Z_{0se}} \tag{7.83}$$

$$(w/h)_{so}=(w/h)\Big|_{R=Z_{0so}}$$

The term $(w/h)'_{so}$ in Equation 7.77 is given as

$$\left(\frac{w}{h}\right)'_{so}=0.78\left(\frac{w}{h}\right)_{so}+0.1\left(\frac{w}{h}\right)_{se} \tag{7.84}$$

After the spacing ratio s/h for the coupled lines is found, we can proceed to find w/h for the coupled lines. The shape ratio for the coupled lines is

$$\left(\frac{w}{h}\right) = \frac{1}{\pi}\cosh^{-1}(d) - \frac{1}{2}\left(\frac{s}{h}\right) \tag{7.85}$$

where

$$d = \frac{\cosh\left[\frac{\pi}{2}\left(\frac{w}{h}\right)_{se}\right](g+1)+g-1}{2} \tag{7.86}$$

$$g = \cosh\left[\frac{\pi}{2}\left(\frac{s}{h}\right)\right] \tag{7.87}$$

The physical length of the directional coupler is obtained using

$$l = \frac{\lambda}{4} = \frac{c}{4f\sqrt{\varepsilon_{eff}}} \tag{7.88}$$

The calculation of effective permittivity constant using odd-mode and even-mode capacitances is detailed in Refs. [1,2].

Design Example

Design and simulate a fifth-order, edge-coupled, half-wavelength resonator BPF with 0.1-dB passband ripple, f_C = 10 GHz, ε_r = 10.2, FBW = 0.15, and dielectric thickness of 0.635 mm.

Solution

The first step is to design the low-pass Chebyshev prototype filter coefficients. They are determined from the design table (Figure 7.7) and are shown in Figure 7.28. The filter coefficients are

$$g_0 = 1 = g_6, \quad g_1 = 1.1468 = g_5, \quad g_2 = 1.3712 = g_4, \quad g_3 = 1.9750$$

The prototype circuit is then transformed into the equivalent BPF using the transformation circuits given in Figure 7.16. The BPF with the final component values is illustrated in Figure 7.29. The frequency response of the filter is simulated with Ansoft Designer, and the insertion and return losses are given in Figure 7.30.

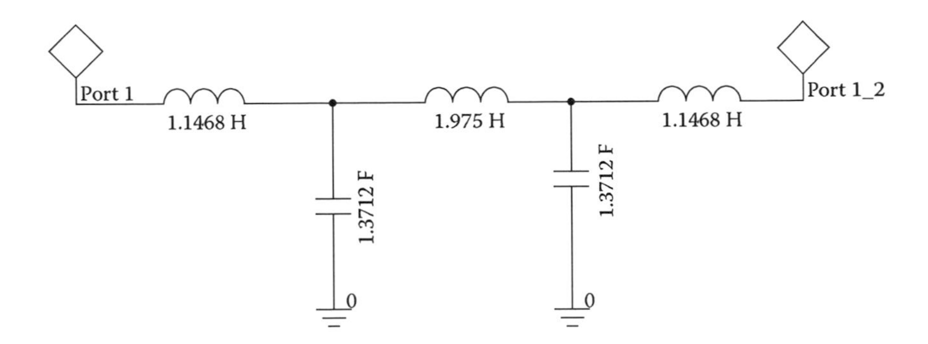

FIGURE 7.28 Low-pass prototype circuit for BPF.

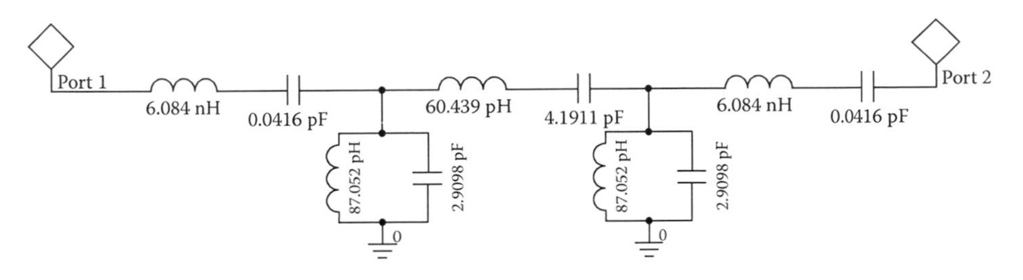

FIGURE 7.29 BPF with final lumped element component values.

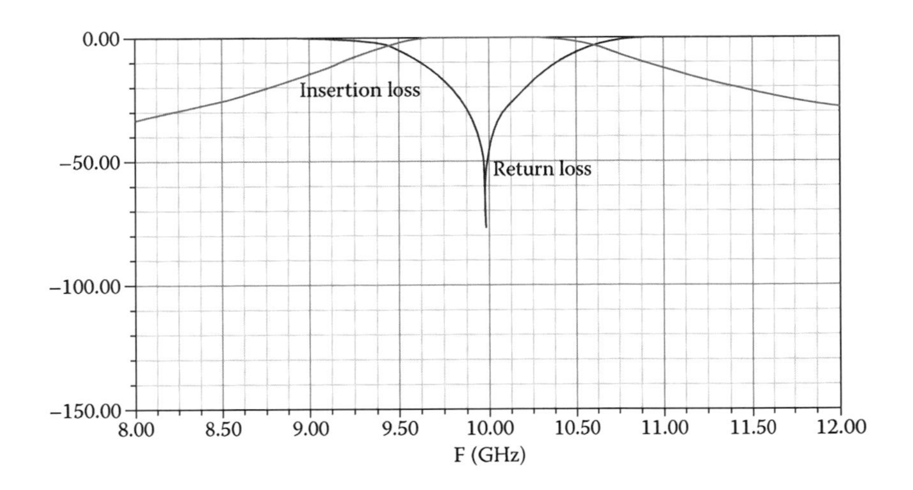

FIGURE 7.30 BPF simulation results with Ansoft Designer.

The MATLAB program has been written to obtain the physical dimensions for the edge-coupled BPF using the formulation given. In addition, the MATLAB program is used to obtain the filter response using *ABCD* two-port parameters. The calculated inverter and corresponding even-mode and odd-mode impedance values are given in Table 7.3.

TABLE 7.3
Even- and Odd-Mode Impedance Values

j	$J_{j,j+1}$	$(Z_{0e})_{j,j+1}$	$(Z_{0o})_{j,j+1}$
0	0.4533	82.9367	37.6092
1	0.1879	61.1600	42.3705
2	0.1432	58.1839	43.8661
3	0.1432	58.1839	43.8661
4	0.1879	61.1600	42.3705
5	0.4533	82.9367	37.6092

The generic MATLAB script that is used to obtain the lumped-element BPF frequency response and physical dimensions of the edge-coupled filter is given below.

```
% Edge Coupled Bandpass Filter Design Program
clc
clear;
close all;

Z0 = 50;          % Charac. Impedance
fc = 10e9;        % Operational frequency
Er = 10.2;        % Rel Perm of the dielectric
BW = .15;         % Bandwidth Percentage
h = 0.635e-3;     % Thickness of the dielectric
n = 5;            % Order of the Filter
c = 3e8;          % speed of light
eps0=8.85e-12;

g(1)  = 1.1468;
g(2)  = 1.3712;
g(3)  = 1.975;
g(4)  = 1.3712;
g(5)  = 1.1468;
g(6)  = 1.0000;

% Conversion from LPF Prototype to BPF Equivalent
Ls1  = (Z0*g(1))/(BW*2*pi*fc);
Ls5  = Ls1;
Cs1  = BW/(g(1)*Z0*2*pi*fc);
Cs5  = Cs1;
Lp2  = (BW*Z0)/(2*pi*fc*g(2));
Cp2  = g(2)/(Z0*BW*2*pi*fc);
Lp4  = Lp2;
Cp4  = Cp2;
Ls3  = (BW*Z0)/(2*pi*fc*g(3));
Cs3  = g(3)/(Z0*BW*2*pi*fc);
Z0J(1)  = sqrt((pi*BW)/(2*g(1)));
```

```
% J Inverter Calculation
for i=1:n-1
    Z0J(i+1) = ((pi*BW)/2)*(1/(sqrt(g(i)*g(i+1))));
end
Z0J(n+1) = sqrt((pi*BW)/(2*g(n)*g(n+1)));


for i=0:1:n
    Z0e(i+1) = Z0*((1+Z0J(i+1)+Z0J(i+1)^2));
    Z0o(i+1) = Z0*((1-Z0J(i+1)+Z0J(i+1)^2));
end
 for i=1:n+1

% the shape ratio for the equivalent single microstrip
correspond to even mode%
whse(i)=8*sqrt((exp(Z0e(i)/2/42.4*sqrt(Er+1))-1)*(7+4/
Er)/11+(1+1/Er)/0.81)/(exp(Z0e(i)/2/42.4*sqrt(Er+1))-1);
% the shape ratio for the equivalent single microstrip
correspond to odd mode%
whso(i)=8*sqrt((exp(Z0o(i)/2/42.4*sqrt(Er+1))-1)*(7+4/
Er)/11+(1+1/Er)/0.81)/(exp(Z0o(i)/2/42.4*sqrt(Er+1))-1);
whso1(i)=0.78*whso(i)+0.1*whse(i);

% space ratio of coupled lines %
sh(i)=(2/pi)*acosh((cosh(pi/2*whse(i))+cosh(pi/2*whso1(i))-2)/
(cosh(pi/2*whso1(i))-cosh(pi/2*whse(i))));
g(i)=cosh((pi/2)*sh(i));
d(i)=(cosh((pi/2)*whse(i))*(g(i)+1)+(g(i)-1))/2;
% shape ratio of coupled lines %
wh(i)=(1/pi)*acosh(d(i))-0.5*(sh(i));

%Calculate the even mode capacitance
if wh(i)<=1
    F(i)=(1+(12/wh(i)))^(-0.5)+0.041*(1-wh(i))^2;
else
    F(i)=(1+(12/wh(i)))^(-0.5);
end
Cp(i)=eps0*Er*wh(i);
epseffs(i)=(Er+1)/2+((Er-1)/2)*F(i);
Cf(i)=sqrt(epseffs(i))/(2*c*Z0)-Cp(i)/2;
A(i)=exp(-0.1*exp(2.33-1.5*wh(i)));
Cf1(i)=(Cf(i)/(1+(A/sh)*tanh(10*sh(i))))*(Er/epseffs(i))^0.25;
Ce(i)=Cp(i)+Cf(i)+Cf1(i);

%Calculate the odd mode capacitance
k(i)=sh(i)/(sh(i)+2*wh(i));
k1(i)=k(i)^2;
k2(i)=sqrt(1-k1(i));
if k1(i)<=0.5
    K(i)=(1/pi)*log(2*(1+sqrt(k2(i)))/(1-sqrt(k2(i))));
else
```

```
    K(i)=(pi)/log(2*(1+sqrt(k2(i)))/(1-sqrt(k2(i))));
end

Cga(i)=eps0*K(i);
Cgd(i)=((eps0*Er)/(pi))*log(coth((pi/4)*sh(i)))+0.65*Cf(i)*
((0.02*sqrt(Er)/sh(i))+(1-1/(Er)^2));
Co(i)=Cp(i)+Cf(i)+Cga(i)+Cgd(i);

%Calculation of effective permittivity constant

Ce1(i)=1/((c^2)*Ce(i)*(Z0e(i))^2);
Co1(i)=1/((c^2)*Co(i)*(Z0o(i))^2);
epseffe(i)=Ce(i)/Ce1(i);
epseffo(i)=Co(i)/Co1(i);
epseff(i)=((sqrt(epseffe(i))+sqrt(epseffo(i)))/2)^2;


%Calculation of length of coupler line
l(i)=(c/(4*fc*sqrt(epseff(i))))*1000;  %in mm

end

freq=8e9:0.0001e9:12e9;
[M,N]=size(freq);
j=sqrt(-1);

for k=1:1:N
    f=freq(M,k);

        % ABCD parameters of Each Component at each frequency
    Zmatrix_Ls1 = [1 (2j*pi*f*Ls1);0 1];
    Zmatrix_Ls5 = Zmatrix_Ls1;
    Zmatrix_Cs1 = [1 (-j/(2*pi*f*Cs1));0 1];
    Zmatrix_Cs5 = Zmatrix_Cs1;
    Zmatrix_Lp2 = [1 0;(-j/(2*pi*f*Lp2)) 1];
    Zmatrix_Lp4 = Zmatrix_Lp2;
    Zmatrix_Cp2 = [1 0;(2j*pi*f*Cp2) 1];
    Zmatrix_Cp4 = Zmatrix_Cp2;
    Zmatrix_Ls3 = [1 (2j*pi*f*Ls3);0 1];
    Zmatrix_Cs3 = [1 (-j/(2*pi*f*Cs3));0 1];

    % ABCD Parameter Conversion
    ABCD=Zmatrix_Ls1*Zmatrix_Cs1*Zmatrix_Lp2*Zmatrix_
Cp2*Zmatrix_Ls3*Zmatrix_Cs3*Zmatrix_Lp4*Zmatrix_Cp4*Zmatrix_
Ls5*Zmatrix_Cs5;

    A=ABCD(1,1);
    B=ABCD(1,2);
    C=ABCD(2,1);
    D=ABCD(2,2);
```

```
    S11(k)=(A+(B/Z0)-(C*Z0)-D)/(A+(B/Z0)+(C*Z0)+D);
    S21(k)=2/(A+(B/Z0)+(C*Z0)+D);
end

fprintf ('Bandpass Filter Lumped Element Component Values : \n
Ls1=% 0.5e\nLs5=% 0.5e\nCs1=% 0.5e\nCs5=%... 0.5e\nLp2=% 0.5e\
nCp2=% 0.5e\nLp4=% 0.5e\nCp4=% 0.5e\nLs3=% 0.5e\nCs3=%...
0.5e\n\n',Ls1,Ls5,Cs1,Cs5,Lp2,Cp2,Lp4,Cp4,Ls3,Cs3);
fprintf('\nInverter Values... :\nZ0J(1)=%0.4f\nZ0J(2)=%0.4f\
nZ0J(3)=%0.4f\nZ0J(4)=%0.4f\nZ0J(5)=%0.4f\nZ0J(6)=%0.4f\
n\n',...
Z0J(1),Z0J(2),Z0J(3),Z0J(4),Z0J(5),Z0J(6));
fprintf('Even-mode impedance Z0e:\nZ0e(1)= %0.4f\nZ0e(2)=
%0.4f\nZ0e(3)= %0.4f \nZ0e(4)= %0.4f \n...
Z0e(5)= %0.4f \nZ0e(6)= %0.4f\n\n',Z0e(1),Z0e(2),Z0e(3),Z0e(4),
Z0e(5),Z0e(6));
fprintf('Odd-mode impedance Z0o:\nZ0o(1)= %0.4f\nZ0o(2)=
%0.4f\nZ0o(3)= %0.4f \nZ0o(4)= %0.4f \n...
Z0o(5)= %0.4f \nZ0o(6)= %0.4f\n\n',Z0o(1),Z0o(2),Z0o(3),Z0o(4),
Z0o(5),Z0o(6));
fprintf ('\nEffective Dielectric Coefficient:\n');
fprintf ('epseff(1)=% 0.4f\n epseff(2)=% 0.4f\n epseff(3)=%
0.4f\n epseff(4)=% 0.4f\n epseff(5)=% 0.4f\n... epseff(6)=%
0.4f\n\n',epseff(1),epseff(2),epseff(3),epseff(4),epseff(5),ep
seff(6));
fprintf ('\nSpacing Ratio for Edge Coupled Microstrip
Lines:\n') ;
fprintf ('s/h(1)=% 0.4f\ns/h(2)=% 0.4f\ns/h(3)=% 0.4f\
ns/h(4)=% 0.4f\ns/h(5)=% 0.4f\ns/h(6)=%... 0.4f\n',sh(1),sh(2),
sh(3),sh(4),sh(5),sh(6));
fprintf ('\nShape Ratio for Edge Coupled Microstrip
Lines:\n');
fprintf ('w/h(1)=% 0.4f\nw/h(2)=% 0.4f\nw/h(3)=% 0.4f\
nw/h(4)=% 0.4f\nw/h(5)=% 0.4f\nw/h(6)=%... 0.4f\n\
n',wh(1),wh(2),wh(3),wh(4),wh(5),wh(6));
fprintf ('Electrical Length (m):\n');
fprintf ('l(1)=% 0.5f\n l(2)=% 0.5f\n l(3)=% 0.5f\n l(4)=%
0.5f\n l(5)=% 0.5f\n ...
l(6)=% 0.5f\n\n',l(1),l(2),l(3),l(4),l(5),l(6));


figure
plot(freq*1e-9,20*log10(abs(S11)),'-mo',freq*1e-
9,20*log10(abs(S21)),'bx')
h = legend('S_{11}(dB)','S_{21}(dB)',2);
title('\bf{Return Loss and Insertion Loss vs Frequency(GHz)}')
grid on
xlabel('Frequency (GHz)')
ylabel('S_{11}(dB) & S_{21}(dB)')
axis([8 12 -100 0]);
```

When the program is executed, the following output giving the physical dimensions of the edge-coupled filter and all the filter-related design values are displayed.

Bandpass filter lumped element component values:

Ls1 = 6.08396e-009
Ls5 = 6.08396e-009
Cs1 = 4.16345e-014
Cs5 = 4.16345e-014
Lp2 = 8.70524e-011
Cp2 = 2.90978e-012
Lp4 = 8.70524e-011
Cp4 = 2.90978e-012
Ls3 = 6.04386e-011
Cs3 = 4.19108e-012

Inverter values:
Z0J(1) = 0.4533
Z0J(2) = 0.1879
Z0J(3) = 0.1432
Z0J(4) = 0.1432
Z0J(5) = 0.1879
Z0J(6) = 0.4533

Even-mode impedance Z_{0e}:
$Z_{0e}(1) = 82.9367$
$Z_{0e}(2) = 61.1600$
$Z_{0e}(3) = 58.1839$
$Z_{0e}(4) = 58.1839$
$Z_{0e}(5) = 61.1600$
$Z_{0e}(6) = 82.9367$

Odd-mode impedance Z_{0o}:
$Z_{0o}(1) = 37.6092$
$Z_{0o}(2) = 42.3705$
$Z_{0o}(3) = 43.8661$
$Z_{0o}(4) = 43.8661$
$Z_{0o}(5) = 42.3705$
$Z_{0o}(6) = 37.6092$

Effective dielectric coefficient:

epseff(1) = 8.5778
epseff(2) = 6.8751
epseff(3) = 6.6707
epseff(4) = 6.7083
epseff(5) = 7.0505
epseff(6) = 9.8710

Spacing ratio for edge-coupled microstrip lines:
s/h(1) = 0.1895
s/h(2) = 0.6365
s/h(3) = 0.8609
s/h(4) = 0.8609
s/h(5) = 0.6365
s/h(6) = 0.1895

Shape ratio for edge-coupled microstrip lines:
w/h(1) = 0.5902
w/h(2) = 0.8875
w/h(3) = 0.9245
w/h(4) = 0.9245
w/h(5) = 0.8875
w/h(6) = 0.5902

Electrical length (m):
l(1) = 2.56079
l(2) = 2.86037
l(3) = 2.90386
l(4) = 2.89570
l(5) = 2.82457
l(6) = 2.38716

The frequency response of the lumped element BPF is also obtained with the program and is shown in Figure 7.31. The results obtained in Figure 7.31 match the results obtained with Ansoft Designer. The physical dimensions calculated with the MATLAB program given are used to simulate the microstrip edge-coupled filter with Sonnet planar electromagnetic simulator, as shown in Figure 7.32.

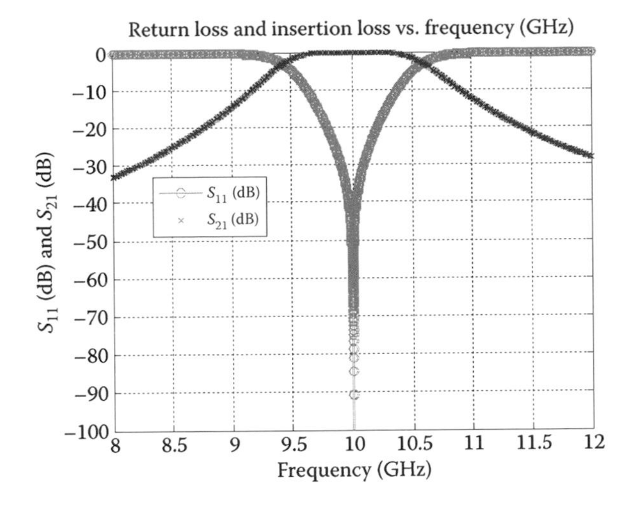

FIGURE 7.31 BPF simulation results with MATLAB.

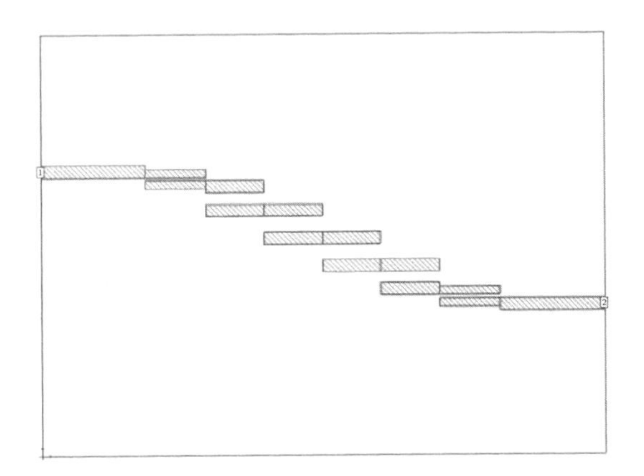

FIGURE 7.32 Simulated edge-coupled microstrip circuit with Sonnet.

The Sonnet simulation results are illustrated in Figure 7.33. Since the material properties are entered, the simulation results are slightly different than the ideal results obtained using lumped elements by Ansoft and MATLAB. Overall, the insertion loss requirement, center frequency, and attenuation profile for edge-coupled filter are achieved with the method used.

7.6 END-COUPLED, CAPACITIVE GAP, HALF-WAVELENGTH RESONATOR BPFs

The general configuration of an end-coupled microstrip BPF is shown in Figure 7.34.

The gap between two adjacent open ends is capacitive and can be represented by inverters. *J*-inverters tend to reflect high impedance levels to the ends of each

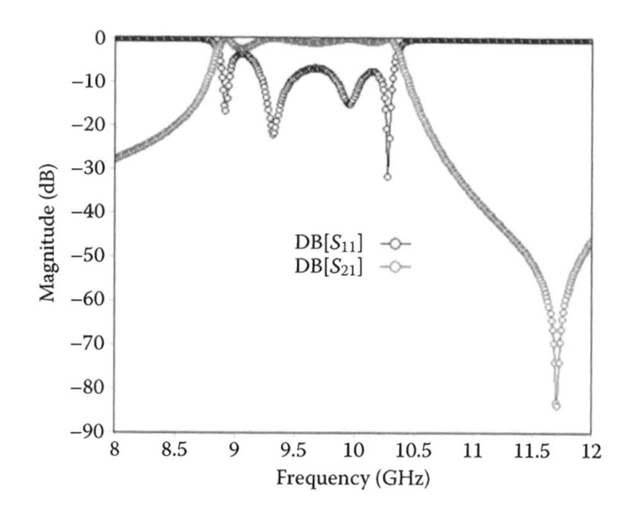

FIGURE 7.33 Edge-coupled BPF simulation results with Sonnet.

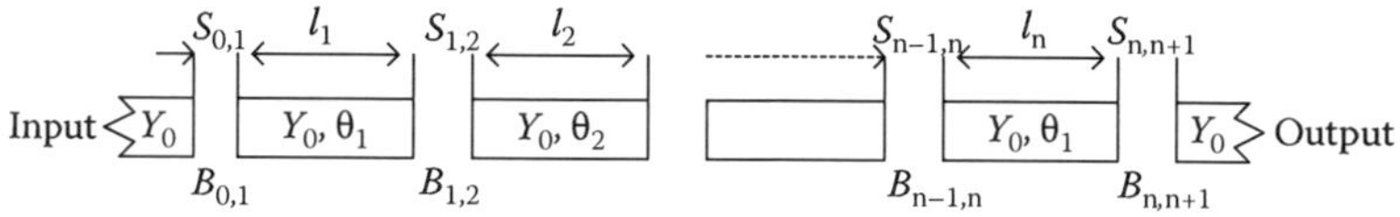

Input Y_0 | $S_{0,1}$ l_1 $S_{1,2}$ l_2 $S_{n-1,n}$ l_n $S_{n,n+1}$ | Y_0, θ_1 | Y_0, θ_2 | Y_0, θ_1 | Y_0 Output | $B_{0,1}$ $B_{1,2}$ $B_{n-1,n}$ $B_{n,n+1}$

FIGURE 7.34 End-coupled microstrip bandpass filter.

half-wavelength resonator causing the resonator to act like a shunt-resonator type of filter. The design equations for the inverters are given by Equations 7.72 through 7.74.

The gap between each resonator can be represented by the equivalent circuit shown in Figure 7.35.

Assuming that the capacitive gap acts perfectly, the susceptance of the series–capacitance discontinuities can be found from

$$\frac{B_{j,j+1}}{Y_0} = \frac{\dfrac{J_{j,j+1}}{Y_0}}{1-\left(\dfrac{J_{j,j+1}}{Y_0}\right)^2} \tag{7.89}$$

and

$$\theta_j = \pi - \frac{1}{2}\left[\tan^{-1}\left(\frac{2B_{j-1,j}}{Y_0}\right) + \tan^{-1}\left(\frac{2B_{j,j+1}}{Y_0}\right)\right] \tag{7.90}$$

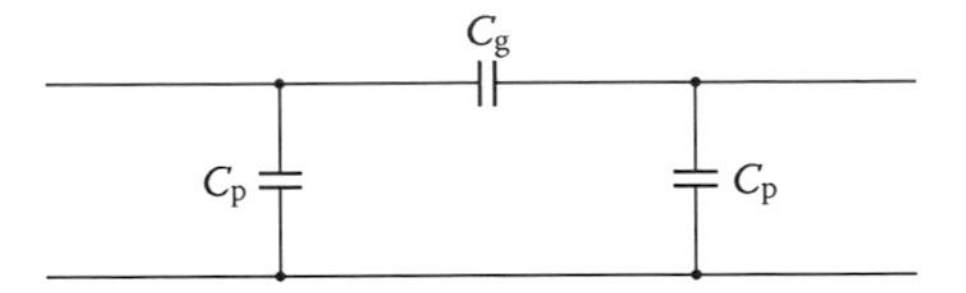

FIGURE 7.35 Capacitive-gap equivalent circuit.

where θ in Equation 7.90 is given in radians. Thus, the final length of each resonator can be found from

$$\ell_j = \frac{\lambda_{g0}}{2\pi}\theta_j - \Delta\ell_j^{e1} - \Delta\ell_j^{e2} \tag{7.91}$$

where

$$\Delta\ell_j^{e1} = \frac{\omega_0 C_p^{j-1,j}}{Y_0}\frac{\lambda_{g0}}{2\pi} \tag{7.92}$$

$$\Delta\ell_j^{e2} = \frac{\omega_0 C_p^{j,j+1}}{Y_0}\frac{\lambda_{g0}}{2\pi} \tag{7.93}$$

The coupling gap between each resonator can be found such that the resultant series capacitance is equal to

$$C_g^{j,j+1} = \frac{B_{j,j+1}}{\omega_0} \tag{7.94}$$

The gap dimensions can be calculated using the closed-form expressions given. Planar electromagnetic simulators can also be utilized to obtain the capacitance values shown in Figure 7.35 with simulation of a two-port microstrip gap shown in Figure 7.36. The two-port parameters can be obtained from the simulation and can be represented in Y parameters as

$$Y = \begin{bmatrix} Y_{11} & Y_{12} \\ Y_{21} & Y_{22} \end{bmatrix} \tag{7.95}$$

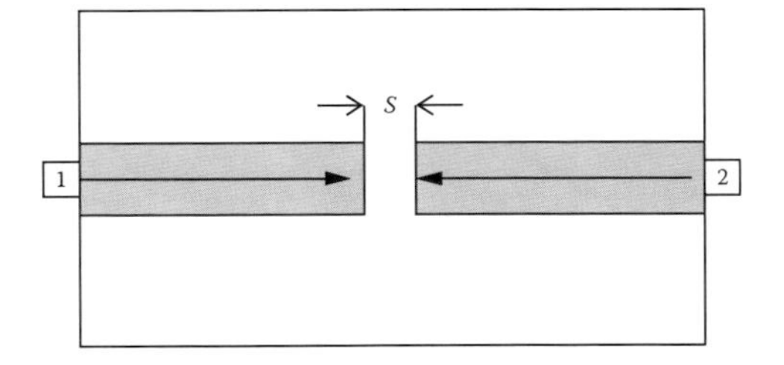

FIGURE 7.36 Layout of microstrip gap for Sonnet simulation.

Using the simulated Y parameters, the following capacitance values are obtained:

$$C_g = -\frac{\text{Im}(Y_{21})}{\omega_0} \tag{7.96}$$

$$C_p = -\frac{\text{Im}(Y_{11} + Y_{21})}{\omega_0} \tag{7.97}$$

Design Example

Design and simulate an end-coupled, capacitive-gap microstrip BPF with the order of $n = 3$ 0.1-dB passband ripple. The center frequency of the filter is at 6 GHz, and the filter has to meet a bandwidth requirement of 2.8%. The filter has to be inserted into 50-Ω characteristic line impedance. For the microstrip implementation, it is given that the dielectric constant is $\varepsilon_r = 10.8$, the thickness of the substrate is 1.27 mm, and the width is 1.1 mm.

Solution

The equivalent circuit of the BPF is derived through the use of the LPF prototype illustrated in Figure 7.37. The *ABCD* parameters of the entire network are found from cascading the *ABCD* parameters of each circuit component. The frequency response of the filter is obtained from converting the network *ABCD* parameters to scattered parameters.

The Chebyshev filter prototype values are determined from the design in Figure 7.7 [1]. The normalized component values of the filter are

$$g_0 = g_4 = 1.0, \quad g_1 = g_3 = 1.0316, \quad \text{and} \quad g_2 = 1.1474$$

The LPF prototype illustrated in Figure 7.1 is used to design the equivalent circuit for the BPF. The low-pass prototype filter is converted to a BPF shown in Figure 7.38 with application of the transformation circuits given in Figure 7.16.

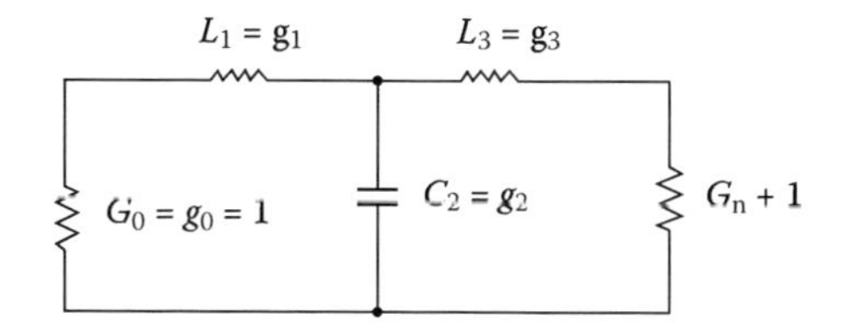

FIGURE 7.37 LPF prototype.

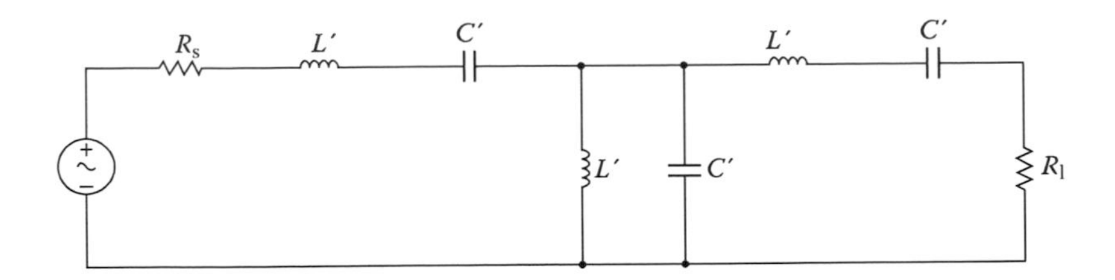

FIGURE 7.38 Equivalent circuit BPF.

In order to determine the *ABCD* parameters of the overall network, the *ABCD* parameters of each component can be cascaded. To obtain the frequency response of the filter, the *ABCD* parameters are converted into scattered parameters as

$$\begin{bmatrix} A & B \\ C & D \end{bmatrix} = \begin{bmatrix} 1 & j\omega L'_s \\ 0 & 1 \end{bmatrix} \begin{bmatrix} 1 & \dfrac{1}{j\omega C'_s} \\ 0 & 1 \end{bmatrix} \begin{bmatrix} 1 & 0 \\ \dfrac{1}{j\omega L'_p} & 1 \end{bmatrix} \begin{bmatrix} 1 & 0 \\ j\omega C'_p & 1 \end{bmatrix} \begin{bmatrix} 1 & j\omega L'_s \\ 0 & 1 \end{bmatrix} \begin{bmatrix} 1 & \dfrac{1}{j\omega C'_s} \\ 0 & 1 \end{bmatrix} \tag{7.98}$$

$$\begin{bmatrix} S_{11} & S_{12} \\ S_{21} & S_{22} \end{bmatrix} = \begin{bmatrix} \dfrac{A + \dfrac{B}{Z_0} - CZ_0 + D}{\psi_7} & \dfrac{2(AD - BC)}{\psi_7} \\ \dfrac{2}{\psi_7} & \dfrac{-A + \dfrac{B}{Z_0} - CZ_0 + D}{\psi_7} \end{bmatrix} \tag{7.99}$$

where

$$\psi_7 = A + \frac{B}{Z_0} + CZ_0 + D \tag{7.100}$$

Insertion and return losses can be obtained from S_{21}. The final component values of the filter are calculated and shown in Figure 7.39.

The insertion and return losses are obtained using MATLAB and are shown in Figure 7.40. The MATLAB script is given below.

```
%Bandpass filter response with equivalent circuit
Z0 = 50;
FBW = .028;
f0 = 6*10^(9);
w0 = 2*pi*f0;
f = [5.4e9:10e6:6.6e9];%5.4e9
syms x;
w = 2*pi*x;
```

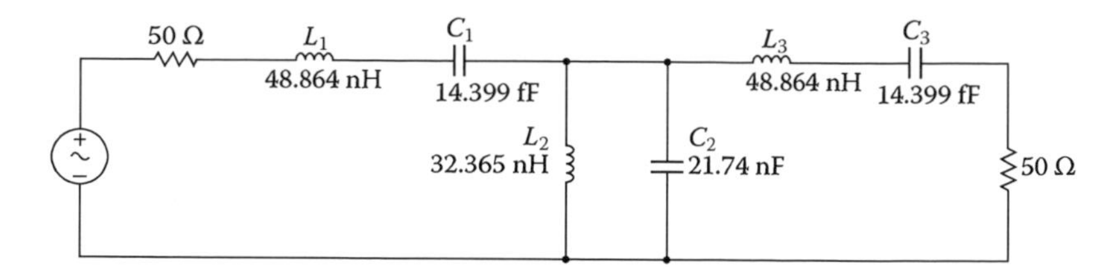

FIGURE 7.39 Equivalent BPF schematic.

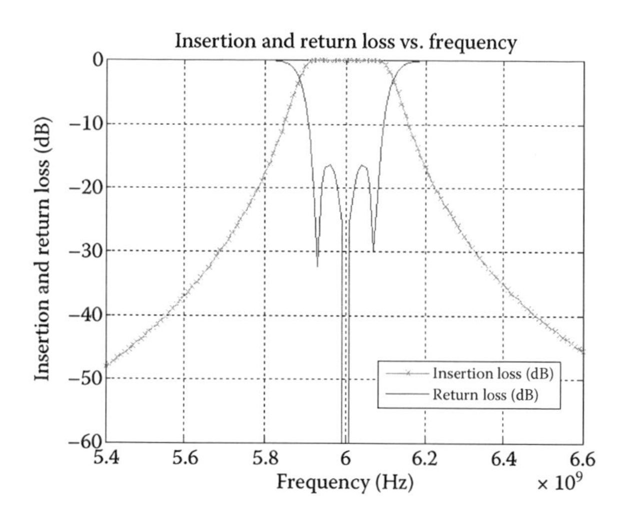

FIGURE 7.40 Insertion loss of the equivalent BPF.

```
g0 = 1;
g4 = 1;
g1 = 1.0316;
g3 = 1.0316;
g2 = 1.1474;
L1 = (g1*Z0)/(w0*FBW);
L3 = (g3*Z0)/(w0*FBW);
C1 = FBW/(w0*g1*Z0);
C3 = FBW/(w0*g3*Z0);
L2 = (FBW*Z0)/(w0*g2);
C2 = g2/(w0*FBW*Z0);

m1 = [1 ((1i)*w*L1);0 1];
m2 = [1 1/((1i)*w*C1);0 1];
m3 = [1 0;(1/((1i)*w*L2)) 1];
m4 = [1 0;((1i)*w*C2) 1];
m5 = [1 ((1i)*w*L3);0 1];
m6 = [1 1/((1i)*w*C3);0 1];
```

```
total = m1*m2*m3*m4*m5*m6;
A = total(1,1);
B = total(1,2);
C = total(2,1);
D = total(2,2);

delta = A+(B/Z0)+(C*Z0)+D;
s21 = 2/delta;
s11 = (A+(B/Z0)-(C*Z0)-D)/delta;
s21 = subs(s21,x,f);
s11 = subs(s11,x,f);
IL = 20*log10(abs(s21));
RL = 20*log10(abs(s11));

figure(1)
plot(f,IL,'-mx')
hold on
plot(f,RL,'b')
h = legend('Insertion Loss (dB)','Return Loss (dB)');
title('Insertion and Return Loss vs Frequency')
xlabel('Frequency (Hz)')
ylabel('Insertion and Return Loss (dB)')
axis([5.4e9 6.6e9 -60 0])
grid on
```

In order to determine the length of each capacitive gap and the value of each parallel capacitor, several simulations of different gap lengths are performed using Sonnet with the configuration shown in Figure 7.35. In the simulation, the width of the microstrip is set to 1.1 mm, and its thickness is taken to be 1.27 mm. The dielectric constant of the material is given to be 10.8. The *Y* parameters of the microstrip, operating at 6 GHz, are extracted for each simulation. In addition, the series and parallel capacitor values are calculated using Equations 7.96 and 7.97. The results are shown in Table 7.4.

Next, the C_g and *s* values obtained from Table 7.4 are plotted as shown in Figure 7.41. The equation of the line is then obtained as displayed in the same figure.

TABLE 7.4
Simulation of Microstrip Gap

s (mm)	$Y_{11} = Y_{22}$	$Y_{12} = Y_{21}$	C_g	C_p
0.05	0.004578	4.412-3	1.1703-13	4.4033-15
0.1	0.003912	3.594-3	9.5334-14	8.4352-15
0.2	0.003286	2.695-3	7.1487-14	1.5677-14
0.5	0.002685	1.466-3	3.8887-14	3.2335-14
0.8	0.002524	8.8508-4	2.3477-14	4.3474-14
1	0.002481	6.4386-4	1.7079-14	4.8732-14

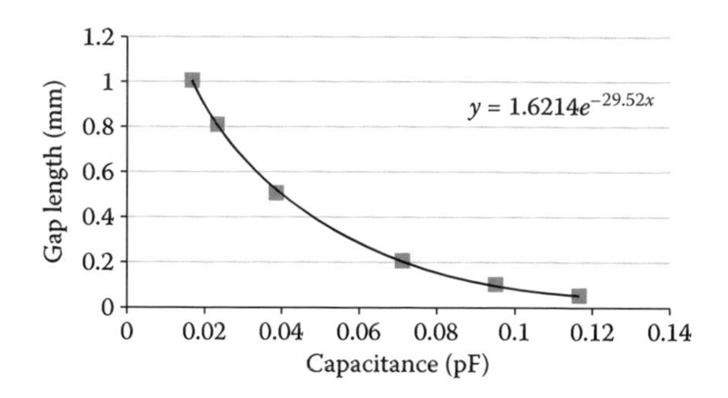

FIGURE 7.41 C_g vs. gap length from simulation.

The equation of the line is used to obtain the length of the gap by interpolation. In the equation on the graph, x corresponds to known capacitance, and y is the corresponding gap length. A similar approach is taken to determine the capacitance C_p terms for each equivalent gap. The interpolation gives the value of the corresponding capacitance for C_p as shown in Figure 7.42.

The summary of the calculated gap lengths and corresponding capacitance values is given in Table 7.5. The end-coupled microstrip BPF is simulated with the

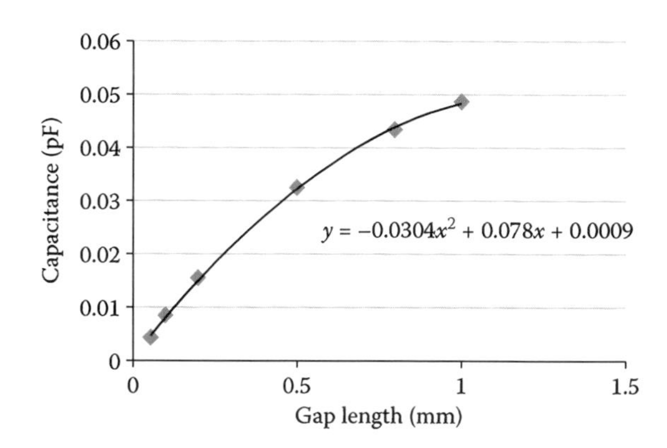

FIGURE 7.42 C_p vs. gap length from simulation.

TABLE 7.5
Summary of the Calculated Lengths and Capacitance Values

Port	C_p (pF)	C_g (pF)	s (mm)
01	0.0051	0.11442	0.055
12	0.0455	0.021482	0.86
23	0.0455	0.021482	0.86
34	0.0051	0.11442	0.055

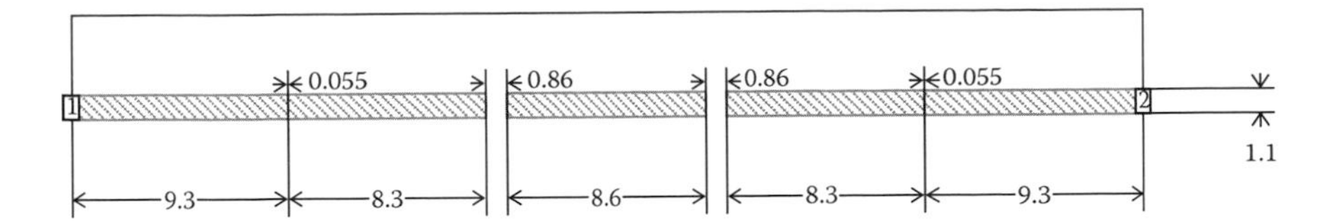

FIGURE 7.43 Simulation of end-coupled microstrip BPF.

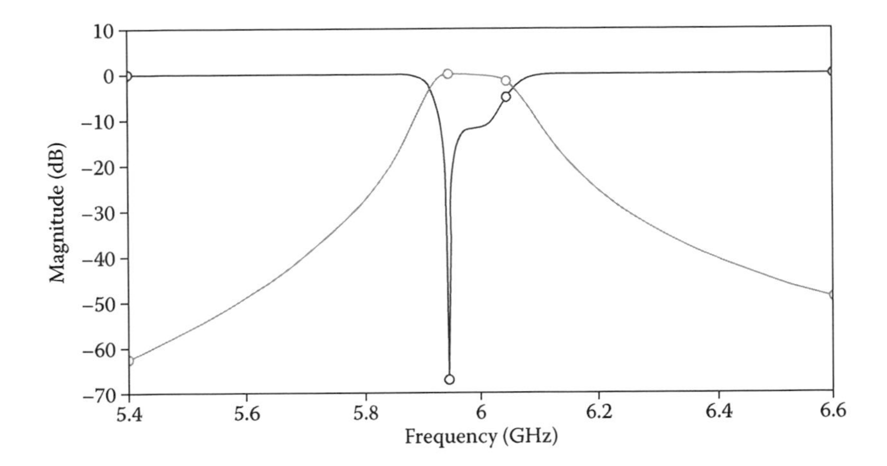

FIGURE 7.44 Simulation results for end-coupled microstrip BPF using Sonnet.

physical values calculated using the generic MATLAB script developed and given below with the results shown in Table 7.5. The Sonnet simulation circuit layout with physical dimensions is illustrated in Figure 7.43. The simulation results for insertion and return losses are given in Figure 7.44.

```
% Generic End Coupled Microstrip Filter Design Program
% Order = 3
% Pass Band Ripple = .1
Z0 = 50;
FBW = .028;
f0 = 6*10^(9);
w0 = 2*pi*f0;

% element values
g0 = 1;
g4 = 1;
g1 = 1.0316;
g3 = 1.0316;
g2 = 1.1474;

% General Design equations
% Jn,n+1/Yo = sqrt((piFBW)/2gngn+1))
J01_Y0 = sqrt((pi*FBW)/(2*g0*g1)); %01 and 34 are equal due to g
```

```
J12_Y0 = (pi*FBW/2)*(1/(sqrt(g1*g2))); %12 and 23 are equal
due to g
J23_Y0 = (pi*FBW/2)*(1/(sqrt(g2*g3)));
J34_Y0 = sqrt((pi*FBW)/(2*g0*g1));
% Bj,j+1/Yo = ((jj,j+1/Y0)/(1 - (jj,j+1)^2))
B01_Y0 = (J01_Y0)/(1 - (J01_Y0)^2);
B12_Y0 = (J12_Y0)/(1 - (J12_Y0)^2);
B23_Y0 = (J23_Y0)/(1 - (J23_Y0)^2);
B34_Y0 = (J34_Y0)/(1 - (J34_Y0)^2);
%theta = pi-.5(atan(2Bj-1,j/Y0)+atan(2Bj,j+1/Y0))
theta1 = pi-.5*(atan(2*B01_Y0)+atan(2*B12_Y0));
theta2 = pi-.5*(atan(2*B12_Y0)+atan(2*B23_Y0));
theta3 = pi-.5*(atan(2*B23_Y0)+atan(2*B34_Y0));
%Cg(j,j+1) = Bj,j+1/wo;
Cg01 = (B01_Y0*(1/Z0))/(2*pi*f0);
Cg12 = (B12_Y0*(1/Z0))/(2*pi*f0);
Cg23 = (B23_Y0*(1/Z0))/(2*pi*f0);
Cg34 = (B34_Y0*(1/Z0))/(2*pi*f0);
% Find right Cg and Cp
%Cg = -im(Y21)/wo
ImY21_01 = Cg01*2*pi*f0;
ImY21_12 = Cg12*2*pi*f0;
ImY21_23 = Cg23*2*pi*f0;
ImY21_34 = Cg34*2*pi*f0;
%------------------------------------------------------------
% Interpolating Cp and s (gap)
% by interpolation from excel spreadsheet
% s01/s34 = .055 mm
% s12/s23 = .86 mm

% by interpolation
% Cp01/Cp34 = .0051 pF
% Cp12/Cp23 = .0455 pF
Cp01 = .0051*10^(-12);
Cp12 = .0455*10^(-12);
Cp23 = .0455*10^(-12);
Cp34 = .0051*10^(-12);
%------------------------------------------------------------
%Microstrip implementation
Er = 10.8;
h = 1.27*10^(-3);
w = 1.1*10^(-3);
Eeff = ((Er+1)/2)+((Er-1)/2)*((1+12*(h/w))^(-.5));
c = 2.99792*10^(8);
vp = c/(sqrt(Eeff));
wav = c/f0;
guide_wav = wav/(sqrt(Eeff));
%------------------------------------------------------------
% Calculate effective length
leff1a = ((w0*Cp01)/(1/Z0))*(guide_wav/(2*pi));
```

```
leff1b = ((w0*Cp12)/(1/Z0))*(guide_wav/(2*pi));

leff2a = ((w0*Cp12)/(1/Z0))*(guide_wav/(2*pi));
leff2b = ((w0*Cp23)/(1/Z0))*(guide_wav/(2*pi));

leff3a = ((w0*Cp23)/(1/Z0))*(guide_wav/(2*pi));
leff3b = ((w0*Cp34)/(1/Z0))*(guide_wav/(2*pi));
%--------------------------------------------------------------
% Calculating physical length
l1 = (guide_wav/(2*pi))*theta1 - leff1a - leff1b;
l2 = (guide_wav/(2*pi))*theta2 - leff2a - leff2b;
l3 = (guide_wav/(2*pi))*theta3 - leff3a - leff3b;
%--------------------------------------------------------------
% Results when Program Executed for design
% l1 = 8.3 mm
% l2 = 8.6 mm
% l3 = 8.3 mm
% s01/s34 = .055 mm
% s12/s23 = .86 mm
% end coupling length are guide wavelength/4 = 4.7 mm
% Cg01 = 1.1442 10-13
% Cg12 = 2.1482 10-14
% Cg23 = 2.1482 10-14
% Cg34 = 1.1442 10-13
% Cp01/Cp34 = .0051 pF
% Cp12/Cp23 = .0455 pF
```

PROBLEMS

1. Design an HPF with a 3-dB equal ripple response and a cut-off frequency of 1 GHz. Source and load impedances are given to be 50 Ω, and attenuation at 0.6 GHz is required to be a minimum of 40 dB.
2. Design a BPF with 5% fractional bandwidth and center frequency of 2 GHz. The filter is desired to have a maximally flat response in the passband and has four sections. The source and load impedances are given to be 50 Ω.
3. Design and simulate a stepped-impedance LPF with cut-off frequency at 2 GHz. The filter is desired to provide a minimum of 30-dB attenuation at 3 GHz. The source and load impedances of the filter are given to be 50 Ω. The ripple is defined to be not more than 0.5 dB in the passband. In addition, it is required to use FR4 as substrate with dielectric constant of 3.7 and dielectric thickness of 60 mils.
4. Design a triple-band BPF with SIR filters. The center frequencies for each band are defined to be 1, 2.4, and 3.6 GHz. Use RO 4003 as a substrate with 32-mils thickness and 3.38 dielectric constant. The insertion loss in the passbands is required to be –3 dB or better. The return loss in the first and the second bands is desired to be –20 dB or lower. The third band stopband attenuation is specified to –30 dB or lower. The ripple in the passband should not exceed 0.1 dB.

REFERENCES

1. A. Eroglu. 2013. *RF Circuit Design Techniques for MF-UHF Applications*. CRC Press, Boca Raton, FL.
2. A. Eroglu, and J.K. Lee. 2008. The complete design of microstrip directional couplers using the synthesis technique. *IEEE Transactions on Microwave Theory and Techniques*, Vol. 57, No. 12, pp. 2756–2761, December.

8 Computer Aided Design Tools for Amplifier Design and Implementation

8.1 INTRODUCTION

Radio frequency (RF)/microwave computer-aided design (CAD) tools have been commonly used in RF power amplifier systems to expedite the design process, increase the system performance, and reduce the associated cost by eliminating the need for several prototypes before the implementation stage [1–3]. RF power amplifiers are simulated with nonlinear circuit simulators, which use large signal equivalent models of the active devices. Passive components used in RF power amplifier simulation are usually modeled as ideal components and hence do not include the frequency characteristics. Furthermore, it is rare to include the electromagnetic effects such as coupling between traces and leads, parasitic effects, current distribution, and radiation effects that exist among the components in RF power amplifier simulation. This is partly due to the requirement in expertise in both nonlinear circuit simulators and electromagnetic simulators.

Nonlinear circuit simulation of RF power amplifiers can be done using harmonic balance technique with the application of Krylov subspace methods in the frequency domain or nonlinear differential algebraic equations using the integration methods, Newton's method, or sparse matrix solution techniques in the time domain [4–6]. Time-domain methods are preferred over frequency-domain methods because of their advantage in providing accurate solutions using the transient response of the circuits.

RF amplifier design therefore can be systematized from the component level to the assembly level and from the assembly level to the system level using CAD tools with the implementation of the design flow diagram given in Figure 8.1. RF amplifier design methodology given in Figure 8.1 consists of three phases: analysis phase, simulation phase, and experimental phase. As seen in the flow diagram, the simulation stage is the bridge to the experimental stage and is required for an optimized successful design.

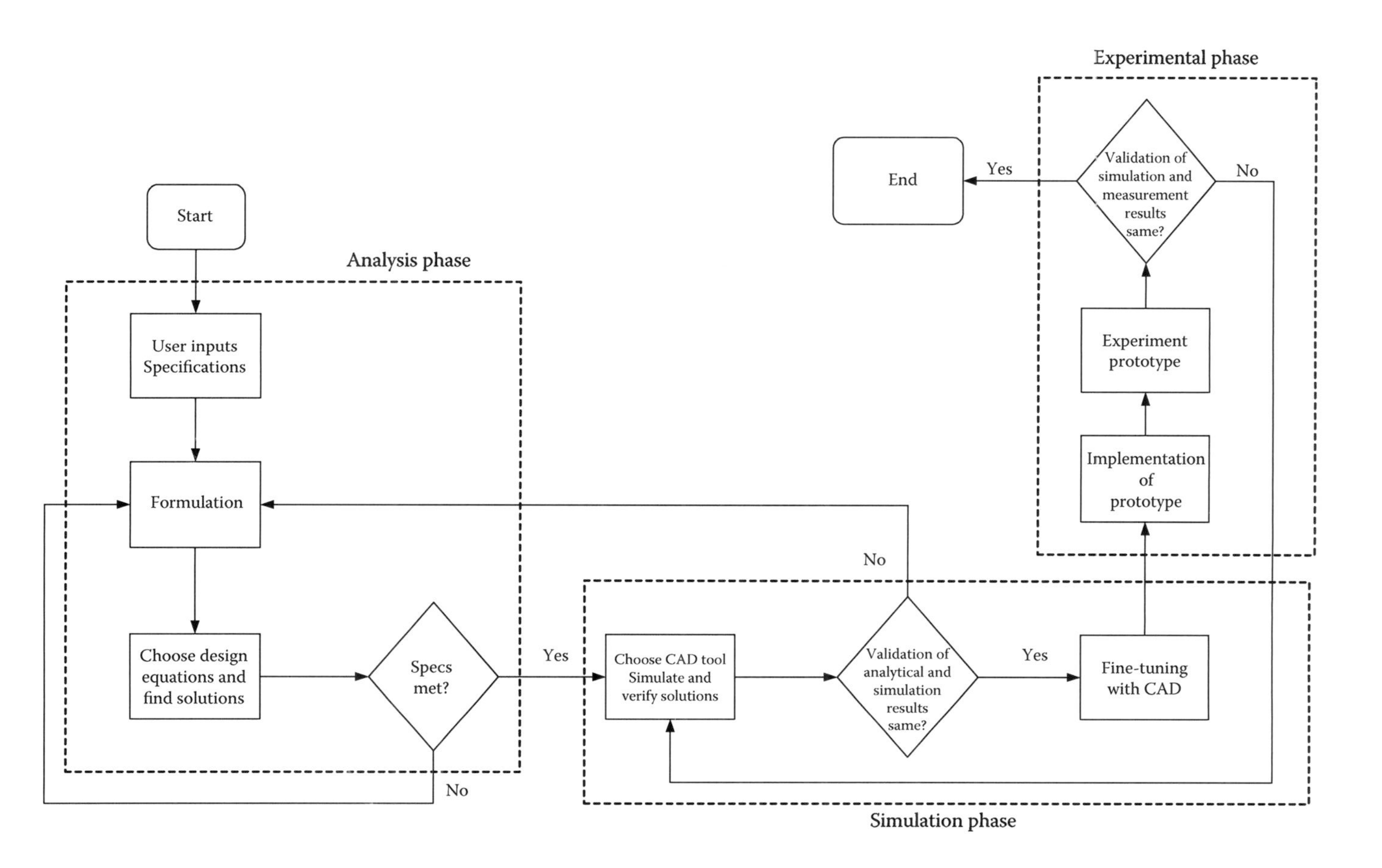

FIGURE 8.1 RF amplifier design methodology using CAD tools.

8.2 PASSIVE COMPONENT DESIGN AND MODELING WITH CAD—COMBINERS

In this section, the design, simulation, and implementation of passive components that are used as surrounding components for RF amplifiers are given. The passive component that will be analyzed and simulated is chosen to be a combiner. The three-phase design detail will be given step by step using combiners as a design example.

Design Example

Design, simulate, build, and measure a high-power combiner using microstrip technology at 13.56 MHz to combine an output of three PA modules with 25 Ω. The output of the combiner is desired to be 30 Ω.

Solution

The design, simulation, and implementation of a three-way combiner will be done using the steps outlined in the flow chart given in Figure 8.1.

8.2.1 Analysis Phase for Combiners

The complete analysis of combiners/dividers has been investigated and given in Ref. [7]. It has been obtained using the formulations in Ref. [7] that a Wilkinson combiner/divider circuit can be represented using the equivalent circuit given in Figure 8.2.

The characteristic impedances of the transmission lines in Figure 8.2 are equal to $Z_{TL} = Z_0 = \sqrt{NR_0}$. The equivalent four-port network for an N-way divider circuit is

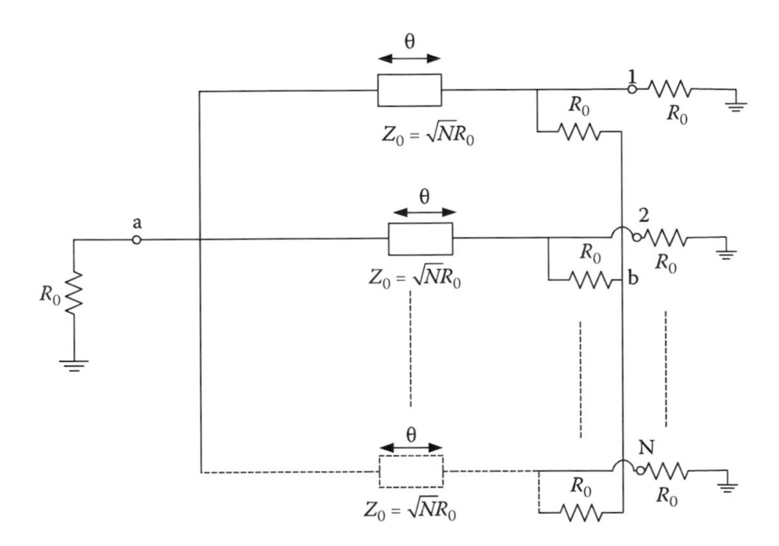

FIGURE 8.2 N-way Wilkinson power divider circuit when source and load impedances are equal.

shown in Figure 8.3. The even mode corresponds to an open circuit at the symmetry plane when voltage source (+V) is placed in series at port 4, whereas the odd mode corresponds to a short circuit when (−V) is placed in series at port 4.

The power delivered to each of the (N − 1) ports with load resistance R_0 is then equal to

$$P = |I_t|^2\left(\frac{R_0}{N-1}\right)\left(\frac{1}{N-1}\right) = |I_t|^2\left(\frac{R_0}{(N-1)^2}\right) \tag{8.1}$$

The power that is available from the excitation port is defined as P_a and is given by

$$P_a = \frac{(NV)^2}{4R_0} \tag{8.2}$$

The isolation between one port and the others is defined as

$$\text{Isolation (dB)} = 10\log\left(\frac{P_a}{P}\right) \tag{8.3}$$

Substitution of Equations 8.1 and 8.2 into Equation 8.3 gives

$$\text{Isolation (dB)} = 10\log\left(\frac{N^2}{4\left|\left[\frac{\cot(\theta)+j\sqrt{N}}{(N+1)\cot(\theta)+j2\sqrt{N}}\right]-\left[\frac{\cot(\theta)+j\sqrt{N}}{\cot(\theta)+j2\sqrt{N}}\right]\right|^2}\right) \tag{8.4}$$

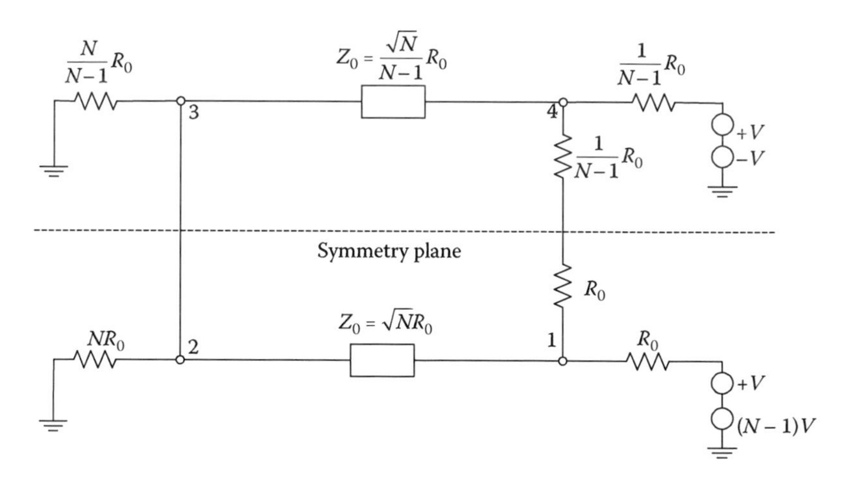

FIGURE 8.3 Four-port network of N-way power divider.

The input voltage standing wave ratio (VSWR) of the system, which is calculated at node a for an N-way divider, is found from

$$\text{VSWR} = \frac{1+|\Gamma|}{1-|\Gamma|} \tag{8.5}$$

where

$$\Gamma = \frac{Z_i - R_0}{Z_i + R_0} \tag{8.6}$$

$$Z_i = \frac{1}{N}\left(Z_{TL}\frac{R_0 + jZ_{TL}\tan\theta}{Z_{TL} + jR_0\tan\theta}\right) = \left(\frac{R_0}{\sqrt{N}}\right)\frac{1+j\sqrt{N}\tan\theta}{\sqrt{N}+j\tan\theta} \tag{8.7}$$

The insertion loss at each port is defined as

$$\text{IL(dB)} = 10\log\left(\frac{N}{1-\Gamma^2}\right) \tag{8.8}$$

The design curves for isolation and insertion loss for combiners/dividers for different number of ports have been given in Ref. [7].

Combiners/dividers might have combiner/divider source and load impedances that are not equal, as shown in Figure 8.4, for some designs.

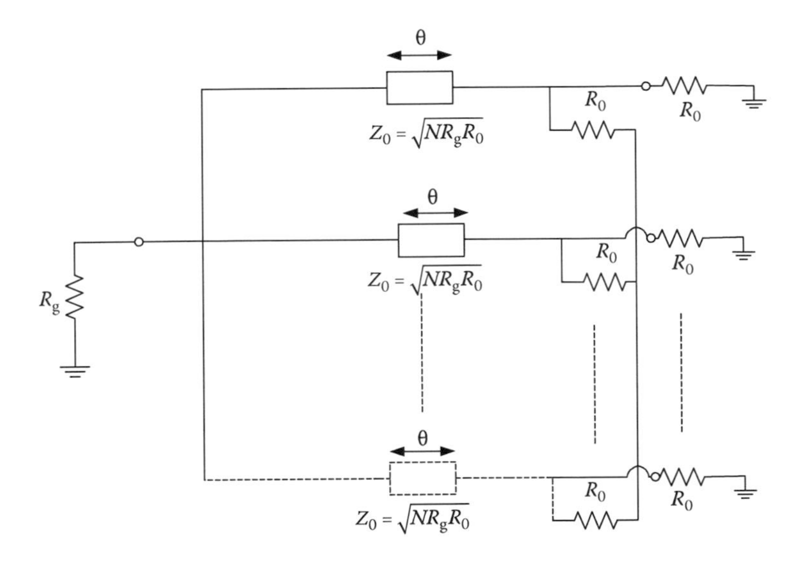

FIGURE 8.4 N-way Wilkinson power divider circuit when source and load impedances are not equal.

Under this condition, the characteristic impedance of the transmission line is then defined as

$$Z_{TL} = \sqrt{nR_x R_0} = \sqrt{nR_g R_0} \tag{8.9}$$

where $R_x = R_g$ is the source impedance.

Isolation and insertion loss are found from Equations 8.4 and 8.7. The input impedance with different source impedance is given by

$$Z_i = \frac{1}{N}\left(Z_{TL}\frac{R_0 + jZ_{TL}\tan\theta}{Z_{TL} + jR_0\tan\theta}\right) = \left(\sqrt{\frac{R_g R_0}{N}}\right)\frac{1 + j\sqrt{N\frac{R_g}{R_0}}\tan\theta}{\sqrt{N\frac{R_g}{R_0}} + j\tan\theta} \tag{8.10}$$

The reflection coefficient at the input is calculated using the equation

$$\Gamma_{in} = \frac{Z_i - R_g}{Z_i + R_g} \tag{8.11}$$

The input VSWR is then equal to

$$\text{VSWR}_{in} = \frac{1 + |\Gamma_{in}|}{1 - |\Gamma_{in}|} \tag{8.12}$$

Isolation is found as

$$\text{Isolation (dB)} = 10\log\left(\frac{N^2}{4\left[\frac{\cot(\theta) + j\sqrt{N\frac{R_g}{R_0}}}{\left(N\frac{R_g}{R_0} + 1\right)\cot(\theta) + j2\sqrt{N\frac{R_g}{R_0}}}\right] - \left[\frac{\cot(\theta) + j\sqrt{N\frac{R_g}{R_0}}}{\cot(\theta) + j2\sqrt{N\frac{R_g}{R_0}}}\right]^2}\right) \tag{8.13}$$

The VSWR at the output ports is calculated using the output reflection coefficient from

$$\Gamma_0 = \left[\frac{2\sqrt{N\frac{R_g}{R_0}}\left(\frac{R_g}{R_0} - 1\right) + j\cot(\theta)\left(N\frac{R_g}{R_0} + 1 - 2\frac{R_g}{R_0}\right)}{2\sqrt{N\frac{R_g}{R_0}}\left(2 + N\frac{R_g}{R_0}\right) + j\cot(\theta)\left(N\frac{R_g}{R_0}(4\tan^2(\theta) - 1) - 1\right)}\right] \tag{8.14}$$

Then, output VSWR is found by

$$\mathrm{VSWR}_{\mathrm{out}} = \frac{1+|\Gamma_{\mathrm{out}}|}{1-|\Gamma_{\mathrm{out}}|} \tag{8.15}$$

The distributed elements can be transformed to the lumped elements for frequencies less than 100 MHz using the for a quarter-wavelength-long transmission line as shown in Figure 8.5.

In Figure 8.5, the element values for the lumped components can be found using the following formulas:

$$L = \frac{Z_0}{2\pi f}, \quad C = \frac{1}{2\pi f Z_0} \tag{8.16}$$

The network in Figure 8.5 also performs impedance transformation from R_1 to R_2 at each distribution port on the combiner. The lumped-element transformation network shown is a π-network, and it consists of three reactive elements. The quality factor for this network can be found from

$$Q = R_1/X_c \tag{8.17}$$

The number of reactive elements can be reduced by transforming the π-network to the L-network, as shown in Figure 8.6. Q of the π-network can be used to obtain the corresponding element values for the L-network when it is transformed. The equivalent L-network is given in Figure 8.6 when $R_1 \geq R_2$.

As a result, each distributed element in Figures 8.2 and 8.4 can be replaced with its equivalent lumped-element L-network shown in Figure 8.6b.

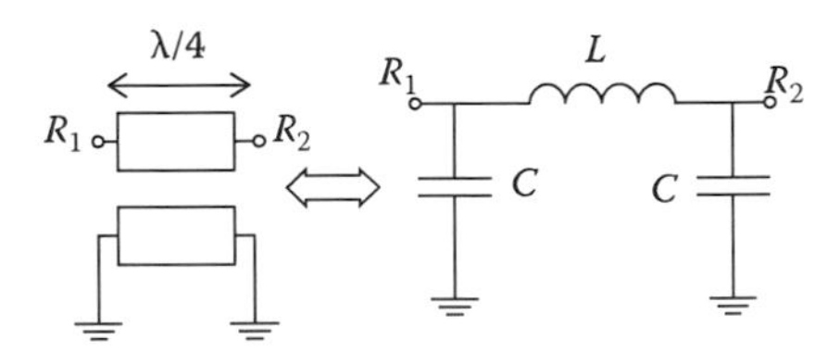

FIGURE 8.5 Distributed element to lumped conversion.

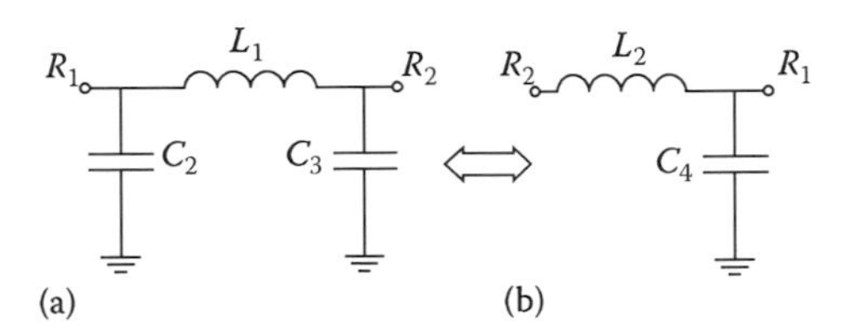

FIGURE 8.6 Transformation from (a) π-network to (b) L-network.

For the design example given, the operational frequency is 13.56 MHz, and the combiner should be capable of providing 12,000-W output power. The combiner is intended to combine the outputs of three PA modules. Each PA module presents 25-Ω impedance to the input of each distribution port on the combiner. The output of the combiner is desired to be 30-Ω. MATLAB GUI in Ref. [7] is used for the combiner to obtain insertion loss, isolation, VSWRs, and characteristic curves, as shown in Figure 8.7. The program does not take into account any imperfection that might exist in the real system and hence theoretically gives perfect isolation when $\theta = 90°$. The design parameters for the three-way combiner are

$$R_0 = 25\ \Omega, \quad R_L = 30\ \Omega, \quad Z_0 = 47.43\ \Omega \tag{8.18}$$

Each PA module is required to provide 4000-W output power under a matched condition. The component values calculated using Equation 8.14 for the π-network given in Figure 8.6a are

$$L_1 = 556.69\ \text{nH}, \quad C_2 = C_3 = 247.46\ \text{pF}, \quad Q = 1.898 \tag{8.19}$$

The corresponding component values of the lumped elements for the *L*-network given in Figure 8.6b are

$$L_3 = 472.2\ \text{nH}, \quad C_4 = 210.5\ \text{pF}, \quad Q = 1.612 \tag{8.20}$$

In both circuits,

$$R_1 = 90\ \Omega, \quad \text{and} \quad R_2 = 25\ \Omega \tag{8.21}$$

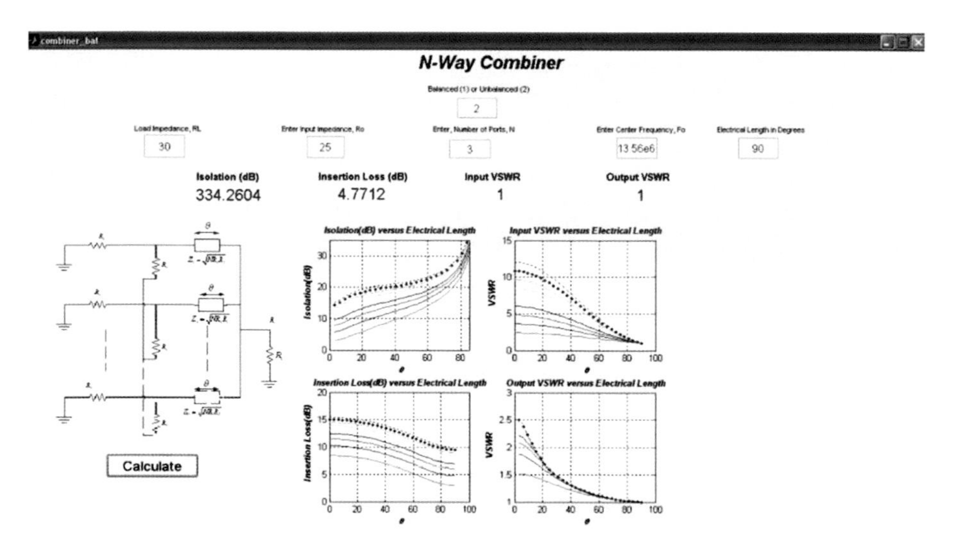

FIGURE 8.7 MATLAB GUI output for three-way unbalanced combiner when $\theta = 90°$.

The lumped-element inductor in the L-network is implemented as a spiral inductor on an alumina substrate having a planar form. The form of the spiral inductor that will be used in our application is shown in Figure 8.8. It is a rectangular spiral inductor with rounded edges versus sharp edges. This type of implementation on the edges increases the effective arcing distance between traces. The physical dimensions for the spiral inductor are the width of the trace, w, the length of the outside edges, l_1 and l_2, and the spacing between the traces, s. The simplified two-port, lumped-element equivalent circuit for the spiral inductor shown in Figure 8.8 is illustrated in Figure 8.9. In Figure 8.9, L is the series inductance of the spiral, and C is the substrate capacitance.

This model ignores the losses in the substrate and the conductor. The lumped-element values for the spiral inductor to perform the required impedance transformation from $R_2 = 25\ \Omega$ to $R_1 = 90\ \Omega$ for the network in Figure 8.9 are calculated to be

$$L = 497\text{ nH}, \quad C = 43.6\text{ pF}$$

The accurate inductance calculation at the high-frequency range can be obtained using Greenhouse's method described in Ref. [8]. The total inductance of the spiral inductor including the effect of mutual couplings is given as

$$L = L_0 + \sum M \tag{8.22}$$

L_0 is the sum of the self inductances for each trace. $\sum M$ takes into account all the mutual inductances in the structure. Equation 8.22 can be written more explicitly for any number of turns for a rectangular spiral inductor as

$$L_i = 0.0002 l_i \left[\ln\left(2\frac{l_i}{\text{GMD}} \right) - 1.25 + \frac{\text{AMD}}{l_i} + \frac{\mu}{4} T \right] \tag{8.23}$$

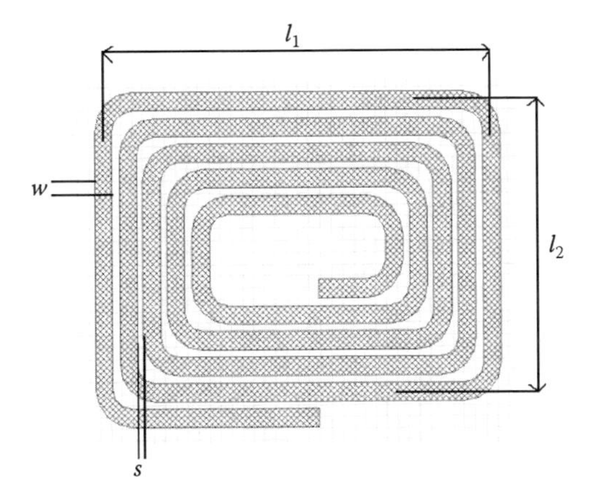

FIGURE 8.8 Spiral inductor model.

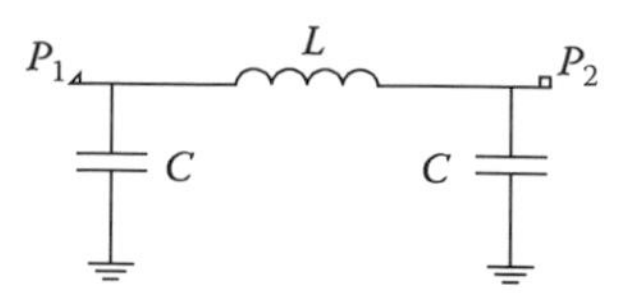

FIGURE 8.9 Simplified equivalent circuit for spiral inductor without loss factor.

and

$$M_{ij} = 0.0002 l_i Q_i \tag{8.24}$$

AMD is the arithmetic mean distance, and *GMD* is used for the geometric distance. C is the capacitance that includes the effect of odd mode, even mode, and interline coupling capacitances between coupled lines of the spiral inductor. The detailed calculation of the capacitances is given in Ref. [9]. The substrate losses and conductor losses are ignored due to the low operational frequency.

The one-port measurement network for the spiral inductor using the model proposed in Figure 8.9 is shown in Figure 8.10. When the equivalent one-port measurement network is used, the effective inductance value is found using $Z = j\omega L_{eff}$

$$L_{eff} = 589.2 \text{ nH} \quad \text{from} \quad Z = j\omega L_{eff} \tag{8.25}$$

The physical dimensions of the microstrip spiral inductor are calculated using the formulation and algorithm developed in Ref. [10] and are given in Table 8.1.

At this point, all the formulation has been done, and solutions are obtained. Based on the analytical values, the combiner meets the specifications. Hence, we are ready at this point to perform the second phase of the design stage.

8.2.2 Simulation Phase for Combiners

The spiral inductor using the dimensions in Table 8.1 is simulated with the method of the moment-based planar electromagnetic simulator, Sonnet. The operational frequency is chosen to be 13.56 MHz. The 3D layout of the simulated structure is illustrated in Figure 8.11. The input port or port 1 is connected via the bridge for inductance measurement. The bridge height and width are given in Table 8.1. The effect of the bridge at the frequency of the operation is minimal due to its increased width.

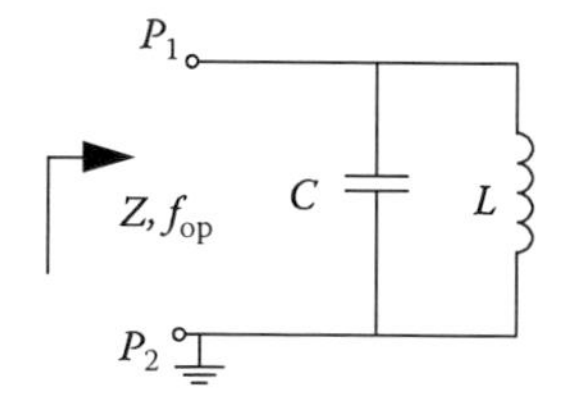

FIGURE 8.10 One-port measurement circuit.

TABLE 8.1
Physical Dimensions of the Spiral Inductor

Trace Width w	Spacing S	Horizontal Trace Length l_1	Vertical Trace Length l_2	Copper Thickness t	
80	30	1870	1450	4.2	
Dielectric Material	**Dielectric Permittivity ε_r**	**Dielectric Thickness h**	**Number of Turns n**	**Bridge Height h_b**	**Bridge Width w_b**
Al_2O_3	9.8	100	6.375	100	350

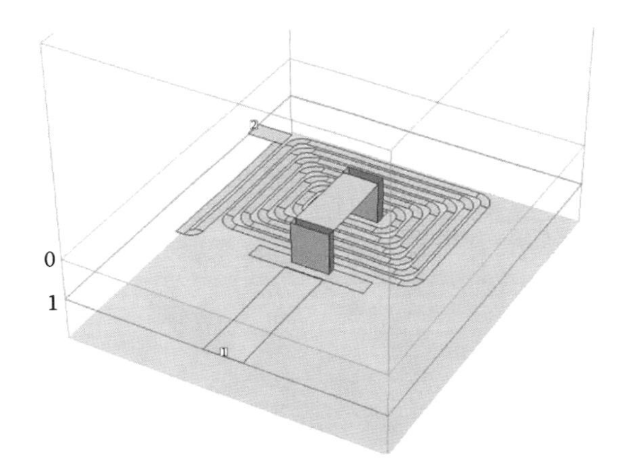

FIGURE 8.11 3D model of the simulated spiral inductor.

The traces, as seen in Figure 8.11, are segmented for parametric study to understand the effect of width and spacing on the self-resonant frequency and the quality factor of the spiral inductor. One of the unique features of the planar electromagnetic simulator is the visualization of the current distribution on the spiral structure. This is specifically important for high-power applications to adjust the necessary spacing between traces to prevent any possible arc during the operation. As seen from Figure 8.12, the current density on the bridge, which is designed to have minimal impact on the device performance and overall inductance, is the lowest. The current density increases as it gets closer to the edges of each trace. It becomes maximum at the edges. This is why during the implementation of the spiral inductor, the corners are rounded to increase the creepage distance to prevent potential arcs.

The inductance value of the spiral inductor is simulated and obtained as 588.6 nH at the operational frequency, which is 13.56 MHz. This is very close to the calculated inductance value.

We are now ready to combine the spiral inductors to simulate a three-way combiner. A three-way planar, high-power combiner is simulated with the method of moment field solver, Ansoft designer, as shown in Figure 8.13. The simulation results for the whole combiner are shown in Figure 8.14.

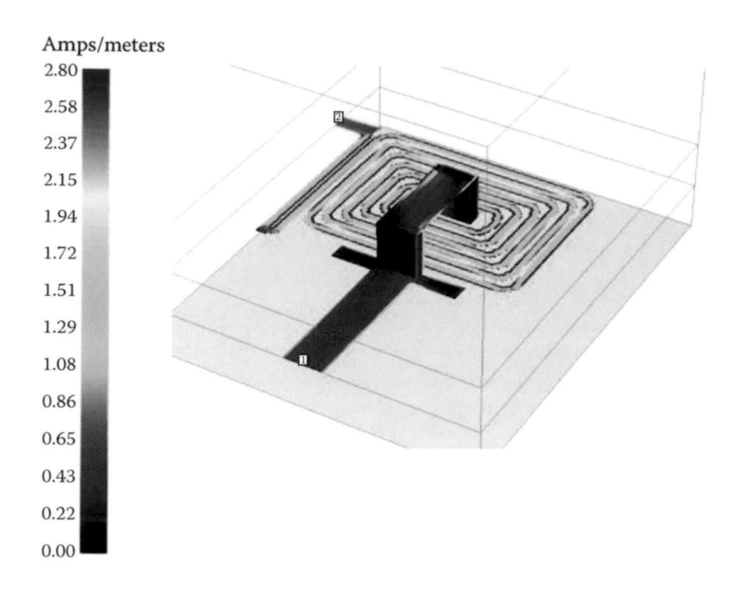

FIGURE 8.12 Current density of the spiral inductor.

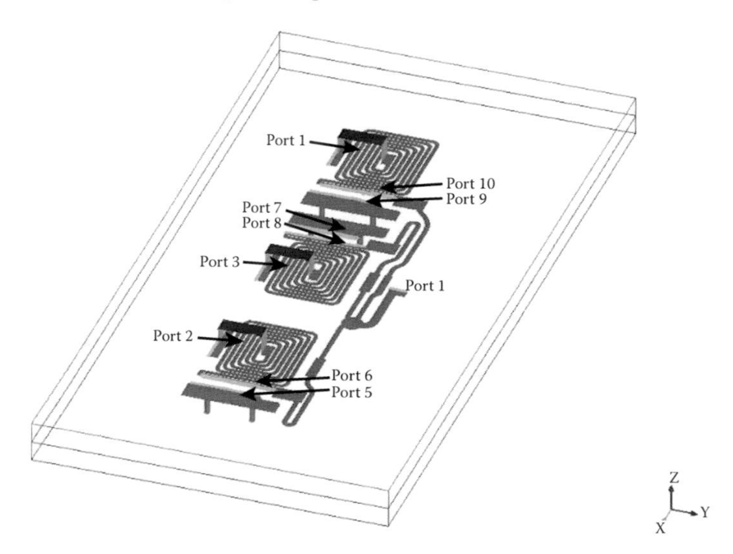

FIGURE 8.13 Simulated three-way combiner in planar form using L-network topology.

Based on the simulation results, the insertion loss and the isolation between each port are found to be –5.15 and –17.21 dB, respectively, at $f = 13.56$ MHz. When the impedance at the output port is measured for the three-way combiner using an electromagnetic simulator, it is found to be $(29.49 - j0.06)\ \Omega$ at the operational frequency as confirmed by the analytical values. The simulated VSWRs at the combiner output vs. frequency are shown in Figure 8.15 and found to be equal to

$$\text{simulated combiner VSWR} = 1.695 \text{ at } f = 13.56 \text{ MHz}$$

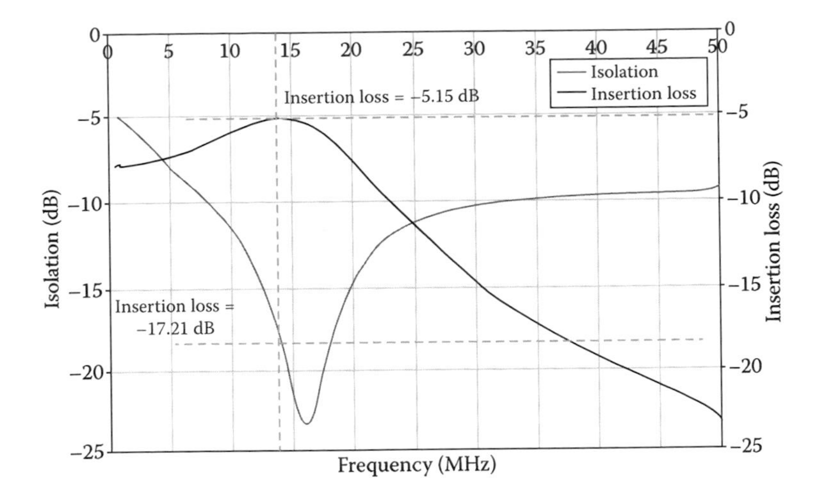

FIGURE 8.14 Simulation results for three-way combiner in planar form using L-network topology.

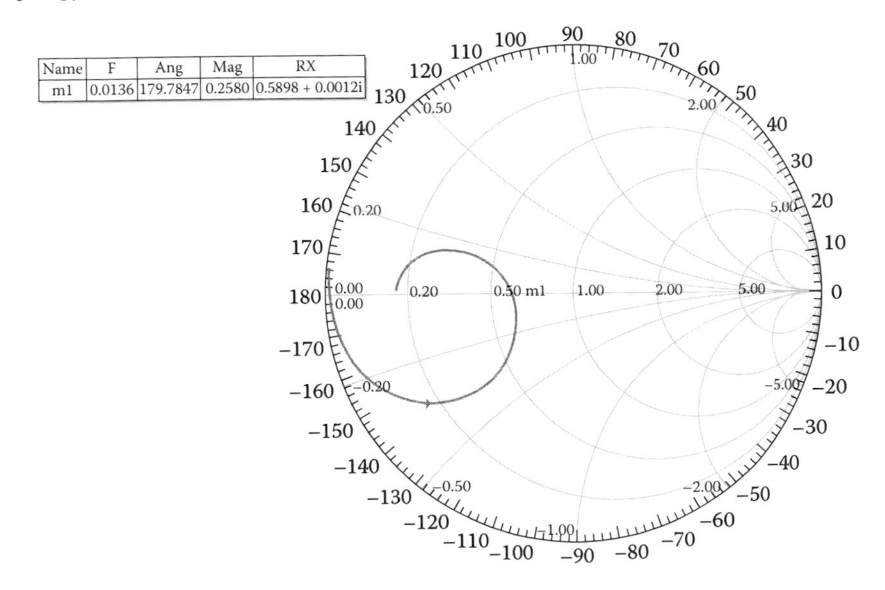

Name	F	Ang	Mag	RX
m1	0.0136	179.7847	0.2580	0.5898 + 0.0012i

FIGURE 8.15 VSWR vs. frequency for three-way microstrip combiner.

Now, we feel more confident to build our prototype and perform the third and last phase in our flow chart given in Figure 8.1.

8.2.3 Experimental Phase for Combiners

The picture of the final constructed combiner is shown in Figure 8.16. The measurements are done using an HP-8504A Network Analyzer. The measured isolation loss and insertion loss for the combiner are illustrated in Figures 8.17 and 8.18 and

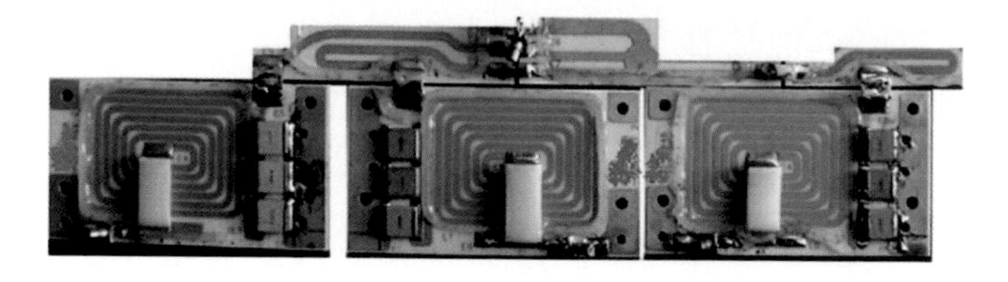

FIGURE 8.16 Three-way combiner implemented in planar.

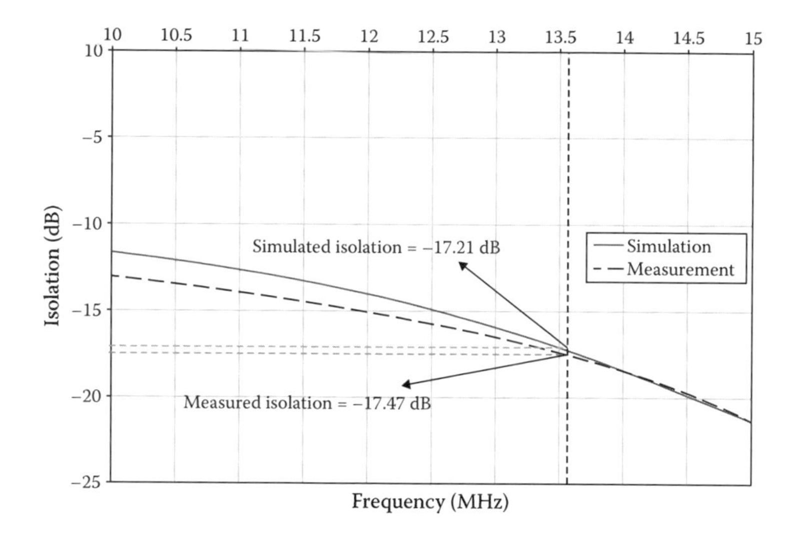

FIGURE 8.17 Three-way combiner implemented in planar.

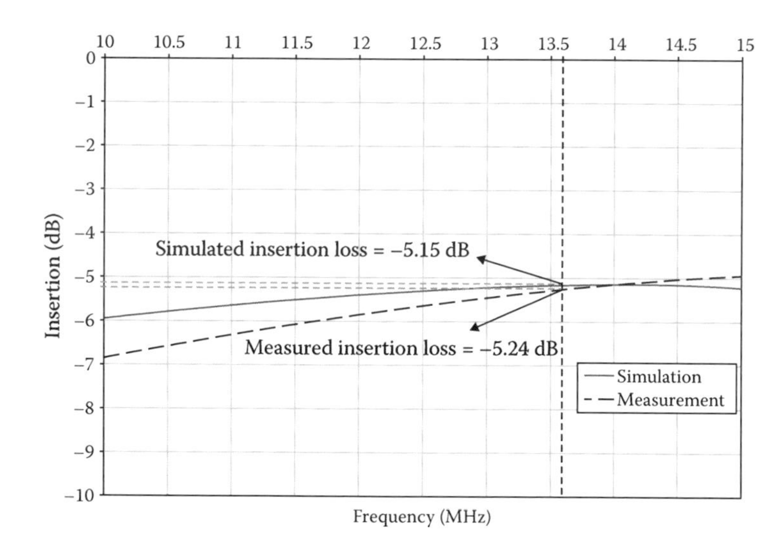

FIGURE 8.18 Measurement results for insertion loss of three-way combiner in planar form using L-network topology.

compared with the simulated results. The isolation loss and insertion loss at the operational frequency are measured to be –17.47 and –5.24 dB, respectively.

The impedance at the output port under this condition is measured to be (30.23 – j0.06) Ω at the operational frequency. This corresponds to

$$\text{measured combiner VSWR} = 1.654 \text{ at } f = 13.56\text{ MHz}$$

This value is very close to the targeted impedance value of 30 Ω at the output port.

The measured insertion loss and return loss are given in Figures 8.17 and 8.18, respectively, with comparison with the simulated results. As seen, the simulated and analytical results are in agreement, and design methodology introduced by the flow chart given in Figure 8.1 leads to completion and realization of a successful design.

8.3 ACTIVE COMPONENT DESIGN AND MODELING WITH CAD

It is a requirement for active devices to be able to handle high power, high gain, and stability for some RF/microwave applications including semiconductor wafer processing and medical resonance imaging. Active devices also have to be rugged, withstand high voltage swing, and have good thermal profile when used as a component in RF amplifiers.

There has been extensive study on failure modes of semiconductor devices to identify the root causes of wire-bond failures and bonding pad fatigue [11]. The investigation of the failure mechanisms of high-power devices led to the inclusion of other sources of failure such as die-attach voids and degradation of the bonding pad. The recent advancement in technology helped to relate some of the wire-bond failures to hotspots created by voids under the silicon die. This initiated research into the growth of these die-attach layer cracks to improve the reliability of active devices that are used in RF/microwave amplifiers. As a result, the design and manufacturing of a die with the desired package that will be used as an active device in RF systems need to have several design standards and the required electrical and mechanical properties to perform without failure.

Hence, it is necessary to have a novel package design for active devices to enable them to perform without failure. In this section, the flow chart in Figure 8.1 will be used to design, simulate, and build a hybrid package for a metal–oxide–semiconductor field-effect transistor (MOSFET). The three-phase design detail will be given step by step using the hybrid MOSFET package design.

Design Example

Design a hybrid package, and obtain package parasitics consisting of four dies to be used in power amplifiers operating at 13.56 MHz as a component to provide the required gain, power output, stability, and thermal profile.

Solution

The design, simulation, and implementation of hybrid package will be done using the steps outlined in the flow chart given in Figure 8.1.

8.3.1 Analysis Phase for Hybrid Package

Consider the layout for the package with a single die given in Figure 8.19. It is intended to use SN96 to attach the die to a metal base using high-pressure reflow. Each die shares a common source and split gate connection.

The proposed package has the following requirements:

- Bond gates are used to interconnect the frame printed circuit board (PCB), as shown in Figure 8.19, three wires per die.
- Stitch bond source leads connect the PCB to the die sources and back to the PCB pad, as shown in Figure 8.19, i.e., stitch A, B, C, D, six wires per die.
- Controlled low loop is used: 0.010 in. minimum, 0.040 in. maximum. Wirebonds shall be bonded within 0.050 in. from the PCB pad edge to minimize the bond length.

Table 8.2 gives the details of the package, the die attach material, and the types of adhesives used.

PCB stratification for the hybrid package is given in Figure 8.20. The calculation of the physical dimensions of the bond wire is done using the wideband bond wire inductance formula:

$$L = \frac{\mu_0 l}{2\pi}\left[\ln\left(\frac{4l}{d}\right) + \mu_r \delta - 1\right] \tag{8.26}$$

where δ is the skin effect factor, which is a function of the wire diameter and frequency and is defined as

$$\delta = 0.25 \tanh(4\, d_s / d) \tag{8.27}$$

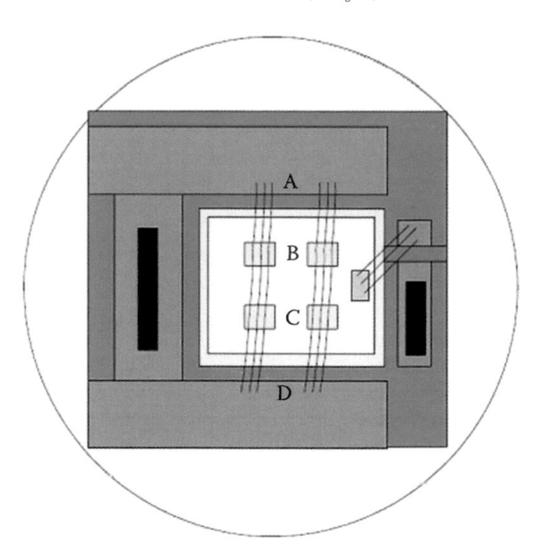

FIGURE 8.19 Single die layout in a proposed package.

TABLE 8.2
Details of the Proposed Package

Package Technology	Die Attach Material	Type
Pressed alumina ceramic	Silver-filled glass	Inorganic adhesive
Laminated alumina ceramic (pin grid array [PGA], ceramic quad flat pack [CQFP], side-braze)	Gold–silicon eutectic	Hard solder
	Silver-filled cyanate ester	Organic adhesive
Molded plastic	Silver-filled epoxy	Organic adhesive

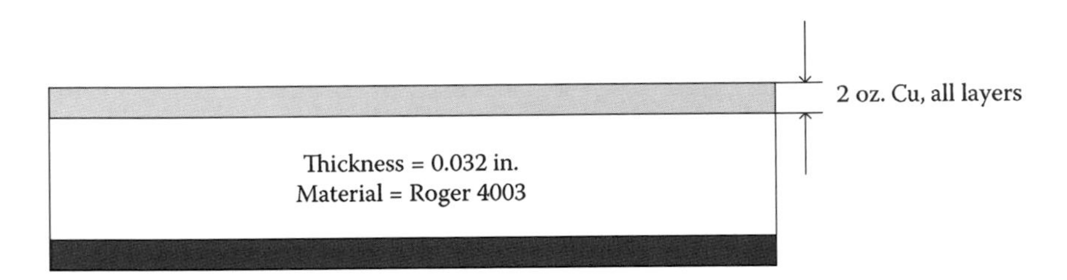

FIGURE 8.20 PCB stratification.

where d_s is defined as

$$d_s = \sqrt{\frac{\rho}{\pi f \mu_r \mu_0}} \tag{8.28}$$

ρ is the resistivity of the bond wire, l is the length, and d is the diameter of the wire. Using Equations 8.26 through 8.28, we find the bond wire inductance as

$$L = \frac{\mu_0 l}{2\pi}\left[\ln\left(\frac{4l}{d}\right) + \mu_r \delta - 1\right] = 3.65\,[\text{nH}] \tag{8.29}$$

The physical length of the bond wire is usually very short, and therefore, the associated resistance is low. The DC resistance of the bond wire is calculated from

$$R_{dc} = \frac{\rho l}{\pi r^2} \tag{8.30}$$

For $d/d_s > 3.4$ using curve fitting, we obtain

$$R_s = R_{dc}\left(0.25\frac{d}{d_s} + 0.2654\right) \tag{8.31}$$

The four dies when placed symmetrically in the final proposed package are illustrated in Figure 8.21.

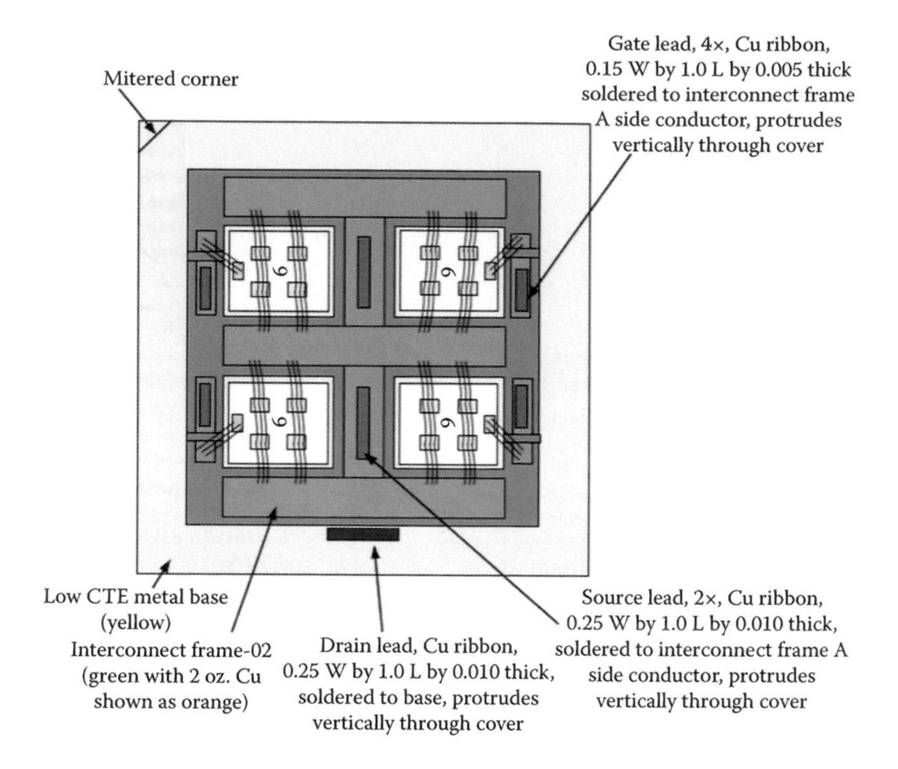

FIGURE 8.21 The complete hybrid package with four dice.

8.3.2 Simulation Phase for Hybrid Package

We can now proceed to the simulation phase in the flow chart. Bond wires are simulated with Ansoft Designer, as shown in Figure 8.22, using the configuration given in Figure 8.19.

The simulation results giving inductance value vs. frequency are illustrated in Figure 8.23. Based on the results, the inductance of the combination of three parallel bond wires is found to be 1.37 nH. This corresponds to 4.11 nH. This is close to the value calculated in Equation 8.29.

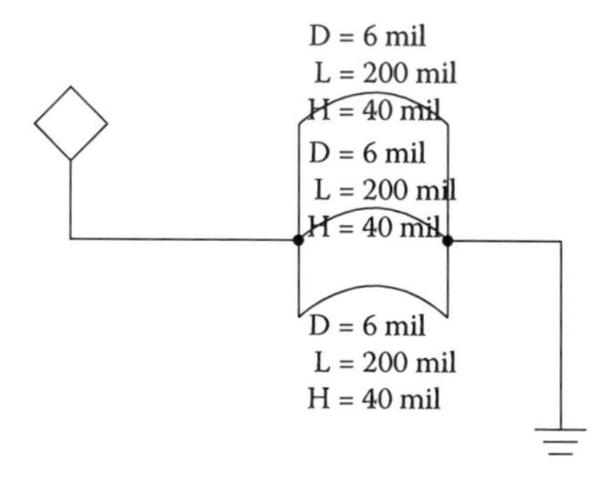

FIGURE 8.22 Bond wire inductance simulation.

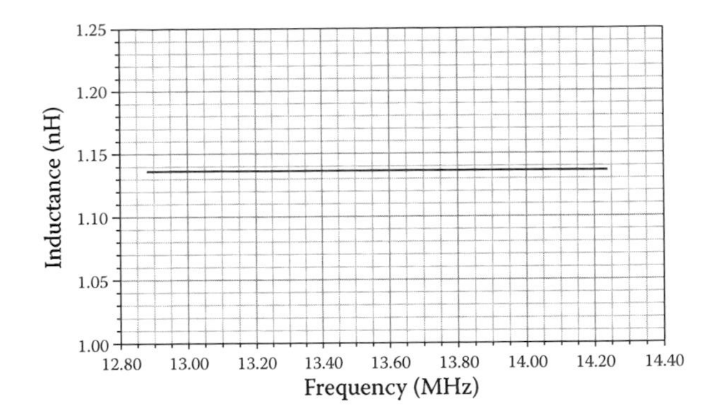

FIGURE 8.23 Simulated bond inductance value for three parallel bond wires.

The planar electromagnetic simulator, Ansoft Designer, is also used to obtain the simulation results for lead and source inductances, as shown in Figures 8.24 and 8.25, respectively.

The complete package shown in Figure 8.20 can now be simulated for all the parasitics since the results are in agreement with the analytical results so far. The

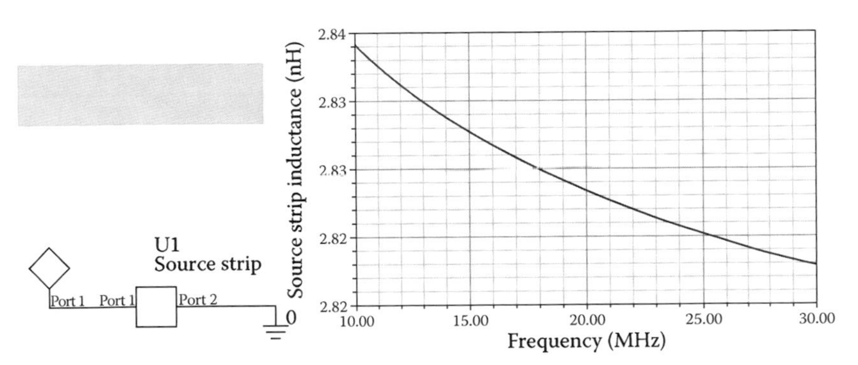

FIGURE 8.24 Simulated source lead inductance vs. frequency.

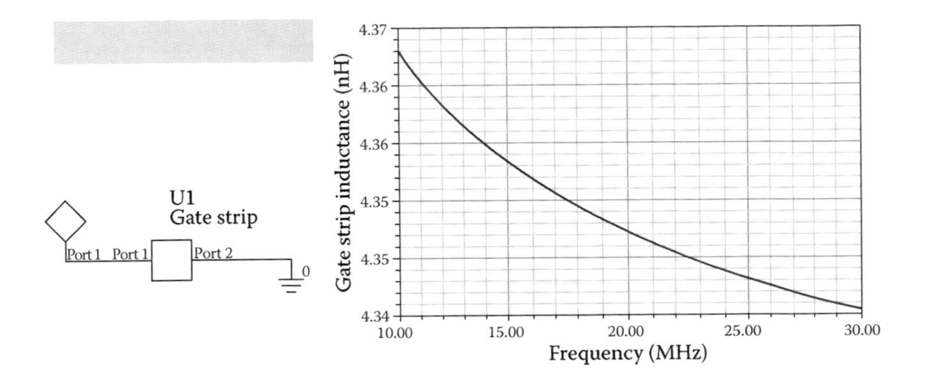

FIGURE 8.25 Simulated gate lead inductance vs. frequency.

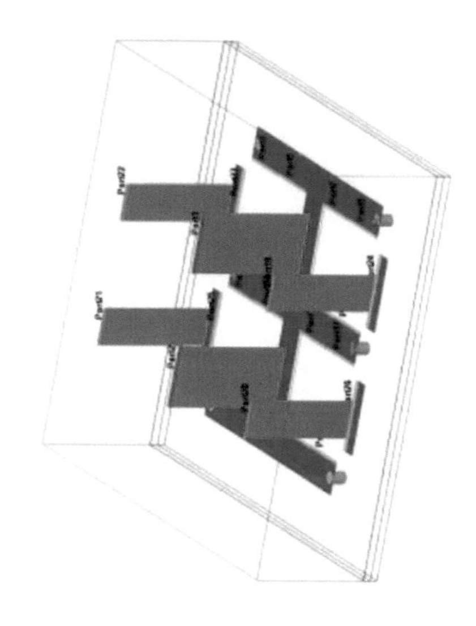

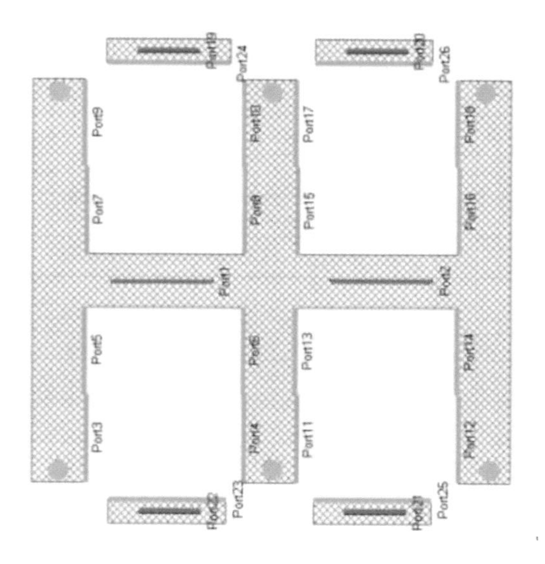

FIGURE 8.26 Simulated hybrid package with four dies.

complete hybrid package simulation is illustrated in Figures 8.26 and 8.27. Figure 8.26 gives the 2D and 3D layout of the structure that is simulated. Figure 8.27 is the circuit when the bond wires are included to the structure that is simulated and is shown in Figure 8.26. Cosimulation technique is used to simulate both electromagnetic structure and bond wires using Ansoft Designer.

The simulation results obtained using a cosimulated circuit are shown in Figure 8.28. Based on the simulated results of the complete four-die hybrid package, the source inductance value is found to be 5.5 [nH].

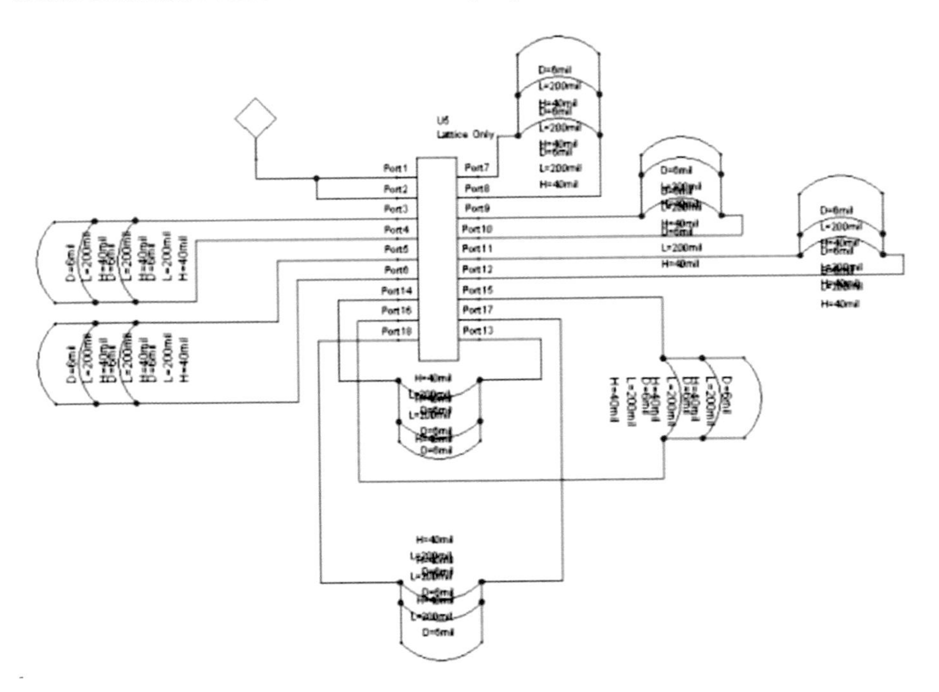

FIGURE 8.27 Cosimulation of four-die hybrid package for parasitics with bond wires.

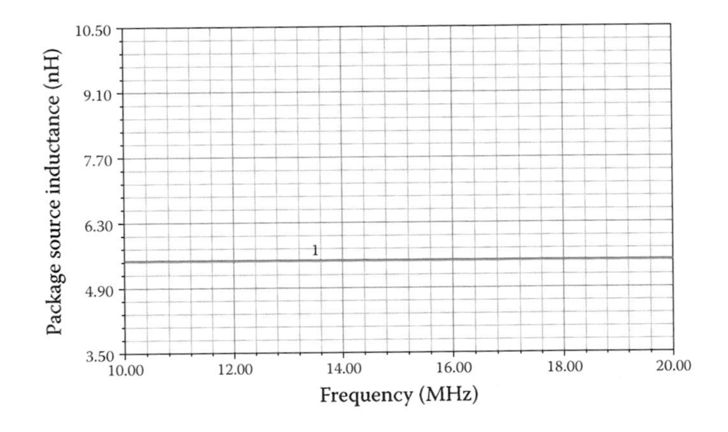

FIGURE 8.28 Simulated four-die hybrid package parasitics.

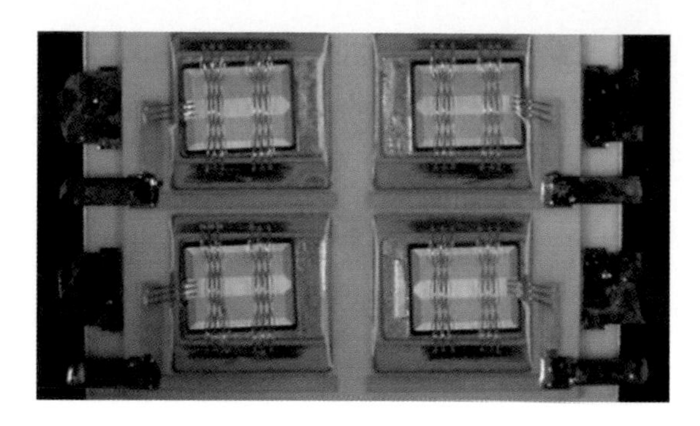

FIGURE 8.29 Constructed four-die hybrid package parasitics.

8.3.3 Experimental Phase for Hybrid Package

The structure is built and measured as part of the third and last phase in the flow chart. The constructed four-die hybrid package is shown in Figure 8.29. Overall, the measured and simulated source inductance values show an improvement over the conventional package parasitic inductance value. For instance, the typical source lead inductance value for TO-247 MOSFET package is reported to be around 7–13 [nH]. The measurement results are found to be in agreement with the simulated results.

REFERENCES

1. E.J. Wilkinson. 1960. An N-way hybrid power divider. *IRE Transactions on Microwave Theory and Techniques*, Vol. MTT-8, pp. 116–118, January.
2. A.D. Saleh. 1980. Planar electrically symmetric N-way hybrid power dividers and combiners. *IEEE Transactions on Microwave Theory and Techniques*, Vol. MTT-28, pp. 555–563.
3. H. Howe, Jr. 1979. Simplified design of high power, N-way, in-phase power divider/combiners. *Microwave Journal*, pp. 43–57, December.
4. X. Tang, and K. Mouthaan. 2009. Analysis and design of compact two-way Wilkinson power dividers using coupled lines. *Asia-Pacific Microwave Conference*, pp. 1319–1322, Singapore.
5. R. Knochel, and B. Mayer. 1990. Broadband printed circuit 0°/180° couplers and high power in phase power dividers. *IEEE MTT-S International Microwave Symposium Digest*, pp. 471–474.
6. K.J. Russel. 1979. Microwave power combining techniques. *IEEE Transactions on Microwave Theory and Techniques*, Vol. MTT-27, pp. 472–478.
7. A. Eroglu. 2013. *RF Circuit Design Techniques for MF-UHF Applications*. CRC Press, Boca Raton, FL.
8. H.M. Greenhouse. 1974. Design of planar rectangular microelectronic inductors. *IEEE Transactions on Parts, Hybrids and Packaging*, Vol. PHP-10, pp. 101–109, June.
9. A. Eroglu, and J.K. Lee. 2008. The complete design of microstrip directional couplers using the synthesis technique. *IEEE Transactions on Instrumentation and Measurement*, Vol. 12, pp. 2756–2761, December.

10. A. Eroglu. 2011. Planar inductor design for high power applications. *Progress in Electromagnetic Research B*, Vol. 35, pp. 53–67, December.
11. A. Hamidi, N. Beck, K. Thomas, and E. Herr. 1999. Effects of current density and chip temperature distribution on lifetime of high power IGBT modules in traction working conditions. *Microelectronics & Reliability*, Vol. 39, Nos. 6–7, pp. 1153–1158, June–July.

Index

Page numbers followed by f and t indicate figures and tables, respectively.

M

N

O

P

Q

R

V

W

Y

Z